Lecture Notes in Computer Science 2456

Edited by G. Goos, J. Hartmanis, and J. van Leeuwen

Springer
Berlin
Heidelberg
New York
Barcelona
Hong Kong
London
Milan
Paris
Tokyo

Roland Traunmüller   Klaus Lenk (Eds.)

Electronic Government

First International Conference, EGOV 2002
Aix-en-Provence, France, September 2-6, 2002
Proceedings

Series Editors

Volume Editors

Roland Traunmüller
University of Linz, Institute of Applied Computer Science
Altenbergerstr. 69, 4040 Linz, Austria
E-mail: traunm@ifs.uni-linz.ac.at

Klaus Lenk
University of Oldenburg, Lehrstuhl für Verwaltungsinformatik
26111 Oldenburg, Germany
E-mail: lenk@uni-oldenburg.de

Cataloging-in-Publication Data applied for

Die Deutsche Bibliothek - CIP-Einheitsaufnahme

Electronic government : first international conference ; proceedings / EGOVS 2002, Aix-en-Provence, France, September 2 - 6, 2002. Roland Traunmüller ; Klaus Lenk (ed.). - Berlin ; Heidelberg ; New York ; Barcelona ; Hong Kong ; London ; Milan ; Paris ; Tokyo : Springer, 2002
(Lecture notes in computer science ; Vol. 2456)
ISBN 3-540-44121-2

CR Subject Classification (1998): K.4, K.6.5, K.5, K.3, C.2, H.5, H.4

ISSN 0302-9743
ISBN 3-540-44121-2 Springer-Verlag Berlin Heidelberg New York

Springer-Verlag Berlin Heidelberg New York
a member of BertelsmannSpringer Science+Business Media GmbH

http://www.springer.de

Typesetting: Camera-ready by author, data conversion by Olgun Computergraphik
Printed on acid-free paper SPIN: 10871160 06/3142 5 4 3 2 1 0

Preface

In defining the state of the art of E-Government, EGOV 2002 was aimed at breaking new ground in the development of innovative solutions in this important field of the emerging Information Society. To promote this aim, the EGOV conference brought together professionals from all over the globe. In order to obtain a rich picture of the state of the art, the subject matter was dealt with in various ways: drawing experiences from case studies, investigating the outcome from projects, and discussing frameworks and guidelines. The large number of contributions and their breadth testify to a particularly vivid discussion, in which many new and fascinating strands are only beginning to emerge. This begs the question where we are heading in the field of E-Government. It is the intention of the introduction provided by the editors to concentrate the wealth of expertise presented into some statements about the future development of E-Government.

The number of subject matters covered by the EGOV 2002 conference proceedings is best illustrated by listing some of them:

- Communication with citizens over the Net: One-Stop-Government, Single-Window-Access, and Seamless Government;
- Frameworks and guidelines for E-Government;
- International and regional projects, case studies, and international comparisons;
- Strategies, implementation policies, and best practice;
- Redesigning cooperation within and between agencies;
- Sustaining business processes, collaborative activities, legal interpretation, and administrative decision making;
- E-Democracy strategies, citizen participation in public affairs, and democratic deliberation;
- Technical implementation aspects (standards for information interchange and processes, digital signatures, platforms, security concepts, and provisions);
- Novel organizational answers and new forms of networks: adhoc cooperation and coalition between public agencies and public–private partnerships;
- Changing legal frameworks, and legal and social implications of new infrastructures and applications;
- Teaching of E-Government.

Many people cooperated over a long period of time to shape the conference and to prepare the program and the proceedings. Our thanks go to the members of the Program Committee (listed below) and to Gabriela Wagner, who heads the DEXA organization. The editors are particularly grateful to Ute Holler for her exceedingly engaged assistance in coordinating the preparation of the program and of the proceedings.

September 2002

Roland Traunmüller
Klaus Lenk

Program Committee

General Chair

Roland Traunmüller, University of Linz, Austria
Klaus Lenk, University of Oldenburg, Germany

Program Committee

Kim Viborg Andersen, Copenhagen Business School, Denmark
Chris Bellamy, Nottingham Trent University, UK
Trevor Bench-Capon, University of Liverpool, UK
Daniéle Bourcier, University of Paris 10, France
Jean-Loup Chappelet, IDHEAP Lausanne, Switzerland
Wichian Chutimaskul, King Mongkut's University of Technology Thonburi, Thailand
Nicolae Costake, Bucharest, Romania
Arthur Csetenyi, Budapest Univ. of Economic Sciences, Hungary
Christian S. Friis, Roskilde University, Denmark
Fernando Galindo, Universidad de Zaragoza, Spain
Michael Gisler, Bundesamt für Informatik und Telekommunikation (BIT), Switzerland
Dimitris Gouscos, eGovLab, University of Athens, Greece
Åke Grönlund, Umeå University, Sweden
Michel Klein, HEC Graduate School of Management, France
Sayeed Klewitz-Hommelsen, Fachhochschule Bonn-Rhein-Sieg, Germany
Friedrich Lachmayer, Republik Österreich BKA-Verfassungsdienst, Austria
Philippe Laluyaux, Clip Card, France
Alan Lovell, UK
Ann Macintosh, Napier University, Edinburgh, UK
Gregoris Mentzas, National Technical University of Athens, Greece
Thomas Menzel, University of Vienna, Austria
Jeremy Millard, Danish Technological Institute, Denmark
Javier Ossandon, Ancitel Spa, Italy
Reinhard Posch, Chief Information Officer, Graz University of Technology, Austria
Corien Prins, Tilburg University, The Netherlands
Gerald Quirchmayr, University of South Australia, Australia
Heinrich Reinermann, Deutsche Hochschule für Verwaltungswissenschaften Speyer, Germany
Reinhard Riedl, Fachbereich Informatik, Universität Rostock, Germany
Giovanni Sartor, CIRSFID Bologna, Italy
Erich Schweighofer, University of Vienna, Austria
Ignace Snellen, Erasmus University of Rotterdam, The Netherlands
Dieter Spahni, Institute for Business and Administration, Switzerland
Efthemis Tambouris, Archetypon S.A., Athens, Greece
Wim van de Donk, Tilburg University, The Netherlands
Mirko Vintar, University of Ljubljana, Slovenia
Francesco Virili, University of Cassino, Italy
Maria Wimmer, University of Linz, Austria

Table of Contents

Framework

Digital Olympics 2008: Creating the Digital Beijing

Knowledge Management

Requirements

Business Process Reengineering

Electronic Service Delivery

Designing Innovative Applications

Electronic Democracy

Information Society Technologies Programme (IST)

Implementing e-Government

Legal Issues

Technical Issues

Varied Contributions

Electronic Government: Where Are We Heading?

Klaus Lenk[1] and Roland Traunmüller[2]

[1] University of Oldenburg, Germany
lenk@uni-oldenburg.de
[2] University of Linz, Austria
traunm@ifs.uni-linz.ac.at

Abstract. In common understanding, Electronic Government focuses upon relatively simple transactions between identifiable customers (citizens, enterprises), on one side, and a multitude of government organisations in charge of particular activities, on the other. Attention is chiefly directed to Electronic Service Delivery. If the promise of e-Government as the principal key to modernising government is to be kept, this concept has to be broadened so as to include the full enabling potential of IT, as well as the complex reality of government and public governance. There is encouraging political support for e-Government, yet implementation problems could inhibit further success.

1 What Electronic Government Is About

Electronic Government (or e-Government) as an expression was coined after the example of Electronic Commerce. But it designates a field of activity which is with us for several decades yet. In many respects, e-Government is just a new name for the informatisation of the public sector, which has been going on for several decades now (Lenk 1994). The use of IT in public administration and in other branches of government (including parliaments and the judiciary) has attained a high level in many countries of the industrialised world. But there was hardly any political interest in this ongoing and almost invisible process of modernising government. Only academics and some far-sighted consultants insisted on the significance of informatisation for public governance and its modernisation (Snellen and van de Donk 1998).

For a long time, their message went unheard (Lenk 1998). Especially New Public Management as the most important explicit movement of government reform hardly recognised the enabling potential of IT for changing the work practices and the business processes in the public sector. Its image of IT was one of an auxiliary tool, to be used for supporting financial management and statistical information.

This situation changed fundamentally with the announcements of the then US Vice President Al Gore in 1993, heralding not only the potential for a renewal of society which an "Information Society" holds, but relating it directly to the need of improving the performance of the public sector. The lesson inherent in harnessing the propagation of an Information Society to public sector modernisation has been learned earlier in Asia than in the European Union, where the Action Plan, based on the Bangemann Report of 1994, concentrated one-sidedly on the private sector of the

R. Traunmüller and K. Lenk (Eds.): EGOV 2002, LNCS 2456, pp. 1–9, 2002.

economy. But the situation has now been redressed also in Europe. The American and Asian examples have greatly stimulated the interest in the potential of e-Government as one important facet of an Information Society.

Although the new impetus presents great opportunities for improving government processes, the underlying concepts of e-Government remain fairly vague. Moreover, they are still driven by analogies from E-Commerce. This may misdirect the attention of governments which are eager to innovate. Things are made to appear as if they were entirely new, and both the achievements and the lessons learned from several decades of using IT in the public sector are neglected by those who are discovering the enabling potential of IT for the first time.

Whilst in the past, IT-support was inward-looking and chiefly brought to bear on typical back office activities, the focus has now moved toward the external relationships of all branches of government. It is on electronic citizen service, on electronic procurement of goods, as well as on electronic democracy including democratic deliberations, citizen information, and electronic voting. Many early projects inspired by the Information Society rhetoric focussed on politically visible fields like online citizen services, without giving much consideration to whether the promised improvements were catering to the most pressing needs of citizens or enterprises. Also in the field of E-Commerce, early projects were launched without caring much about what the potential customers would actually need. Up to the "dot.com" crisis of the year 2000, market research and target group identification had been largely absent. But in the private sector, market forces quickly taught the right lessons.

The errors of e-Government are much harder to detect, and incentives to correct them do not always exist. A case in point is the assumption that online access "24 hours, 7 days a week" would meet the prime concern of most citizens when they have to approach a public organisation for services delivery or other reasons. The results of research on citizen-government relations, which were accumulated over decades, have been totally neglected in order to present things in such a way as if existing "online" solutions (which fall short even of involving interaction via videoconferencing) would hold the key to solving all problems. The political wish to announce serious actions and quick solutions has also led to focusing on transactions like registering a car or applying for an identity card, which citizens mostly do not consider as a service but rather as a nuisance. Many governments hoped to speed up the diffusion of Internet use within the population by offering relatively simple government "services" over this channel.

So it was less a desire to enhance back office productivity and service quality which prompted governments to embark on the type of Electronic Service Delivery, which is so prominent now. The overriding political concerns seemed to be of an economic nature, the state assuming a forerunner role in entering the Information Society. Administrative modernisation was piggybacked by economic policy. In federally funded projects in Germany it was at no time clear what the priorities actually were: propagating an electronic signature deemed necessary to make the country become a player in the world league of E-Commerce, or improving service quality and productivity in the public sector.

Among proponents of e-Government, this concentration on Electronic Service Delivery has contributed to a fairly distorted view of the whole machinery of government and of public governance. The system of public governance is now changing in many respects, and states as the former key players are re-positioning themselves with respect to global corporate players and a civil society which is discovering new ways of self-organisation. And again, quite like in the New Public Management wave of administrative modernisation, which is now ebbing off, the role which IT can play in this context is grossly underestimated.

Most of the present endeavours at promoting e-Government fall short of acknowledging two things, the complexity of government and of governance, and the potential of IT beyond what is cast on the market so far. But both are extremely important. Where these two circumstances will join, they will make e-Government a fascinating experience.

The convergence of new forms of governance and future ventures in IT will transform the ways in which we work, communicate, deliberate and negotiate. To set the stage, we have to broaden the concept of e-Government (Lenk and Traunmüller 2001). If we do not succeed in showing a way out of the narrow corridor of improving access to simple and highly automated business processes within a given institutional frame, e-Government might soon become another example of exaggerated "hype".

In order to broaden the concept of e-Government beyond Electronic Service Delivery, the German Society for Informatics (GI) published, together with the German Society of Electrical Engineering (VDE), in September 2000 a Memorandum on "Electronic Government as a key resource for modernising government. This memorandum dealt with the great prospects of a wider usage of information technology for a lasting modernisation of the state and public administration. With the same intention the OCG (Austrian Computer Society) established a "Forum e-Government" (www.gi-ev.de/informatik/presse/presse_memorandum.pdf and www.ocg.at).

If the promise of e-Government as the principal key to modernising government and governance in more than a superficial sense will ever materialise, a clearer view of the agenda of modernising public services should come to prevail. This view should not be tainted by considerations of applying readily available solutions to problems which are not sufficiently investigated. Knowledge of the administrative domain has to be harnessed to a good understanding of the opportunities opened up by technology. And the technological perspective should not be restricted to the present state of development of the technology and to what is on the market now.

In what follows we therefore deal, beyond the immediate perspectives of Electronic Service Delivery, with both the requirements for a pervasive e-Government flowing from the complexity of the public sector and the enabling potential of IT for e-Government. To conclude, we shortly address the significance of political support and successful change management, as critical factors for success. In so doing we draw on a theoretical foundation of e-Government which we published earlier (Lenk and Traunmüller 1999; see also Lenk and Traunmüller 2000 and 2001; Lenk, Traunmüller and Wimmer 2002).

2 Electronic Service Delivery: The Immediate Perspectives

The delivery of services over the Internet has attracted most of the attention devoted to e-Government. A "virtual administration" is more and more taking shape. Public administrations will eventually appear no longer as a set of independent agencies which have to be approached separately, but as a collective unit with which contact can be made via one and the same "portal", or "window". Such a common access structure will reduce neither their intrinsic complexity nor the required precision of their work. No institutional reform is required to make this happen. A One-Stop Government or "Single Window Service" will alleviate many burdens, for individuals and businesses alike. At the same time, it will make public administration more transparent.

One has to admit: this picture is still a more a vision than reality. Most administrative bodies are committed to first steps at improving their own services, without looking at what their neighbours do. Yet the integration of their business processes across organisational boundaries will become very attractive, at least from the addressees' standpoint. In order to make such a vision of "seamless government" come true, specifications of future service delivery arrangements have to be elaborated with great care. These have to take into account a multitude of dimensions:

- Addressing the needs of target groups (e.g. professionals, taxpayers, the elderly)
- Allowing for a multichannel access mix (one-stop shop, online, letter, fax etc.)
- Taking into account service complexity (which varies according to the categories of business processes supported)
- Establishing the required level of service integration (eventually single-window access to all services regardless of government level and organizational unit)
- Providing the required level of security (user identification; authentication, cancellation and non-repudiability of documents and communications)
- Implementing a data protection policy and transparency measures
- Making reliability and usability a prime concern (creating user interfaces which match existing skills, incentives, culture)

These requirements cannot be discussed in detail here (Lenk 2002). We will instead draw attention to the role of portals in an architecture for service delivery which is predicated on a clear separation between front offices and back offices. In such an architecture, front offices can be customer-centered, whilst back offices are task-driven. Some sort of middleware (or "mid office") is required to link front offices to back offices. Typically, front offices comprehend portals which may either directly be accessed by the citizens or used in a physical one-stop shop or a call center where citizens are served.

Portals cover a wider range of functions, and they can be designed to cater for specific demands of their users. Among their most prominent functions is the provision of information about services. This comprises basic information as to which services are available (or which duties citizens have to comply with), as well as information about how to get in contact, which evidence should be presented, etc. For a citizen, this may be of help in preparing a contact at a physical (front) office or in deciding about further steps on the Internet. Next, the download of forms is often of great help to citizens, even if such forms are filled in by hand and mailed back offline.

A great advantage of portals is the possibility of accessing information which is not confined to the range of services offered by a specific agency. At one single access point, citizens can obtain information about all levels of government, and it is of secondary importance which government level or which other (private or ssemi-public) actor runs the portal. Canada is perhaps the leading country with regard to single-window access to government. www.help.gv.at is the address of the Austrian national portal, where easy access to information concerning 55 life events (getting a passport, marriage, change of address etc) is provided. Similarly, the French Portal http://www.service-public.fr provides not only information but holds over 1,000 forms for download, covering the French national administration as well as local and regional government. Moreover, both portals have recently created specific entries for business situations, in addition to typical life events. The work on their extension into a gateway providing access to electronic transactions is in progress.

Since many contacts of ordinary citizens with public bodies are not so frequent, we predict that unassisted online transactions will fulfill only part of the needs of citizens. When it comes to more complex transactions, a citizen may need personalised help, or want personal contact and explanations. This should be made possible without the citizen having to go in person to the back offices where the service is produced or clerical work is done, and it can be achieved either indirectly via physical one-stop shops or call centres or directly through the portal.

Perhaps quite soon will we see the emergence of "telepresence" of a human agent based in a back office somewhere, via advanced forms of videoconferencing, in a front office service situation or directly on a screen at home. We also expect that more and more physical one-stop offices will be created, which rely on Internet-based (and Intranet-based) services. In an "administration à accès pluriel" (Rapport Lasserre 2000), multichannel access involving multifunctional front offices will bring the advantages of the Internet also to citizens who for one reason or another do not use the Internet personally. This option is not yet pursued in countries which still hope to encourage the diffusion of Internet access by making it attractive through online public service offerings. May it be recalled at this point that France, over 20 years ago, first embarked on administrative "Télé-services" in order to market the Télétel (Minitel) system. After some deceiving pilots, this strategy was quickly abandoned in favour of encouraging the (successful) diffusion of Minitel through zero-cost access to a nationwide phone directory. But in a world dominated by players from other world regions, European experiences obviously do not count, even not for the European Union.

To sum up, portals provide an ideal leverage for the modernisation of public services. Yet many problems still have to be solved. Opening a "window to the outside" will lead to considerable rearrangement inside government. This could be limited to re-arranging the interface of business processes so as to make them candidates for Web Services. But it can also amount to a momentous "E-Transformation" inside government, which would be attuned to the equally momentous changes in public governance which are already beginning to be felt worldwide. But, as we said before, narrow views about e-Government can prevent us from facing this challenge.

3 Broadening the View: Government and Public Administration from a Systemic Perspective

In order to apprehend the wide range of opportunities which e-Government now opens up for improving the public business, we have to recall that public governance is not just about delivering services. It includes democratic policy formulation, the execution of these policies, and the evaluation of their results so as to improve policy making in the future. The ways in which branches of government work: in policy-making and planning, in deciding cases and in settling conflicts, are manifold and often quite different from what can be found in the private sector.

The diversity of work practices, business processes, organisational structures and institutional settings in state, politics and administration cannot easily be described. It also depends on legal and political preconditions, which furthermore differ considerably from country to country. Especially at the level of the European Union, this diversity is seldom acknowledged. Moreover, vendors have an interest to downplay it in order to sell their products to more than just one country. But the governmental systems that have evolved on the European continent have a very complex structure. The French, the German and also the former Austro-Hungarian model of public administration are still very influential, and they are quite different from British or American governmental traditions. Many EU-financed European pilot projects have stumbled over these differences.

Most important is a look at what are the typical results, or products, of the executive branch of government, including local government and including also mixed forms of production like public-private partnerships. Only to a minor part are the products of administrative activity typical services, where individual "customers" can be identified. More often than not, they benefit a multitude of addressees, providing public goods e.g. in the form of common infrastructures. Examples include road construction and maintenance, or police patrols. Other kinds of government services consist in financial transfers, which aim at a redistribution and at social justice by giving to some and taking from others.

But the most important type of a administrative action is regulation through authoritative decision making, which takes place e.g. in granting licenses, in allocating rights and duties. Administrative decision-making has mostly to do with more or less complex situations where political and legal regulations are applied to distinct cases. To give an example, a building permit will not only benefit a houseowner, but compel the neighbours to tolerate building activities.

Finally, among the activities of the public sector we should not forget information management activities, especially in the large field of building up and maintaining fundamental data bases about inhabitants, land and economic activities. Such data bases include civic registers, geographic information systems, official statistics etc. They support administrative action and, at the same time, they provide services to the economy.

A closer look at the types of processes and products which are characteristic of the public sector is required for assessing which type of information system could support them. To complete the picture, we should also mention the policy making side, e.g. in the legislative branch of government. For many situations, there is no possibility of importing ready-made systems from the private sector. A case in point is "E-council": a system to support the deliberations and the work of local government council members (Schwabe 2000). Such systems are specific to the public sector, and there is not much willingness as yet to spend public money for their development.

4 Taking the Potential of Information Technology Seriously

Not only is government and public governance an uncharted field for many of those who presently pay lip service to e-Government. Also the vast enabling potential of IT, beyond what is to be found on the market so far, remains largely unacknowledged. It could be brought to improve many processes and structures in the public sector.

In our book, we conceive of the history of IT use in the public sector as a series of application generations, reflecting the respective state of advancement in hard- and software (Lenk and Traunmueller 1999, p.21ff.). e-Government is no exception to this. Here, the most relevant feature is communication and world wide information access over the Internet. Each generation of IT carried some general guiding ideas about what could be done with the technology. An example is provided by the idea of creating huge data banks (as well as that of regulating their use through data protection legislation), which took shape in the wake of disk storage devices. Another example is the "paperless office" as a guiding idea which was prompted by the advent of the PC. Each IT generation suggested new applications, and the practice of business was perceived principally in the light of what the latest generation of computers or information systems could do to support it. The general pattern is that problems always tended to be perceived in the light of available solutions.

The development of administrative informatics can thus be understood in a sort of dialectical movement. New applications suggested by new waves of technology seemed to arrive just in time so that problems besieging a field of practice could be tackled. The new generation of technology seemed to hold the ultimate solution for all problems. Yet when the new perspective was put into use it soon appeared that its promise was only partial. It became clear that under the spell of a central guiding idea its promise was overstated.

e-Government is firmly anchored within this trend. The current fashion of looking at administrative processes from a citizen-interaction perspective is just a continuation of the temptation to seek inspiration from technological progress and to derive from it guiding principles for good practice. To stress the positive side: now that, with the help of Internet technology, new forms of electronic service delivery appear possible, the problems of citizens in their dealings with administrative agencies seem to be taken seriously for the first time. But there is the other side of the coin as well: the interaction is interpreted in a way so as to make technology-driven solutions appear as valid solutions to them. It is seldom question of *social* innovations in administrative or political practice, which are IT-mediated or IT-enabled (Hoff et al. 2000). Not surprisingly, many truly important policy fields have not got yet advanced IT support. Providing services to handicapped persons, providing neighbourhood social services, or dealing with people with immigration status are hardly given a thought in e-Government strategies. A large part of the population seems to be simply absent in political statements about the E-Society.

5 Political Support – A Window of Opportunity

But nevertheless, the fast growing political interest in e-Government arouses great hopes. In order to prevent such hopes from dissipating, we now have to look for quick and tangible success for important groups of stakeholders. There are encouraging signs. Inter-organisational cooperation which is of vital importance to the innovation alliances has considerably increased. Even in a very complex polity like Germany where local governments compete with each other, and moreover are extremely jealous of anything the Land or the Federation does, cooperation is progressing. Still, the lacking willingness of many agencies to make investments in long-range projects, as well as the reluctance to spend money for qualifying staff, are points of distress.

Another point of concern is the lack of clear visions of what a modern public sector should look like. Among the central questions that have to be answered is the following one: Under which conditions do we want our public organisations to function in the future? Which products and services do we want them to provide? And should these be produced and/or delivered by public organisations themselves or from external sources or in partnership with others? The lack of well-founded visions of a modernised public sector becomes obvious when actors trying to promote e-Government find it difficult to figure out viable business models for new IT-based administrative services.

If such visions are not developed, the temptation will persist to look at daily practice only in the light of what the technology can do to improve it. In the end, therefore, strategic thinking will be required, Only if well-founded visions of the future work of state and administration will be developed, will e-Government become a lasting success.

6 Implementation – The Hidden Threat to e-Government

According to a recent management brief issued by OECD (OECD 2001), the inability of governments to manage large public IT projects threatens to undermine efforts to implement e-Government. In a climate of euphoria, it is easy to overlook hindrances on the way to a lasting improvement of governmental and administrative practice. Political discourse tends to lose contact with the reality of what can be achieved with given resources and in a reasonable period of time.

Action has to be taken to improve the conditions for successfully implementing e-Government projects. On one hand, thanks to the evaluation of past technical inventions we already have considerable knowledge about the success factors for projects and their diffusion. On the other hand, again and again we forget the lesson learned. One reason is that too many experiences made during implementation are generally scattered and not communicated. Also there is a widespread inclination to ascribe implementation difficulties to an immature state of technology.

Furthermore, there is a gap between those making concepts and those who have to implement them. The technical and logistic implementation of solutions is usually under the responsibility of field organisations and their management. In adapting software to the structures and working processes of the organisation they often miss adequate support for planning and implementing the required organisational changes. Software suppliers tend to provide technical solutions to complex socio-technical problems. Theirs is the role of an engineer, but there seems to be no architect in charge of the overall human-machine interaction system. Procedures of systems design will have to evolve toward holistic methodologies, balancing the technology package and the complex socio-technical work reality (Lenk and Traunmüller 1999, p.93ff.).

There is thus a real danger that e-Government will glide down the slope from a mountain of euphoria into a valley of deception. Only in broadening the concept and in recalling its basic tenets will we steer it toward lasting success.

References

1. Hoff, J., Horrocks, I, Tops, P. (eds.). Democratic Governance and New Technology. Technologically mediated innovations in political practice in Western Europe. London, New York: Routledge, 2000.
2. Lenk, K. Information systems in public administration: from research to design. In: Informatization in the Public Sector. 3, 1994, pp. 307-324.
3. Lenk, K. Reform Opportunities Missed: Will the innovative potential of information systems in public administration remain dormant forever? In: Information, Communication & Society. 1, 1998, pp. 163-181.
4. Lenk, K. Elektronische Bürgerdienste im Flächenland als staatlich-kommunale Gemeinschaftsaufgabe. In: Verwaltung & Management 8, 2002, pp.4-10.
5. Lenk, K., Traunmüller, R. Öffentliche Verwaltung und Informationstechnik (=Schriftenreihe Verwaltungsinformatik 20) Heidelberg: Decker, 1999.
6. Lenk, K., Traunmüller, R.. *Perspectives on Electronic Government.* In: Galindo and Quirchmayr (eds.). Advances in Electronic Government, Proceedings of the IFIP WG 8.5 Conference in Zaragoza, 2000, Zaragoza: University Press, 2000, pp. 11-27.
7. Lenk, K., Traunmüller, R. Broadening the Concept of Electronic Government. In: J.E.J. Prins (ed.), Designing E-Government, Amsterdam: Kluwer, 2001, pp.63-74.
8. Lenk, K., Traunmüller, R., Wimmer, M.A. The Significance of Law and Knowledge for Electronic Government. In A. Grönlund (ed.), "Electronic Government - Design, Applications and Management", Hershey (PA): Idea Group Publishing, 2002, pp. 61-77.
9. OECD PUMA Public Policy Brief No.8, March 2001. The Hidden threat to E-Government. Avoiding large government IT failures.
10. Rapport Lasserre. L'Etat et les technologies de l'information et de la communication: vers une administration à accès pluriel. Paris: La documentation francaise, 2000.
11. Schwabe, G. E-Councils – Systems, Experiences, Perspectives. In: A.M.Tjoa et al., Proceedings of the 11th International Workshop on Database and Expert Systems Applications, 4-8 September, 2000, Greenwich. Los Alamitos: IEEE Press, pp.384-388.
12. Snellen, I.Th.M.; van de Donk, W.B.H.J. (eds.). Public Administration in an Information Age. Amsterdam: IOS Press, 1998.

Centralization Revisited? Problems on Implementing Integrated Service Delivery in The Netherlands

Jeroen Kraaijenbrink

University of Twente, P.O. Box 217, 7500 AE Enschede, The Netherlands
j.kraaijenbrink@sms.utwente.nl

Abstract. In the Netherlands, the development of integrated public service delivery has been an important topic for over a decade. Despite the investments, the results are meager. In the literature, an overwhelming and contradictory amount of conceivable problems is mentioned that can explain these lagging results. Four case studies were carried out to find out which of these problems are most pressing in the particular context of integrated public service delivery. These are found to be: (1) indistinct and subdivided responsibilities, (2) focus on the autonomy of the own organization, and (3) insufficient scale. Given these problems, and given their different importance in the four cases, it is argued that the effective development of integrated public service delivery in the Netherlands requires more centralization.

1 Introduction

Enhancing the level of public service delivery has received much attention in the Netherlands but has not lead to substantial results yet [2], [4], [7], [9]. This applies in particular to the integration of service delivery (ISD). In short, ISD means that multiple public organizations cooperate to deliver their services in an integrated way, usually by means of an integrated counter. By doing so, these organizations try to offer a solution to the problem of fragmentation. Fragmentation of service delivery is seen as an important problem for at least thirty years in the Netherlands [8], [10], [11], [27]. It is considered problematic for both citizens and government. Whereas citizens cannot find their way in the bureaucracies, government does not reach its citizens, and public policy remains ineffective.

This paper argues that three characteristics of ISD explain why results are meager. First, to realize ISD, interorganizational *cooperation* is a necessity. Organizations must tune work processes, create new services together and mutually adapt their applications. This implies a major *change* for participating organizations, which can be problematic on its own. The third characteristic is the use of *information and communication technology* (ICT). Because public services are to a large extent information services [1], exchange of data and information is one of the crucial elements of ISD. It is unthinkable that this exchange of information can take place without the support of ICT.

R. Traunmüller and K. Lenk (Eds.): EGOV 2002, LNCS 2456, pp. 10–17, 2002.

Given the literature, there are numerous theoretical reasons for the failure of interorganizational cooperation, organizational change and information system development, e.g. [5], [14], [15], [16], [19], [22], [23], [24], and [25]. Moreover, literature on success factors and solutions is innumerable as well, e.g. [3], [12], [13], [18], [19], [23], and [30]. Adding only the findings of this limited number of authors, at least 150 different reasons for success and failure appear [17]. However, it is not evident which of these are most important in the particular context of ISD in the Netherlands. Therefore this literature does not provide organizations with helpful insights in problems associated with ISD and situations in which these are more probable to occur. Because the literature is not well adapted to the particular situation of ISD the question remains:

Which are the problems hampering integrated service delivery in the Netherlands and why do these problems exist?

This paper reports about a research carried out to answer this question. It has the following structure. The next section discusses the research method. Section 3 presents the main problems and answers the first part of the central question. Section 4 answers the second part by discussing probable causes. Finally, Section 5 discusses the research and proposes topics for further research.

2 Research Method

To answer the research question, four case studies were conducted. Because the number of ISD projects that have passed the planning phase is limited in the Netherlands, the selection of cases was mainly based on convenience sampling. However, as Table 1 shows, a certain spread in domain, approach, results, and duration was achieved. The following cases were selected:

1. Counter for (starting) companies: cooperation of mainly chamber of commerce, tax office, and municipalities.
2. Counter for (starting) companies: combination of physical and virtual counter.
3. Health counter: cooperation of nursing and old people's homes and a municipality.
4. Counter for the unemployed: cooperation of municipal social service department, job centres, and social security.

Table 1. Short Description of the Cases

	Case 1	**Case 2**	**Case 3**	**Case 4**
Start	1999	1999	1996	1996
Approach	Mainly decentralized	Mainly decentralized	Decentralized	Decentralized, then centralized
Integration	Little	Little	Virtually none	Some
Electronic communication	Little	Little	Virtually none	Much

The case studies consisted of an extensive documentation analysis and additional interviews with project leaders. A structured list of 150 theoretical reasons for success and failure was used to systematically check which of these were present and which were most pressing in each individual case. The period of data collection was September to December 2001.

3 Main Problems in the Cases

For answering the first part of the research question, this paper only presents cross-case results. The individual case studies are discussed in [17], but are not needed for the purpose of this paper. The main problems that were identified in the cases can be split up in three categories:

- Problems on indistinct and subdivided tasks and responsibilities
- Problems through a focus on the autonomy of the own organization
- Problems on scale

Problems on Indistinct and Subdivided Tasks and Responsibilities

These problems root in the legally defined organization structure of the public sector in the Netherlands. Within this structure, tasks and responsibilities are often not distinctly divided between organizations – both in a horizontal and a vertical way. As a result of this, the decision-making processes involve multiple organizations.

A. *Division of authority and responsibilities:* ISD asks for cooperation, tuning, and rethinking tasks and responsibilities. In the cases organizations emphasized their own – often legally based – responsibilities, and did not entrust them to other organizations. In Cases 1 and 3 parties tried to set up a foundation to overcome this problem. However they did not come to an agreement because none of the parties was willing or able to hand over responsibilities to this foundation.

B. *Confining role of national public organizations:* The cooperating local organizations are dependent on decisions of their national counterparts. Local organizations that try to cooperate perceive that these organizations often do not support them in their attempts. Therefore they are limited in their scope. In Case 2 the national public organizations were even called 'the common enemy' (translated from Dutch).

C. *Legal constraints and uncertainty:* A sizable part of public organizations' positions and tasks is legally defined. When parties cooperate, they must act within these constraints, which can lead to severe restrictions for ISD. In the Netherlands, tax offices are for example not allowed to make their systems accessible online. This makes electronic communication in Cases 1 and 2 difficult.

   The frequent number of small policy changes was perceived as a problem as well because it causes uncertainty. In Case 4 the concerning project leader experienced the change that was promoted as major by government just as one out of many others.

Problems through a Focus on the Autonomy of the Own Organization

Each of the participating organizations has its own identity, culture, and way of working. When they cooperate, adaptations must be made on each of these. However, organizations experienced this as problematic and therefore kept their internal focus.

A. *Giving up identity:* Placing part of organization's services behind an integrated counter makes an organization less visible to its citizens. With losing this visibility, parties in the cases also feared they would loose part of their identity. This went hand in hand with a reserved attitude toward collective responsibilities and tasks. Case 1 illustrates this with its business cards: the logo of every participating organization is on it (see Fig. 1).

B. *Tuning of work processes:* Every participating organization has its own way of working. Tuning and adjusting work processes can streamline activities. However, the cases indicate that this asks for major efforts of all organizations, which lead to nearly no adaptations at all. The project leader of Case 2 illustrated this with stating that accumulating registration forms of the seven participating organizations would lead to a seventy centimeters' pile of paper. He remarked about this: 'when we interfere with that, we just have a very sour life and tiring discussions that lead to nothing' (translated from Dutch).

C. *Lack of standards:* Because of the large diversity of applications and formats in use in the participating organizations, electronic information exchange between them is far from easy. However, parties did not overcome this problem in other ways than by sending faxes, making phone calls, sending e-mails, and retyping data. There was also a lack of standards in definitions. In Cases 1 and 2 every participating organization had its own definition of entrepreneur on which they did not agree till date.

Fig. 1. Example of a Business Card

Problems on Scale

Because of the local level of the ISD, the projects have a relative small scale. Due to this small scale, organizations do not have sufficient resources and power to fulfill the needs for a sound integrated counter.

A. *Insufficient personnel:* During the ISD-projects organizations had to continue their regular activities. This lead to capacity problems for the projects. Moreover, because some important functions were only occupied by one or two persons, illness and dependence on single persons was reported as a problem. As a result of this, in Case 1 the communication about the project was stopped for half a year because of a long illness of only one of the project members.

B. *Insufficient financial capacity:* Parties indicated that the development of the necessary infrastructure and mid-office functionalities largely exceeded their bearing power. According to them these investments should be made at a higher aggregation level. The project leader of Case 2 indicates that the costs of building a suitable mid-office are estimated at about ten million Euro, with a budget of about one million Euro for this complete individual project.

C. *Ensuring privacy, authenticity, and security:* Electronic information exchange is an important enabler for ISD. Because of sensitivity of information however, exchange should be secure. This exchange is not limited to the local participating organizations because some databases are owned at a national level. Therefore, this can hardly be arranged at a local level. Case 4 illustrates that it is also difficult at a national level. Although electronic integrated information exchange was enabled, and security was claimed, the system was already hacked in its first few days of use.

Next to the nine problems in total, there were other problems mentioned as well. However, considering the four cases together, these nine appear to be the most pressing problems out of the 150 reasons for success and failure mentioned in theory. Yet, they do not appear in the same extent in each case. The next section discusses these differences.

4 Situations in Which They Are More Probable

Now that an answer is given to the first part of the research question, this section discusses the second part, which is about causes of the two types of problems. This section compares the approaches followed in each of the four cases.

Cases 1 and 2 seem to be most typical for the current way of developing public integrated counters in the Netherlands. As Table 1 shows, these cases have attained relative little integration and electronic communication. These two cases came also with most identified unsolved problems. The relative recent start of the projects could explain this. However, project leaders did not expect that the current approach would indeed lead to better integration and electronic exchange of information. According to them, solutions are needed on a higher organization level.

In the third case less problems were identified, but this project was also relatively unsuccessful in terms of level of integration and electronic communication. Because of the relative autonomous position of participating organizations, almost no problems were mentioned on tasks and responsibilities. Problems on scale were also scarcely mentioned, but this could be explained by the fact that ICT investments and attempts to really integrate were postponed.

In the fourth case all of the above-mentioned problems were identified during the project. This was particularly true in the stadium in which the project shifted from a decentralized to a centralized approach. However, after taking radical measures like changing laws and reorganization of the social security sector, this lead to relative good results on integration and electronic communication, as shown in Table 1.

Considering the different approaches of the four projects, it seems that the current most common approach in the Netherlands comes with most identified unsolved problems. This approach has centralized and decentralized elements in it, which means that none of the organizations on both national and local level has complete responsibility. It seems that as a result of this not much really happens. Although other causes exist, based on these limited number of cases I state that the current mixed approach is a major cause of the problems on integrating service delivery in the Netherlands.

Cases 3 and 4 support this statement. Both a centralized (Case 4) and a decentralized approach (Case 3) lead to fewer identified unsolved problems. In Case 3 few problems were reported, but as mentioned above, the results were relatively limited, partly because of limited resources. Considering the better results and solution of problems in Case 4, it seems that a centralized approach fits better to the needs of ISD. Conceivable reasons for this are that sufficient resources, capacity and authority are present to really change situations and reorganize towards a more integrated way of service delivery.

5 Towards More Centralization?

At first sight the suggestion that ISD requires a more top-down approach may seem outdated and in contrast with the current literature on organizational change and cooperation that favors participation and a bottom-up approach e.g. [5], [14], [23]. Considering the fourth case it appears indeed that forcing parties to cooperate just leads to frustration, stagnation, and a rigid attitude of parties and no results. However, after taking more radical measures like changing laws and redefining parties, it seems to lead to better results at the end. Therefore, it is not the obligation that leads to better results, but the reorganization that takes place. I state that problems exist because the current organization of the Dutch public sector does not fit the needs of ISD. ISD is more likely to succeed when this mismatch is eliminated. This is in accordance with interaction theory, which states that organizational problems should be fixed before introducing systems and that the organization should be in line with these systems [20], [21].

The preference for a top-down and discontinuous approach resembles much of what is said by Hammer and Davenport about Business Process Reengineering as well [6], [12]. Although BPR is about single private organizations and has not been very successful in the public sector [16], [28], it seems that at least some principles of this method are valuable in this particular context.

Although no cases were analyzed in which a centralized approach was taken from the beginning, some experiences in other public electronic information exchange projects indicate that it can be a successful approach. An example is the Kruispuntbank Belgium (enabling electronic information exchange in social security). Compared to the RINIS initiative in the Netherlands – which aims at similar results – it appears that in Belgium a centralized and drastic approach, involving changes in law, has lead to better results years ago than achieved with RINIS at this moment [30].

Of course these results are not a sufficient proof of the value of a centralized approach for ISD in the Netherlands. Therefore more and quantitative research is necessary to explore under which circumstances more centralization is advisable.

Although this paper suggests that a centralized approach has benefits for ISD, it is not a pleading for centralization in general. It is only due to the current relative decentralized approach of ISD in the Netherlands that more centralization is worth considering. This implies that the benefits that Simon relates to centralization – coordination, expertise, and responsibility – at this moment outweigh the costs, including higher workload of higher paid personnel, higher communication costs, and less available information [26].

Acknowledgements

The author would like to thank dr. Ronald Leenes, prof.dr. Robert Stegwee, dr. Jörgen Svensson, and dr. Fons Wijnhoven of the University of Twente for their comments, suggestions, and corrections, which greatly improved this paper.

References

1. Broek, M. van der: Informatievoorziening in de Excellente Gemeente. Elsevier Bedrijfsinformatie (2000)
2. Cap Gemini Ernst & Young: Web-based Survey on Electronic Public Services. Brussels: European Commission DG Information Society (2001)
3. Cavaye, A.L.M. & P.B. Cragg: Factors contributing to the success of customer oriented interorganizational systems. J. of Strategic Inf. Systems, vol. 4 (1995), no. 1, 13-30
4. CCTA (Central Computer and Telecommunications Agency): Benchmarking Electronic Service Delivery. A Report by the Central IT Unit (2000)
5. Daft, R.L.: Organization Theory & Design. West Publishing Company, Minneapolis/St. Paul, fifth edition (1995)
6. Davenport, Thomas, H.: Process Innovation: Reengineering Work through Information Technology. Boston, Massachusetts. Harvard Business School Press (1993)
7. Deloitte & Touche: Overheid oNLine. Trendonderzoek naar Gemeenten en Provincies op Internet. Onderzoeksrapport (2000)
8. Derksen, W. et al.: De Blik naar Buiten: Geïntegreerde Dienstverlening als Structuurprincipe. Wetenschappelijke Raad voor het Regeringsbeleid. Den Haag (1995)
9. Duivenboden, Hein van, & Mirjam Lips: Klantgericht Werken in de Publieke Sector. Inrichting van de Elektronische Overheid. Uitgeverij LEMMA BV, Utrecht (2001)
10. Eenmalige Adviescommissie ICT en Overheid: Burger en Overheid in de Informatiesamenleving: de Noodzaak van Institutionele Innovatie. Den Haag. September (2001)
11. Frissen, P.H.A.: De Virtuele Staat: Politiek, Bestuur, Technologie: een Postmodern Verhaal. Academic Service, Schoonhoven (1996)
12. Hammer, Michael: Reengineering Work: Don't Automate, Obliterate. Harvard Business Review (1990), July-August, 104-112
13. Hammer, Michael & James Champy: Reengineering the Corporation: a Manifesto for Business Revolution. New York. Harper Business (1993)

14. Jarrar, Y.F. & E.M. Aspinwall: Business Process Re-engineering: Learning from Organizational Experience. Total Quality Management, vol. 10 (1999), no. 2, 173-186
15. Kern, H. et al.: Building the New Enterprise: People, Processes, and Technology. Sun Microsystems Press. USA (1998)
16. Kock, N.F., Jr. & R.J. McQueen: Is Re-engineering Possible in the Public Sector? A Brazilian Case Study. Business Change and Re-engineering, vol. 3 (1996), no. 3, 3-12
17. Kraaijenbrink, Jeroen: De Lange Weg: een Onderzoek naar Knelpunten bij Interorganisationele Samenwerking rond Geïntegreerde Loketten. Graduation Essay. University of Twente, Faculty of Technology & Management. Enschede (2002)
18. Kurnia, S. & R.B. Johnston: The Need for a Processual View of Inter-Organizational Systems Adoption. Journal of Strategic Information Systems (2000) no. 9, 295-319
19. Lawrence, Peter: Workflow Handbook 1997. John Wiley & Sons. Chichester (1997)
20. Markus, M. Lynne: Power, Politics, and MIS Implementation. Communications of the ACM. Vol. 26 (1983), no. 6, 430-444
21. Markus, M.L.: Systems in Organizations: Bugs and Features. Pitman (1984)
22. Rose, G., H.Khoo & D.W. Straub: Current Technological Impediments to Business-to-Consumer Electronic Commerce, In: Communications of the Association for Information Systems (1999). http://members.aol.com/grose00000/cais/article.html. Visited. March 9, 2001
23. Sabherwal, R. & J. Elam: Overcoming the Problems in Information Systems Development by Building and Sustaining Commitment. Accounting Management & Information Technology, vol. 5 (1995), no. 3/4, 283-309
24. Senge, Peter et al.: De Dans der Verandering: Nieuwe Uitdagingen voor de Lerende Organisatie. Academic Service. Schoonhoven (1999)
25. Shaw, M. et al.: Handbook on Electronic Commerce. Springer-Verlag. Berlin (2000)
26. Simon, Herbert A.: Administrative Behavior. A Study of Decision-Making Processes in Administrative Organizations. Fourth edition, The Free Press, New York (1997)
27. SUWI: Structuur Uitvoering Werk en Inkomen: eerste Voortgangsrapportage. 13 Oktober (2000). http://www.suwi.nl, visited Oktober 18th 2001
28. Thaens, M., V.J.J.M. Bekkers & H.P.M. van Duivenboden: Business Process Redesign and Public Administration: a Perfect Match? Paper presented at the Conference of the European Group of Public Administration. Rotterdam. September 6-9 (1995)
29. Tidd, Joe, John Bessant & Keith Pavitt: Managing Innovation: Integrating Technological, Market and Organizational Change. John Wiley & Sons. Chichester (1997)
30. Zuurmond, A. et al.: Dienstverlening Centraal. De Uitdaging van ICT voor de Publieke Dienstverlening. Pre-advies Raad Openbaar Bestuur. (1998)

From Websites to e-Government in Germany

Dieter Klumpp

Alcatel SEL Stiftung, Lorenzstr. 10, 70435 Stuttgart, Germany
d.klumpp@alcatel.de

Abstract. European neighbors wondered at Germanys relatively slow start into Electronic Government. With a certain anxiety they had looked at this nation of eighty millions with its five million of public employees - the level of entire countries like Norway, Finland or Denmark. Consequently Europe was not surprised that Germany in 1997 took over a pole position presenting the first law on digital signature, an important element for Electronic Government purposes. But it turned out that this early law meant only a pre-dawn. It turned out that Germany - as it is undoubtedly big - needed much time and energy to get itself in motion. The presumed "giant" evolved to be a specter of mostly isolated actions and projects. Since two years - beginning with a memorandum on Electronic Government triggered by the two major IT associations - there are more and more actors striving for joint action. German Federal Government set up the program "BundOnline2005" putting comprehensive targets and respectable money into the necessary transformation of government processes, aiming at - as Chancellor Schroeder stated - "not citizens but data have to run". A public-private initiative "D21" is backing jointly strategies. Germany now appears to reach necessary pace to play the expected visionary role as well as to overcome the obstacles every innovative infrastructure at the beginning is confronted with.

1 Introduction

Germanys administration system is somewhat admired all over the world. But looking a bit closer there are rather the secondary virtues like "punctuality", "correctness", "quality", "thoughtfulness" and even "Prussian", which are connected with the image of Germany since times. Germans themselves of course hate their "bureaucracies" as all other nations do with theirs, but Germans are convinced that there exists some special kind of "reliability" within the German administration system. Without going deeper whether the German image is reflecting merely historical inheritance or today's reality, it appears to international observers that Germanys ranking in the Electronic Government sector has been on a quite surprisingly humble level for years. In early 2001 the French Prime Minister after presenting the first three years of "AdmiFrance" was dazzled and proud to learn from German Minister of the Interior that he felt Germany some years behind of France concerning the Electronic Government implementation.

Even after two years of a dedicated modernization program led by German Ministry of the Interior Germany has not found yet its specific role within the European movement towards Electronic Government. But this role is needed –

R. Traunmüller and K. Lenk (Eds.): EGOV 2002, LNCS 2456, pp. 18–25, 2002.

Germanys contributions to structural modernization are estimated to be very valuable. The outstanding example was the development of the national rooted ("accredited") Digital Signature coming up with the first worldwide Federal law 1997 which became a pattern for implementation plans of many countries [1]. European countries in this early stage were somewhat reluctant with such German innovations because they feared that Germany was building up "too exhaustive" quality standards which could hamper the e-Business sector.

The "career" of the topic "Electronic Government" in Germany over the last five years shows some quite different roots, out of which some of the most important shall be outlined hereafter.

2 Cities "Internet Projects" As Website Collections

In mid Nineties the first worldwide parliamentarian commission on multimedia in the federal state of Baden-Wuerttemberg had the vision of a "tele-administration" being possible with the forthcoming broadband era. Visionary members of the group as Franz-Josef Radermacher predicted that Electronic Government would not arise out of technical possibilities but it would be compelled by the necessity "that we will have do 120% of public sector work with 80% of the clerks". Such forceful hints to Electronic Government procedures as a necessary means for rationalization vanished with the rising expectations of the public awaiting "millions of workplaces to be created by multimedia" (former EU commissioner Martin Bangemann).

German cities discovered Electronic Government predominantly participating in regional or EU programs [5]. As all over the world cities are confronted directly with the citizen some local administrations felt obliged to use the internet as an immediate opportunity to close an information gap between citizens and the governing authorities by offering electronic leaflets. The city projects tried to set up internet based "services for all" which in fact turned out to be not much more than the electronic access to the habitual paper-stuffed and always-occupied bureaucracy allowing to choose the license plate number for cars, asking for a paper form by mail up to clicking the Lord Mayor presented in cute colour photos. Cities, villages and their associations began to set up workshops on the new media "internet", mainly backed and funded by big players from the IT sector. The worldwide wave of public-private initiatives like the "Bay Area Initiative" slowly swapped over to German circles without raising the urge to modernize with all the power given [10].

With all respect to the early adopters one could witness the false step [7] towards Electronic Government: Simply "cloning" the procedures and patterns of e-Business for Electronic Government purposes leads to isolated public websites. This "experimental way" meant a tremendous loss of time and money for German cities. Each "successful" project overlapped with such of hundreds of 24.000 German cities leaving everywhere the problem of updating websites with too few people and resources. The most influential case for years was the city of Mannheim[1], where the two main modernizers, the Lord Mayor and his Communication Director succeeded in implementing a varied website with only a handful of contracted young IT specialists. There were very few cities following this "best principle" way - each one claimed to

[1] see www.mannheim.de and Blumenthal, J. in Alcatel SEL Stiftung [12]

have a “best principle” of its own. Similar leapfrog competitions[2] arose between the 16 German states [8].

3 Digital Signature and the Media@Komm Lead Projects

Another important root can be discovered in the broad German Science specter reaching from Public Administration Competence Centers like Speyer [9] to Knowledge Management and Organization Systems like the Fraunhofer Society Institute in Stuttgart. In 1996 one of Germanys most influential IT-specialized jurists, Alexander Rossnagel, hold the guest professorship funded by Alcatel Foundation at the Technical University of Darmstadt using this fruitful period to write down a first draft of the German Law on Digital Signature which was brought to final decision in 1997 [10]. It lined out an infrastructure based on public keys and long-term certified roots given by public or private trust centers. This innovation had to pass from the beginning a nightmare of misapprehension or better: a lack of understanding. The different actors in Germany and EU directories discussed the “security of the 128 kb key”[3], some businessmen confused signature with “a digital stamp imposed by authorities for every electron3ic move on the internet” and a few even suspected that the visible hand of the state was grasping for new tax revenues threatening to paralyze Internet economy”.

German Ministry of Research meanwhile had built up a multimedia department promoting all kind of research for the “Information Society”, which was after the 1999 election transferred to the Ministry of Economy and Technology. The “German” Digital signature was in worldwide discussions consequently milled to an “European” family of “electronic signatures”, out of which the “self-determined user” should freely choose between worthless short-term software products and a long-term guaranteed infrastructure service[4]. One of the major tasks of the Media@Komm Lead Projects, funded with 30bn should be the rollout of digital signature as infrastructure reaching the “critical mass”. But from the beginning the “critical mass” could not be reached because the projects had to undergo a “full competition” way to find out ten winners out of 120 proposals. After many months of finding out of the ten the three winning proposals, even the winners (Bremen, Esslingen and Nuremberg-Erlangen) were not enabled to co-operate for another year. Consequently every project group had to begin from the bottom, EU project organization insisted even on the founding of private companies to deal with the private funding corresponding to the portion of public money.

The city of Bremen - which represents also the small state of Bremen - after two years climbed at the top of German communities with a broad interactive Electronic Government website and with a strong focus on standards. The impetus in Bremen emanates undoubtedly from scientific knowledge built up in years around the universitarian chair of Herbert Kubicek, supported by reform-oriented politicians.

---

[2] benchmarking is one of the favored marketing tools of the ever-so-present consulting firms

[3] a topic of kryptology

[4] countries like Japan and even European countries meanwhile discovered the charming of a nation-rooted Public Key Infrastructure and are settling a knowledge transfer with German Root Authority. Germany itself will need some time to join the way it opened years ago

4 The Memorandum "Electronic Government"

The lead projects on Electronic Government lacked cooperative structures. In early 2000 representatives of German Information Technology Society (ITG) and German Informatics Society (GI) met in Frankfurt to set up a Working Group on Electronic Government comprising more than 40 representatives of science, administrations, companies, unions and politics. After six months of bi-weekly meetings Klaus Lenk (Head of Administration Informatics of the Oldenburg University) presented the policy paper to State Secretary Brigitte Zypries and gathered experts in September 2000 stating: "With the pervasive success of the Internet, many opportunities are now emerging which together can be seen as leading to an "Electronic government". This expression is not meant to designate improved citizen services only. Instead, "government" is taken in a wide sense to denote governing and administrating in a democratic system including democratic processes of policy formulation. (...) The challenges addressed by the new motto are at least as great as those of electronic commerce. In view of the diversity of tasks involved in government and administration, they are even much broader and more varied" [2].

This vision came too early to prevent the specter of Electronic Government actors from falling into the e-Commerce depression a year later. The Memorandum quickly became the silent referential point in German discussion, it is still now present with hundreds of copies in all websites from associations, politics, science, cities foundations, trade unions and consulting companies. Being the presumably most important paper in German discussion it nevertheless has never been translated into another language. The latter is paradigmatic for the German way: the 42 who signed it up got a lot of applause, but none of those who applauded felt empowered to put it on the international scene simply translating it. All agreed to the recommendations inclusively the admonishing hint to begin immediately with reshaping of administration processes, "because there is open a time slot for Germany now".

The 1999 Memorandum established some important "Leitbilder" (models) to reshape the Electronic Government strategies: "From the point of view of the individual citizen, contacts in administrative offices can in the future be established via Internet portals and service shops, saving both time and effort. But the improvement of relations between citizens and the administration by means of new access structures constitutes just the tip of an iceberg. In order to get clearer view we have to add three further viewpoints which (...) are not looking at government and administration from the outside, but dealing directly with the machinery of public services and governance which constitute the larger part of the iceberg. These three viewpoints are:

- A reorganization viewpoint, consistently taking everyday (business) processes as the starting point.
- The perspective of telecooperation, which makes it clear that not only cooperation in routine affairs, but also complex and controversial negotiations may be carried out over distance, exhibiting ever less need for the persons involved to be present at one and the same place.
- A knowledge viewpoint, illustrating how much information technology can contribute to make processes more effective and transparent through support-ing knowledge as the most important asset of government and administration.

- In the triangle of relations between citizens (including companies), politics, and administration, the interplay between all of these perspectives is evident. But there is an urgent need to develop reference models and pilot projects in order to reveal the full extent of the possibilities at hand" [2].

Following Klaus Lenk the critical success factors include a strategic thinking instead of the "attitude of curious but indiscriminate trying out of different approaches". Without doubt a financing initiative is needed but very difficult to realize due to Germanys budget burdens during the still ongoing unification process. "People who drive the realization of electronic government are perhaps the most essential factor. In order to develop the requisite human resources, an unprecedented qualification effort has to be undertaken. A large number of the people working in the public sector need to become aware of the potential which information technology holds and also become able to estimate how they may better be able to structure their own working processes with this potential. This is closely related to a further requirement, namely a competent change management, which places people in the foreground"[2].

The Memorandum stated that a suitable IT infrastructure will be required providing the full range of availability and security. This clear request for "safer" infrastructures needed some time to trickle down from science and companies to the public sector including the politicians. One of the thresholds could be described as a traditional individualistic "non-cooperation attitude" and the incurable German passion to await visions from abroad, especially from overseas.

5 Thresholds Hampering Cooperation

To understand the slow start of Germany albeit being well equipped with all the necessary skills in "classic" administration, having leading scientists on Administration and well-trained professionals in all government sectors, one must examine the thresholds that hamper cooperation within Germany. As Federalism is an unchangeable constitutional basis of Germany, there are considerable difficulties to get along with common standards and procedures, which always mean a extensive degree of co-operation. After reunification German states still suffer several kinds of divides which could afford the necessary joint strategies towards Electronic Government. As even after 1989 there is a relatively sharp divide between the so-called "A states" (governed by Social Democrats) and "B states" (governed by Christian Democrats), there have grown more remarkable divides like between the "richer" (southern) and "poorer" (northern) states, between metropolitan states like Berlin versus the surrounding state of Brandenburg and last but not least the divide between the "old" (western) and the "new" (eastern) states after reunification. But all of them join immediately whenever Federal Government is suspected to interfere, because "all evil is central". Some German federal institutions and parties still try to maintain divides on such precedent topics like Electronic Government. Actors do have all types of explications why there is no need to extensively co-operate across the inner-German borders. As above mentioned, it is quite difficult to get a big mass to speed up.

Germany with its 80 Millions of people is not surprisingly the biggest public IT-market in Europe. KableNET [4] estimated Germany with (EUROS) 13,313 billions ahead of UK with 12,118 and France with 10,006 billions. With roughly 5 Millions of public employees, thereof 15% on federal level, 40% in the 16 states and 45% all over the communities, Germany counts on a big workforce in the public sector and can in future still count on it due to the mighty unionists of ver.di who clearly exclude major cutoffs, but are backing the modernization process with emphasis. Small and medium companies are discovering that Electronic Government business is much more than a niche and present products which are apt to transform paperwork to electronic processes. German SAP - spread over the globe - has reached to be No. 1 worldwide in offering comprehensive workflow systems which meet the demand of the public sector for reliability. German politicians - who are proud of their national car products - are still somewhat hesitant regarding and fostering export opportunities of German IT services, although the eastern countries are demanding for it in their phase of preparing EU membership. Russia dedicated its version of the D21 initiative, "Electronic Russia" exclusively to Electronic Government [12].

Electronic Government actors in Germany since 2001 begun looking forward to leaving individualistic approaches behind. Now it seems that after being despaired with the growing number of excellent projects that did not produce the necessary spark to the building up of common infrastructures, there are appearing new structures of cooperation and joint strategies.

6 Shared Strategies towards Implementation

In the 2002 Accenture research report, "e-Government Leadership - Realizing the Vision" [3] is perceived that "governments are, albeit slowly, realizing their visions. More importantly, there is a growing recognition that e-Government is not just about technology - but about harnessing technology as just one of the tools to transform the way governments operate". The report is finding an astonishing jump: "Germany significantly improved its overall ranking from 15th in the 2001 report to 9th this year, and was placed sixth among the Visionary Challengers. Its improved performance reflects a greater emphasis on bringing Government services online and further development of its strong base of mature services in the Revenue and Postal sectors. Following Chancellor Gerhard Schroeder's unveiling of Germany's e-Government vision, BundOnline 2005 (9/2000), the Government has implemented a range of measures to accelerate the implementation of its vision. The German Government 's focus is on the modernization of federal administrative structures that will deliver speedy, service-oriented, approachable and cheaper electronic administration by the year 2005. The Government identified 18 pilot projects over to lead its early efforts in delivering online services to citizens and businesses. They include the repayment of student loans, electronic tax declarations, and the processing of customs matters. The public procurement project, Öffentlicher Eink@uf Online is intended to combine the authorities' entire contract award process, from defining their requirements to delivering products".

The new strategic lines of Electronic Government in Germany are already visible. First of all, Electronic Government must be prevented from budget cuts that are threatening the smaller projects, most of them in communities. Second, the co-operations between cities and between German federal states must be fostered dramatically. The most likely move will be the creation of "Centers of Competence"[5]. Such centers are arising on the Federal Level, where the BSI (Federal Agency for IT Security) developed a "Handbook Electronic Government", which enables communities to develop secure administration processes, a valuable work which provides a manual for first steps and which will be equipped soon as a knowledge platform for best practices. The Mainz outlet of German Regulatory Authority presents its nation root signature organization every week to another international delegation weaving a worldwide network of partnership. In June 2002 Ministry of Interior launched SAGA[6] as important offer to the states and cities to shaping joint infrastructures [11].

Large states like North Rhine-Westphalia and Bavaria are constructing competence centers, the little (especially the eastern) ones intensify their talks about co-operation from technical infrastructure standards to "Shared Services Centers". Stuttgart Universities are creating modularized scientific entities to offer interdisciplinary approaches[7] to Electronic Government which will provide the necessary push for a Competence Center" funded by public and private partners offering Electronic Government training. Science already is acting along an European approach with partners.

"Shared Services Centers" (or as the author prefers: "Overlay Administration") could be the most important Shibboleth for Electronic Government in Germany. The discovery of co-operation is driven by the insight to concentrating the ever-too-small-budgets, to avoiding long trial-and-error, to creating new "highly interactive" and "automated" administration processes instead of passing on medieval structures to the 21st century Electronic Government. Like in other European countries the way to Electronic Government is not the easiest in Germany too. Modernizers in all sectors are confronted with their rivals who reject any change and are leaving the big tasks and burdens to the grandchilds. Local and regional politicians still prefer the "quick" internet projcct to sow up with "modernization", for their interests "rationalization" lies athwart popularity. Interaction between the European partners is the adequate remedy for the obstacles yet recognized, realizing a forceful push for innovative Electronic Government infrastructures.

References

1. Schwemmer, J., Elektronische Signaturen: praktische Erfahrungen, Beobachtungen, Schlussfolgerungen, in: Kubicek/ Klumpp/ Büllesbach et al. (Eds.), Innovation@ Infrastruktur. Zur Gestaltung der Informationsgesellschaft, Heidelberg 2002
2. Lenk, K./ Klumpp, D. (Eds.), Electronic Government als Schlüssel zur Modernisierung von Staat und Verwåltung, Memorandum des FA Verwaltungsinformatik der Gesellschaft für Informatik e.V. und des FB 1 der ITG im VDE, Bonn/Frankfurt September 2000
3. Accenture (Ed.), eGovernment Leadership - Realizing the Vision, April 2002
4. Annual Public Sector Information Technology Spending in Europe, 2000/2001, Kable Ltd., London 2001

---

[5] German term "Kompetenz-Zentrum" comprises "building up own skills" as well as "bundling administration tasks" and "knowledge transfer"

[6] The acronym for **S**tandards and **A**rchitectures for e**G**overnment **A**pplications reminds ironically to old Germanic very long stories, see: www.bund.de/saga

[7] see www.alcatel.de/stiftung (English version under construction)

5. Hagen, M./ Kubicek, H. et al.: One-Stop-Government in Europe - Results from 11 National Surveys, COST Action A 14 - Government and Democracy in the Information Age - Working Group "ICT in Public Administration", Bremen 2000
6. Klumpp, D., Von der Veränderung der Schnittstelle Mensch-Bürger, in: Kubicek et al. (Eds.), Multimedia@Verwaltung. Jahrbuch Telekommunikation und Gesellschaft, Heidelberg 1999,
7. Masser, K. / Gerhards, R., WEB-TEST II, Bundesländer im Vergleich, in: Innovative Verwaltung 5/97
8. Reinermann, H./ von Lucke, J. (Hrsg.): Portale in der öffentlichen Verwaltung, Forschungsbericht, Band 205, Forschungsinstitut für öffentliche Verwaltung, Speyer 2000, 1-6
9. Roßnagel, A.: Die digitale Signatur in der öffentlichen Verwaltung, in: Kubicek, H. et al.. Jahrbuch Telekommunikation und Gesellschaft 1999, Heidelberg 1999, 158
10. Klumpp, D./ Schwemmle, M., Wettlauf Informationsgesellschaft. Regierungsprogramme im internationalen Überblick, Berlin 2000
11. Bundesministerium des Innern (Ed.) SAGA - Standards und Architekturen für eGovernment Anwendungen im Rahmen der Initiative BundOnline 2005, Berlin, 04.06.2002 (draft v 0.9)
12. Alcatel SEL Stiftung (Ed.), Verwaltung und Region im Electronic Government, Tagungsdokumentation Brandenburg, Stuttgart 2002

BRAINCHILD, Building a Constituency for Future Research in Knowledge Management for Local Administrations*

Martin van Rossum[1], Daniele Chauvel[2], and Alasdair Mangham[3]

[1] The Hague University of Professional Education
vanrossum@thehague.nl
[2] European center for Knowledge Management
daniele.chauvel@free.fr
[3] London Borough of Camden
alasdair.mangham@camden.gov.uk

Abstract. The overall objective of this network of excellence is to train the BRAINCHILD change masters, champions and activists (called "Chief Knowledge Officers) in knowledge management for local administrations,. Applying the action learning method for a C.K.O. Graduate Course, a pilot group of C.K.O.'s will develop at the same time the strategic roadmaps for future applied research in Public Admin e-Work & Next Generation KM Systems. After validation the C.K.O. Graduate Course will be offered on a large European scale to the members of the founding BRAINCHILD networks (Telecities & Elanet) of advanced local authorities. The course will have official accreditation, from ecKM – Groupe ESC Marseille Provence and other participating academic institutes.

1 Relevance[1]

Today we are experiencing a transformation of the public administration into a demand driven public service organisation. Driving this is the citizen's need for information and services in their search for a better life as responsible and caring human beings. From this point of view, the way in which services are offered at the moment is often insufficient or difficult to access. In order to obtain the desired information /service, it is very common that several public departments and institutions must be contacted, that the information required is not always available and that consequently the process of collecting the information or requesting the service is very time-consuming.

---

* This article is a compilation of inputs from many contributors to the BRAINCHILD network of excellence, of which explicitly should be mentioned Daniele Chauvel - the Director of ecKM , the European center for Knowledge Management, at the Graduate School of Business Marseille Provence (France) and Alasdair Mangham - programme manager for the Camden Connect Team (London Borough of Camden, UK)

[1] Based on input from Alasdair Mangham

R. Traunmüller and K. Lenk (Eds.): EGOV 2002, LNCS 2456, pp. 26–32, 2002.

At the forefront of applied research today is the development of demand driven information systems as a solution for the improvement of citizen services. This line of development includes a number of research challenges. Most important is the question of how to handle the dynamic aspects of demand driven information systems. The service offered as a response to specific questions is not static but operates within a dynamic growing service context. The citizens are looking for a service which will improve the quality of daily life. From the citizens perspective it is of no interest whether the information originates from the private or public sector in arrange of activities including: childcare, education, cultural activities, tourism, sports and health care. The citizen is interested in a service that will help in the specific situation, that is quick and responsive Current trends in Electronic Government are focused at delivering seamless services for citizens. This has involved Government agencies examining the service from a citizen centric point of view and then building electronic delivery platforms based around the service that the citizen requires. The most obvious manifestation of this approach is in the creation of portals where a number of services are joined together to give the citizen the impression of a holistic service.

What often lies behind the portals is a collection of Local and Central Government departments and/or private institutions / corporations who are feeding information out of their respective silos into the portal. The raised expectations of the citizen of the joined up-front end are very often not matched by the service that is delivered. Part of the solution is undoubtedly to re-engineer the back office to match the service paradigms being dictated by the new citizen focused front end. However, it will not be possible to re-engineer the back office to match every joined up service delivery scenario.

The creation of joined up service delivery models also raises expectations in the citizen of a joined up policy between the government actors. There is a common expectation that the joined up front end will be able to deal directly with up to 80% of routine transactions, with the 20% that require expert intervention being routed through to the back office.

The concentration of the tools tends to be on ways of managing the 80% of routine enquiries. Whilst these tools are important for the delivery of a seamless service to the citizens they do not enable either the experts or the policy makers to act in a joined up way. One important aspect of dealing with the more complex enquiries is the ability to join up data held in back office systems. However, data without context is meaningless. Data that can be viewed in context becomes information. Information that is analyzed and can be applied is knowledge. When this knowledge is distilled, organized, stored and redeployed according to specific user needs, then a corporation / organisation is employing Knowledge Management. For these reasons future applied research, for which BRAINCHILD network of excellence intends to design the strategic roadmaps, should target both the development of e-Work Systems as well as organisational knowledge management tools for accomplishing process improvements in public administrations. The alternative approach will be taken into account as well, that is to exploit artificial intelligence techniques to represent bureaucratic rules and procedures. Represented declaratively, bureaucratic rules and procedures, are not concealed in the black box of computer code, but made manageable by other (knowledge base) software tools.

The specific focus in that case will be to address the problem where a client must deal with the rules and procedures of several different agencies simultaneously to solve his/her problem. For political (etc) reasons, it is not always likely the diverse agencies will integrate their operations. Instead, we will alternatively also take into account a technical solution, in the form of *virtual integration* of procedures. This relies on the procedures being represented declaratively ,through utilising the new web services such as WSDL/UDDI and SOAP so other software can provide *on-the-fly re-engineering* to deliver a customised, integrated view of the procedure to the client.

To benefit from every customer or partner interaction, public administrations must give employees opportunities to record what was learned. Then, other employees must have access to the data and information and the means to understand it in context. Knowledge management solutions help an administration to gain insight and understanding from their own experiences. When employees use this KM system, best practices are perpetuated throughout the organization, and each employee has the tools to be as effective as the best employee. To the extent knowledge and information are the core assets of a public organisation, knowledge and information management become the means for ensuring success. From this perspective, the knowledge-related challenges facing the public and private sectors are similar. Public administration must develop a critical awareness of the value of its stock of knowledge, manage this resource optimally, and utilize it for more proficient service delivery (and policy making). However the role played by change masters, champions and activists is crucial for successful change in organisations. The overall objective of this network of excellence is to (develop training courses for) train the BRAINCHILD change masters, champions and activists (called "Chief Knowledge Officers) in knowledge management for local administrations,. Applying the action learning method for a C.K.O. Graduate Course, a pilot group of to be trained C.K.O.'s will develop at the same time the strategic roadmaps for future applied research. After validation the C.K.O. Graduate Course will be offered on a large European scale to the members of the founding BRAINCHILD networks (Telecities & Elanet) of advanced local authorities. The course will have official accreditation, from e[c]KM – Groupe ESC Marseille Provence and other participating academic institutes.

The research should be based on results from earlier & current studies in European IST projects within the area of "smart government" In that sense BRAINCHILD will "cluster" ongoing IST project experiences. One of the requirements for the C.K.O. Graduate Course applicants will be that they play – on behalf of their local authorities – a key role in current "smart government" projects. It will be our intention to include from the outset local authorities from NAS-countries.

2 New Paradigm

During the past decade the source of economic competitiveness has shifted from the industrial paradigm to one that is knowledge-based. The value of know-how, knowledge and the intangible assets embraced by an organisation has become increasingly more important in comparison to tangible assets. The social reality within the industrial paradigm is conditioned by a high degree of formality, rule-based hierarchy, standardisation and an evasion of confrontation and criticism. The new paradigm, which is now emerging on the business landscape, calls instead for transversal and flat organisational structures, fast-innovation processes, managerial agility and interpersonal effectiveness.

3 Knowledge Management[2]

Knowledge Management (KM) is an emerging phenomenon variously at the centre of global economic transformation, organisational success, the eventual demise of private enterprise capitalism, new forms of work and the forthcoming paradigm shift from InfoWar to Knowledge Warfare (K Warfare). Competitive advantage is located in "learning organisations", "brain-based organisations", "intellectual capital" and the "economics of ideas." Knowledge has assumed this centrality in conjunction with sweeping changes in organisational forms and the dawning post-industrial and information revolutions. Definitions of and approaches to KM show that the meanings associated with "Knowledge Management" are multiple rather than singular, and the field has a set of intellectual roots which are neither incongruous, not consistent, but certainly different in their understanding of the matter. Nonetheless, actors and observers now agree that Knowledge Management is an important domain of study and application that has the ability to deliver significant business benefits.

The domain of Knowledge Management is becoming more sophisticated and extending its reach, reflecting the impact of emerging technologies as well as the influx of specialty domains (psychology, sociology, the organisational sciences) and new fields of endeavour (Public Administration, not-for-profit organisations). KM is no longer a systems issue, but rather a holistic initiative that understands organisations as socio-technical phenomena whatever their core activity.

As new-paradigm organisations become more knowledge-based, work becomes more knowledge-intensive. This truism applies to private and public sector organisations. It quickly becomes critical to understand the way people think, learn, use knowledge, and how these knowledge processes are embedded in the systems, structures and routines of an organisation. The knowledge assets in play include those that are explicit, easily accessed and codified, but especially those that remain implicit and locked in a psychosocial context. Knowledge Management is focused on the systematic development of all such assets.

Experience shows that successful organisational change requires (1) thorough diagnosis and scenario analysis, (2) clear change objectives, (3) organisational commitment that stems from the unfailing support of top management, (4) change masters or champions who provide leadership, and (5) careful, incremental reinforcement of any progress achieved. These principles apply to all types of organisations, whatever their sector of activity or change objective. A decade of experience with Knowledge Management change efforts similarly confirms these guidelines for action.

4 Public Administration

Public administration has pressures and constraints that are similar to those of any business enterprise. The private sector's primary objective is wealth creation, and the primary means for doing so increasingly centre on knowledge and knowledge processes within an organisation. The primary objective of public administration is social stability and effective public service, and historically the primary means for doing so has centred on legislation, procedural and information management.

[2] This and following paragraphs based on input from Daniele Chauvel

Public administration is now being called upon to act in an increasingly prosperous and knowledge-rich society while addressing the problems of the "digital divide." A radical transformation in the way public administration manages its knowledge is on the horizon, both within its organisations and in its external relationships with citizens. The knowledge-related objectives for today's public administrations include:

- Effective public services that address the public's issues and requirements relevantly, competently and timely, while consuming minimal resources.
- The preparation of citizens, organisations and public agencies to be effective policy partners: create sound public opinions; develop constructive public debates and policy formation; conceptualize, plan, decide and implement public actions
- Assure an acceptable quality of life through building, maintaining and leveraging commercial and public intellectual capital.
- Ensure a prosperous society by developing citizens to become competent knowledge workers, and institutions that are competitive.

5 C.K.O.s in Public Administration

To the extent knowledge and information are the core assets of a public organisation, knowledge and information management become the means for ensuring success. From this perspective, the knowledge-related challenges facing the public and private sectors are similar. Public administration must develop a critical awareness of the value of its (ever growing) stock of knowledge, manage this resource optimally, and utilize it for more proficient service delivery.

Change masters, champions and activists.

Peter Drucker has written that, "...unless an organisation sees that its task is to lead change, that organisation--whether a business, a university, or a hospital--will not survive.

Since the 1960's the discipline of Organisation Development has taught that organisational change requires steadfast leadership and management that is best materialised by dedicating a person or a team to the effort. In the late 1990's Nonaka, Krogh & Ichijo introduced the idea of a Knowledge Activist: "...someone, some group that takes on particular responsibility for energizing and coordinating knowledge creation efforts throughout the corporation."

The Knowledge Activist described by Nonaka et. al. fills the task description developed by Organisation Development some 50 years ago: a person who leads the knowledge-related change effort in an organisation. Knowledge Activists facilitate the creation and dissemination of knowledge, ensure interconnections between people and leverage innovation. This individual is a "merchant of foresight", developing a "bird's eye perspective" on the change required for knowledge development in a company.

6 The Need

The aim of BRAINCHILD Network of Excellence is to develop a program that addresses the theoretical foundations of knowledge management in public administrations, together with organisational change and development specific to public organisations, in parallel with the knowledge management system tools. This will lead to the development of best practices and technologies that allow public institutions to identify, share, disseminate and create or reuse knowledge & knowledge based systems.

This will form the pedagogical foundation on which a training program for public administration C.K.O.'s is developed. This educational program will benefit from the decade of Knowledge Management experience already acquired in the private sector, but will be specifically tailored to the particularities of the public sector.

BRAINCHILD will focus on the following priority thematic area:

Strategic Roadmaps for Applied Research Addressing Major Societal and Economic Challenges (1.1.2.i)

Objectives: To prepare the ground for RTD activities under FP6 by investigating future research challenges, roadmaps and associated implementation models in the domain of e-government/e-governance and e-work. A constituency of RTD stakeholders will be created by the joint networks of Telecities & Elanet.

Challenges will be identified in the following areas:

- Future e-work systems, "e-business" in public administrations, to facilitate seamless joint-up service delivery to the citizens;
- Organisational knowledge management including context- and location-sensitive solutions for acquisition, sharing, trading, and delivery of knowledge (including next generation knowledge management systems and artificial intelligence techniques [3]) to support public sector employees in their roles as seamless joint-up service delivery agents (**1.1.2.iv. Knowledge and interface technologies**);
- Models and scenarios to shape future policies for a knowledge-based economy in conjunction with the MUTEIS project (socio-economic research on the transition to a knowledge-based economy)

Focus: Activities of the C.K.O. Graduate Course participants will include studying the following aspects:

- Building and strengthening RTD communities that bring together research, business and user organisations with the aim of developing shared visions, scenarios and objectives and facilitating the integration of European research resources to address major future business and work challenges (clusters of IST research projects, national or industrial initiatives in which the C.K.O. Graduate Course participants play a key role themselves);
- Identifying research tasks for both objective-driven and exploratory research. Work should also help to identify and explore the set of complementary activities required to improve RTD impact. These include links to other research frameworks, innovation and take-up actions, training and mobility, standardisation, dissemination activities and the integration of international efforts.

[3] Actions to design & develop next generation knowledge management systems:

- Knowledge management of streaming and archival media resources for the open sharing of content"
- Digital media management, indexing / analysis and search / retrieval in the broadcast domain"
- km3 - Knowledge Monitoring, Maintaining, Mastering"
- Flexible Controls and Learning In Bureaucratic Systems"

- Identifying the key actors in the field, stimulating interest and achieving broad-based consensus on the way forward to meet the research challenges.

Addressing eEurope and eEurope+ Objectives

Objective: To support the broad adoption of IST solutions for "e-commerce" and e-work in public administrations, including the security of on-line transactions, greater flexibility in work organisation and better access to e-work facilities for local and virtual communities, and for SMEs, thus contributing to the realisation of eEurope and eEurope+ objectives.

Focus:

- Best practice actions for trust, security, e-work, organisational knowledge management and process improvement in public administrations (including next generation knowledge management systems);
- Encouraging the participation of Accession States (eEurope+), e-inclusion, implecations of European integration and enlargement for governance.
- Knowledge-based Society and social cohesion (**1.7.1.**).
- Citizenship, democracy and new forms of governance (**1.7.2**).

Clustering of Projects

Objectives: To facilitate synergy between existing projects that see an added value in working together on common objectives. The PACE, EDEN, PRELUDE, KEeLAN, EUSlanD and MUTEIS projects - in which Telecities & Elanet members play a key role - in particular will offer opportunities to share efforts with BRAINCHILD when designing strategic roadmaps for future applied research.

Organizing for Online Service Delivery: The Effects of Network Technology on the Organization of Transactional Service Delivery in Dutch Local Government

Marcel Hoogwout

Department of Public Administration and Public Policy (BSK) University Twente,
PO box 217, 7500 AE Enschede, The Netherlands
m.hoogwout@bsk.utwente.nl

1 Introduction

Dutch central government has, like many other governments, set high aims to offer government services on line. In 2002 about 25% of all services should be online. Because over 60% of all government services are provided by local governments, the challenge is to help these relatively less powerful local government organizations to realise this ambition. Local governments have to operate in an environment where investing in e-government is not evident. To overcome the problems they encounter in improving (online) government services new forms for organizing the service delivery emerge. This paper explores the problems local governments encounter in improving their transactional (online) service delivery and investigates the organizational solutions that arise to overcome these problems. The central question will be to what extent the new organizational forms contribute to the central government aims to realise the high e-government ambitions.

2 The Service Delivery Paradox

In spite of the relatively overwhelming political attention for client oriented government, Dutch local government is relatively slow in implementing electronic service delivery. There seems to be a structural contradiction, between the moral point of view in which it is generally accepted that citizens should be treated as clients of governments, and the practical point of view that makes local governments hesitant to invest in client convenience. Local politicians hesitate because there is little incentive for municipalities to improve their service delivery. Implementing improvements not only requires considerable financial resources, but also radical organizational changes and process redesign [1]. Legal problems have to be solved. Employee attitude has to change in order to promote citizen-oriented service delivery. Furthermore service integration requires that various organizations co-operate and share authority. This is difficult to achieve without a strong external incentive [2]. For most services, government not only has a monopoly, but also from an economic perspective, investing in service delivery improvement on a municipal level makes little sense. Costs will rise while turnover and income will not rise accordingly. The number of transactions for

R. Traunmüller and K. Lenk (Eds.): EGOV 2002, LNCS 2456, pp. 33–36, 2002.

most services is so low that most Dutch local governments simply are too small to warrant the investment. Pressing for improvements in service delivery is also not very rewarding for politicians. The relation between election behaviour and the experience of service delivery is, if it can be proven, very weak. On top of that Dutch citizens are relatively content with the quality of service delivery. Citizens award their local government a 7,2 on a 1-to-10-scale for service delivery. About 80% of all respondents affirm that they are more or less content about the transactional service delivery [3]. Besides these practical arguments also more fundamental dilemmas prevent local governments to embrace the citizens as client consumers. Transactional service delivery touches dilemmas like for example:

- Defending collective interest versus defending the interests of the individual: Sometimes government has to disappoint individuals on behalf of the society as a group.
- Maintaining rules and controlling citizens versus helping them.
- Privacy protection versus transparency through connecting governmental databases for optimal proactive service delivery: Big Brother versus Soft Sister [4];
- Neutral civil servants versus servants committed to the citizen's interest. The Weberian bureaucracy model which puts the chosen politicians forward as the sole clients of the civil servants collides with the concerned street level bureaucrat that cares for its clients and acts that way.
- Raising and educating citizens instead of spoiling them by treating them like a king client.
- 'Consumer democracy' versus 'representative democracy' [1]. Asking citizens for their needs through monitoring instruments and analysing consumer patterns compete with policy articulation based on chosen representatives.
- Hierarchy versus market allocation: by choosing for government as the sole entity to deliver a certain service and not leaving it to the market, one also chooses to accept the negative side effects of the hierarchy system of allocation.

In the light of the arguments and dilemmas shown above, it is understandable that most local politicians don't prioritise service delivery improvement. Given the efforts of central government of improving transactional service delivery this hesitation in local government is felt to be highly unsatisfying.

3 The Changing Organizational Landscape

Central and local governments are more and more aware of this service delivery paradox. To break through the hesitating attitude they are experimenting with new organization models for arranging transactional service delivery, models that avoid the paradox and benefit from the economies of scale and market incentives. An enabling force in this respect is the rapid development of network technology in recent years. Five dominant trends are visible in the changing landscape of the public sector in the Netherlands:

1. Back offices (production) and front offices (client contacts) are increasingly separated in time, space and organizational responsibility.
2. Back offices responsible for the production of the original legal task look for opportunities to co-operate and merge with other back offices to benefit from economies of scale.

3. Front offices are placed in competition. The incentives of the market economy are used to improve service delivery to clients, while avoiding the service delivery paradox.
4. Front office tasks are not only placed in a competitive environment with other governmental front offices but are also distributed over non-governmental organizations that are used to compete in the market.
5. Transactional service delivery tends to be modularised and standardised. The various transaction modules enable (physical and internet) front offices both in the public and private sector to combine a set of modules around the demand patterns (life events) of the client groups they want to serve. This also means that these front offices can shop among different governmental and non-governmental back offices to choose the optimal set of governmental and non-governmental transaction services for their own means. The modularisation eventually can lead to direct one-to-one transactions between the back office organizations and the citizens as government clients.

The five trends have in common that they all lead to an increasing variety of organization models and a fading of the organizational borders. Traditional government organizations exist next to more hybrid forms, while also powers and responsibilities are shared with competing private organizations. The Dutch central government recognizes the trends and the hazards that come with it. To a certain extent they embrace the change and are willing to experiment. The fading of the borders of governmental organizations creates however new challenges for the legislator to guarantee the legitimacy of governmental service transactions. The Dutch government has, for this reason, announced the constitution of a new law that will be designed to facilitate the great variety of new organizational arrangements (Wet Bestuurlijk Maatwerk).

4 Cases of New Organizational Forms in The Netherlands

Three examples of newly developed service delivery concepts in local governments illustrate the existence of the five trends mentioned above.

- **Rental Subsidy by the Ministry of Domestic Housing (VROM):** The department of collective housing is responsible for the allocation of rental subsidy to over one million households in The Netherlands. The development of a completely new modularised and automated transaction process has led to an experiment in which the intake is done not only by local governments alone, but also by real estate brokers and housing corporations as a natural complement to their services around rental housing. The administrative processing of the applications is done by the ministry of VROM. The local governments are put in a position that they have to compete with non-governmental organizations while the centralisation and modularisation of the transaction service enables to profit from economies of scale. (trends 1, 3, 4 and 5).
- **BV Woonnetwerk-Noord**: Woonnetwerk-Noord is an initiative of the City of Groningen with several housing corporations, real estate brokers, project developers, the utility company and the regional and central government. They intent to offer services both through the internet and through a walk-in singe window shop in the center of town (trends 1, 4 and 5).

- **WoningNet Utrecht**: This co-operation between the city of Utrecht and the 12 housing corporations in the Utrecht region has taken over the municipal tasks connected to the allocation and distribution of social housing in the City of Utrecht. Next to that it does the (on line) matching between supply and demand of social housing. The organization also provides the intake for financial rent support, all from the perspective of a single window for the citizen that is in need for social housing. The organization is also planning to facilitate both physical and digital front offices of local governments and housing corporations with their modularised intake and matching services (trends 1, 2, 3, 4 and 5).

5 Conclusion

The examples illustrate that the five organizational trends are increasingly redesigning the organizational landscape of service delivery in The Netherlands. They all rely heavily on the use of network technology to operate their service delivery concepts. For the citizens, as clients of governmental transaction services, dealing with government appears to improve. They enjoy the integration of associated services in a single window while the number of outlets and the individual choice increase. All examples show that mainly through the bundling of resources the necessary investments can be warranted. Especially the rental subsidy and the WoningNet example contribute in their new organizational form to the central government aims to realize the high e-government ambitions. Although the organizational models that were discussed above indeed seem to help to overcome the existing service delivery paradox, time will tell if these new models are viable enough to survive also in the long run.

References

1. Bellamy, C., and Taylor, J.A. (1998). *Governing in the information age*. Buckingham ; Bristol, PA: Open University Press
2. Kraayenbrink, J. Back to the Future: Centralization on its Revival? Problems in the Current Organization of Public Integrated Service Delivery. Paper presented at *DEXA 2002, submitted*.
3. Hoogwout, M. (2001). Leuker kunnen we het niet maken, maar willen we het wel makkelijker? Waarom overheden geen haast hebben met het verbeteren van de dienstverlening. In H.P.M. Van Duivenboden, and M. Lips (Eds.), *Klantgericht werken in de publieke sector: Inrichting van de elektronische overheid*. 149-166. Utrecht: Uitgeverij LEMMA BV.
4. Frissen, P., Koers, A.W., and Snellen, I.T.M. (1992). *Orwell of Athene? democratie en informatiesamenleving*. 's-Gravenhage: Sdu Juridische & Fiscale Uitgeverij.

Public Sector Process Rebuilding Using Information Systems

Kim Viborg Andersen

Department of Informatics; Copenhagen Business School,
Howitzvej 60, DK – 2000 Frederiksberg
andersen@cbs.dk

Abstract. Ongoing modernization of the public sector, gray-zone/ semi-public organizations, and virtual/ teleworking/ Internet use are among the organizational features that need consideration for reorganizing the work processes using information systems. Although politics is not to be ignored, organizational and institutional changes alter the face of the public sector and pave the road for what we call Public Sector Process Rebuilding (PPR).

1 Introduction

With annual governmental expenditures on information systems (IS) sprouting, it is evident that the stakeholders seek a return from their efforts in implementing new IS in their organizations. For the government the motives can also include issues such as: did the R&D expenditures lead to higher economic growth, did the computerization of the public offices lead to savings in manpower, better service, and more services? Did interaction with citizens and companies improve? However, the impacts of using IS often do not live up to the expectations which is a continual source of vexation for the government.

Accordingly, one direction is to *include still more variables* to find valid explanations for *why IS does not always match the intentions, expectations, and needs*. The interest in factors leading to a successful implementation has stemmed from technical issues, to include concerns on environmental factors, organizational factors, and in turn political issues. Another route is to *refine the measurements of what constitute successful implementation*. Whereas, the managerial perspective and the workers' perspective have earlier dominated the literature on IS, recent studies have brought attention to the *internal and external work processes at both macro and micro-level* in the organizations. The process orientation of the studies of IS and organizational transformation is still a more popular route to follow.

There are indeed *various obstacles for reorganizing government*, such as the extreme *openness of the public sector organizations*. Also, public *organizations rarely change in any rapid nor top-down manner*. Rather they change in an incremental manner and only partly top-down. While this is true in a large part of the public sector, the increased use of teleworking, virtual organizations, and quangos/ semi-governmental organizations, only make us more optimistic on the possibility of reorganizing the "business processes" in the public sector.

R. Traunmüller and K. Lenk (Eds.): EGOV 2002, LNCS 2456, pp. 37–44, 2002.

2 Reorganizing the Processes

The very idea of reorganizing processes originates from the basic question: are we doing our business in the most optimal way? Are we doing our job well enough? Are we giving all we've got? (Osborne & Gaebler, 1993). This last question is just as crucial for the public sector as any other large organization, though it may be disputable whether this is contemplated enough by its members. For example, OECD (1995) emphasized the following significant areas in facilitating the exacting of maximum organizational benefits from IS:

- enhancing management, planning and control of the IS functions
- using technology to redesign and improve administrative processes
- providing better access to quality information
- harnessing the potential of new technologies
- developing and applying standards
- attracting and retaining high-caliber IS professionals
- increasing research into the economic, social, legal and political implications of new IS opportunities; and
- assessing experiences

Within the private sector, concepts of reorganization, redesigning and reengineering the processes have gained enormous popularity and – some argue –valuable impacts on the actual practices. Besides the initial articles and books (Davenport & Short, 1990; Hammer, 1990; Hammer & Champy, 1993), numerous books were published showing how IS was affecting dramatic and radical changes in an organization (e.g., Caudle, 1995; Champy, 1995; Davenport, 1993; Davenport & Stoddard, 1994). By contrast, the public administration community did not applause the reengineering concepts as breaking new ground. Rather, they jeered that, BPR was at best old wine in new bottles. At worst, BPR applied in the public sector would lead to misjudgment and actions inconsistent with the 'spirit' of the public sector.

The BPR approach emphasizes that changes in processes are to be drastic rather than incremental. Also, the approach points to broad, cross-functional processes and, if needed, a radical change towards such processes. This of course points to high risk for failure as well. Is risk part of the rationale of the public sector? Traditionally speaking, it would be a quick, resounding NO, but in 1997 we wouldn't be so fast to say taking risks is uncharacteristic of the public sector. First of all, one part of BPR tradition true enough emphasized short term efficiency and longer-term strategic advancement. However, as noted by Coombs and Hull (1996), during the 1990s a "Soft BPR" emerged emphasizing human costs and benefits rather than just organizational shape. Second, the face of the public sector is changing radically not only with rightsizing and downsizing, but also in organization and leadership. It is old news that public services do not inevitably have to be performed by the peers employed in the public sector. By all means, contract them out. Form quangos. Farm out, but keep a short leash on the administrative power. And don't forget to take steering and delegating seriously. These are just some of the central shifts in the public sector management during the 1980s and mid-1990s that makes it uncertain whether the public sector is overall risk wary.

Within the public sector, scholars affiliated with the *US National Academy of Public Administration* further define reengineering as: "...a radical improvement approach that critically examines, rethinks and redesigns mission product and service processes within a political environment. It achieves dramatic mission performance and gains from multiple customer and stakeholder perspectives. It is a key part of a process management approach for optimal performance that continually evaluates, adjusts, or removes processes."

The BPR concept argues likewise, that "researcher...and managers.. must begin to think of process change as a mediating factor between the IT initiative and economic return" (ibid.cit., p. 46). Thus, information technology is not seen as the sole factor that can lead to a miracle outcome. Instead, process reengineering is. However, IT is given the role as the almighty enabler connecting individuals, work groups and departments. Davenport formulated it as "..to suggest that process designs be developed independently of IS or other enablers is to ignore valuable tools for shaping processes" (p. 50). In Davenport's work, the impacts from IT on process innovation is grouped in nine categories: *automational* (eliminating human labor from a process), *informational* (capturing process information for purposes of understanding), *sequential* (changing process sequence), *tracking* (closely monitoring process status and objects), *analytical* (improving analysis of information and decision-making), *geographical* (coordinating processes across distances), *integrative* (coordination between tasks and processes), *intellectual* (capturing and distributing intellectual assets), and *disintermediating* (eliminating intermediaries from a process). Although these impacts differ slightly from the findings our research has identified for the public sector, these areas form a solid basis for addressing the use of IS in the service development, fulfillment and logistical functions of the public sector.

Using IS to increase the ongoing innovation of the work processes can for example include the use of computers to capture the work processes in case-handling within the social welfare administration and use these data to carefully examine the work processes. Within this area is also IS that enables one to track case status. It could mean that when Peggy Sue submits an application for housing and/ or subsidy to a public housing authority it will trigger actions by several public agencies. Also, to generate integrative organizations and have dialogue with agencies and various administrative levels of government spread over *different geographical locations*, IS is a powerful tool. In this area of IS applications, we have also placed forecasting and models. The use of the Internet or Intranet, can help co-working in designing new procedures, without long flights and wasteful commuting time. Finally, knowledge workers in the public sector often need to access same kind of data for designing new work procedures. IS can be used to rationalize the use of the *intellectual assets* in the organization by providing easy access to frequently used data. For example, easy access to the budget for the organization, the current account, documents describing the work processes, etc. can be granted.

IS can also be used to *reorganize the service fulfillment*. Meeting the citizens' needs is essential for the public sector, yet, quite often the employee has to choose between several solutions to match the needs specified by the citizen. E- procurement and the *one-stop services* for the citizens are examples of how IS reorganizes service fulfillment. Likewise, *voice communication* is essential to being responsive to the voice of partners. Computer-based voice response mail and answering machines are still rather uncommon in public offices. IS application here, that is virtual components and IS interorganizational communication tools, can eliminate the lack of transparency of each individual's work. Unfortunately, e-mail is still not an integrated part of the public sector, although studies show benefits in using e-mail in enhancing service quality and effectiveness.

The use of *logistical systems* is more than desirable in all parts of the public sector. For example, scheduling nurses' work week is a highly complicating matter, optimizing the financial issues, while considering labor market regulations etc. Logistical planning systems help keep a record of the laws, stock of workers etc. Also, IS allows remote monitoring of items/ processes. For instance, cars passing through a toll booth in Singapore's highways do not need to toss money in a receptor or in the hands of an attendant. Instead, smart cards with bar codes are read rapidly using telemetry. In addition, Singapore's central government is in charge of exchanging data on various import/ export issues, such as international trade bodies, traders, intermediaries, financial institutions, and port and airport authorities. Exchanging data involves for example collecting, manipulating and transmitting. By implementing new IS and reorganizing the work processes in the Trade Development Board, they were able to handle more cases, reduce the staff, and increase efficiency as well (Teo, Tan, & Wei, 1997).

3 Indicators of Change in Institutional Setting for IS Use

During the 1980s and 1990s governments have been committed to de-bureaucratization and have viewed privatization as a means to achieve this goal. The governments in most first world countries have behaved in line with this strategy. In fact, in recent years the concept of privatization has been used in different ways. Privatization can mean the sale of assets, contracting out, introduction of user charges, voucher schemes, deregulation and government withdrawal from public obligations.

The main aim of the modernization programs has been to improve public services for citizens and companies, to use resources more efficiently, and to halt the expansion of the public sector. Also, we have seen numerously budget reforms, the introduction of management techniques from the private sector, de-bureaucratization, decentralization, experiments with more independent local governments, and a more flexible job structure in the public sector.

For example, Andersen, Greve and Torfing (1996) found that *budget reforms* during the 1980s simplified many of the financial procedures within public institutions. Instead of practicing some highly complex procedures of budgeting and accounting, many public institutions are now free to act within a given financial limit.

Management techniques from the private sector are now more widely applied. Corporate management techniques are adopted, the skills of the chief executives upgraded, and focus on actual output receives more emphasis. *De-bureaucratization* has legitimized the task to minimize the number of public rules and regulations. This has been done by merging different public institutions, privatizing public tasks, and simplifying the remaining rules and regulations. However, the results have been viewed with some skepticism.

Decentralization has been a major component in the strategy to reorganize the public sector. Decision-making responsibility has been shifted downwards from the central government agencies to local government and public institutions. Within local governments the politicians have more power due to the extended use of framework laws, but this power is shared with street level bureaucrats and professionals, and they allow that users exerted more direct influence on the day-to-day operation of public institutions. Within public institutions overall, executive managers have been given greater freedom to formulate strategies for future development. A more *flexible job structure* in the public sector has been introduced. In fact, they allowed that salaries varied according to individual skills and performance.

These changes were all part of a responsive state strategy that aimed at increasing the efficiency of the public sector. A marketization strategy chaperoned this responsive state strategy which promised to intensify the use of market forces and competition within the public sector, hereby demonstrated by the current telecommunication industry, bus-services, airports, national train-companies. In most countries, we have noticed an increase in quarrels over who shall produce, organize, and finance the public services. The overall trend seems to be: democratic principles of the public sector will be left to the politicians. Outsourcing of producing and financing will stimulate the application of BPR in the public sector. The public sector's view on service information has accordingly been revised substantially. Thus, a recent OECD study on public administration arrived at the following objectives for information service made by the public sector (Arnberg, 1996):

- increase democratic legitimacy
- enable clients to claim entitlements
- improve clients' opportunities to influence service content and participate in the provision of services
- manage the expectations of clients about service levels and service quality according to the resources available
- facilitate and create conditions for client choice
- enforce performance on the providers
- restore the confidence of clients in the public sector and its agencies

Thus, IS is at the meso-political level seen as a tool that might help "to reinvent" democracy through various areas involving public contact and citizen access to public held data, one-stop shopping, interactive electronic service, point-of-contact data entry, expert systems, information storage, and revenue generation. Although we have seen this shift at the political agenda towards increased commitment in using IS in the public sector, we still have to see empirical studies that can demonstrate that such impacts are achieved.

3.1 Quangos and Outsourcing: Reorganizing the Incentives and Management

The use of new organizational forms for the public sector has taken its most radical form with the widespread use of quangos. In 1992, the quangos in the UK used about 30% of the total public budget, involving about 5,500 organizations. Similar importance is seen in the more socialist countries, such as the Netherlands and Denmark.

Also, outsourcing of computing activities has escalated during the past 10 years. In a study of computing in Japanese local government, Sekiguchi and Andersen (1999) found that the number of "outsourced" personnel has increased by 126% during the period 1985-1995. During the same period, the total number of computing personnel grew by about 88%. The expenses on computing for local government skyrocketed by about 175%. The high percentage of outsourced employees shows that the computer capability of the existing in-house staff lags far behind the increase in the need of computing skills.

3.2 Virtual Organizations, Teleworking, Internet: Reorganizing the Face of the Public Sector

Virtual organizations is a very popular term and has been connoted with some of the elements in rebuilding the organizations. Although there are varying definitions on what virtual organizations are, networking (e-mail, voice-mail, facsimile, instant messaging, M2M, etc.), restructuring of the organization (outsourcing jobs/ functions, downsizing, transformation), and team culture (geographically and functional), are three elements one finds in most examples (Dutton, 1997). Using such virtual construct is facing major challenges since:

- most users have so far rejected major innovations in telecommuting (or tele-access)
- managers are often unwilling to relinquish control and supervision
- trust and commitment can be undermined if IS is used as a substitute for person-to-person communication
- privacy of users and consumers are potentially threatened

Within the public sector, many teachers, researchers etc. work from home often in an arrangement that comes close to what can be termed virtual organizations. Although this group is important and numbers to a substantial group, few IS applications have aimed at stimulating this process and also to reorganize the work processes. Even less has been accomplished when we look at the much larger fraction of the public sector that is employed as case-workers (social security, housing, etc.).

Similarly, the use of the Internet to allow remote work and contact with the public sector's partners and citizens/ politicians, is an element that is rebuilding the public organization. So far, the knowledge of the impacts of this development is very limited. However, our studies on the organizational changes associated with the use of Internet is that they are driven not from top-down nor in a dramatically patterns. Our studies also show, however, that most Internet applications in government rarely allow two-way communication nor direct access to public employees.

4 Conclusions: Re-building the Work Processes

Above we have pointed to an overall modernization of the public sector, including quangos and a (semi-) virtual construct. We see this development as indicator of a public organization moving towards the technology based public organization. This does not mean that politics no longer is an issue. Nor does it imply that technocratic factors are no longer influencing the capability or opportunities more or less than before. Neither that information systems or new organizational forms is eroding power and politics as being an element of the public sector.

Thus, the overall message with our concept on PPR is to glean for some useful parts of BPR and process innovation ideas while giving ear to criticisms of the concept's application in the public sector. Though we in principle believe that governments should be as small as possible and contract out tasks as much as possible, the core of the public sector above all needs to be in optimal working order. The public organization of today and tomorrow is not easier to manage than older forms of organizations. On top of that, if all existing work procedures are merely transferred to remote workers without reorganizing the work processes, little would be accomplished and counter productive results may lurk.

Areas beyond the central government, but also local government, semi-governmental and other areas of governmental areas, can benefit from our concept. We must remember that political processes include a wide range of activities, such as limiting student enrollment at universities, hiring personnel, or setting the level of welfare service. These are essential parts of (implementing) general political decisions, and extremely important for the content of the public policy. The one million dollar question is: can IS be applied here along with reorganizing the processes? In other words, can we help graduates from vocational training schools get a job faster by using information technology, and yet not expand the number of state tasks? We believe the answer is yes.

In the public administration we face institutional powers with respect to checks and balances, power distribution and professional training. Whereas the low-risk automation has been ongoing in the public administration during the 1980s and 1990s, we believe that more high-risk re-engineering and paradigm-shifts will appear. Researchers at the US National Academy of Public Administration formulated six starting points for reengineering in the public administration. We have adopted their insights and adjusted their list of factors critical for a successful reengineering in the public sector.

Equally important to setting specific goals, is the rebuilding of the structures to support these goals along with implementing the new IS. This requires that we know the work processes. Although this is the case in a large part of the public sector, the flow of information, the share of information, and the manipulation of the information are just some of the items where our knowledge is in fact quite limited. However, without such knowledge prior to rebuilding the structures, the outcome will depend more on luck than professional responsibility, commitment, and involvement.

Also, the keywords "measurement" and "expectations" should be considered carefully. Within the public sector, it is difficult, but not impossible, to measure the processes (including the input and outcome of them). Likewise, the expectations from the "stakeholders" must be identified and tied to the performance management. This is naturally complicated by the change of a political cabinet after a national election and by the often rigid systems for the customers' to impose their influence on the content of the public service. Nevertheless, our message here is that rebuilding public organization is not successful if the only thing accomplished is increased satisfaction for the employees, or information systems that has a better user interface. The clue is that the expectations have to be known and that the important ones are not the expectations of the employees, whether they are shot-term or long-term.

References

1. Andersen, K. V., Greve, C., & Torfing, J. Reorganizing the Danish Welfare State 1982-93: A decade of Conservative rule. Scandinavian Studies, 68(2), 161-187. (1996)
2. Arnberg, M.: Informing clients: statements of service information and service standards. In OECD, Responsive government: service quality initiatives (pp. 245-64). Paris: OECD, PUMA. (1996)

3. Caudle, S. L.: Reengineering for results. Keys to success from government experience. Washington, DC: National Academy of Public Administration. (1995)
4. Champy, James: Reengineering management. The mandate for new leadership. New York: Harper Collins. (1995)
5. Coombs, Rod, & Hull, Richard: The politics of IT strategy and development in organizations. In William H. Dutton (Ed.), Information and Communication Technologies: Visions and Realities. Oxford: Oxford University Press. (1996)
6. Davenport, Thomas, & Short, J. E.: The new industrial engineering: Information technology and business process redesign. Sloan Management Review, 31, 11-27. (1990)
7. Davenport, Thomas: Process innovation. Reengineering work through information technology. Harvard Business School Press. (1993)
8. Donk, Snellen, & Tops: Orwell in Athens. A perspective on informatization and democracy. Amsterdam: IOS Press. (1995)
9. Dutton, W. H.: Virtual organizations. Unpublished manuscript. Seminar at CRITO. (1997)
10. Hammer, Michael: Reengineering work. Don't automate, obliterate. Harvard Business Review, 90, 104-12. (1990)
11. Hammer, Michael, & Champy, James: Re-engineering the corporation. A manifesto for business revolution. New York: Harper Business. (1993)
12. National Academy of Public Administration: http://www.clearlake.ibm. com/Alliance/regodata.html#concepts. (1996)
13. OECD: Governance in transition. Public management reforms in OECD countries. Paris: Author. (1995)
14. Osborne, David, & Gaebler, Ted: Reinventing government: How the entrepreneurial spirit is transforming the public sector. (1993)
15. Sekiguchi, Y., & Andersen, K.V. Information systems in Japanese government. Information Infrastructure and Policy, 6, 109-26. (1999)
16. Teo, Hock-Hai, Tan, Bernard C.Y., & Wei, Kwok-Kee: Organizational Transformation Using Electronic Data Interchange: The Case of TradeNet in Singapore. Journal of Management Information Systems, 13 (4), 139 - 166. (1997)
17. Thaens, Bekkers, & Van Duivenboden: BPR in the public administration. Conference for the European Group of Public Administration (EGPA). Permanent Study Group on ICT. Budapest, Hungary. (1995).

What Is Needed to Allow e-Citizenship?

Reinhard Posch

Secure Information Technology Center Austria
Inffeldgasse 16a, A-8010 Graz
Reinhard.Posch@a-sit.at

Abstract. e-citizenship is used as a term for participation of citizens in e-technologies. This includes all levels of e-government as well as e-commerce. However, e-government and e-commerce exhibit quite different characteristics. Whereas e-commerce has the clear target to intensify usage and turnover, e-government seeks its goals in availability and comfort not in augmented frequency. This paper discusses some aspects resulting from these facts using the Austrian approach as a model.

1 The Austrian View to Approach e-Citizenship

This report is presenting the Austrian view as to what is facilitating the use and the participation in e-government. E-government must be properly designed to allow for e-citizenship. Homogeneous approaches including all levels of administration are the key to acceptance. Such e-government consists of many elements that need to fit nicely together.

On one side such fitting together will be the basis for an efficient implementation, on the other hand efficiency is one of the effects. The targets resulting in this effect have to comprise increase of comfort and convenience of use as these two are the main factors yielding frequency. Therefore we have to agree on underlying principles that govern the implementation and use of information and communication technologies in the administration.

1. Citizens shall be encouraged to use e-technologies at their will.
2. Efficiency shall be looked at from a global rather than a local viewpoint.
3. We have to constitute the right for electronic proceedings for citizens.
4. Security and data protection have to be an overall principle.
5. Administration must be as transparent as possible.
6. We have to care for interoperability at all levels.
7. Open standards and freely available interface specifications have to be deployed wherever possible.
8. Appropriate change management has to enable further development at all times.
9. Competition, analogue to the private sector, shall enable long term efficiency.

E-government needs some basic elements to result in benefits adequate to the effort needed when implementing:

1. The structure of the e-government approach is a critical element. As the main effort is enabling change management, importance of such structure is often considered too late.

R. Traunmüller and K. Lenk (Eds.): EGOV 2002, LNCS 2456, pp. 45–51, 2002.

2. The importance of open specifications of interfaces exceeds the importance of open source.
3. A further critical element is signature tokens. Speedy deployment of such tokens (the Bürgerkarte) is needed. As these tokens are placed in the hands of citizens, this is the element which should be introduced first. Thus, it can also act as a tool to raise awareness [1].
4. Applications have to form a structured back office that uses the specified interfaces.
5. Physical and virtual front offices that comply with the standard interfaces specified complement a modular e-government.
6. The whole system gains a European dimension both through electronic signatures [2] and through standards for forms that use XML/XSL [3] [4] and thus exhibit a potential of being language independent.

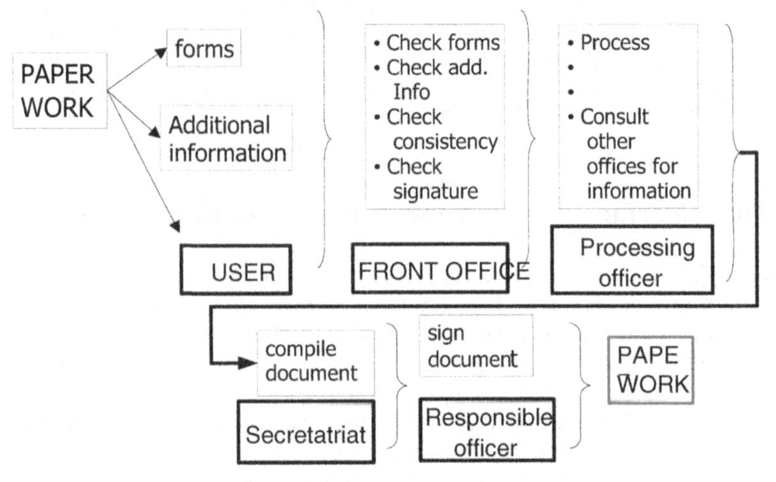

Fig. 1. The way paper works

Today many pieces of administration processes have electronic components. In general, these do not fit together automatically. A complicated chain of semi automated processing results from this situation. We have various roles and transitions that make the work difficult even if a set of tasks can be fully automated in many cases.

2 Structured e-Government

Many of the tasks shown in figure 1 are operations to transfer and transform information. In many cases this is the major part of the work to be done and exhibits a high potential of automation.

As a consequence e-government can be restructured (see figure 2). By introducing electronic signatures and thus avoiding all possible unwanted manipulation, there is a potential to let many administrative processes happen automatically. It is only necessary to interfere, when a decision is to be made which does not follow exact rules which are mapped into the workflow.

The process designed in figure 2 will be implemented in the financial administration (FinanzOnline [5]) by start of 2003.

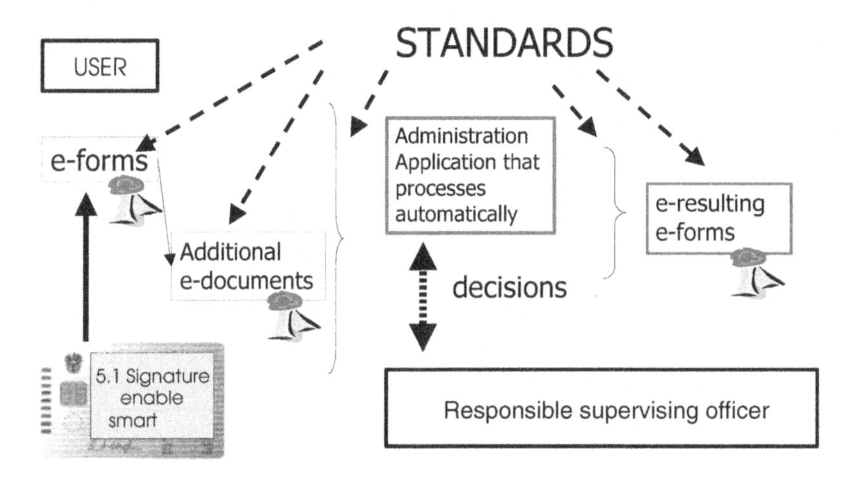

Fig. 2. The way e-government could work

Such modification of government processes is only acceptable for citizens, when it presents a uniform picture of the administration. To this end there is need for a well formed structure of the whole e-government process.

A well formed structure allows the administration to focus on the core business and to increasingly let private competition participate in the electronic and face to face services of the administration.

To achieve large acceptance help and information systems taking the special context into account for optimum comfort are crucial elements. These systems can integrate many elements of multimedia assistance. IP-telephony etc. for optimum assistance are just some of the examples.

For the resulting e-government strategy electronic signatures are the underlying principle. Today smart cards form the infrastructure so that citizens have a tool to "electronically present themselves" to the administration. Compatibility and interoperability are to be assured on an as broad as possible level [6].

An appropriate structure allows for gradually introducing electronic procedures both, when accessing the e-government system and when accessing the back office. In some cases the system or back office can even be a simple workstation operated manually interfacing to the defined standards in the first approach.

3 Enabling Change Management

Efficiency is the effect that should be the result. Convenience and comfort as well as security are the tools to reach this goal. By implementing well defined structures and interfaces the tools available can be dynamically adopted.

For practical reasons there is a need for enabling a dynamical change of all parts within the system. This need is fulfilled by well defined interfaces. Security interfaces (signature [7], identification, etc.) and structural interfaces (person record [8], payment record, portal communication ...) have to form part of such systems.

Such a structure enables the implementation of semi automated or fully automated back office applications according to cost benefit conditions.

In all cases citizens can access government applications independent of the technology used for the back office application. This access can be face to face or fully electronically at the citizen's preference.

As a first step in this direction, Austria has decided to implement a web access through a generic applicable signed form for all federal ministries. Through this, proceedings can be managed that do not explicitly require special form handling.

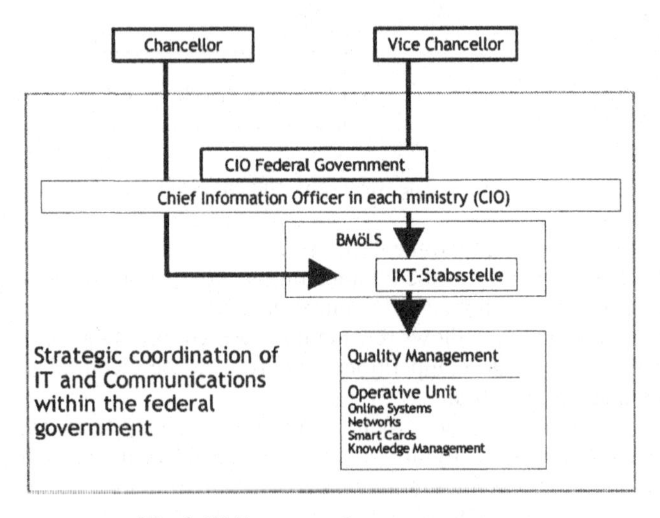

Fig. 3. Well structured e-government

4 Interoperable Documents

The use of appropriate formats is able to solve language and accessibility problems [9]. A form (document) can be prepared in Braille in English and it can be presented in German on the screen. The electronic signature is unchanged and valid in all cases by applying appropriate transformations.

New generations of browsers support such transformations directly. For older browsers automatic gateways can perform the transition in a web-based application for the administration. It is important that structure and formal description are interoperable among the various applications within the administration. This interoperability on a higher level can contribute to move from an e-member state to an e-Europe state in administration.

An "open interface initiative" is needed to allow interoperability on a transnational basis. Various levels of transformation can be used to cope both with standard formats and with individual approaches on a national level.

Not only is it important to provide standards for formats, electronic documents or at least their automated transformation must enable automated processing. There is a need for standard parsing mechanisms so that the source of a document need not be part of an application. Rather the document will itself tell what it can be used for, as it contains recognized tags that characterize the inform

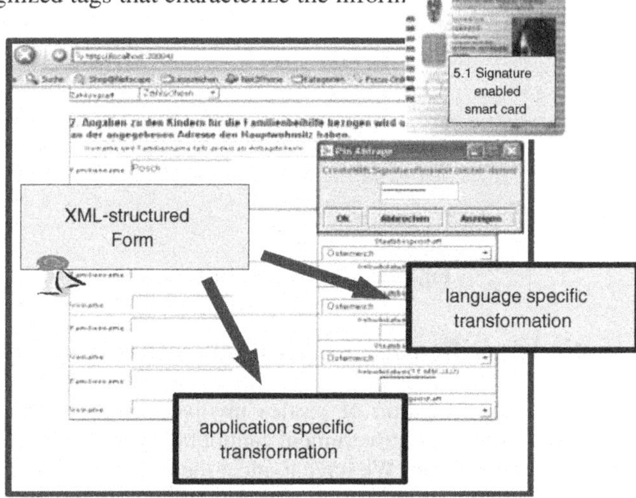

Fig. 4. Interoperability of formats and forms

Austria is currently building public accessible databases [10] that offer specifications, interoperability conditions as well as other information needed for the various institutions (public or private) involved in e-government.

Standard formats with formal description not only make it possible to process electronic forms automatically. By choosing XML documents it is also possible to include electronic signatures directly in the document [11] [12] – a benefit when implementing e-government applications. When designed properly, such forms can be identically rebuilt from the screen or from a printout. This makes it possible to include electronic signatures also on printed forms like tax bills etc. even when these are sent as printouts. A forgery is impossible.

Authentic, automatic reconstruction even from printout is crucial to cope with the various media and products that are and will be in place.

5 Integrated Signature

Figure 5 demonstrates such a form which is signed. In this case it constitutes a payment record issued by a bank for a specific administrative task. For tuition fees and many other applications such a technique makes the payment itself technologically independent from the administrative proceedings.

This same technique can be applied for issuing documents by the administration. More importantly on a European level this technique could enable independent interoperable administration still allowing fully electronic access.

With electronic signatures, with an appropriate structure of the e-government access, and with the necessary compatibility of forms citizen can govern their own data needed for administration and yield compatibility not only on a national but on a European level.

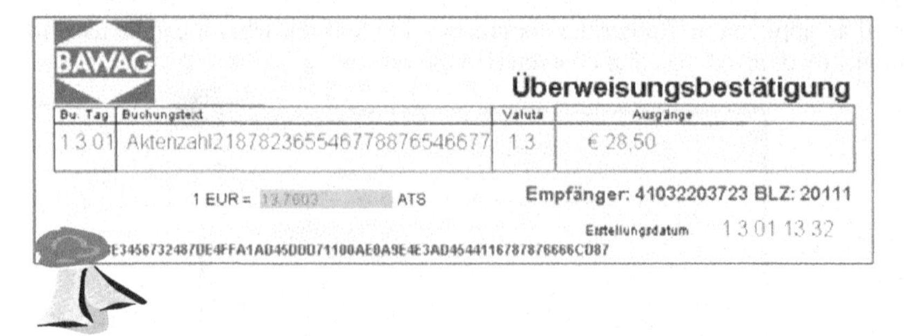

BAWAG

Überweisungsbestätigung

Bu. Tag	Buchungstext	Valuta	Ausgänge
13.01	Aktenzahl218782365546778876546677	1.3	€ 28,50

1 EUR = 13.7603 ATS

Empfänger: 41032203723 BLZ: 20111

Erstellungsdatum 13.01 13.32

E3456732487DE4FFA1AD45DD71100AE0A9E4E3AD4544116787876666CD87

Fig. 5. Paying for administration

6 Automatic Identification of All Parties Involved

A crucial step is the identification of parties involved in government proceedings. This is especially important as e-government cannot be based on prior registration and identification during registration. We have to meet the following demands:

1. Identification must cope with people having an identical name.
2. Identification must be possible without prior contact with the administrative unit.
3. Identification must be based on adequate cryptography.
4. Identification must also comprise corporate bodies.
5. Identification must meet data protection requirements.
6. Identification must be capable of serving as a basis for confidentiality.

In the Austrian context these problems have been solved with a cryptographic binding of the ZMR (the number used in the registry of residents) to a certificate [13]. By application of a domain specific one way function this serves the above goals without infringing anonymity where appropriate.

7 Finally Structuring Organization

It is a key issue that the various appearances of e-government toward citizens present a coherent look and feel. This must be the case without respect to where the administrative application is located or whether it is federal or local. In a huge organization like governments this requires a clear organization and adherence to standards.

The information and communications technology board installed for the federal government in Austria is a means that serves as a tool to achieve such coordination which can also be communicated to and coordinated with the various levels of administration.

As a summary we can state that we need structures to enable e-citizenship. We need structures on the document level, structures on the security level comprising identification, signatures and confidentiality, and structures on the organizational level. Unless we reach such a structured approach it is unlikely that we meet the citizen's need in a mid and long term view.

References

1. Posch R. and Leitold H.: "White Book Bürgerkarte", first version by order of Ministry of Public Service and Sports, June 2001.
2. Directive 1999/93/EC of the European Parliament and of the Council on 13. December 1999 on a community framework for electronic signatures.
3. Bray T., Paoli J., Sperberg-McQueen C. M., Maler E.: "Extensible Markup Language (XML) 1.0 (Second Edition)", W3C Recommendation, October 2000.
4. Clark J.: "XSL Transformations (XSLT)", W3C Recommendation, 1999.
5. FinanzOnline, Ministry of Finance, http://www.bmf.gv.at/, preview demo available at http://fondemo.datakom.at/
6. eEurope Smart Card Initiative, Trailblazer 1, "Public Identity", http://eeurope-smartcards.org/
7. Hollosi A., Karlinger G., Posch R: "Security-Layer Specification", February 2002, http://www.buergerkarte.at/konzept/securitylayer/spezifikation/20020225/
8. Reichstädter P. and Hollosi A.: "PersonData – an XML Specification of Personal Data", April 2002, http://reference.e-government.gv.at/
9. Chisholm W., Vanderheiden G., Jacobs I.: "Web Content Accessibility Guidelines 1.0", W3C Recommendation, May 1999.
10. Austrian e-government reference portal of federal, provincial, and municipal administrations, http://reference.e-government.gv.at/
11. Eastlake D., Reagle J., and Solo D.: "XML-Signature Syntax and Processing", W3C Recommendation, 2002.
12. ETSI SEC: "XML Advanced Electronic Signatures (XAdES)", Technical Specification ETSI TS 101903, 2002.
13. Hollosi A.: "Identity Link – XML Specification", March 2002, http://www.buergerkarte.at/konzept/personenbindung/spezifikation/20020411/
14. Leitold H. and Hollosi A.: "An Open Interface Enabling Secure e-Government", forthcoming.
15. International Organization for Standardization: "Information technology - Security techniques - Evaluation criteria for IT security", ISO/IEC 15408-1 to 15408-3, 1999.
16. ETSI SEC: "Electronic Signature Formats, v.1.2.2", Technical Specification ETSI TS 103733, 2000.

Private Sanctity – e-Practices Overriding Democratic Rigor in e-Voting

Åke Grönlund

Umeå University, Informatics
gron@informatik.umu.se

Abstract. The discussion on electronic voting has so far mostly focused on technical issues, mainly concerning security and privacy. This paper reports an empirical study on how the symbolic values of democracy, as manifested in the act of voting, are considered by e-voters. The study found that the voters in a student election in actions as well as in stated views gave priority to convenience over security and privacy. They voted electronically from home despite uncertainty about the security of the technical system. We argue that this is an indication that the view of the principles of democratic practices will change, and that what might be called an "e-practices mode of thinking" will to some extent prevail over a "rigid democracy mode".

1 Introduction

Electronic voting has been discussed for some time, most intensely so immediately after the US president election debacle in Fall 2000. The topic for the debate has mostly been technical issues relating to security, authentication, and privacy, but also turnout and, to some extent, economic considerations (CIVTF, 2000; Cranor, 2000; IPI, 2001; Rubin, 2001).

This paper reports a study that takes the issue a step further. Suppose we had perfectly secure systems for electronic voting – then what? What would they be used for? Voting is an act that is surrounded with a nimbus as it embodies the essence of democracy. The rules and procedures framing voting express the view of what a vote is, and hence what democracy is. Although different in details across countries and democratic variants (e.g. Barber, 1984; Åström, 2001), what is sometimes referred to as "the sanctity of the act of voting" (Statskontoret, 2001: 49) is usually described as coming from the following characteristics:

- The vote is an individual choice, and the result of individual deliberation. Although public debate is of course essential you should make up your mind yourself, at least without pressure from anyone else.
- The vote is at the same time an expression of a collective – you are part of a people who has decided to make decisions in public matters this way. Considering the alternatives, you should feel proud of the fact that you can indeed go to the polling booth. This means the way you view your voting is important – it is the expression of freedom of choice. The vote is a manifestation of the "sacred" mode of rule, democracy.

R. Traunmüller and K. Lenk (Eds.): EGOV 2002, LNCS 2456, pp. 52–60, 2002.

- The content of the vote, or more precisely the link between the vote and you, is your private property. It is up to you whether or not you want to disclose your political preferences.
- The voting procedures, hence, must be considered as reliable in all aspects by everyone – even those who do not vote. This includes surrounding events such as the preceding and ensuing public debates. (CIVTF, 2000; Johansson, 2001; Statskontoret, 2001).

Examining current voting procedures we find that precincts are physically set up so that people are not too hard pressed to individually honor the above principles. For instance, when you actually cast your vote, usually by dropping a piece of paper in a ballot box, there is a curtain or some other arrangement to prevent others from seeing what you do.

The briefest thought about conditions in the home reveals that this physical guarantee for privacy cannot be arranged without Orwellian measures. So, if we had secure electronic voting from the home, privacy would have to be guaranteed by the individual rather than government. In fact, not only the voter herself but also, often, by her family who would have to agree not to shouldersurf. Would we do that? Or would we change or minds about the principles?

To gain at least some preliminary empirical findings about people's views of the act of voting in an e-voting context, we made an investigation of the first Swedish Internet election, arranged by a student union at Umeå University in May 2001.

2 The Election

Umeå Student Union, representing some 12 000 students in social sciences, humanities, and teacher education, is one of three unions at Umeå University, Sweden. Its Council is elected in annual public proportional elections. The procedures are quite similar to those in Swedish public elections. Voters vote for parties and are allowed to strike out candidates from the list provided by the party of their choice.

Turnout in student elections is generally very low across Sweden. In Umeå, the 2000 election attracted only 11,5 % of the electorate. The low turnout was the main reason for choosing to use Internet in 2001. Eventually sponsoring from the Department of Justice required broader aims, as the government was interested in a pilot test for voter behavior. This led to the goals eventually being the following:

- Increase turnout by about 50 % (from 11,5 % to ca 17 %).
- Investigate voters' attitudes to Internet elections (concerning privacy as well as symbolic values).
- Investigate strengths and weaknesses generally with Internet elections.
- Improve the election process.
- Get more people active in the democratic process.
- Make democratic debate more available to people.

It was hoped that improved marketing in combination with the "improved election process" would achieve this. The improvement of the election process included not only electronic voting, also an electronic discussion forum was set up. The voting technology used was developed by the US company Safevote (www.safevote.com).

The discussion forum was developed by Vivarto technologies (www.Vivarto.com), a Swedish partner to Safevote.

Voting could be done in three ways, by mail, at a precinct, or over the Internet. Voting by mail and by Internet was open from April 27 to May 10. Precinct voting was open May 10-11. It was possible to vote more than once over the Internet. It was also possible to override a mail vote by precinct voting. The voting periods and the option to re-vote were chosen so as to minimize the impact of technical failure, vote selling and changed conditions like some political scandal appearing during the early voting period (re-voting is allowed also in Swedish public elections –votes cast at post offices can be overridden by voting at a precinct).

These considerations led to that votes were given different priority: A precinct vote overrides an Internet vote, which overrides a mail vote.

3 The Investigation

On the initiative of the Department of Justice, an evaluation group was appointed. This group designed a set of measures for evaluations. The security of the technical system and of the logistics used was assessed by expert review. Voters' attitudes and behavior were investigated by a large telephone interview study designed by the evaluation group and executed by Sweden Statistics (http://www.scb.se/).

From the member file of Umeå Student Union (11 859 people), 2 500 names were randomly selected. Interviews were conducted from May 21 to June 6. 80 % responded, 2026 individuals.

3.1 The Findings

This investigation assessed voters' attitudes to critical issues of e-voting: how they view the "sacred act" of voting, and how they feel about privacy and integrity. As the possible effect of Internet voting on the rate of participation in elections has been on the agenda, we also tried to assess voters' reasons for participating or not, and their view of how practical it is to vote electronically from the home.

3.1.1 Participation

The marketing activities were considered successful. For instance, 87 % of those who did not vote knew that there was an Internet voting option. Despite that, the expected increase in turnout did not happen - on the contrary, it sank from an already low 10,4 % in the 2000 election to only 9,3 %.

3.1.2 Voting Method

Internet voting was by far the most popular method (Table 1). The low turnout may seem an indication that Internet voting does not help increase turnout, but the issue may be more complicated. There were some other factors that may have contributed to the decrease. One was that a rather big party (200 voters in the year 2000 election), also the only party directly targeting the students at the teacher education, did not run this time. Also, there were fewer precincts in the 2001 election. We therefore tried to find some further guidance by means of the interviews as to why people chose to vote or not.

Table 1. Voting by method

Voting method	Male %	Female %	Total %	Total number
Internet	66	60	63	678
Precinct	16	13	14	181
Mail	19	27	23	244

3.1.3 Reasons for Voting

Voters were asked about their reasons for choosing to vote or not. Table 3 below shows that the most common reason was that they saw it as natural to use their right to vote. Only 12 % were interested in the Union's activities. 24 % mentioned the Internet voting option as a reason. It should be noted that the voter population is very fluid. Only 28 % of those who voted in 2001 and were entitled to vote in the 2000 election also voted then. The Union's hypothesis was that Internet voting would make a positive difference - in a population where interest in the body to be elected is low and participation is not necessarily by routine, an easily accessible vote method may make a difference

There is indeed some support in the study that the Internet method appealed more to new voters. Dividing voters into two groups depending on whether or not they voted in the previous election shows that voting by the Internet was more common among the new voters, 67 % compared to 55 % (Table 2).

Table 2. Voting method 2001 compared to participation in the year 2000 election (%)

Voting method 2001	Voted 2000		Not entitled to vote 2000
	Yes	No	
Internet	**55**	**67**	63
Precinct	15	13	14
Mail	30	18	21

Further, 40 % of those who voted 2001 but not 2000 mentioned the Internet option as the reason for voting (Table 3). This should be compared with the corresponding figure for the total population, only 24 %.

Table 3. Main reason for voting among those who participated in the 2001 election (%)

	Interested in the Union	The Internet option	It is natural to vote	Other
All voters	12	**24**	62	4
Voted 2000	12	12	74	3
Did not vote 2000	10	**40**	43	7
Not entitled 2000	13	19	67	1

For one fourth of the voters, the main reason was that they felt attracted by the Internet voting alternative. For people who participated 2001 but not 2000, there was a greater share that felt this way. Viewed in this perspective, Internet voting could have increased the participation to a considerable extent, especially concerning attracting new voters to the booth.

3.1.4 Reasons for *Not* Voting

Some 90 % of the total population did not vote. Table 4 summarizes their stated reasons for not voting.

Table 4. Main reason for not voting (%)

	Not interested in the Union	Didn't get to doing it	Other
All non-voters (%)	48	24	27
Voted 2000	22	43	36
Did not vote 2000	54	21	25
Not entitled 2000	48	23	29

As Table 4 shows, close to half the population claims to have refrained from voting due to a lack of interest in the student union and its activities. About one fourth mentions that the main reason for not voting that "it just did not happen". Among those who participated in 2000 but not in 2001, there was a greater share who mentioned that they just did not make it and comparatively few, less than one fourth, that stated a lack of interest in the student union which caused them to not to vote.

The figures indicate a volatile electorate, with a high number "just not making it" and with a general democratic mindset ("it is natural to vote") by far overriding the interest for the object of the election, the student union. Even the attraction of the voting method was twice as high as the interest in the union. This seems to give further support for the hypothesis that the Internet option has indeed helped in preventing turnout to fall even more. Internet would then be an enabler of turnout, although not a determining factor.

3.1.5 The Sacred Act of Voting?

As discussed above, the "sanctity" of the act of voting can be expressed by several parameters: the *reasons* for voting, the *actual privacy* in the voting act and during the ensuing handling of the vote, the *integrity* of the vote itself after it has been cast, the *view* of just how this privacy should be implemented, and the general *credibility* of the election as a whole.

As for the *reasons for voting*, as we discussed above people seemed to be quite serious about it, as the most prominent reason for voting was that it is natural to use one's vote.

Privacy during the voting act when performed in the home obviously has to be arranged by each individual voter. 90 % said nobody else saw what they voted for. When someone did see it, it was a family member (8 %) or a friend (2 %). This indicates that the voters approached the privacy of voting much the same way as in a precinct voting. On the one hand, it might be claimed that 10 % seeing the voting is too much. On the other hand, probably more than 10 % of the population voluntarily reveals their political preferences to other people that are close to them.

It seems fair to say the voters in this election maintained what might be called a "personal sanctity" which is comparable to what the imposed privacy of the voting booth achieves.

It thus seems that Internet voting does not change voter's view of how the voting situation should be arranged.

3.2 Privacy and Integrity

The technology in the Umeå election was not completely safe. It used standard operating systems and standard web browsers. It did not use active components such as JavaScript, but it required this feature of the browser to be enabled[1]. There were also security problems in the handling of the codes used for identifying voters, as these were distributed by ordinary mail.

There was ambiguous information to voters, as the system was announced as simply "safe". As the setup clearly made use of standard technology, at least the somewhat knowledgeable user should raise some doubts about this statement. We were interested in how voters perceived the system's ability to guarantee privacy and integrity, and their view of how any (fears for) deficiencies in this respect would affect them.

During interviews, the respondents were asked to rate their agreement or disagreement with some statements relating to privacy and integrity matters by rating them on a 5-grade scale: strongly agree, partly agree, partly disagree, strongly disagree, and don't know. The results were as follows.

23 % agreed partly or strongly to the statement My vote may be disclosed to some unauthorized person:

	Strongly agree	Partly agree	Partly disagree	Strongly disagree	Don't know
All Internet voters (%)	7	16	18	48	10

Almost half of the Internet voters agreed partly or strongly to the statement *My opinion can be registered:*

	Strongly agree	Partly agree	Partly disagree	Strongly disagree	Don't know
All Internet voters (%)	24	22	15	31	8

44 % felt more or less strongly that My vote may disappear somehow:

	Strongly agree	Partly agree	Partly disagree	Strongly disagree	Don't know
All Internet voters (%)	15	29	16	32	7

Only 42 % agreed strongly to the statement *I feel confident that my vote will be counted.* This means a clear majority was not so sure about it:

	Strongly agree	Partly agree	Partly disagree	Strongly disagree	Don't know
All Internet voters (%)	42	33	12	3	8

Only 34 % agreed strongly to the statement Safety *is enough to protect voting secrecy:*

	Strongly agree	Partly agree	Partly disagree	Strongly disagree	Don't know
All Internet voters (%)	34	35	11	5	14

[1] More precisely, the actual voting system did not require this but the Web site from which the voting took place did. This meant in practice voters did in fact have to enable the JavaScript option

Altogether, it seems fair to say that the above answers reveal a great skepticism as to the privacy and integrity of Internet voting. Against this background it appears strange that they did in fact vote – note again that these are the views of those who actually cast their vote over the Internet.

A possible explanation could be that convenience is more important to them, so we asked a couple of questions about that.

3.3 Convenience

To assess the students' view of the convenience of Internet voting we asked them to assess fourstatements, again on a similar 5-grade scale as above. 90 % considered it very good or rather good not having to go to a precinct:

	Very good	Rather good	Rather bad	Very bad	Don't know
All Internet voters (%)	74	16	3	3	3

Although a solid majority of 81 % found Internet voting easy, it may come as something of a surprise that 6 % found it not so easy. After all, these were students with enough Internet experience to choose Internet as voting method. There were in fact a lot of technical problems - 33 % of the voters experienced such. This usually meant they had to reload the page as the connection timed out (instructions for this were on the screen as the possibility was foreseen). It is therefore likely that the 6 % who found it not so did so because of the actual system.

	Strongly agree	Partly agree	Partly disagree	Strongly disagree	Don't know
All Internet voters (%)	54	27	5	1	13

59 % felt that restriction of voting to precincts would indeed affect turnout by disagreeing partly or strongly to the statement *It does not affect turnout if voting is restricted to precincts with a higher degree of security:*

	Strongly agree	Partly agree	Partly disagree	Strongly disagree	Don't know
All Internet voters (%)	12	14	17	42	14

More speculatively 54 % thought that the Internet voting option had increased the turnout in this particular election: *What influence do you think the Internet voting option has on turnout in a student election?*

	Probably increased it	Probably not changed it	Probably decreased it	Don't know
All Internet voters (%)	54	33	5	7

Taken together it seems fair to say that these answers show that respondents found Internet voting practical. Indeed – because they actually voted – so practical that convenience overrides the perceived deficiencies in privacy and integrity.

4 Discussion

The figures presented above clearly tell that in this particular election, convenience was considered more important than security. Is this also valid more generally?

Obviously this investigation cannot give a clear answer to that. Students are a special group, a student election is not really considered as important as a national or municipal one, and e-voting is not yet an established practice so views may be tentative.

Still, when it comes to privacy issues, or the more particular right to keep your political opinion to yourself, if your political views are disclosed on the Internet it doesn't really matter if this happens in a small-scale and politically unimportant election or a national, important one. Therefore, the case should represent the views expressed on these issues.

A question that this investigation cannot answer is whether this is an expression of Swedish naiveté ("it doesn't happen to us in safe Sweden"), or if it is indeed a more profoundly considered active priority. To the support of the latter it may be argued that frequent Internet users do have to make some such prioritization if they are going to be able to use the Internet for shopping and such. Perhaps Internet voting is viewed as just a kind of Internet shopping? On the other hand it appears not unlikely that people in non-democratic countries, and probably young democracies, as well as Swedish immigrants from such countries, would think differently.

Another issue that is not clearly answered by this investigation is whether the answers reflect emerging opinions or residues of old ones. For instance, is the students' view of how the voting should be done – alone – a residue of the "manual" procedure, which is deep rooted? Are the views of privacy an example of technological naiveté leading to over-optimistic assumptions of the (low) probability that the kind of things that they state "might" happen actually will occur?

The students were experienced technology users - all used computers at work, and most probably also at home, most were their early 20s and thus from a generation which have used computers also in school. They were also experienced Internet users – all have student email addresses, and an estimated 80 % have broadband connection in their home. Altogether, it does not seem unfair to say that this indicates that the answers represent new views, those of "pragmatic Internet users" rather than anything else.

The implications of the findings, if they are valid also for a wider population, are that voting procedures will have to change:

1. Internet voting was the priority one voting method, yet turnout did not increase. Perhaps the optimistic view that e-voting will increase turnout should be reformulated: e-voting will become necessary so as to prevent turnout to fall even more.
2. If e-voting from the home is used on a large scale, the idea of the "voting day" will probably have to change, for convenience as well as because of technical issues. Attackers will have to go on for a longer period of time to make any difference, and this will increase the risk/chance of disclosure, and technical problems will not have such a disastrous effect.
3. If e-voting from the home is allowed, the task of guaranteeing privacy at the moment of voting will have to be delegated to the individual voter. The system will have to accept "private sanctity", that is ideologically and/or psychologically rather than physically enforced privacy. The Umeå experiences indicate that at least this population was up to the ethical standards necessary, but trusting this is probably the hardest challenge for the democratic system.

References

Barber, B (1984) *Strong Democracy*. Berkeley: University of California Press.

CIVTF (2000) *A Report on the Feasibility of Internet Voting*. January 2000. California Internet Voting Task Force (http://www.ss.ca.gov/executive/ivote).

Cranor, L.F. (2001) *Voting After Florida: No Easy Answers* (http://www.research.att.com/~lorrie/voting/essay.html).

IPI (2001) *Report from the National Workshop on Internet Voting*. Internet Policy Institute (http://www.netvoting.org/Resources/InternetVotingReport.pdf).

Johansson, S (2001) *Teknik och administration i valförfarandet.*(Technology and administraton in voting procedures) Slutbetänkande från Valtekniska utredningen (SOU 2000:125). http://justitie.regeringen.se/propositionermm/sou/pdf/sou2000_125.pdf

Olsson, A.R. (2001) E-*röstning. En lägesbeskrivning*. (E-voting. State of the art) Stockholm: IT-kommissionen. Observatoriet för IT, Demokrati och medborgaskap, Observatorierapport 35/2001.

Rubin, A. (2001) *Security Considerations for Remote Electronic Voting over the Internet* AT&T Labs – Research, Florham Park, NJ (http://www.avirubin.com/e-voting.security.html).

Statskontoret (2001) *Utvärdering av kårvalet vid Umeå studentkår*. (Evaluation of Umeå Student Internet Election). Stockholm: Report 2001:26.

Åström, J. (2001) Should online Democracy be Strong, Thin or Quick? *Communications of the ACM*, Jan 2001.

Reconfiguring the Political Value Chain: The Potential Role of Web Services

Francesco Virili[1] and Maddalena Sorrentino[2]

[1] Dipartimento Impresa e Lavoro, Università di Cassino
Via Mazzaroppi, 1 - 03043 Cassino (FR) - Italy
francesco.virili@eco.unicas.it
[2] Dipartimento Scienze dell'Economia e della Gestione Aziendale
Università Cattolica del Sacro Cuore di Milano
Via Necchi, 5 - 20123 Milano - Italy
mso@mi.unicatt.it

Abstract. A new technological standard, called 'Web services' has recently made its first appearance in the Web technologies arena. Our question here is: what is the role of Web services for eGovernment? In the present contribution, the concept of 'political value chain' is introduced and the process of value reconfiguration is illustrated, evidencing one of the potential roles of IT on administrative activities: the facilitation of 'citizen value' creation activities connection. A brief illustration of the Web services technology is then given, finally exploring its potential for e-Government activities and the related research issues.

1 Introduction

A new technological standard, called 'Web services' has recently made its first appearance in the Web technologies arena. Our question here is: what is the role of Web services for eGovernment?

In the following sections, the concept of 'political value chain' is introduced (Section 2), and the process of value reconfiguration is illustrated (Section 3), evidencing one of the potential roles of IT on administrative activities: the facilitation of 'citizen value' creation activities connection. In Section 4 the Web services technology is illustrated, finally exploring, in Section 5, its potential for e-Government activities and the related research issues.

2 The Political Value Chain

In a recent contribution about the use of IT for business process reengineering in the public administration, Anderson [1] maps on a value chain model, based on the classical Porterian concept [10], some examples of IT applications in the public sector. Anderson calls this scheme, inspired by [5] and depicted here in Figure 1, 'political

R. Traunmüller and K. Lenk (Eds.): EGOV 2002, LNCS 2456, pp. 61–68, 2002.

value chain'. Actually, the transposition of a typical concept of industrial business strategy (that of value creation) to the Public Administration, would require more caution: the author notices, a bit superficially, that 'in the public sector there is typically no financial margin of value to be added by innovation. Instead, the public sector can partly add value by shaping the business environment and helping companies be more efficient and effective. In part, too, the public sector is legitimised by its political actions in the democratic domain. So the margin of value in Figure 1 is cast as some combination of the economic, the democratic and the technical'. The idea of value production in the public administration would deserve a deeper analysis, and also the notion of purely 'financial margin' in the industrial business doesn't give justice to the more complex and comprehensive concept of 'customer value' discussed by Porter, that certainly includes many immaterial and qualitative aspects the customer is willing to pay for.

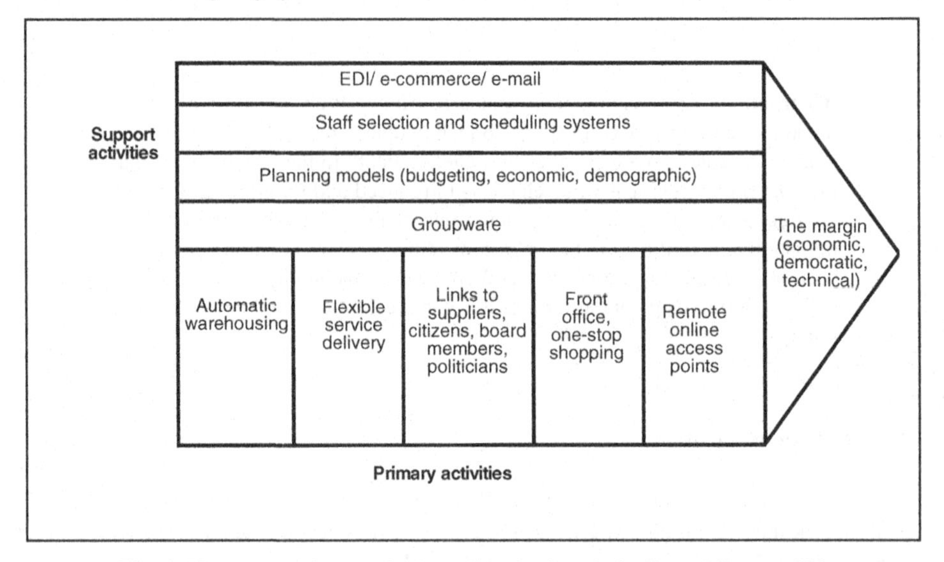

Fig. 1. IT opportunities within the political value chain. From [1], page 320

Moreover, the value chain is better used to represent the value creation process in the industrial production than in the service sector. From this point of view, the analysis of Stabell and Fjeldstad [14] certainly deserves attention: the two authors observe, not without reason, that there is a great difference between the value production processes of a typical industrial business, that inspired Porter's value chain proposition, and that of a service provider, like a bank or a hospital. For example, in Figure 1 the phase of 'inbound logistic', usually referred to the 'raw materials' of the production process, is somehow kept in the model to figure out an 'automatic warehousing' IT application. It is hard to say that inbound logistic is a value production phase in most of the typical public administration activities: immaterial services often constitute the main output, and there is no relevant inbound logistic process. In such cases, different value configurations may be considered as a starting point, like the 'value shops' and the 'value networks' discussed by Stabell and Fjeldstad.

Nevertheless, the political value chain is a simple and interesting starting point for our present work, and we would use it as it is, knowing that it may deserve some deeper analysis and (re)definition. For our purposes, we take for granted the existence, in the public sector, of a sequence of value creation activities, aimed to the production of what Anderson calls 'margin', that we would still call 'value'. We may better say 'citizen value', instead of 'customer value'. We won't investigate, in this contribution, on the specific characteristic of the 'citizen value' and on the details of the value creation activities description, categorization and sequence.

3 The Value Reconfiguration Process

Taken for granted the existence, in the public administration activities, of several value creation activities, we may point the attention to a process that is becoming increasingly common in several industrial and service sectors, generally called 'value reconfiguration', that is well described, for example, by Malone, Yates and Benjamin, in [8] using the framework of the transaction costs theory [17]. According to the authors, ICT may significantly reduce the overall transaction costs, inducing organisations to externalise some activities of the value production process without loosing control (value disaggregation). Moreover, an extensive use of Information and Communication Technologies would then allow the integration of third parties value production activities, creating new inter-organizational value configurations. By this value reconfiguration process, the organisation at the final end of the value creation system may develop and manage a wider and more articulated offer, integrating products and services from several other organisations. For example, Seifert and Wimmer [14], describe the value reconfiguration process focusing on the financial industry. They analyse the case study of a German mortgage bank, the Rheinische Hypothekenbank (Rheinhyp), that externalised the division of 'direct customers' (mortgages distributed via Internet) to a new joint venture company, 'Extrahyp'. Extrahyp was involved in the value production activities related to the new distribution channel; moreover, it was used to develop a richer product/service offering: in addition to the basic Rheinhyp mortgages other products and services issued by third parties were introduced. Finally, Extrahyp started issuing IT services to other banks.

Is this concept of value reconfiguration applicable to the public administration activities? The framework used by Malone, Yates and Benjamin [8] is based on Oliver Williamson's theories, that were later extended by the same author to the governance mechanisms [18], with some important modifications: a significant new concept is that of 'inefficiencies by design' (see also [16] for an application). Basically, we should now take into account, besides the classical transaction costs, also the cost of political consensus. In facts, some degree of governance inefficiency may be accepted (and even introduced on purpose) in order to 'buttress weak political property rights' ([18], page 199) extending consensus with compromising governance choices. The existence of this efficiency/consensus trade-off should not affect the potential role of ICT as transaction costs reducer and driver of value reconfiguration processes [2][8][10], though some research work should be devoted to deal with the enhanced complexity of the modified framework, with its peculiar aspects, and also to some recent criticisms like [3], based on the ambivalent effects of IT externalities on transaction costs.

In the next section we are pointing the attention to a new technology that may potentially play a central role in the value reconfiguration process.

4 Web Services: An Emerging Standard

In April 2001, some 52 IT companies and 'power users', participating in the W3C consortium (including Microsoft, IBM, HP, Sun, SAP, Boeing, ...) conveyed to a workshop in San Jose (California) to advise the W3C on the further actions to be taken with regard to Web services. All of them published their 'position papers', (http://www.w3.org/2001/01/WSWS), discussing their peculiar view and means of implementation of the new technology.

Web services are self-contained, modular business process applications that Web users or Web connected programs can access over a network via a standardized XML-based interface, in a platform-independent and language-neutral way [4] [5]. This makes it possible to build bridges between systems that otherwise would require extensive development efforts. Web services are designed to be published, discovered, and invoked dynamically in a distributed computing environment. By facilitating real-time programmatic interaction between applications over the Internet, Web Services may allow companies to more easily exchange information, leverage information resources, and integrate business processes.

In practice, a Web service is a software reusable component (i.e. a small functionality, a little 'piece' of an application) that can be written by anybody (for example a software vendor), and published to be later retrieved and dynamically used within an existing application by anyone (for example an IS developer). Adopting this framework, companies in the future will be able to buy their information technologies as services provided over the Internet, rather than owning and maintaining all their hardware and software (Hagel, 2001). The functionalities that can be implemented by Web services have virtually no limits, ranging from major services as storage management and customer relationship management (CRM) down to much more limited services such as the furnishing of a stock quote and the checking of bids for an auction item.

Users can access some Web services through a peer-to-peer arrangement rather than by going to a central server. Some services can communicate with other services and this exchange of procedures and data is generally enabled by a class of software known as middleware. Services previously possible only with the older standardized service known as Electronic Data Interchange (EDI) are now likely to become Web services. Besides the standardization and wide availability to users and businesses of the Internet itself, Web services are also increasingly enabled by the use of the Extensible Markup Language (XML) as a means of standardizing data formats and exchanging data.

Through Web services systems can advertise the presence of business processes, information, or tasks to be consumed by other systems. Web services can be delivered to any customer device - e.g., cell phone, (PDA) and PC - and can be created or transformed from existing applications. More important, Web services use repositories of services that can be searched to locate the desired function to create a dynamic value chain. Web services go beyond software components, because they can describe their own functionality, look for, and dynamically interact with other Web services. They provide a means for different organizations to connect their applications with one another to conduct dynamic e-business across a network, no matter what their application, design or run-time environment.

By this new software layer, it's possible to build applications without having to know whom users are, where they are, or anything else about them. Users of these applications can source them as easily as they would be able to source static data on the Web, with complete freedom and no concern about the format, platform, or anything else.

So the revolutionary aspect of using Web services is that they are self-integrating with other similar applications. Until now, using traditional software tools to make two e-business technologies work together required lots of work and planning, to agree on the standards to pass data, the protocols, the platforms, etc. Thanks to Web services, applications will be able to automatically integrate with each other wherever they originate, with no additional work.

4.1 The Web Services Architecture

The Web Services architecture, depicted in Figure 2, is based upon the interactions between three roles: service provider, service registry and service requestor[4]. The interactions involve the 'publish', 'find' and 'bind' operations. Together, these roles and operations act upon the Web Service software module and its description. In a typical scenario, a service provider hosts a network-accessible software module (an implementation of a Web service).

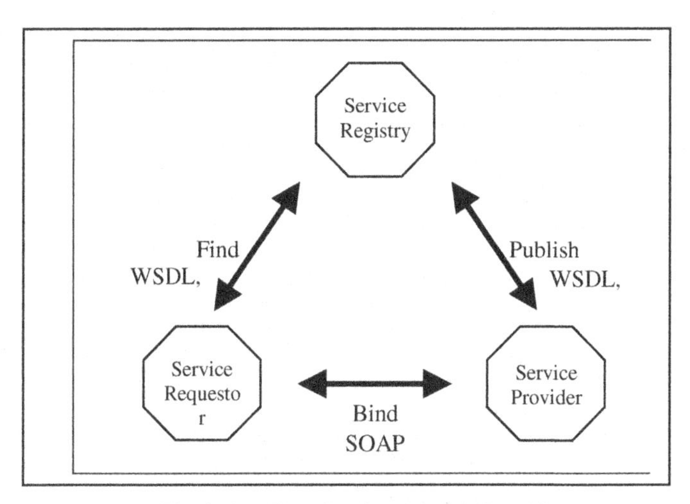

Fig. 2. The Web Services Model (from[4])

The service provider defines a service description for the Web service and 'publishes' it to a service requestor or service registry. The service requestor uses a 'find' operation to retrieve the service description locally or from the service registry; it uses the service description to 'bind' with the service provider and to invoke or interact with the Web service implementation. Service provider and service requestor roles are logical constructs and a service can exhibit characteristics of both.
The foundation of Web Services is represented by new standard technologies which meet the basic requirements for their implementation:

- SOAP (to communicate) a standard mechanism for sending requests to services and receiving responses;
- WSDL (to describe) a standard way to describe services, with a input/output interface specifications and some meta information (copyright, version, update URL, etc.);
- UDDI (to advertise and syndicate) a standard means of locating relevant services with the desired characteristics.

4.2 Building Web Services

Using Web services, as usual in componentized software applications, the system development process may be fractioned in two major parts: the standard components development and the integration into the target system.
Even if, given the novelty of the platform, a consolidated methodology does not yet exist, we may figure out, in the development process of the standard components, that approximately four major phases might be followed: building, deployment, running, management.

The 'build' phase includes the development and testing of the Web service implementation, and the definition of the descriptions for both the service interface and the service implementation. Web services implementation can be provided by creating new Web Services, transforming existing applications into Web Services, and composing new Web Services from other Web Services and applications.
The 'deploy' phase includes the publication of the service interface and service implementation definition to a service requestor or service registry and deployment of the executables for the Web service into an execution environment (typically, a Web application server).

During the 'run' phase, the Web service is available for invocation. At this point, the Web service is fully deployed, operational and network-accessible from the service provider. Then the service requestor can perform the find and bind operations.

The 'manage' phase covers ongoing management and administration of the Web service application. Security, availability, performance, quality of service and business processes must all be addressed.

On the user side, the deployment of Web services in existing systems should not require any effort or resources for application integration. This fact would surely have a significant value to developers, and we may figure out it could have a dramatic impact on the way of designing and implementing Information Systems that may be dynamically adapted to new business needs or organizational changes. The whole IS development process may be radically transformed, as foreseen by (Lyytinen et al, 1998): '... the distinctions between 'internal' and 'external' applications have greyed. The impact of this greying is both the altering and the broadening of design considerations such as availability, security, support and access for all applications. In response to these issues new mechanisms and methods of application assembly are emerging. [...] These are a far cry from the application-oriented, data flow diagramming, functional design and bespoke application days of yore. Against these changes, the role of the software developer necessarily changes. Some will manufacture components; the majority will facilitate their adaptation, choice, understanding and use'. (page 248).

5 Preliminary Conclusions: Exploring the Potential Role of Web Services

What is the potential role of Web services in the value reconfiguration process? John Hagel III, in [5], writes:

Two and a half years ago, Marc Singer and I wrote 'Unbundling the Corporation' [6]. In that article, we described [...] how the Internet would facilitate the unbundling [process], leading to much more tightly focused companies. The rise of the Web services architecture will not only speed this unbundling but will spur the growth of the new companies by letting them mobilize a greater range of resources to reach a broader set of customers (page 113).

Obviously, this statement is only a hypothesis that should be confirmed by evidence and better investigated. If we transpose this hypothesis to the Public Administration sector, the peculiar aspects of governance [10][11] and the higher complexity of the resulting framework would obviously require some additional efforts. The resulting research agenda would encompass theoretical aspects like transaction cost theory investigation and application, management aspects like the definition of the new organisational assets, and applicative aspects like the development of a security infrastructure to ensure the required level of trust.

References

1. Andersen, K.V. Reengineering public sector organisations using information technology, in 5, pp. 313-329.
2. Ciborra C., Teams, Markets and Systems, Cambridge University Press, 1993.
3. Cordella, A., Does Information Technology Always Lead to Lower Transaction Costs?, Proceedings of ECIS 2001, Bled (Slovenia).
4. Kreger, H., Web Services Conceptual Architecture, white paper, IBM Software Group, May 2001.
5. Hagel III, J. and J.S. Brown, Your Next IT Strategy, Harvard Business Review, October 2001, 106-113.

6. Hagel III, J. and M. Singer, Unbundling the Corporation, Harvard Business Review, March-April 1999, Vol.77(2) pp.133-141.
7. Heeks, R. (ed.), Reinventing Government in the Information Age: International Practice in IT-Enabled Public Sector Reform. Routledge, 2001.
8. Malone, T.W., Yates, J and R. Benjamin, Electronic Markets and Electronic Hierarchies. Communications of the ACM, 30(6), 1987, 484-497.
9. Moreton, R. and M. Chester, Transforming the Business: The IT Contribution, McGraw-Hill, 1996.
10. Lenk, K. and R. Traunmüller, Perspectives on Electronic Government, IFIP WG 8.5 IS in Public Administration Working Conference on Advances in Electronic Government, 10-11 February 2000, Zaragosa, Spain.
11. Lenk, K. and R. Traunmüller (eds.), Öffentliche Verwaltung und Informationstechnik - Perspectiven einer radikalen Neugestaltung mit Informationstechnik. Heidelberg Decker, 1999, German.
12. Picot A., Bortenlänger C. and H. Röhrl, Organization of electronic market: contributions from the new institutional economics, The Information Society, 13, (1997),107-123.
13. Porter, M.E. Competitive strategy: techniques for analyzing industries and competitors, NY: Free Press, 1998 first edition 1980.
14. Seifert, F. and A. Wimmer, Towards networked banking: the impact of IT on the financial industry value chain, Proceedings of the European Conference on Information Systems (ECIS), Bled, Slovenia, 2001, pp.474-84.
15. Stabell, C.B. and Ø. Fjeldstad, Configuring Value for Competitive Advantage: On Chains, Shops, and Networks, Strategic Management Journal 19 (1998), pp. 413-37.
16. Virili, F., The Italian e-Government Action Plan: from Gaining Efficiency to Rethinking Goverment. Proceedings of DEXA 2001 International Workshop on e-Government, Munich 2001.
17. Williamson, O.E., Markets and Hierarchies: Analysis and Antitrust Implications, Free Press, 1975.
18. Williamson, O.E., The Mechanisms of Governance, Oxford University Press, 1996.

The E-GOV Action Plan in Beijing

Xinxiang Chen

Capinfo Company Limited
Beijing Network & Multimedia Lab

1 IT Brings New Opportunities and Challenges

Information and Communication Technology (IT) is one of the most magic forces. It heavily affects people's daily life, learning and work, even the working style of the government in civil society. IT will make sure that each individual in the society can common share other's knowledge and ideas. By means of IT, people can collaborate with each other without limitation of time difference and geographic location. It will help people to exploit their potential and accelerate the development of the society.
Also IT can bring large challenges, such as innovation, productivity and efficiency, in front of enterprises, firms and government. They will face the competition not only in the special local area, but also all over the world.
IT is fast becoming a vital engine of growth for the world economy. So IT reforms the earth. The preparation of manuscripts, which are to be reproduced by photo-offset requires special care. Papers submitted in a technically unsuitable form will be returned for retyping, or canceled if the volume cannot otherwise be finished on time.

2 Internet in China

2.1 The Growth of Internet Users in China Shown in Figure 1

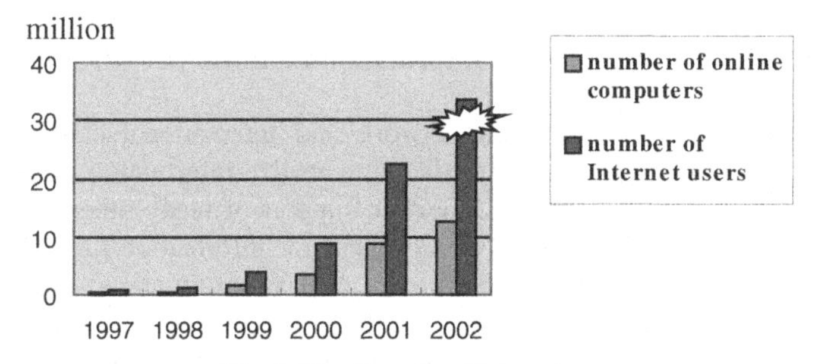

Fig. 1. The Growth of Internet user

2.2 Here Is the Statistics/Investigation from CNNIC in 01/2002

- Internet users in China: 33.7 million
- Online computers: 12.54 million

R. Traunmüller and K. Lenk (Eds.): EGOV 2002, LNCS 2456, pp. 69–74, 2002.

- Registered domain names in CNNIC: 127,319; Beijing accounting for 33.8%
- Registered Web Sites in CNNIC: 277,100; Beijing accounting for 20.6%
- Total capacity for international traffic: 7,597.5M

2.3 Penetration of IT in Beijing(01/2002)

- Internet users: 4.18 million, 12.39%
- PCs: 110.5 per thousand people, the ratio of the computers for family use has propagated to 28.4%
- Mobile phones: 5.95 million, 4.00%
- Telephones: 5.16 million, 2.84%
- TV sets: 148.9 per hundred houses
- 234 thousand employees in IT area

3 The E-GOV Action Plan in Beijing

- First Stage (1998 ~ 2000): Launching the supporting infrastructure, including information database and the portal website. Starting the E-GOV process in Beijing
- Second Stage (2001 ~ 2002): Developing, examining and managing the One-Stop, online interactive public services, available to the enterprises and citizens in Beijing
- Third Stage (2003 ~ 2005): to establish a systematic, well-structured, electronic service delivery system based on the high-speed, broadband municipal network, to build one of the best E-GOV in China

4 The E-GOV Action in Beijing: First Stage (1998-2000)

4.1 Establish the Capital Public Information Platform (CPIP)

Based on information transmission network and international standard protocol, CPIP, a public facility, has been constructed to realize information exchange and to provide various kinds of services. Under the support of many other existing public network platforms, it can create an environment for information gathering, processing, distribution, exchanging and sharing, as well as information service providing and information management. It is a pivot of Beijing public information gathering, and a gateway for inter-communication and information exchange of professional industry. Covering all the 10 districts in Beijing, the CPIP is interconnected with key information resources such as CHINANET, CSTNET, CERNET, and CHINAGBN. Today the CPIP, providing the fundamental services for cyber Beijing, is one of the largest IP-VPN based metropolitan area networks in Beijing and utilized by most of the government entities. It is also an important port for information exchange between Beijing and other cities both home and abroad. It also acts as a joint point, facilitating the inter-connection between cable TV network and multi-media broadband telecommunication network. The network structure of CPIP is shown in figure 2.

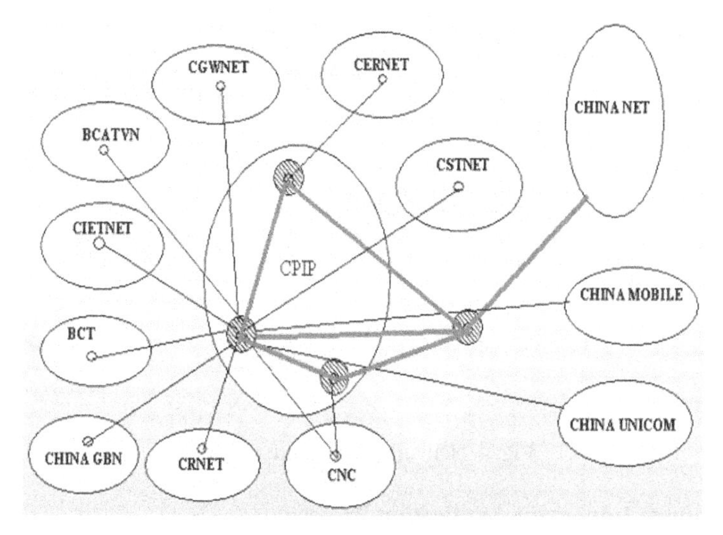

Fig. 2. Network structure of CPIP

4.2 Develop the Portal Website for the Municipal Government (http://www.beijing.gov.cn)

Beijing-China project - "the window of the capital" opened in 07/01/1998. As a key component of "Digitalization of Beijing" program, it is a uniform and unique portal for E-GOV, currently linking over 100 websites of different government departments in Beijing. The multi-language versions of the portal are shown as below.

Fig. 3. Chinese Simplified Version

4.3 Construct the Internal Network for the Government Departments and Computerize the Civil Services

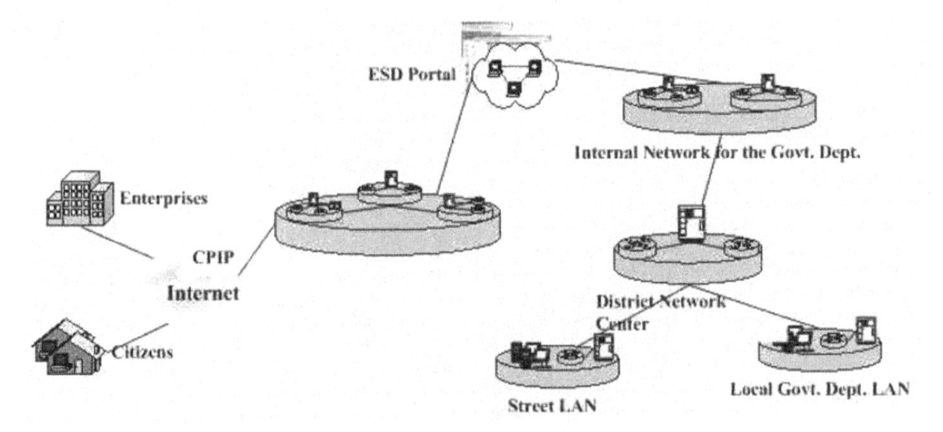

Fig. 4. Network architecture of E-GOV

4.4 Promote Important Application Projects

- Beijing Medical Insurance Information Subsystem & Beijing Citizen Card: the system will incorporate personal data of about six million insured people in Beijing by the end of 2002. Upon completion, it will be one of the most advanced social security information systems in China.
- Beijing Community Services Information System: the system is designed to cover all the 18 districts in Beijing and to set up a hotline call center system incorporating the 149 street-level service stations.
- Beijing Municipal Government e-Procurement System: it is an online bidding system enabling the Beijing Municipal Government, by inviting both domestic and foreign suppliers, to carry out procurement activities on a more open and transparent basis.

5 The E-GOV Action in Beijing: Second Stage (2001~2002)

- Upgrade the Organization, Reform the management modes and optimize the business process
- Integrate all information resources by building and updating a batch of administration database
- Launch the ESD (Electronic Services Delivery) schema to realize the electronic transaction between enterprises, citizens and the government. The ESD platform will provide a unified authentication and authorization service for government department, allowing the users to conduct their business with their counterparts in a secure manner, using any application, at anytime and from anywhere
- Continue to perfect the CPIP and speed up establishing the broadband network administration system

- Continue to improve the portal website – "the window of the capital", bringing innovative, interactive and efficient delivery to the Beijing government as well as high quality online services to the public
  1. Realize Single Sign-On for all departmental services, one identity across all departmental services.
  2. As part of EDS, an application integration platform was implemented, featuring the following aspects:
  3. Integration with the BackOffice Transaction System
  4. Reliable data transmission based on the XML integration
  5. Routing and dispatching of the workflow
- Carry out overall planning and security protection, formulate unique standard, relevant policies and regulations, and create a good soft-environment for implementation of E-GOV

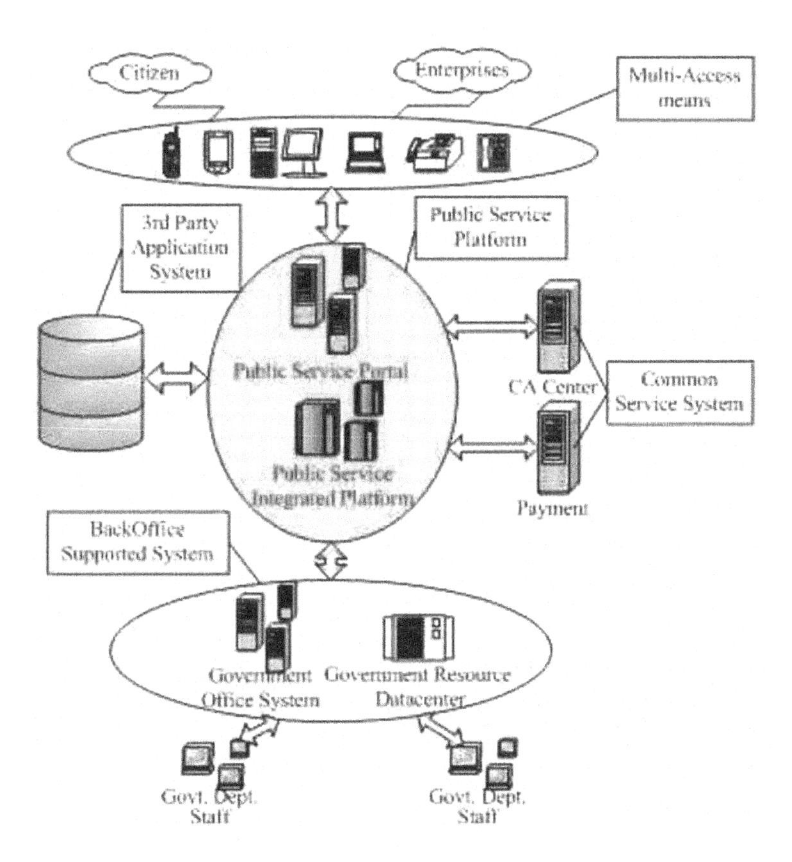

Fig. 5. Framework of ESD

6 The E-GOV Action in Beijing: Third Stage (2003-2005)

To establish a systematic, well-structured electronic services delivery system based on the high-speed, broadband municipal network, build one of the best E-GOV in China.

7 Conclusion

Over the past five years, the Beijing municipal government provided a lot of financial support for the programs related to the E-GOV Action plan, and will continue to offer help. But other than the money problem, the most important and difficult part involving the implementation of the action plan is the mindset change that is a requirement for all the citizens. Beijing will go ahead with its efforts to carry out E-GOV and hope to play a leading role in this particular field, providing more and better services for the industries and the citizens in the information era. There is still a long way to go.

The POWER-Light Version: Improving Legal Quality under Time Pressure

Tom M. van Engers and Radboud A.W. Vanlerberghe

t.m.van.engers@acm.org, ravanl@oimp.nl

Abstract. The Dutch Tax and Customs Administration conducts a research program Program for an Ontology-based Working Environment for Rules and regulations (POWER). In this research program that was started in 1999 and is sponsored by the European Commission (E-POWER) since September 2001 an ICT-based methodology has been developed that enables the formalization of legal sources and finally the design of legal knowledge-based systems. The full-scale POWER-method however although much less time consuming than normal software design methodologies is still too elaborate especially if we want to apply this method in legal drafting or policy making processes. We therefore created the POWER-light version, a variant of the POWER-method that helps to improve legal quality and can be used with relatively little effort and in short time. Although the POWER-light version lacks many of the advantages of the regular POWER-method (e.g. its verification, simulation and knowledge-based component generation abilities) it offers a first step. The POWER-light approach offers the tools to get the best possible legal quality given the time restrictions.

1 Introduction

One of the goals of E-Government is providing citizens with means to access the governmental body of knowledge. This knowledge is based upon legislation, but also incorporates the business policy and interpretation that is added to the explicit knowledge corpus as it is reflected in the many legal documents like the different laws, regulations, case law etc.

In the POWER research program (Program for an Ontology-based Working Environment for Rules and regulations) a method and different supporting tools have been developed that support the chain of processes from drafting to implementation. Central to the POWER-method is a formalization process in which the legal knowledge sources are captured and translated into formal models, which we refer to as POWER-models (see Van Engers and Glasséé 2001). These formal models are the basis for the systems development process(in which we create knowledge-based components) that in many cases follows the modeling process. The POWER-models are also used to detect (potential) defects in the knowledge sources e.g. inconsistencies and circularities (see Spreeuwenberg et al. 2001). The formal models can be used for simulation of the effects of (new) legislation as well.

It is obvious that the initial legal quality has great impact on the quality of the (e-) governmental services that are based upon it. Simulation of legal effects and verification of the (legal) knowledge sources helps to improve legal quality.

R. Traunmüller and K. Lenk (Eds.): EGOV 2002, LNCS 2456, pp. 75–83, 2002.

In this paper we will give a brief description of the POWER-method.

While this method already reduces implementation time and improves legal quality, it is still too time consuming if the focus of the POWER-user is limited to the political decision making process. In this process legislation drafters, sometimes working closely together with knowledge groups in the public administrations, have to produce drafts under enormous time pressure. This leaves almost no time to integrally apply the POWER-method. Therefore we designed a POWER-light version. This method is specifically suited to conduct a quick scan of legal quality and can be applied with very little effort. That way even in the pressure cooker of the political decision-making process it is still possible to perform a quality check on draft legislation. In this paper we will explain the POWER-light method and show some experiences with this method in a recent legislation drafting process.

2 Managing Corporate Knowledge

If we want to model the knowledge of public administrations or other organizations that execute regulations, it seems best to focus on the existing documentary knowledge sources first before eliciting experts in order to model the knowledge of a certain domain. In most cases these documents contain the 'rules' in the form of an informal (or pseudo-formal) representation, e.g. the income tax law. Interpretation of these informal expressions is needed. This interpretation reflects the opinions of the public administration and consequently influences the business policy.

We consequently have to capture the expert knowledge to establish the correct interpretation of these documentary knowledge sources. The experts are also consulted to understand the processes in which the domain knowledge is used. This process knowledge is expressed in process models.

Usually experts from different disciplines and backgrounds are involved in the knowledge specification processes. Their knowledge is made explicit with help of the knowledge engineers, knowledge that would otherwise have remained implicit. The knowledge can furthermore be specified in a way that makes it easier to establish its validity. In addition to improving the efficiency of constituency treatment in its operational units, the knowledge-based systems serve primarily as a dissemination vehicle allowing the DTCA to make more effective use of the knowledge of its sparse experts, while improving the quality of law enforcement[1] in its operational units. The POWER program elaborates on that insight. The focus of the POWER program differs from 'traditional' knowledge engineering approaches (see for example Sudkamp 1988). In POWER we focus primarily on the knowledge specification. This specification can be used to create knowledge-based systems but this is just one application form. The POWER knowledge specifications are also used for enhancing the quality of legislation as well as for e.g. policy-impact analysis. The position of the POWER method is depicted in figure 1.

[1] Quality of law enforcement is defined as satisfaction of the constituency with the adoption of the principles of equality before law, predictability of law enforcement and proper use of authority by law enforcement agencies.

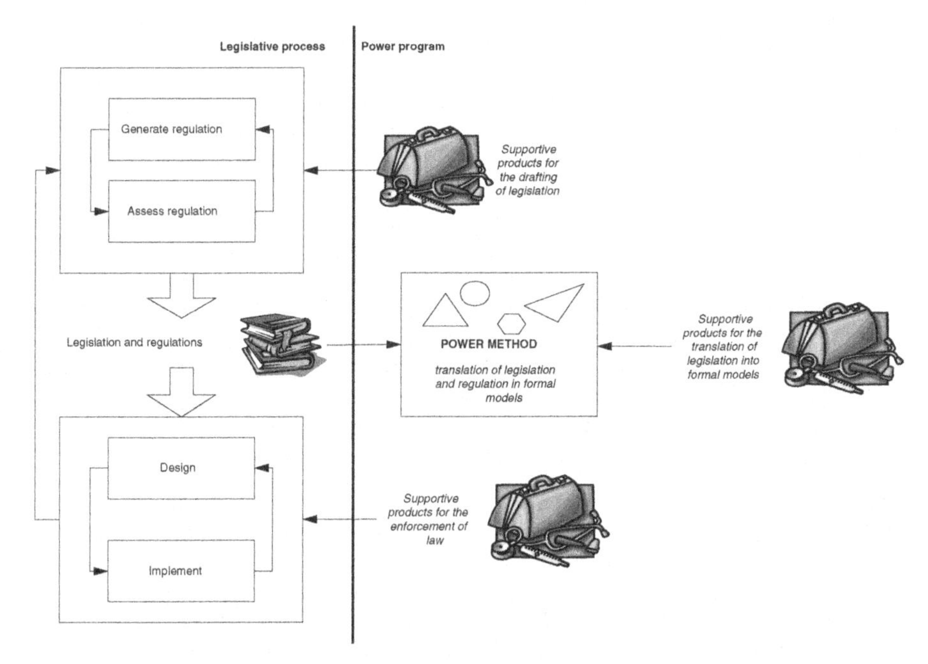

Fig. 1. The POWER method and tool in the legislative process

In common modeling approaches used in traditional information science process and domain knowledge are intermingled, if not in the different models or design documents that are used than in many cases in the software code. In the AI (expert systems or knowledge-based systems approach) the knowledge is separated from the inferences that are made upon it. But although a knowledge-based approach already has some advantages compared to traditional modeling methods we could even further improve the strength of knowledge modeling by besides distinguishing and separating process from domain knowledge also distinguishing the different knowledge sources and include explicit references to the original sources (and even specific distinct parts of these sources) in our models. By creating knowledge models this way, organizations get a means to manage their knowledge.

3 The POWER Modeling Process

The POWER modeling process first legal sources, such as law, executive measures both from the authorized ministry and the executive bodies, but also from expertise groups from such bodies that are active within the field of application, are translated into formal specifications. These formal specifications are termed conceptual models. These initial conceptual models closely reflect the structure and nature of the legal sources.

In most cases, however, this will result in weakly structured models due to the nature of the political processes that produced the legal source. Their emphasis is on reaching an agreement between the negotiating parties. Hence a second process that combines and refactors dependent legal sources into a consistent and well-structured conceptual model is required.

These processes need not be conducted as sequential steps. Typically, the modeler will select a legal source, most likely a few chapters of law, he/she knows is relevant to the domain, conduct the translation process and then partially refactor the specifications.
Verification of those resulting specifications will expose missing specifications, and perhaps inconsistencies. Additional specifications may be translated from executive measures, and inconsistencies may be resolved through executive decisions of the authorized expert group. Refactoring the specifications with these newly added specifications results in a more consistent and complete conceptual model. Eventually, the verification process demonstrates the consistency and completeness for the task at hand. The design of the enterprises software systems for enabling integration of the task into the executive business processes may be well underway by that time, in order to achieve the throughput time-to-execution required.

The modeling process is also likely to take advantage of existing specifications that can be reused for new executive tasks, and needs to take into account the process of reaching a political agreement, resulting in frequent changes. Tractability will be required both to verify the proper authority for the specifications, as well as for effectively maintaining this correspondence in the face of changes in the legal sources.

The legislation-drafting process often takes place under an enormous time pressure. This is due to the political consensus making process that leaves almost no time for the drafter to incorporate the outcomes of this process in a new draft. We therefore developed a POWER-light version that is based upon the original POWER-method, but consists of a more global approach. The consequence is that no formal models are produced and consequently the formal verification process, simulation of effects as well as the generation of knowledge-based components are not supported. The purpose of the POWER-light version is to support a global first scan of the draft's quality within a very short period of time. This can also be the first step to a full scale POWER-modeling and analysis process.

4 POWER-Light

The POWER-light method is comprised of two separate analysis steps; Structure analysis and Domain modeling.

The first step of the POWER-light method consists of a structural analysis of the legal knowledge sources (documents). Legal documents typically have an invariant structure that is used to construct meaning (e.g. lex specialis is constructed from the order of the members in the law). Another characteristic of legal texts is the use of references, both absolute references (e.g. 'article 3.11') or relative (e.g. 'the previous member'). Problems often occur during the drafting process due to the time pressure. Many drafters may be involved in a drafting process and even when special attention is paid to co-ordination legislative texts contain structural errors. Correcting these errors after a new law has become effective (i.e. after the political decision-making process has taken place) is costly (and in practice only happens when additional changes have to be made). Detecting these errors prior to subjecting them to a political decision may eliminate the need for these costly corrections.

The second part of the POWER-light method deals with possible defects regarding the system of the legislative text on a material level. Co-operation of different legislative drafters under time and pressure occasionally leads to inconsistencies or redundancies within the text. By trying to model (parts of) the legislative text some of these defects may be detected and consequently corrected before the legislative text is passed.

4.1 The Structural Analysis

Most legislative texts have a distinct (and invariant) structure (the invariance may be law-family dependent). Since legislation drafting is a dynamic process, often executed under time pressure with involvement of a group of people, it is prone to structural defects. These defects may cause serious problems in the operational units of the governments bodies. An example of such structural defect is a reference to a missing member (e.g. a reference to a concept that should be defined in a previous member.

We developed a software tool (Juridical Editor and Working Environment for Legislation drafters, in short JEWEL) that supports the structural analysis of legal documents.

JEWEL supports the detection of the most common structural defects:

Discontinuity in the numbering of structure blocks.

Legal texts are composed of structural elements (such as sections, article, members, sentences etc.). We call these structural elements structure blocks. To be able to refer to these structure blocks legislation drafters use a form of numbering. While creating and updating legislative texts structure-blocks are frequently edited, replaced and/or added. This may result in e.g. a numbering of the structure-blocks that is not continuous, structure-blocks that are placed out of sequence etc. Maintenance of the numbering is part of legislation drafting.

Incorrect use of references.

References in a text that have been heavily modified can be defective in two ways. The first possibility is that a reference refers to a structure-block that no longer exists. This type of reference defect is easy to detect and the automated tool will show that specific reference to be 'empty'. The other type of reference defect is harder to spot. This defect involves references to existing structure-blocks that do not have content relating to the issue addressed by the reference. Although it is possible for analysts to identify a few of these defects, only experts can truly expose these defects.

Inconsistent hierarchy.

Most legislative texts possess a structure that encompass a specific hierarchy. In the co-operation between different departments the use of two different hierarchy-types will diminish the readability of the legislative text and hampers future adjustments. Automated detection is possible, because as a result of the use of two different hierarchy-types, the structure-analysis will not function properly.

Defects found in this part of the analysis are, apart from the content-reference defect, self-apparent and have no need for validation by a legal expert.

4.2 Domain Analysis

During legislative drafting different people work on small parts of a legislative text. As a result it is difficult for the final editor to maintain an overall consistency throughout the entire text. In a first attempt to model a legislative text, many of these potential systematic defects come to light. By not only restricting oneself to the translation of legislative texts into conceptual models but also take into account integration and process-design issues that become apparent during translating, even more potential defects are identified in an early stage of the legislation design process.

From this analyses the following potential defects can be identified:

Definition of concepts and terms. Defects regarding the definition of concepts can take several forms.

First and foremost is ambiguity. Through attempting to model the definitions in a formal language, it often becomes apparent that more than one way of modeling the text is possible. By pointing out these options legislative drafters can be made aware of those passages in the draft that need further or different specification.

Conflicts in definitions may arise from having several people work on the same legislation. By trying to define the constraints regarding a single concept, these conflicts are brought to light.

Another important potential defect is incompleteness of the text. By defining the scope of each definition and combining these it is possible to check whether all cases that are to be addressed by the prospective legislation are accounted for.

Redundancies. By constructing a simple decision model it is possible to see whether all segments of the legislative text are used in enforcing the law.

Vagueness. Due to the time pressure legislative drafters often do not have the time to work out all legislative rules out in detail at first, but intend to do so in a later stage of the drafting process. Sometimes however certain of these "rough" rules are overlooked in these later stages. By using a simple decision model knowledge engineers check whether all conclusions enclosed in the legislative text can be reached through the use of common knowledge/sense.

5 Operating Procedure (Putting the Quick-Scan into Practice)

The first step in the POWER light procedure is to have the text parsed by an automated tool built for just that purpose called JEWEL. The parser brings to light structural defects in two different manners. It is able to recognize defects that pertain to incorrect numbering of structure-blocks. The other way that defects in the structure become apparent is when the parser is unable to recognize parts of the text and fails to mark these parts as structure-blocks. The defects in the latter category are usually the result of existing legislative texts being modified over the years.

Additional to the automated detection of structural defects the text is scanned manually to determine whether the structure-blocks that are referred to exist, and to check wether the concept that is linked to the reference can be found back in that structure-block. This ensures that whenever a reference refers to an existing structure-block, it refers to the correct structure-block.

The next step is to export the structured text to a case tool for further modeling. This is supported by another tool developed in POWER: OPAL (Ontology-based Parser for the Analysis of Legal texts). This tool automates the building of a prototypical conceptual model that is used in the regular POWER-method. OPAL analyses the text using a lexicon and grammar to identify concepts and their relation to one another. The concepts and relations are translated into types and associations. During this translation OPAL presents different modeling options to the knowledge analists. These options may indicate that in certain cases two or more different ways of reading the text are present.

As with the structure analysis the automated translation is followed by a manual one.

At this point the POWER light method starts to differ from the regular POWER method. Normally the normative expressions in the domain are translated into formal representations (in OCL-invariants). In the POWER-light version we stop when a basic conceptual model containing the concepts and relationships (associations) is produced. Consequently the entire text is scanned for issues that would pose difficulties in the translation of the normative expressions, integration of partial models and task mapping. This is done by attempting to construct both partial integrated models and constructing a simple decision model based upon the text and its intended meaning.

All potentials defects found in the entire procedure are then grouped by the structure blocks in which they occur (allowing us to refer to the original knowledge source). At the present time these potential defects are specified in two ways. Where the text poses problems for modeling the potential defect it is presented in the form of a question. (e.g. "Is it correct that with regard to this specific legislation investment costs may be deducted?). When potential defects do not qualify as ambiguity in the text these defects are presented using specific cases that address the defect. (e.g. In this legislation a boat would be considered to be a truck for the purpose of taxation. Is this the intended meaning?) The report is presented to the legislation drafters. They will than reflect on these issues and clarify possible misunderstandings. This results in a report with the actual defects instead of potential defects.

Depending on the time available several versions of draft legislation can be reviewed this way.

6 Benefits of POWER-Light

Knowledge engineers can add a fresh look at the same issues legislative drafters work on. By critical reading and asking for clarification and questioning decisions made by drafters, knowledge engineers force legislative drafters to reconsider and reweigh issues, possibly resulting in adjustments of that legislation. Not only will this clarify explicit choices that have been made, but more importantly it will also expose implicit assumptions used by the drafters, assumptions that might needlessly lead to restrictions on the set of possible implementation scenarios considered.

POWER-light offers a systematical approach that enables us to enhance the quality of the legislation. It provides drafters with an extra control that helps clearing defects.

Further more POWER-light helps to create legislation that will be more easily implemented by considering enforcement issues (in the task-mapping step) in an early stage of the legislative process.

7 Experiences with POWER-Light So Far

Although the POWER-light method has only recently been developed and is still in an early development stage, it has already been put into practice in two different legal domains. The first concerned the adaptation of an existing law (succession, in Dutch successie), the second concerned the development of new legislation (mileage taxation, in Dutch kilometerheffing).

In the first case we focused on three constitutes of the legal domain, 'partner begrip' (partnership), 'algemeen nut' (general profit) and 'bedrijfsopvolging' (business succesion). A new element was added; 'constructie van conserverende aanslagen' (construction of preliminary taxation). In this case the POWER-light method was applied when most of the legislation had got its shape and was already approved by the Ministry of Finance. However the application of the light version did yield several benefits. The analyses pointed out areas where further legislation was desirable. Furthermore it provided the enforcers of that legislation with an insight into the system of that legislation, thereby improving their ability to implement the legislation.

The second application of the POWER light method is still a continuing project (mileage taxation, in Dutch kilometerheffing). Even though the project has not been finished yet, benefits of POWER light can already be identified. In this case the POWER light method was involved in an early stage of new legislation. The major contribution of the analyses was the reconsidering of the basis assumptions of the legislation drafters. By basing themselves upon existing legislation the legislation drafters had overlooked the system of that legislation. They than proceeded to use parts of that legislation without including other parts that were crucial to the legislation as a whole.

Furthermore an implicit assumption was that by using parts of one legal text all relevant case law regarding that text would be included in the new legislation. By making that assumption explicit the accompanying text with the legislation was revised and as a result it was ensured that judges would support the assumption. As a final result several concepts within the text where elaborated upon in order to ensure unity of policy in the several units responsible for enforcing the legislation. At this point the legislation has not been passed yet and is expected to undergo several more changes. The legislation drafters involved have requested that the POWER-light method will be applied to the future versions of the legislation.

8 Conclusions

The POWER-light version of the POWER-method helps to improve legal quality and can be used with relatively little effort and in short time. Although the POWER-light version lacks many of the advantages of the regular POWER-method (e.g. its verification, simulation and knowledge-based component generation abilities) it offers a first step. Legislation drafters and people from the knowledge groups involved in drafting new legislation often have to work under enormous time pressure. It is evident that legal quality has great impact on the citizens' compliance. The POWER-light approach offers the tools to get the best possible legal quality given the time restrictions. By consequently applying the regular POWER-method when in the implementation stage the time pressure has been a little reduced, we can create knowledge-based applications that provide better E-services to the citizen.

References

1. Van Engers, T.M., Glassée, E., 2001, Facilitating the Legislation Process Using a Shared Conceptual Model, in IEEE Intelligent Systems January/February 2001 p 50-58.
2. Van Engers, T.M., Kordelaar, P.J.M., Ter Horst, E.A., POWER to the E-Government, in 2001 Knowledge Management in e-Government KMGov-2001, IFIP, ISBN 3 85487 246 1.
3. Spreeuwenberg, S., Van Engers, T.M., Gerrits, R.,2001, The Role of Verification in Improving the Quality of Legal Decision-Making, in Legal Knowledge and Information Systems, IOS press, ISSN 0922-6389.
4. Sudkamp,T. A., 1988, Languages and Machines; An Introduction to the Theory of Computer Science, Reading Mass, Addison-Wesley Publishing Company,

Intranet "Saarland*Plus*" – Enabling New Methods of Cooperation within the Ministerial Administration

Benedikt Gursch, Christian Seel, and Öner Güngöz

Institute for Information Systems (IWi)
at the German Research Center for Artificial Intelligence (DFKI),
Stuhlsatzenhausweg 3, Bd. 43.8, 66123 Saarbruecken, Germany
{Gursch,Seel,Guengoez}@iwi.uni-sb.de

Abstract. The potentials of the information and communication technology become more and more important for the optimization and support of the administrational work processes. The Saarland state government sets a special focus on the usage of the new technologies within the public administration and attaches value to it in the context of a global E-business strategy.
The presented document gives a short overview on the implementation of an intranet as a platform for ministerial comprehensive communication.

1 Introduction

The usage of the Information and communication technologies (ICT) creates beneficial potential of optimisation for the necessary restructuring of the public administration. In general, the digitalisation of information and communication as well as the rapid rise of the internet as a central resource for information exchange facilitates the integration of value-added chains independent from time and space. From the internal perspective, this enables new potentials of co-operation within and between administrative authorities, for example by creating virtual project rooms.[1]

According to this trend, an E-Business Strategy was developed by the state government of Saarland – one of the smaller federal states of Germany – as far back as the beginning of the year 2000. Included are different strategic projects, which give impulses for the efficient arrangement of administration procedures and which should set milestones for the modernisation of administration. The spreading of ICT, new forms of work and education, as well as the creation of jobs within the technical and service sector are in focus along with modernising administration [2].

The project "Intranet Saarland*Plus*", a subsidiary project of the strategic project "Networked State Government" aims at the optimization of the information and knowledge exchange within and between the different ministerial administrative authorities. This is achieved by the implementation of an intranet solution as a departmental comprehensive communication platform. Using this project as case-study, the presented paper provides in the following a short overview of the gained implementation polices and experiences.

R. Traunmüller and K. Lenk (Eds.): EGOV 2002, LNCS 2456, pp. 84–87, 2002.

2 Implementing the Intranet Saarland*Plus* – Experiences within the Ministerial Administration

2.1 Business Case and Challenges

The ministerial administration as the area of application for the intranet Saarland*Plus* is characterized by a dual role. On the one hand, duties of a supervising and intervening administration have to be fulfilled. On the other, the ministerial administration provides services for specific groups of citizen. As a result, the addressees can emerge at the same time as recipients of the administration's output and as subject to the supervision of the ministries. There is only a rare contact to the citizen, moreover the main target group of the services performed is built by commercial and social partners who play the role of a intermediary to the primary beneficiaries [3].

The state government of Saarland consists of eight departments and employs an overall amount of 25.000 people. One can see, just from the number of employees involved, that a radical reorganisation of work processes in the sense of a Business Process Reengineering (BPR) would be problematic, the more so as the existing procedures have to correspond to certain formal criteria according to the described duties. Furthermore, cultural barriers as well as resistance against organisational changes represent a vital obstacle. However, to improve work processes and the flow of information, a Continuous Process Improvement (CPI) approach was pursued to implement the system. Consequently, the Intranet "Saarland*Plus*" will enable that

- the knowledge is available at the location where it is needed, and
- the knowledge is stored in the place where it originated.

In addition, the Intranet "Saarland*Plus*" permits the adaptation of the internet technologies already used for the state government's web presence to the internal structures.

2.2 Conception and Implementation

The vision for Intranet "SaarlandPlus" was to provide an integrated communication platform that would incorporate the existing isolated intranet applications within the state government and that would support the co-operation within and between the various ministerial departments. To achieve this goal an integrated framework was developed that anticipates the various requirements within the state government. Considering the organizational structure described above it was necessary to realize the intranet in three expansion stages.

The first expansion stage is available for all state employees accessing the intranet. Information relevant to the whole workforce are stored there. Lists of telephone numbers and addresses, directories of legal regulations, work distribution plans, job advertisements, ICT related information as well as information concerning departmental vocational trainings fall under this category.

The second expansion stage contains the design of department specific sections. It is directed at the staff of each department and is only accessible for the particular employees. Announcements of the executive board, information of the staff council, important internal appointments or the internal departmental employee newspaper can be accessed here.

With *the third expansion stage*, cross-departmental project teams can use the intranet as a virtual project room in order to exchange relevant information for their team, e.g. appointments, protocols and other internal data.

Each of the described stages follows a procedural model containing four phases – requirement analysis, main and detailed concept, implementation and evaluation. During the *requirements analysis*, the requirements referring to the contents and functional range of the intranet are determined by the employees and the decision makers. This parallel top-down and bottom-up proceeding is necessary to detect any differences in perception at an early stage. The framework concept results from an iterative process, which has to take the results of the requirements analysis into consideration. It comprises a *main and a detailed concept* of hardware and software as well as the contents to be implemented. This is extended through a maintenance concept, a marketing strategy and a training concept. The *implementation* contains the technical and organisational implementation of the specifications which were compiled in the framework concept. The results of the implementation are then aligned with needs of the users during *evaluation*. As a consequence, this guarantees that the intranet is adapted to the user requirements and that possible barriers of use are anticipated.

The described phases were encapsulated by an *integrated framework* focusing on the change management process, the organizational structures, the technical infrastructure and the contents (information and knowledge). This ensures the active participation of the executives and employees during all phases.

The *change management process* was one of the most important parts of the project. The essential element of a successful change management is the commu-nication. Therefore several efforts were taken to enforce an open-minded information policy. The deployed communication tools ranged from kick-off-meetings in every department, e-mail-newsletters, circular letters and workshops to training, roadshows and marketing measures at the official starting date of the intranet. This integrated approach of treating the employees like customers and integrating them in every stage and phase of the development ensured a successful change management process.

Nevertheless, change management and communication are not sufficient to change given organizational structures. Therefore it was necessary to define *areas of responsibility* and assign them to specific employees. This areas of responsibility consisted in:

- *Organization:* Support for the intranet activities and promotion of the intranet in the departments and building an internal support-network for the intranet.
- *Content:* Support for the content provider and training of the employees. Designation of a contact person in every department for content-specific questions and a coordinator of the content providers.
- *Technique:* Support for technical issues and especially for the installation and support of the software needed to access the information via the intranet.

The process of assigning responsibilities was strongly enhanced by the fact that the project was residing in the staff office of innovation, research and technology that is assigned to the director of the state chancellery.

An important premise for the technical implementation of the intranet was represented by the existing content management system (CMS). RedDot was already in use for the state government's internet presence and had to be also utilised as the platform for the intranet. This led to several technical and organizational challenges:

- The chosen CMS is a very powerful tool, but it is not very useful for integrating content of different departments. As the integration was a main part of the conception, it was quite a challenge to solve this problem.
- The RedDot-technology which is used to maintain the content is quite simple, but requires a basic understanding of the difference of a web editor and a word processing software. It was another challenge to train the employees according to their knowledge in the use of the RedDot system.
- There was a bottle-neck in the RedDot implementation, caused by an internal service provider who was also responsible for the internet presence using the same platform.

To achieve a broad acceptance and usage of the intranet, the content has to be useful for the work process of the employees and, in addition to that, the installation of content has to be very simple. Using the existing CMS was one step in this process, another consisted in providing a general mandatory procedure for evaluating, converting and providing the contents for the intranet. Finally the accessibility was enhanced through usability-studies and corresponding measures.

3 Conclusion

The chosen procedural model has proven itself to be very successful. Due to the active integration of the employees, executives and stakeholders during the requirement analysis, it was possible to achieve consensus among all those involved. In addition, the knowledge resulting from the already existing isolated intranet applications was very useful. During the different phases, the integration of the employees and anticipation of cultural restraints were quite significant. The chosen expansion stage concept proved to be quite advantageous here. Through step-by-step expansion, employees had the opportunity to become familiar with the system and to actively participate in the formation of their own information and communication platform. The next steps of the projects will consist in the continuous expansion of the intranet's contents as well as the sustainable institutionalisation of intranet related activities within the ministerial administrations.

References

1. PwC: Die Zukunft heißt E-Government - Deutschlands Städte auf dem Weg zur virtuellen Verwaltung, Frankfurt (2000), p. 10.
2. Staatskanzlei des Saarlands: Szenario für eine eBusiness-Landschaft im Saarland, http://www.staatskanzlei.saarland.de/innovation_2121.htm
3. Breitling, M.; Heckmann, M.; Luzius, M.: Nüttgens, M.: Service Engineering in der Ministerialverwaltung. IM Information Management & Consulting, 13 (1998), pp. 91-98.

e-Learning for e-Government

Michel R. Klein and Jacques Dang

HEC School of management, 78350 Jouy-en-Josas, France
{Kleinm,Dang}@hec.fr

Abstract. The present paper presents some trends in educational activities for central and local government in France. It describes briefly the main characteristics of a brokerage platform used to import and export digitalized educational materials while protecting intellectual property rights.

1 Trends in Education for Central and Local Government

Central and local governments are facing the same kind of problems that other educational organisations, in particular private business organisations, are facing with respect to the continuous education of their personnel.

- the competence of the local government personnel is, as anywhere else, a prerequisite for improving the efficiency of administrative services.
- the persons in charge of providing continuous education are subject to budget constraint as anywhere else.
- there is a trend toward organising mix seminars where some teaching is done face to face but there are periods during which the participants interact with instructors from the distance.
- they have to help their personnel to receive training concerning new topics and software.

To these standard constraints are added others which are known to anybody who has been working for local government. Financial incentive used to reward improvement in competence is nearly non existent since promotion is essentially linked to age. Of course, there is always the reward of doing more interesting things, but in spite of the existence of this motivation for the personnel, it is unfortunately true, in our experience, that persons in authority , in our case elected members of the town council and the Mayor , are usually not very keen on professional education and put their energy on other matters.(like politics, for example!)

Another problem is that the public of potential learners in local government are spread all around the country and so it is natural that the educational centres are also spread over the country.

In France continuous education for local government is provided essentially by the Centre National de Formation des Personnels Territoriaux (CNFPT) which is offering seminars in about ten different educational centres and longer educational programs in four schools in Anger, Strasbourg, Montpellier and Nancy. Towns and other local governments can also use services of the local universities and private companies. The CNFPT is itself a public institution financed through a compulsory

R. Traunmüller and K. Lenk (Eds.): EGOV 2002, LNCS 2456, pp. 88–92, 2002.

tax paid by towns its seminars are free and towns rely mostly on its service for their basic continuous educational needs.

As a consequence once content has been developed by an expert on a particular topic it is necessary to educate and support local instructors who are going to teach this content in decentralised centres.

This constraint is all the more important since the decentralised centres have a fairly high degree of autonomy. They select their own instructors who are usually local professionals. The standard procedure is to educate local instructors to the content and let them use it in courses or seminars run locally.

One example in France concerns the educational needs which were generated by the new accounting plan M14 for local government which became mandatory a few years ago in all French towns.

The content of the course to introduce accounting personnel to the new accounting plan was developed by a group of experts. Then it was necessary to familiarise the local instructors with this content. These instructors were usually financial managers of medium size towns. For some specialised topics such as the impact of the change of the accounting plan on the financial management of towns it was a need to train instructors in the use of specialised software. In such seminars it is usually greatly appreciated to have an expert taking part in the seminar who can give the benefit of his experience.

In certain domains the trend is moving fairly fast. The development of training required with respect to office systems and Internet is a good example of such a domain. The type of competence to acquire is not different from that of any other organisation private or public. In such a situation it is often good strategy to re-use existing course material and learning resources which have been developed elsewhere. In such a situation it is useful to have outside support in order to:

- import and export educational material in electronic form
- support instructors in mastering the learning resources
- make it possible for experts to participate from the distance under the form of a live lecture through video (Klein,1998,2001).

2 Learning Resources, Course Description and Learning Activities

A learning resource in our context is something which is used to transfer knowledge during a course. It can be a case study, a case studies guide, a set of slides to introduce basic concepts of a topic of teaching, a chapter of a book on a given topic, an article, a collection of exercises with solutions etc…

The basic descriptors of a learning resource (LR) are: discipline to which it belongs, its type (case studies, complete course, syllabus,…), the key words which helps to define its topic, the language used, the level of dealing with the subject (elementary, medium, advanced..), the author, the educational institution , etc…
Persons in charge of teaching a topic are usually looking for a complete set of LR to teach this topic or a specific LR.

A learning activity is something dynamic such a live lecture , the coaching of a group of students working on a project, a business game session , etc…

3 The Use of a Brokerage Platform for Learning Resources

We shall now describe briefly the characteristics of one particular brokerage platform, which can be use to help solve some of the problem we have mentioned above. This software was developed within the framework of the European Project UNIVERSAL For more detail the reader is referred to [Brantner et als,2001].

3.1 Main Function of the System

The main functions of the system can best be introduced by presenting the main window of the system.(see fig 1). These functions are:

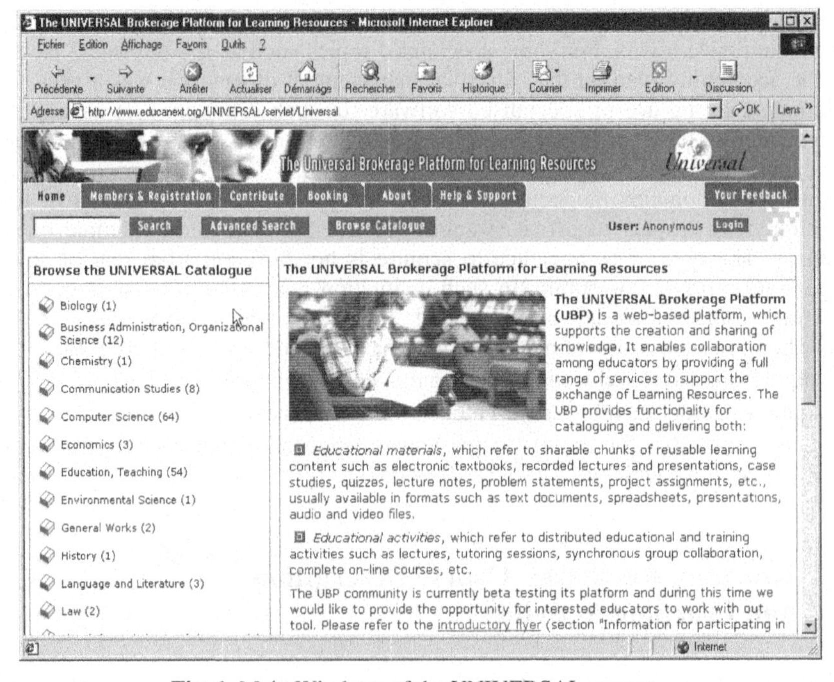

Fig. 1. Main Windows of the UNIVERSAL system

Contribute: this function allows a content provider (expert or a faculty member) who has been developing a course or just a given Learning Resource (LR) to add the **description** of the course or of the learning resource to the catalogue and to up-load in electronic form the learning resource itself: on the platform.

Booking: this function allows the instructor to download a LR once he has accepted the terms of use. These terms of use take the form of a contract. It is a this

Members and Registration. This function allow a potential user (content provider, instructor, student) to register to the system, to use its functions and to communicate with other authorised persons.

Search: allows a teacher or student to search for a given learning resource according to various criteria and to access the syllabus and document used in a seminars or course.

help & assistance: Provides access to a short introduction to the use of the system, in the future to its user manual.

3.2 Description of the Functions

3.2.1 Registering to the System

This function allows an institution to provide a list of persons belonging to the institution (faculties, students, ..) who are authorised to use the system. The registered user is provided with a user name and password. The system then knows to which institution the user belongs.

As a consequence a registered user of a given institution can be authorised to access the LR put together by a group of educational institutions. For example all Business Schools of a given alliance can decide to allow their faculties to access each others LR.

3.2.2 Searching for a Course Description or a Learning Resource

This function allows a registered user to search for a specific learning resource: for example an introduction to the town accounting plan or a complete course or seminar to a topic at a certain level such as an introductory seminar to the accounting plan of French town. In this latter case the user may wish to search for course descriptions on a given topic, ie their syllabus, describing the course with the topics treated at each class and the associated LR. Two types of search are allowed in UNIVERSAL simple and advanced.

3.2.3 Booking and Delivery of a Learning Resource

This function allows a registered user to book and import a learning resource. At the present stage two types of contracts are available to book a LR.. free access or restricted access to the members of a given organisation or association of organisations. The contract is there to provide protection of intellectual property rights and is a key element for decision by potential providers of LR.

3.2.4 Contribute

This function allows a content provider to:

- enter the description of the learning resources he wishes to make accessible.
- define the terms of the contract he wishes to have with potential users
- make the learning resource accessible from his own Universal interoperable server or upload it to the Universal Brokerage Platform (UBP) to make it accessible from the UBP directly.

4 Supporting Educational Activities

One important feature of UNIVERSAL is to provide a multi point IP- based video function called ISABEL. This function allows the teacher to teach to his students face to face and from the distance to other students through video. This service is presently running in the Linux environment. This function was, for example, used in a joint virtual course with students at HEC near Paris and Wirtschaft University in Vienna UNIVERSAL can automatically set up video conference sessions on demand and monitor the status of the live connection. It provides a convenient interface for managing on line class sessions.

5 Conclusions

The system we have presented is not specifically designed to support the exchange of learning resources for government personnel. This system was designed to support educational activities for higher and continuous education in general. It can be used for the education in the governmental sector. The fundamental goal of such a system is to make more readily available what is existing to interested persons and to protect intellectual property rights of authors.

Bibliography and Web Sites

1. For ISABEL see http://Isabel.dit.upm.es
2. For UNIVERSAL see http://www.ist.Universal.org
3. Brantner Stephan, Enzi Thomas, Guth Suzanne, Neumann Gustaf, Simon Bernd, UNIVERSAL-Design and Implementation of a highly Flexible E-Market Place for Learning Resources, Infonova and Vienna University of Economics and Business Administration, Information System Dept., 2001
4. Klein Michel, Borgman Hans, PC-based video as a tool for supporting collaborative work in teaching and research in Management, Proceedings 3 rd International Conference , Louvain- la –Vieille, May 7-9, 1998.
5. Klein Michel, Gauthier Valérie, Mayon-White William, Rajkovic Vladislav, Developing Synergies between Faculty and Students of European Business Schools through Telecommunication and Computer Supported Cooperative Tools , Proceedings Fourteenth Bled Electronic Commerce Conference , O'Keefe Bob, Gricar Jose et als (Eds), Moderna Organizacija, 2001.
6. Klein Michel, Keravel Alain, Impact des recherches sur les technologies de l'information et de la communication et leurs usages sur la formation à la gestion , Recherche et Enseignement de la Gestion, B.Moingeon (Editeur) , L'Harmattan, 2002.

Multi-level Information Modeling and Preservation of eGOV Data*

Richard Marciano[1], Bertram Ludäscher[1], Ilya Zaslavsky[1], Reagan Moore[1], and Keith Pezzoli[2]

University of California San Diego, 9500 Gilman Drive, San Diego, CA 92093 USA
[1] San Diego Supercomputer Center, MC 0505
{marciano,ludaesch,zaslavsk,moore}@sdsc.edu,
http://www.sdsc.edu
[2] Urban Studies and Planning Program, MC 0517
kpezzoli@sdsc.edu, http://regionalworkbench.org

Abstract. This paper addresses the issue of long-term preservation of and access to digital government information. We show how the preservation process is enhanced by storing an infrastructure-independent representation of the raw data, together with a model dependency graph (an executable graph of database view mappings). This allows for the design of decision-support tools and services for improving government transparency and promoting citizen access to eGOV data. A case-study, the *Florida Ballots Project,* is used to illustrate the approach.

1 Introduction and Approach

A common demand is that e-Government services promote citizen access to government information [1], such as official records kept at an archival institution [3]. Today, thanks to the ubiquitous Web, access to digital data is often less of a problem than actual information content. We argue that a multi-level information or "deep" modeling approach combined with an appropriate infrastructure independent representation mechanism can greatly enhance the value of eGOV data to the interested public, future researchers, and "digital archeologists/historians". We use the 2000 U.S. Presidential Election as an example of the deep modeling approach.

"On behalf of the State Elections Canvassing Commission and in accordance with the laws of the State of Florida, I hereby declare Governor George W. Bush the winner of Florida's 25 Electoral Votes," said Florida's Secretary of State, Katherine Harris, as she certified George W. Bush the winner over Al Gore, on November 26, 2000. The National Archives and Records Administration (NARA), went on to record this 25-Vote result by collecting two documents for permanent retention:

- *Certificate of Ascertainment,* containing the proposed Electors: http://www.nara.gov/fedreg/elctcoll/2000/certafl.html

---

* Work partially supported by NSF/NPACI ACI-9619020 award (National Archives and Records Administration / NARA supplement) and National Historical Publications and Records Commission / NHPRC award ("Methodologies for Preservation and Access of Software-dependent Electronic Records").

R. Traunmüller and K. Lenk (Eds.): EGOV 2002, LNCS 2456, pp. 93–100, 2002.

- *Certificate of Vote,* capturing the winning Electors: http://www.nara.gov/fedreg/elctcoll/2000/certvfl.html

More recently, two election media studies, started rethinking the entire process:

(1) *USA Today / the Miami Herald* on April 4, 2001, http://www.cnn.com/2001/ALLPOLITICS/04/04/florida.recount.01/

(2) The NORC *Florida Ballots Project*[1], on November 12, 2001, http://www.norc.org/fl, the results of which we use in our case study.

These studies present parameters under which either candidate could have won.[2] They suggest that with the 25 Votes, one should consider the retention of a parameter space that captures greater context. Fig. 1 depicts a model dependency graph we derived from examining NORC, and tries to formally define such a suitable parameter space as an example of our "deep modeling" approach.

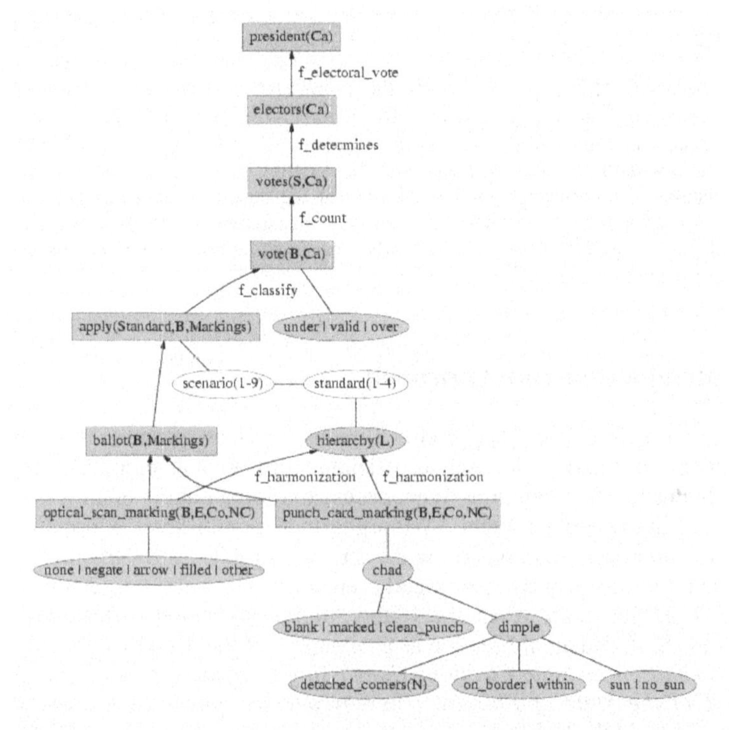

Fig. 1. 2000 Presidential Florida Election Model Dependency Graph, derived from NORC.

---

[1] At NORC, the National Opinion Research Center, at the University of Chicago, a consortium of news organizations including: the New York Times, the Wall Street Journal, the Washington Post Co. (The Washington Post, Newsweek), Tribune Publishing (LA Times, Chicago Tribune, etc.), CNN, Associated Press, and others. While the first study only looked at the ***Undervote*** (ballots rejected because no vote was recorded for the president), the second study looked at all 180,000 uncertified ballots in Florida's 67 counties, including the ***Overvote*** (ballots rejected because more than one presidential candidate was selected).

[2] Interactive analysis of the USA Today study at: http://usatoday.com/graphics/news/gra/gvote/frame.htm.

2 Multi-level Model Dependency Graphs

For several reasons, the depth of information modeling that corresponds to news reports and even official archival records is quite limited (e.g., the top three nodes in Fig.1), with NORC being a notable exception. One of the findings was that, depending on the specific scenario or applied standards, different outcomes can result.

The multiple possible outcomes can be made precise through "deep modeling" using a *graph of database mappings* as follows. The graph in Fig. 1 is an abstraction of such a graph (i.e., a network of "views" in the database sense). A view is a relational table that is defined by a query expression. Note that views can be layered and defined on top of other views, resulting in a graph of mappings. The overall graph itself defines a (complex) view, mapping the ("raw") input data to the final result (the President Elect in Fig.1). In general, a deep modeling approach using database views comprises:

1. ***relational schemas*** for all relevant entities and relationships (in the figure: parameterized entities and attributes)
2. ***view definitions*** (= database queries) precisely defining the mappings from one schema to another
3. ***constraints*** (logic formulas) over the relational schemas (to express, e.g., which standards can be applied to which ballot type)

Thus, in the actual graph, nodes correspond to relational views defined on top of other views or base tables. In our abstraction in Fig.1, nodes stand for *parameterized entities* (boxes) and *attributes* (ovals), while directed edges denote *database mappings*, i.e., functions between relational schemas.

Together with the raw data, the graph of database mappings can then be *executed* as a (complex) database query with a *verifiable* and non-controversial output. Of course this does not prevent a political controversy from happening, but it can be dealt with at a less superficial and more informed level: In Fig. 1, the *scenario* (see *Appendix E*) and/or the specific *standard* (see *Appendix D*) being applied to specific sets of ballots, uniquely determine the database tuples in the views above; in particular, the topmost tuple, i.e., which president should be named president elect. Thus, the only degree of freedom and non-determinism that such a graph of mappings allows is in the input data, in this case, the raw ballot data and the scenario/standard to apply.

NORC did everything to guarantee that the raw data was as objective as possible – in particular the coders did *not* compute the function *f_classify* themselves, i.e., they did not determine the votes. Instead every coder just described the markings seen and the *f_classify* determines the vote (under/valid/over) as a function of the standard and the markings on the ballot (see *Appendix B*). The crux is that those functions can be expressed and implemented as *database queries.* For example, the edge:

f_electoral_vote: *electors(S,Ca)* $\rightarrow$ *president(Ca)*

means that whether candidate *Ca* is elected president is a function (called "electoral vote") of the electors (of all States *S*) of *Ca.* The latter is itself a function of ***votes****(S,Ca)*, i.e., the votes that candidate *Ca* received in state *S.* Clearly, given the corresponding relational tables, the result of *president(Ca)* or *electors(Ca)* can be represented as a database query on a table representing the votes per state and candidate (=*votes(S,Ca)*).

Some citizens may be interested in the top-most node only: who is the president elect? Others may choose to study the reports from the news agencies and study how many electors each candidate could win or how many certified votes per state each candidate had. The extremely close outcome of the presidential race (the differences in votes between candidates were below the statistical error margin in the state of Florida – the state which ultimately determined the election, 271 to 266 Electoral Votes for Gore) sparked an enormous controversy about the official outcome of the election and almost led to a constitutional crisis.

As a result NORC conducted a thorough study aimed at resolving the issues. Translated in our framework, this means that one can resolve the controversial issues in a precise and for the interested citizen, verifiable way (depending on the available raw data of course). The model dependency graph shows that at the lowest level, the raw data consists, e.g., of *optical_scan_markings* and *punch_card_markings*. For example, *B,E,Co,NC* means that on the ballot with identifier *B*, the coder *Co* has described the element *E* (e.g., a specific chad or specific box for a candidate) to have a marking NC (=NORC Code, e.g., *dimpled chad with two detached corners*). In the graph, the node *ballot(B,Markings)* then provides a convenient way to represent the information: the ballot *B* with all of its markings (including, for each element *E* and each coder *Co*, the observed markings encoded as *NC*). One point of the controversy was which *standard* should be applied to determine the intent of the voter.[3] Depending on the county or even precinct, and the type of ballot (see *Appendix A*) different standards could be applied. For convenience, NORC created *scenarios*, where each scenario explicitly states which standard is applied to which set of ballots. The markings coming from different ballot types (optical and punch card) were "harmonized" (see *Appendix C*) so that one could easily express standards even if applied *across* different ballot types. This "harmonization mapping" was itself documented but was added only as another convenience: one can still apply standards directly to the markings of a ballot (but without harmonization one needs to do this for each ballot type individually).

Thus, under the assumption that the raw data is uncontroversial, by using a graph of mappings, the dispute can be localized to the specific choice of scenarios/standards being applied. In this way, transparency and verifiability of the process can be guaranteed for every interested citizen.

3 Preservation Issues

The modeling of the study as a network of database transformations also has advantages for the preservation information. The NORC study provided all raw data online and precise descriptions of the mappings as part of the accompanying documentation. Moreover, a "Scenario Manager" (see Fig. 2) has been developed that allows the user to inspect the outcome of applying different scenarios. From an archival point of you, however, the specific choice of system (Microsoft Access)

[3] Of course if other more *reliable* technologies were available that would lead to unambiguous voting results and avoid the discussion about which standard to apply – however, this is not the point here: even if this specific controversy was caused by an anachronistic voting system, many other eGOV data issues (e.g., redistricting) will always present a "deep modeling challenge".

introduces an infrastructure dependency: *"Will a researcher be able to evaluate and experiment with the study 5, 10, or 50 years from now?"*[4]

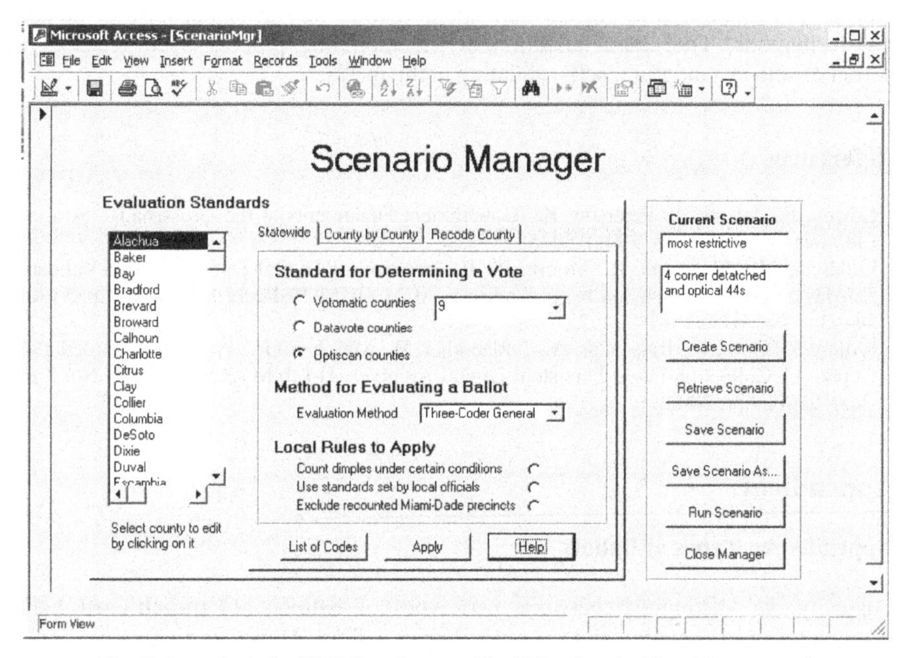

Fig. 2. Snapshot of a NORC tool created by Elliot Jaspin (Cox Newspapers)

A more robust and generic solution is to explicitly model the functionality of the Scenario Manager in an infrastructure independent way. First, mappings in a standard such as SQL. In this way, an "archival package" can contain the raw data together with all mappings and can be run on any future SQL engine, provided the standard is still supported.

One can go even one step further and create a *self-instantiating, self-validating archive* [2]. As in the case of SQL, we store an *executable* version of the constraints and mappings of a scenario. Moreover, we also package the *execution engine itself.* If the engine is SQL, this may not be a viable solution. Instead, for our running example, one could express all mappings as logic programs (Datalog/Prolog queries) and archive the complete execution engine (some complete Prolog system are smaller than the raw data of the NORC study) as part of the archive. Assuming that the logic engine is implemented in an infrastructure independent way (e.g., a Prolog engine in Java Byte Code), the complete analysis can be unrolled in the future by instantiating the graph of database mappings, and validating its integrity constraints. If the mappings satisfy certain properties, the analysis can in fact be *reversed,* i.e., one could try to solve the inverse problem and ask under which scenarios/standards a specific outcome is obtained.

[4] In fact, the problem occurs today: the Scenario Manager crashed several times during our experiments.

4 Conclusion

Multi-level (or "deep") information modeling provides a mechanism for capturing process information in a formal and unambiguous way as a network of database transformations. The characterization of the modeling process itself leads to the notion of self-instantiating, self-validating archives [2].

References

1. Cowell, E., Jacobs, J., Peterson, K.: Government Documents at the Crossroads. American Libraries. Infotrac. (Sep. 2001), 52–55
2. Ludäscher, B., Marciano, R., Moore, R.: Preservation of Digital Data with Self-Validating, Self-Instantiating Knowledge-based Archives, ACM SIGMOD Record, Vol. 30, No. 3 (2001) 54–63.
3. Moore, R., Baru, C., Rajasekar, A., Ludaescher, B., Marciano, R., Wan, M., Schroeder, W., Gupta, A.: Collection-based Persistent Digital Archives, D-Lib Magazine Vol. 6, No. 3 & 4, (Mar. 2000, Apr. 2000)

Appendices

Appendix A: Types of Ballots

Logically, we distinguish between two types of ballots: (1) **punch-card** ballots (*Votamatic* and *Datavote*), and (2) **optical-scan** ballots. However, there were really 5 types of voting systems in use in Florida. **Votomatic**, where a hand-held stylus was used to punch the pre-scored paper or *chad*, **Datavote**, where voters use a mechanical punching machine, **Optical Scan**, where ovals are filled in, or arrows connected, **Lever**, where the *Datavote* process was followed, and **Paper**, where the *Optical Scan* process was followed.

Voting system	*Number of counties*	*Undervotes*	*Overvotes*	*Total number of uncertified ballots*
Votomatic	15	53,215	84,822	138,037
Datavote	9	771	4,427	5,198
Lever	1			
Optical Scan	41	7,204	24,571	31,775
Paper	1			
Total	**67**	**61,190**	**113,820**	**175,010**

Appendix B: Visual Markings and NORC Codes Used

Most of the uncertified ballots were due to *overvotes*, and most of the problems came from the *punch-card* ballots. It is easy to see why, by looking at the following animation of a *Votomatic* voting machine (Doug Jones, University of Iowa):
http://www.cs.uiowa.edu/~jones/voting/votomat/animate.html

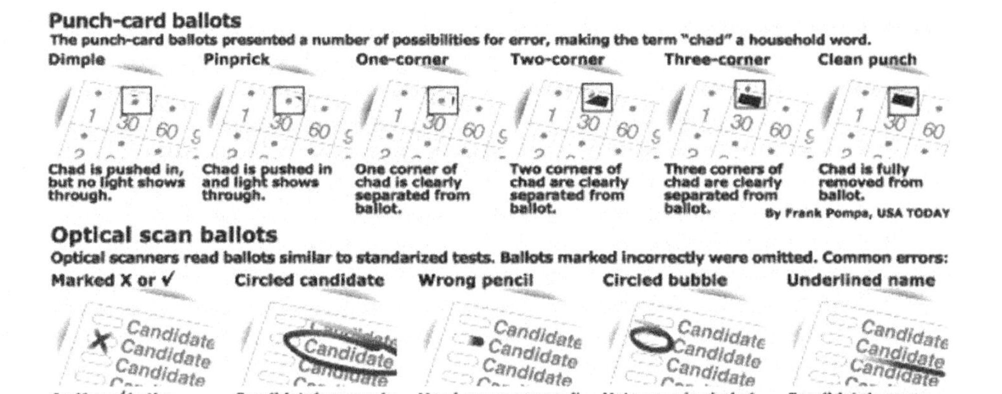
Punch-card ballots
The punch-card ballots presented a number of possibilities for error, making the term "chad" a household word.
Dimple
Pinprick
One-corner
Two-corner
Three-corner
Clean punch
Chad is pushed in, but no light shows through.
Chad is pushed in and light shows through.
One corner of chad is clearly separated from ballot.
Two corners of chad are clearly separated from ballot.
Three corners of chad are clearly separated from ballot.
Chad is fully removed from ballot.
By Frank Pompa, USA TODAY
Optical scan ballots
Optical scanners read ballots similar to standarized tests. Ballots marked incorrectly were omitted. Common errors:
Marked X or ✓
Circled candidate
Wrong pencil
Circled bubble
Underlined name
An X or ✓ in the vote area.
Candidate's name is circled instead of filling in the vote area.
Used a pen or pencil that the machine could not read.
Vote area is circled instead of filled in.
Candidate's name is underlined.
By Frank Pompa, USA TODAY

Examples of Undervote ballots for punch-card and optical-scan from the USA Today study.

NORC Codes used to classify each mark on **punch-card** and **optical-scan** ballots.

Punch-card ballots		Optical-scan ballots	
Code	**Meaning**	**Code**	**Meaning**
0	blank, no mark seen	00	blank, no mark seen
1	1-corner of chad detached	11	circled party name
2	2-corners of chad detached	12	other mark on or near party name
3	3-corders of chad detached	21	circled candidate name
4	4-corners of chad detached, clean punch	22	other mark on or near candidate name
5	dimpled chad, no sunlight	31	arrow/oval mark other than fill: circle, x, /, check, scribble
6	dimpled chad, with sunlight	32	other mark near oval/arrow
7	dimple within chad area, off chad, with or without sunlight	44	arrow/oval filled
8	dimple on border of chad area, with or without sunlight	88	arrow/oval filled or marked other than fill, then erased or partially erased
9	chad marked with pencil or pen	99	negated mark: scribble-through, cross-out, "NO", and similar

Appendix C: Harmonized Codes

Equivalence Classes	NORC Codes	
	Punch-Card	**Optical-Scan**
0	0	00 / 99 / 88
1	8	
2	7	
3	5	11 / 12 / 21 / 22 / 31 / 32
4	6	
5	1	
6	2	
7	3	
8	4	44
9	9	

Appendix D: Standards

Standards to specify evidence of voter intent.

Standards	Equivalence Class Codes	
	Punch-Card	Optical-Scan
1. Dimple or better	>= 3	>= 3
2. One-corner detached	>= 5	>= 3
3. Two-corner detached	>= 6	>= 3
4. Dimple (if rest of ballot is dimpled)	(>= 6) && (3, 4, 5)	>= 3

Appendix E: Scenarios

Scenarios
1. Prevailing statewide standard
2. Supreme Court "simple"
3. Supreme Court "complex"
4. 67-county custom standards
5. Two-corners-detached statewide
6. "Most inclusive" statewide
7. "Most restrictive" statewide
8. The Gore 4-county recount strategy
9. "Dimples when other dimples present"

For example, *Scenario 5.*, *Two-corners-detached statewide*, is based on arguments made by George W. Bush's attorneys during the 36-day period following Election Day, where Standard 3. is applied statewide.
Also, *Scenario 8.*, *The Gore 4-county recount strategy*, is based on early post-election results, where the Gore camp requested hand counts in 4 heavily Democratic counties: Miami-Dade, Broward, Palm Beach and Volusia. Standard 2 is applied to some of Miami-Dade precincts.

e-Government and the Internet in the Caribbean: An Initial Assessment

Fay Durrant

Head Department of Library and Information Studies
University of the West Indies, Mona, Jamaica
claudette.durrant@uwimona.edu.jm

Abstract. Effective e-government requires cultural change, the incorporation of inter-organizational teams, identification and evaluation of knowledge management assets, and incorporation of facilitating information and communication technologies. Government services must harness this range of information resources. Several governments in the Caribbean have recognized the importance of consolidating and exploiting their dispersed knowledge resources. The objectives of e-government are being analysed with a view to determining the most appropriate means of delivering services via electronic means. The paper examines particularly communication with citizens over the Internet, the delivery of Internet based government information, and aids to the citizenry in using these new facilities. Telecentres located in libraries and community centres, in Jamaica and other parts of the Caribbean, demonstrate the early development of facilities for enhancing government communication with citizens over the Internet, and interaction between citizens and those providing services

In the context of policies for "access to information" and delivery of government services, Caribbean governments have initiated programmes for electronic communication with citizens and the delivery of some services. While the process is still at an early stage it is instructive to assess the advances which have been made and to identify and evaluate the opportunities for further exploitation of the existing human, material and technological resources.

Governments have adopted the electronic and telecommunications facilities which have become available in the region. The Internet, with special applications, is now one of the visible signs, which has allowed people to be connected, and to exchange information. It therefore provides one of the basic requirements - electronic networking - for delivery of information services.

The United Nations Public Administration Network (UNPAN) defines e-government as "a permanent commitment by government to improve the relationship between the private citizen and the public sector through enhanced, cost-effective and efficient delivery of services, information and knowledge."

Electronic government can begin with making specific pieces of information available, and can subsequently integrate information sources to make interactive services available electronically. Most governments in the Caribbean are at the first stage and have websites for ministries, departments, corporations, and agencies. These websites serve to communicate information to citizens, and in some cases to receive feedback. It is anticipated that these websites will provide the bases for government and organizational portals, and the full scale provision of electronic services.

R. Traunmüller and K. Lenk (Eds.): EGOV 2002, LNCS 2456, pp. 101–104, 2002.

For e-government to be successfully implemented, an important step would be the identification of a single point of initial contact for the citizen to locate government services and or information on these services. Portals and gateways are being developed by governments to facilitate "one-stop shopping". and pathways and links to services and related websites. These are provided by the government information services, and in some cases the Office of the Prime Minister. These sites must of course be linked to the information held by other government organizations, so that a citizen's consultation is seamless and efficient.

As governments in the Caribbean continue to use the Internet to develop websites, to inform the public of policies, missions, and services, it seems opportune to undertake an assessment of these websites, and to identify their stages of development, and the incorporation of features considered essential for effective information services.

A pilot survey of government websites in the Caribbean shows that there are now sites of the central banks, the government information services, ministries of finance, ministries of agriculture, and the offices of disaster preparedness, statistical offices, and the Registrar General's Department. Of the twenty sites surveyed, all permitted or encouraged citizens and others to make contact with the organization, while fifty percent, provided mission statements of the organizations, and provided online publications.

Darrall West (2000) used twenty-seven features as the basis for undertaking a survey of government websites to determine their roles in providing information and services. These include:

Access features, design features such as tables of contents, site and subject indexes, frequently asked questions, interactive online services, multimedia, feedback, payment facilities, security and privacy. This researcher considers that additional features may include statements of government policies and statements of government services offered.

Of the twenty sites reviewed in this pilot survey, it was found that these sites can be classified as follows:

Statements of government policies and services	85%
Street Address, Phone, Fax, Email	75%
Comment and feedback	55%
Links to other sites	70%
Online publications	60%
Online databases – library or other	20%
Table of contents, site map	60%
Interactive online services	10%
Privacy and security policy statements	5%
Frequently asked questions	50%
Search engine, search engine	25%
Chat and instant messaging	0%
Date of last update	60%
Online payments	0%

While the sites do not necessarily state that they are part of an e-government strategy, the majority of the designers have clearly included statements of government policy and services. The majority are actively updated and permit comments and feedback.

There is however much further to go in meeting the demands of citizens and the challenges of knowledge management. Interactive access to information held by governments will require an integrated set of communications networks, which can result from interconnection of the communications channels and the content held on various databases and legacy systems. Offering online services such as the payment of taxes , passport services, or the issuing of certificates are important first steps. The Government of Jamaica, for example, has made some important advances in this area and the Registrar General's Department now enables citizens to download application forms or birth, death and marriage certificates.

From this preliminary survey there is also an indication that the features relating to presentation and delivery of services still need to be further developed and analyzed. The services identified so far include, for example, provision by the Central Banks of current information on the exchange rates. This information can be seen as very useful to the general population. Several of the central banks show the daily exchange rate with a range of currencies, running along the bottom of the screen. There can be further study as to how this type of information is accessed and used by a selection of business people and users from the general public.

Such features are the beginning of online government services, and to be effective, they must be based on the integration of the various sources of data held by different departments and agencies.

At this time the most important issue is assurance that integration of content and technology provide the basis of "one stop shopping " for government services. Our short or medium term vision shows citizens being able to access a portal to obtain information on government services, to transact business with the government, and to purchase licenses or permits, In the long term the objective will be to reduce the number of departments a citizen has to contact to obtain a required service. Interesting developments are taking place in the websites of the Gateway to the Dominican Republic, the Bahamas, St Kitts Nevis, Belize and the Jamaica Information Service.

The websites studied show that they are developed by individual ministries, and departments without always being linked to a "gateway" to government organizations and their services, which is generally recognizable by the general population. Some governments have used systems of naming URLs to provide some means of easy recognition by the public. The Country Gateway project being implemented by the World Bank globally has two instances in the Caribbean. The portal for the Dominican Republic has been established on the Internet, and in Jamaica a similar project is well advanced.

The development of e-government must involve each ministry department or corporation focusing on its own strategy within it priority areas. An important difference between business and government is that the number of citizens who are likely to use the services. Everyone citizen potentially has interest in accessing government information. The Registrar General's Department in Jamaica is an example of a site which all citizens are likely to access, as their need for certificates and other documents arises. The portal sites must therefore have the capacity for continuous use by multiple users.

In the content of e-government services in the Caribbean, there will be need for the citizenry to have public access to government portals. Telecentres such as those set up in the public libraries in Jamaica by the Jamaica Library Service and by the National Library of Jamaica provide examples of access points where information on government services can be accessed via the Internet. In addition to providing access to the

Internet these centres can provide training to the public and have begun providing access to local information resources.

The telecentres established by the Jamaica Sustainable Development Network in collaboration with the Government of Jamaica, the United Nations Development Programme and the University of the West Indies also provide opportunities for citizens to access government services via the Internet, and to take advantage of the services as they develop.

A survey of telecentres in Jamaica (Durrant 2001) showed that the network of public libraries and related information centers, are an example of facilities which citizens can use to access e-government websites. They indicate an area for further research to determine best practices in the development of the effective websites, as well as type of site and information service which satisfies the needs of citizens in certain contexts.

The data gathered so far shows that the Caribbean governments are making efforts to bridge the "digital divide". There is need, however, for more systematic study to inform and advise how further development would most effectively be done. Specific recommendations are for:

- the development by each government of a clearly identifiable website as an entry point for that government's presence on the Internet;
- use of URLs for the government websites based on a pattern which presents mnemonic features for easy recognition:
- incorporating information from the existing databases, to facilitate seamless access to a range of government services.
- development of access to online publications
- development of privacy and security policies and procedures
- implementation of online payment systems for government services.

References

United Nations Public Administration Network http://www.unpan.org

West, Darral M: Assessing E-Government: The Internet, Democracy, and Service Delivery, by State and Federal Governments. September 2000.

Durrant, Fay: Telecentres in Jamaica: report prepared for the Jamaica National Commission for Unesco. November 2001.

Durrant, Fay: Report on the establishment of two multipurpose telecentres: report prepared for the Jamaica National Commission for Unesco. February 2002.

Towards Interoperability amongst European Public Administrations

Alejandro Fernández

Ibermatica, IT Services, Partenon 16, 28042 Madrid, Spain
a.fernandez.martinez@ibermatica.com

Abstract. In this paper we present the technical approach followed in the InfoCitizen project. Its novel approximation to Public Administration necessities lies in the combination of Web Services as providers and Intelligent Agents as consumers of public services. The distributed nature of a Web Service network requires a similarly distributed architecture, that in turn requires an assurance of availability and transparency achieved with the use of Intelligent Agents.

1 Interoperability in European Public Administrations

The goal of this paper is to present the technical solutions the Infocitizen [1] development team is implementing to cope with the goal of providing a software solution to the interoperability of European public services. InfoCitizen attempts to conduct electronic transactions in multi-agent settings – e.g. multi-country involvement – in a manner as transparent as possible for the citizen. In the following we will elaborate on the requirements posed by InfoCitizen regarding:

1. transparent public service provision for the citizen and
2. multi-agent setting of public service provision.

The conceptual model for the project has already been presented in [2]. This document is organized as follows: we will start by providing a brief overview of agent technology in the scope of the requirements of the project, followed by the generic architecture that will be implemented in order to achieve the goals of the project. Finally we will introduce the IberAgents architecture, which provides a generic multi-agent and distributed [3] framework, which does not only fulfill InfoCitizen requirements, but also provides a generic integration and distributed framework to be used in any kind of distributed scenario.

2 Agent Responsibilities in the Scope of InfoCitizen

As the delegate of a user or a Public Administration in the system, an agent is responsible for the negotiation on behalf of the entity it represents. Thus, it will identify the parties it communicates with, and will not disclose information marked as "private".

If some further information is required to perform a service, e.g. some user information, it will try to obtain it using the available resources. If this step fails, the

R. Traunmüller and K. Lenk (Eds.): EGOV 2002, LNCS 2456, pp. 105–110, 2002.

Agent will then notify the requesting party, which must then decide which steps to take. Otherwise, the achieved result is communicated and the operation marked as successful.

In what follows, two kinds of agents will be discussed. The representative of the user in the system is the *Personal Agent*, and thus the only element in the system that communicates directly with the user, either on demand or on its own initiative. It can locate other components, but does not necessarily contain the know-how to fulfill a user necessity.

The second type is the *Interoperable Agent.* It is in these components that process information resides, and have the capacity to perform a task if the necessary input is provided.

Two main issues agents have to take into account are security [4] and anonymity of the citizens they will interact with. Regarding security, a Personal Agent must authenticate the citizen or PA employee requesting an operation. In the first case, where access to the system is public, a secure mechanism (such as Digital Signatures) must be used. In the special case where only PA employees can access Personal Agents, password authentication can be enough to enforce security.

An Interoperable Agent is often the unique entry point for administrative requests. It is thus its responsibility to keep access information private and anonymous, including access logs, identity of service consumers, and citizen profiles.

It is extremely important that all privacy concerns must be considered when implementing these components. One of the crucial goals of the project is to prevent third parties from keeping track of user activities in the system; in this regard, even the corresponding PA must see the system as an opaque layer.

At the same time, an audit trail must be stored, for the sake of data protection and integrity. Only if the user provides his or her consent, will the trail be revealed to an administrator, to undo an unwanted operation or correct any accidental errors.

3 InfoCitizen Architecture

In Infociticen [1], components are categorized in the following levels

Table 1. Top-down organization of component levels

Level	Component	Responsibilities
Personal	Personal Agent	presents a web interface to the user checks user authentication locates necessary services accesses interoperable agents contacts the user if necessary
Interoperability	Interoperable Agent	abstracts a process converts information between formats ensures anonymity of service consumer
Registry	Global Registry	registers components locates agents and services
Service	Service-supply Component	gives access to a legacy system
Legacy	Legacy System	performs the actual task

The topmost level is the only one visible to the user: Personal Agents provide a web interface, as the external access point to the system. Either accessed by an authenticated citizen or PA employee, it has the ability to perform any administrative task contained in the system.

If the Personal Agent cannot do the task on its own, it will delegate on an Interoperable Agent. On this level, the abstract information that describes a process is made specific (either for an administrative unit, such as a country, or for a special situation). Additionally, Interoperable Agents keep access data private, to ensure the anonymity of the users.

To locate available resources, Agents access the *Global Registry*. This special component contains a directory-like listing of Agents and Services, and can perform searches for specific components.

In the Service level, Agents access *Service-supply Components* in a uniform way. These are components that encapsulate access to legacy systems, and provide a SOAP interface within InfoCitizen. They should not access any other system components.

The lowest level is the Legacy level, where *Legacy Systems* reside. These include existing infrastructure in the PA, such as registration databases, and all other components external to InfoCitizen (e.g., an e-mail server). The PA can use the Legacy System in the traditional fashion, but the existence of a Service-supply Component ensures accessibility from within InfoCitizen.

Component Hierarchy

The top-down approach assures that components above can only access those below them; Service-supply Components, being on the lowest level, are just passive providers (they cannot access other system components, only external Legacy Systems).

Location of Components

In UML terminology, a *node* is the physical location of a component. It represents a computer or a cluster of computers.

The topology of the InfoCitizen system should be kept as flexible as possible, so that components can reside on different nodes.

In the Personal and Interoperability levels, there are no *a priori* requirements for node assignment. However, a good strategy is to locate all Interoperable Agents for the same PA in a common node. Personal Agents might thus reside in different nodes, depending on the number of users expected.

In the Service level, the usual approach is to place Service-supply Components in the same nodes as Legacy Systems, i.e. within the PA network. An external access point is provided for the rest of InfoCitizen nodes.

4 IberAgents

All the functionally and requirements highlighted in the previous paragraphs will be implemented by a multi-agent and distributed platform which is being developed by Ibermatica. It is called IberAgents.

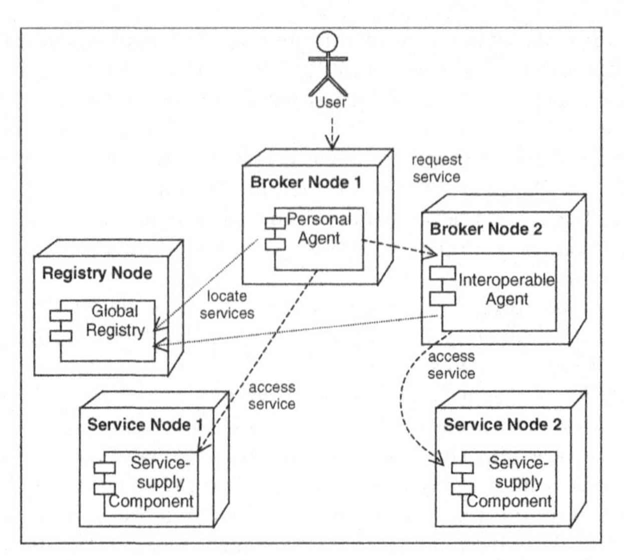

Fig. 1. Possible node organization of components. The Legacy Systems are not shown, since they usually reside in the same nodes as the Service-supply Components (Service Nodes)

The Agent Platform is organized in a distributed fashion, with several *components* (either *agents* or *services*) running on a number of *nodes*. At any given node there must be only one Java Virtual Machine (JVM) executing the agent platform.

Agents

In this architecture an *agent* has the following functionalities:

- *Interoperable*: agents can communicate with other agents.
- *Adaptive*: the path of execution of some tasks is not fixed, but depends on previous runs.
- *Delegated*: agents can make other processes run tasks assigned to them.
- *Proactive*: the agent is able to initiate operations on its own.

The features below are desirable, but not essential for an agent.

- *Distributed*: the agent can execute code on several machines.
- *Customizable*: agents can represent users, and respond to their customization.
- *Interactive*: agents can communicate with users via a graphical interface.

Services

In InfoCitizen, a Service-supply Component wraps a Legacy System for SOAP [6] access.

Correspondingly, IberAgents uses the term *web service*. A web service is a component for creating open distributed systems. Their main advantage arises from their availability via lightweight protocols like HTTP, and XML standards like WSDL and SOAP.

Directory

Component discovery is performed using a directory of existing services and agents.

One of the design principles of IberAgents is that *developers do not have to know or care about what node a service is running in.* As exposed in [3], this poses a number of problems, which are being addressed with the use of Intelligent Agents.

Communication

Agents must be able to communicate with other agents, and with services located on any node. This communication among agents and services is performed via SOAP, being completely **asynchronous** whenever possible (i.e., when a service does not return anything).

A **broadcast** is also possible, in order to allow agents to communicate with multiple components at the same time; and to collect answers to any message from multiple sources.

A provision for communication with other agent platforms is also considered essential, to enable cooperation with other environments.

Supervised Running

In IberAgents, all the agents are proactive, which means that they can start their tasks without external calls. Therefore, they must keep running, even if an exception or error happens at runtime. The agent manager will restart the agent if necessary.

For system administration purposes, agents can be stopped and started from any node in the platform, via a web interface.

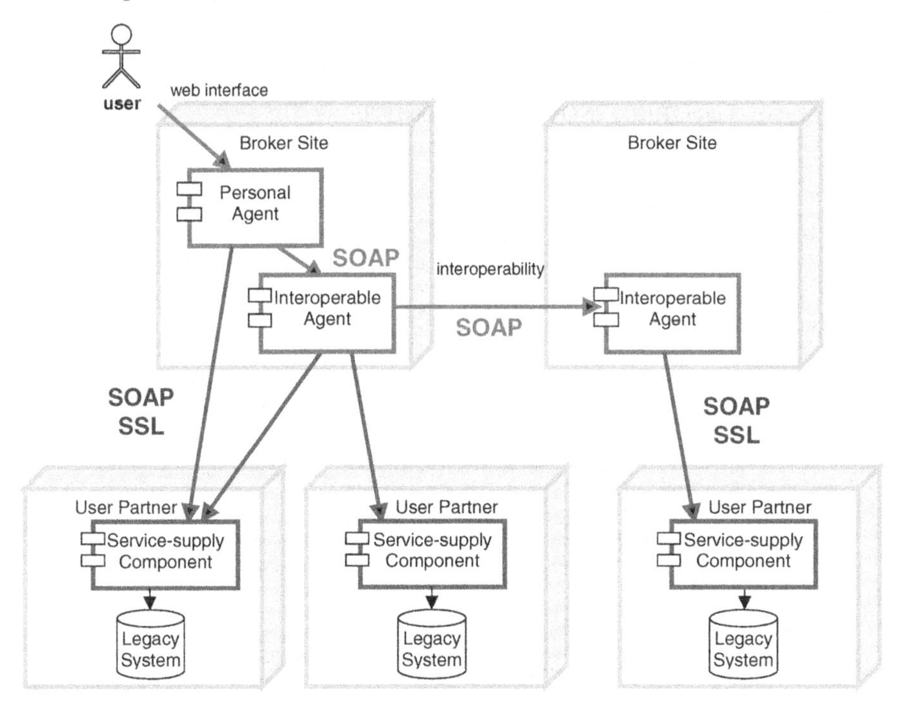

Fig. 2. InfoCitizen Integration Scheme. The *broker sites* can be developed using IberAgents, while the *user partner* sites might be implemented using other SOAP platforms

5 Integration of IberAgents with Other Modules

Just building a closed agent platform would hardly satisfy the needs of the InfoCitizen project. What is really needed is a way to interact with any external broker site, implemented using any underlying technologies.

In the spirit of web services, IberAgents publishes a SOAP interface for each of its accessible components. To the developer, there are no differences between an IberAgents site and any other InfoCitizen broker sites.

This openness leaves the door open to different implementations, and is essential to enable cooperative or competing solutions on other platforms.

As an example inside InfoCitizen, Service-supply Components will be developed using whatever technologies are suitable to the PA existing infrastructure. The standard SOAP interface will make Legacy Systems accessible from IberAgents.

References

1. InfoCitizen Consortium: IST-2000-28759 Annex 1
2. Tarabanis, K., Peristeras, V.: Towards Interoperability amongst European Public Administrations. Submitted to DEXA EGov (2002)
3. Waldo, J., Wyant, G., Wollrath, A., Kendall, S.: A Note on Distributed Computing. Sun Microsystems Laboratories, ref. SMLI TR-94-29 (1994)
4. IBM Corp., Microsoft Corp.: Security in a Web Services World: A Proposed Architecture and Roadmap (2002)
5. http://www-106.ibm.com/developerworks/webservices/library/ws-secmap/
6. W3C Note: SOAP: Simple Object Access Protocol 1.1 (2000). http://www.w3.org/TR/2000/NOTE-SOAP-20000508/

Assessing e-Government Implementation Processes: A Pan-European Survey of Administrations Officials

François Heinderyckx

Université Libre de Bruxelles, Dept of Information and Communication,
Av. F.D. Roosevelt 50/123, B-1050 Brussels, Belgium
Member of the Academic board of e-Forum
(Forum for European e-Public Services, www.eu-forum.org)
fheinder@ulb.ac.be

Abstract. A survey was conducted on behalf of e-Forum among 150 high ranking officials involved in e-government development in all 15 European Union countries. The results provide a unique pan-European examination of perceptions of officials driving the process of shifting towards what is generally referred to as 'e-government'. Issues covered in the survey include authentication techniques, financing e-government investments, benefits and fears among citizens, businesses, administrations and government, barriers and facilitators, priorities within the adminstrations.

1 Methodology

From December 2001 to May 2002, the Forum for European e-Public Services (e-Forum is a non-profit organization funded, for its start-up, by the European Commission) conducted a face to face survey among 150 key representatives of the public sector in all 15 EU countries. Interviewees were identified as key actors in e-government development and belonged mostly to general administration, be it at the national, regional or local level.

The questionnaire included closed and open questions covering various aspects of e-government with particular attention to perceived fears and expectations among citizens, businesses, administration and government as well barriers and potential catalysts in this area.

2 Results

The reader is reminded that respondents were all high ranking public servants, so that all indications collected are to be understood as these officials' perception of those indications.

One should also note that the sample size and the sampling method did not aim at providing a representative sample of the targeted population, but rather to collect information helping to assess broad trends in e-Government development in European administrations. As a result, breaking down the data even at the level of nations would

R. Traunmüller and K. Lenk (Eds.): EGOV 2002, LNCS 2456, pp. 111–115, 2002.

bear no relevance. Only two break-downs of the data will be considered, by groups of nations and by level of government (central versus local or regional).

2.1 Benchmarking by the European Commission

In 2001, EU member States agreed on a list of 20 basic public services (12 for the citizens, 8 for businesses) likely to offer e-government solutions. The Commission will monitor progress in implementation of those services on a half-yearly basis. The survey indicates that a significant number of people (about one third in our sample) in the administration are not fully aware of this new benchmarking initiative. Although about half of the respondents are uncertain whether the benchmarking will actually measure their progress towards e-Government, an overwhelming majority feel that it will increase their motivation to progress faster and will impact their plans or priorities.

2.2 Authentication

One of the central issues in e-Government projects remains authentication techniques. According to our survey, identification by user id and password is clearly the leading approach. Future developments include various means at medium and long term. PKI (Public Key Infrastructures, i.e. certificates sent by e-mail) are well represented, either currently or in future plans. Smart cards technology are significantly considered, but mostly within a few years time. Most administrations have no plans to resort to biometric recognition technologies. There is no spectacular difference in tackling authentication of citizens and businesses.

2.3 Financing e-Government Investments

Almost all respondents indicate that e-Government investments are included in their normal budget. A number of countries in Northern Europe, particularly in the British Isles and Ireland, do report special budgets on top of their usual departmental budgets. Very few countries report a possible co-financing by the private sector. Few respondents rely on benefit from cost reductions induced by e-Government implementation.

2.4 Priorities in Creating Benefits for the Citizens and the Businesses

The improvement of the quality of services was most often ranked as the top priority in developing e-government services for the citizens. Second in importance is to improve citizen's access to administrators and information (even more so in central administrations), followed by goals of improving efficiency, transparency and providing access 24 hours a day, 7 days a week (this is significantly more marked in Southern Europe). Improving cost-effectiveness is ranked higher in Northern Europe. Improving participation of citizens in democracy appears more crucial at the local and regional levels.

As for businesses, improving the quality of the services ranks at the top of priorities, particularly so in central administrations, and significantly more so in Northern Europe. Second to that priority come a group of 4 goals, namely continuous access to services (particularly in Southern Europe), enable services to be provided more cost-effectively (particularly in Northern Europe and in local administrations), improve the efficiency of administrative operations (particularly in Southern Europe and in local administrations) and improve business access to administrators and information (more so in central administrations and in Southern Europe).

2.5 Benefits for the Administration and the Government

As for the administration itself, the main goal is clearly to improve customer satisfaction, thus making the job easier (this is even more the case in Northern Europe). Also quite important is the expected increase in flexibility in working conditions, particularly so in local administrations, and even more in Northern Europe. Other high ranking priorities include personal development in new technologies (particularly in Southern Europe) and improved autonomy in the job (particularly in local administrations and in Southern Europe).

Respondents were also asked what they thought were their government's most important goals and objectives in developing e-government. It appears that administrations, quite homogeneously perceive their government's most important goals as seizing an opportunity to rationalize administrative procedures. Second in importance are improvement of citizens' well-being (particularly in local administrations and in Northern countries) and reduction of cost of administration (significantly more so in Northern Europe).

2.6 Fears Induced by the Development of e-Government (As Perceived by Administrations)

Administrations clearly (and with homogeneity) perceive 3 areas of concern among citizens: loss of information confidentiality, loss of human contact and digital divide (not all citizens will have access to the new services, and not all citizens will be able to use technologies properly). There is also a perceived concern following an increased control of citizens by the government (particularly in Southern Europe).

Regarding businesses' fears, as perceived by administrations, the major area of concern lies with the loss of information confidentiality and increased control by the government (particularly as perceived by local administrations for the latter). There is also concern about the fact that not all businesses will have access. But the loss of human contact and the issue of the ability to use the technologies is viewed as much less a concern for businesses than it is for citizens.

As regards the administration's perception of their own staff's fears, highest ranking concerns include inability to use new technologies properly, increased pressure from users/customers and inability to cope with increased speed (significantly more so in local administrations and in Northern Europe for the latter). Possible job cuts and increased control on individual performance are, to a lesser extent, other areas of concern. Loss of personal contact is perceived by administrations as much less a concern for their own staff than it is for citizens.

Finally, when asked about their perception of concerns induced within their government, administration officials clearly identify 4 areas of anxiety: failure of e-government projects (particularly in Northern Europe), digital divide, high cost of implementation (particularly among respondents working in central administrations) and risk of attacks and fraud by hackers (particularly in Southern Europe). The risk to end up with no real change ('window dressing') is also identified as a concern, particularly by respondents in Northern Europe and in local administrations.

2.7 Barriers and Facilitators

When asked to assess the importance of various barriers in the development of e-government, administration officials rank most often concerns about security and confidentiality as most prevalent. Second to that main concern, issues of lack of access among citizens, high set-up costs, lack of co-operation among administration departments and lack of political will and drive (particularly in Northern Europe for the latter) are most commonly identified as barriers in developing e-government.

When it comes to factors likely to facilitate implementation of e-government, one single factor stands out across Europe: strong leadership from the government. To a lesser extent, other factors also call upon political action: dedicated budgets, appropriate legal framework and availability of approved standards (particularly in Northern Europe). Also quoted are better internet penetration in households (particularly quoted within local administrations) and appropriate skills within the administration (significantly more quoted in Southern Europe).

Moreover, respondents were asked what their expectations were about an international association such as e-Forum. These questions provide valuable information as to what high ranking public servants involved in developing e-government are lacking in doing so. The strongest demand is, by far, on sharing experiences and best practices. This indicates that although each e-government project is clearly unique in its setting and constraints, administration officials in charge of their development are seeking experiences and practices elsewhere to feed into their set-up process. The prevalence of their demand in that respect indicates the absence of efficient structures in sharing such information at the European level.

Other salient expectations include: offer a repository for best e-government related documents in Europe (particularly among central administrations), provide an opportunity to develop informal network of colleagues (particularly in local administrations and in Northern Europe) and have a permanent up-to-date list of existing e-government services in European countries.

2.8 Priorities within Administrations

Face to face interviews allowed respondents for more spontaneous and open comments about the various issues related to the implementation of e-government. Some of these recurrent comments indicate patterns of opinions which appropriately supplement the main questionnaire.

A number of respondents believe that businesses as well as citizens expect e-government to provide a single point of access to administration and public services. Moreover, there is a recurrent view that e-government interfaces should be thought of as complements to existing, traditional systems rather than as substitutes, not only to accommodate those who can't access the new services, but also for those who do not want to. Regarding the issues of security and confidentiality, many consider that it is up to governments to build up people's trust and confidence. A number of respondents also stress the fact that efforts to develop e-government solutions should concentrate on back-office issues. It is also the case that many think that too much attention is focused on technical matters at the expense of considerations for a wide array of issues related to the more human aspects, i.e. the various problems to be solved regarding the people both as users and as administration employees. Many also expressed both their conviction that cross-departmental work was to be developed, and their skepticism that such change could really be achieved within the foreseeable future. Further along those lines, provided that e-government development is inseparable from administrative procedures' simplification and, broadly speaking, from a thorough business process reengineering, the transition can only be considered within a long term process which unfortunately exceeds usual political mandates and planning.

3 Perspectives

In spite of the necessary caution in using results from a survey conducted on a limited sample, converging views of these hand-picked key officials do provide some very relevant facts about the process at hand within administrations as perceived by those involved. Overall, it appears that officials managing the process of implementation have a well framed view of barriers, catalysts, fears and expectations associated with such process. The issues they raise clearly call for ample reforms which they appear dedicated to undertake. As for the way to achieve these e-government driven reforms, they seem to have developed a rather clear view of the priorities, although their agenda may appear to not necessarily match that of their political leaders, be it in nature or in timing. The shift towards e-government can be seen, to some extent, as the continuation of an on-going process which started with the implementation of computers and data-processing, so that administrations' experience in that area should be seen as a real asset. However, given that e-government consists in developing automated tools to directly interface with citizens and businesses, there is little doubt that the reforms and transformations at hand significantly differ in nature and exceed in amplitude that which lead to computerization.

A One-Stop Government Prototype Based on Use Cases and Scenarios

Olivier Glassey

IDHEAP and INFORGE, University of Lausanne,
1015 Lausanne, Switzerland
olivier.glassey@idheap.unil.ch

Abstract. In this paper we show the methodology we used to build a prototype for One-stop Government. We started by defining ten simple use cases, and then we developed scenarios, business rules and sequence diagrams for each of them. This work was based on a conceptual model for One-stop Government we developed in a previous research. We also explain why the use cases and the scenarios proved very helpful for the conception and the development of the prototype. Last we show the software architecture, based on distributed components, and the operation of the prototype with a few examples.

1 Introduction

In previous work we defined a conceptual model of a One-stop public administration, putting the accent on both the structural and behavioral aspects of such a system. We decided to work on such a model because we believe that there is a need for conceptual methodologies in that field, where various researches have already been conducted, such as the GAEL (Guichet Administratif En Ligne) project in Switzerland [3] or the BTÖV (Bedarf für Telekooperation in der Öffentlichen Verwaltung) method in Germany [6]. Indeed many researchers and public administrations need stable models of processes that can help public administrations managing the complexity and the rapid evolution of new technologies. This is also what a Swiss working group on e-Government appointed by the federal government believes [5]. Furthermore this group identifies the definition of "patterns of applications" as one of three essential action domains for e-Government.

During our modeling work, we took into account the internal processes of an administration, its relations with the customers and the expectations of the citizens: we made a six month survey in the "Administration Cantonale Vaudoise" (ACV), a large public administration at the cantonal level in Switzerland. There we met project managers, domain managers, departmental managers and users representatives for a total of 28 formal interviews. We also had many casual talks with different civil servants. Moreover we conducted two online surveys (1998 and 2000) in order to discover the expectations of the citizens in the field of electronic administrative services and we had close to 500 questionnaires to analyze. We also took into account the different targets that a One-stop public administration can have (citizens, businesses, civil

R. Traunmüller and K. Lenk (Eds.): EGOV 2002, LNCS 2456, pp. 116–123, 2002.

servants, other administrations, etc.) and different distribution canals (Internet, wireless, public kiosks, Intranet and so on). Using these field data we filled in "summary cards", inspired by CRC cards (Class-Responsibilities-Collaborators). For more information on this technique, we recommend [1]. First these cards helped us collecting the input from the various people we met and transforming it into a structural model through an iterative abstraction process. We defined this class model using the Unified Modeling Language (UML), developed by [2]. Then we used the cards to create use cases and scenarios following the methodology developed by [7]. We will not go into the details of this model because it is not the aim this paper where we want to explain how we built a prototype based on this model. In order to validate and to refine the conceptual model we decided to build a small system of a One-stop public administration and throughout this paper we will explain the different steps of this work. It has to be noted that it is only small prototype will limited functionalities because we lacked the resources to develop a full-scale system. However we think that the interest of this paper is focused on the methodology we used to build the prototype rather that on the final system. That is also why we will not show any implementation details and we will stay at the general software architecture level.

2 Use Cases and Scenarios

We wanted to illustrate rather typical Information-Communication-Transaction administrative services and we selected ten functionalities that we wanted to be part of the prototype. For each one of these we built use cases, scenarios and sequence diagrams. Here we shortly explained why we chose these particular functionalities even though they may seem quite simple:

– **Content publication:** this functionality demonstrates problems of content management, validation, update and access rights, as well as the ability of non-technical persons to publish content without specific knowledge.

– **Directory:** we found during our surveys of the citizens that it was one of the most expected services.

– **Calendar:** this tool illustrates typical problems of dynamic content management and of interfaces with databases.

– **Appointments and personal agenda:** the ability for a user to have his own agenda is a limited personalization capability in the prototype; a client can propose an appointment to a civil servant or receive reminders for various events.

– **Validation and acceptation of an appointment:** this shows limited workflow functionalities and interactions with a human actor; the system checks for conflicts in appointments and a person in charge validates it before the user receives a confirmation.

– **Discussion forums:** this service is an interesting feature because many forums are third party software and sometimes remotely hosted, so it allowed us to show the integration of commercial components or application in a One-stop public administration.

– **Chat:** basically we chose to add a chat to the prototype for the same reasons as above, with the idea of a future integration with the agenda, allowing the client to make an appointment for a official chat with a civil servant, which would provide advanced Communication services.

– **Search engine:** this is a very useful feature to find one's way in the mass of information provided by online government and it shows the ability to integrate distant functionalities within our prototype; a search engine is quite simple, but we can imagine to add a payment or a certification "engine" in a real system, which are necessary tools to support complex Transaction services.

– **Official forms:** official forms are the bases of a public administration's operations and they are at the convergence of Information, Communication and Transaction services; they are also the most requested feature that came out of our surveys.

– **Tax declaration:** this is the only really Transaction service of the prototype and we added it because it helped us illustrate the integration of legacy systems (we developed a prototype of tax declaration in 1997).

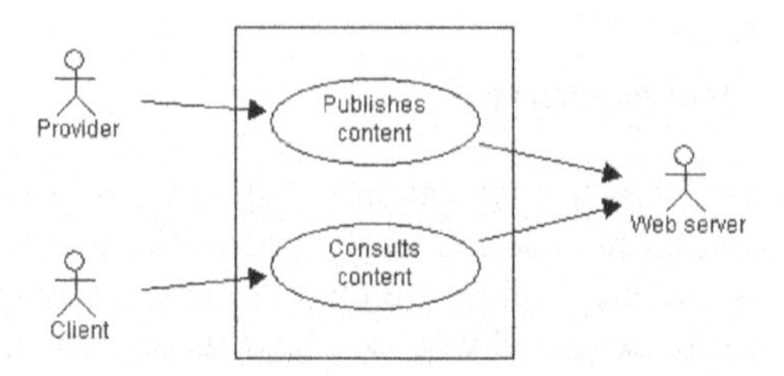

Fig. 1. Simple use case diagram

For each of these ten functionalities we created use case diagrams such as shown in Fig. 1 and we developed basic scenarios (Table 1), but here we will only show them for the first feature of the prototype. The use case diagrams and the scenarios are very useful to have a common language between technical persons, users, managers and leaders. They also provide a solid basis for a posteriori tests and validation of a system.

In order to complete the scenarios, which show in what order a procedure is accomplished, we defined business rules (an example is given in Table 2) to explain how a particular activity is conducted, using the methodology developed by [8].

When a scenario is a textual representation of a procedure or a service, UML also provides a graphical way of describing it. Indeed the sequence diagrams (Fig. 2) show the different steps of a procedure, but they also add information by introducing classes, actors and messages. They provide a solid basis for developers who can later on work on activity and state diagrams before they begin to actually work on the code. Furthermore there are many CASE tools that do code generation and reverse engineering, thus reducing greatly the workload of the software engineers.

Table 1. Basic scenario

Use case name	**Content publication**
Abstract	The public worker (or service provider in our terminology) publishes content on the Web: news, job offers, etc. The client can consult this content online after it as been validated by a person in charge.
Normal sequence of events	1. The provider publishes new content 2. The content is validated 3. The content is made available online 4. The client consults the content he needs
Alternate sequence	1. The client does not consult any content
Exception handling	At step 2, the content does not go online if it is not validated. A specific procedure begins.
Triggers	
Assumptions	
Preconditions	New content has to be published.
Postconditions	Content is accessible on line.
Authors and date	Olivier Glassey, February 12th, 2002

Table 2. Business rule

Business rule	**Job offer publication**
Short description	All job positions to be filled must be made publicly available, event if they are filled internally later on.
Source	Job regulation in public administrations
Type	Fact

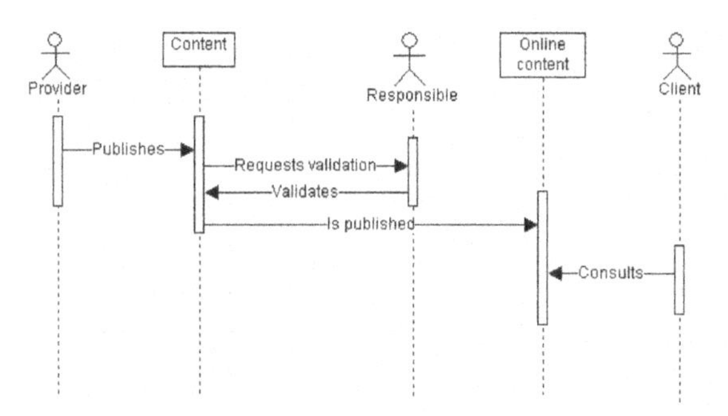

Fig. 2. Sequence diagram

As we mentioned above, we created such diagrams for each of the ten functionality of the prototype, then we used our structural model in conjunction with a CASE tool to create the class models of our prototype. These allowed us to generate the software "backbone", but we also needed to define logical software architecture before we started to really implement this system.

3 Architecture of the Prototype

We needed a software architecture that allowed us to take into account the following constraints, that we had found were very strong in public administrations:

- Ability to access heterogeneous applications and platforms
- Simple user interface and openness to the Internet
- Use of standard and robust technologies
- Modularity and scalability

To create the middleware layer we chose a distributed component architecture because it fitted our needs best and because there are really strong standards on the market. We will not present the characteristics of this type of architecture here, but we recommend [4] which is one of the foundation papers of this model. The three main components communication models are CORBA from the Object Management Group, COM+ (and now .NET) from Microsoft and Java/RMI for Sun Microsystems. Neither will we discuss these here, as there is an excellent comparison in [9]. Let us just say that for reasons of resources and simplicity we built our prototype using the COM+ model for the back-office applications. To be able to distribute these services to the clients we decided to build a dynamic Web interface based on the PHP/MySQL couple (respectively a server script language and a database, both of which are Open Source). The output displayed in the end-user interface is pure HTML, which means that it can be read by any browser such as Internet Explorer, Netscape or Opera.

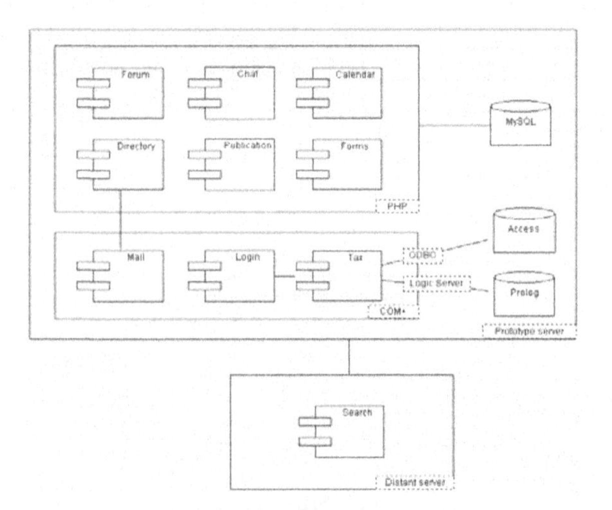

Fig. 3. Architecture of the prototype

Fig. 3 shows the general logical architecture of the prototype and of its functionalities. For most of the system, we adapted existing PHP applications (which can be seen also as components): forum, chat, calendar, directory, content publication tools and access to the official forms. On the other hand, the tax declaration application we built in 1997 was written in Delphi and used the ODBC gateway to connect to a Microsoft Access database and a Logic Server to retrieve calculation rules in Prolog. We simply wrote a COM component that encapsulates this application and is able to be distributed over the Internet using the ASP scripting language. We also added a commercial component called ASLogin that allowed us to secure the tax declaration pages. Finally we added the Google search engine in our prototype. We will not go into anymore details regarding the implementation of this prototype as this is rather a "toy" compared to real systems and because we think the interesting concept here is the general architecture: it allowed us to integrate various commercial and Open Source components and applications into what appears to be a unified and transparent system with a single interface.

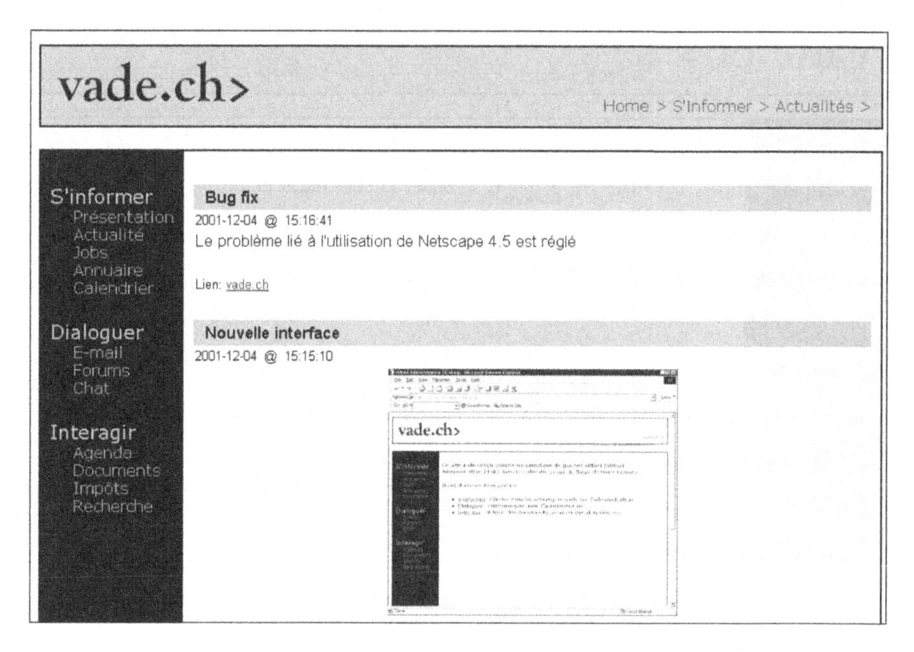

Fig. 4. Interface to access content

4 Operation of the Prototype

The prototype can be tested at http://uts.unil.ch/vade/, but it is only available in French. Here we will briefly describe its operation and show how users with different access rights can accomplish various tasks within a single interface. Let us mention that we made no particular work on the ergonomics of the prototype as it was not our intent. We used very classical navigation techniques with a left menu based on the Information-Communication-Transaction typology and an upper menu that shows the hierarchy of the pages. We realize that this is not entirely satisfying and we will work on a different interface in the future, most likely based on the "life events" metaphor.

The first screen (Fig. 4) shows the interface that a regular client (with no particular rights) sees when he chooses to see the latest news. The second one (Fig. 5) shows how a user with specific access right can publish new content within the same interface. Without any HTML knowledge he or she can publish text, links or images that will be validated before going online. Another example (not shown here) of clients with different access rights using a similar interface is the directory: a regular client can only browse or search the directory when selected civil servants have the access to add, edit or delete entries, in that case with additional links in the interface.

The prototype in itself does not show anything very impressive, but we think that its strong point is the concept of a single and universal interface that can be used via any distribution channel and by any type of client.

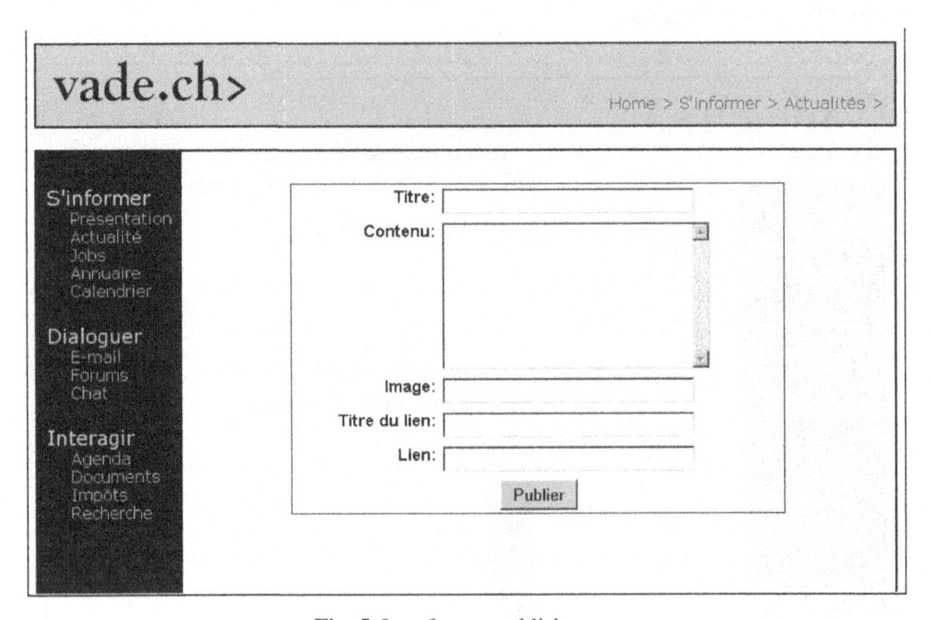

Fig. 5. Interface to publish content

5 Conclusions

What we have shown in this paper is the application of a conceptual model and methodology to a practical case of One-stop Government. The result of this work is a simple but functional prototype. We now want to emphasize what we believe are the strong points of this work. First we think that it is almost necessary to find a common language between technical persons, users and managers, and we found that the use of graphical models developed in UML provide a good way to achieve that. We also found that scenarios and sequence diagrams, although they are simple enough to be understood by anyone, are a great basis for software engineers to conceptualize and implement information systems that satisfy the needs of their users. We also realized that we needed a solid and standard logical architecture in order to integrate heterogeneous applications and software components into a seemingly unique One-stop Government system. Furthermore we think that, in order for this system to be accessible from anywhere and by anyone via a standard user interface, the best choice was to use a middleware layer interfaced with the Internet. In the future it is also very likely that this middleware will have to be interfaced with wireless technologies. Thus the architecture we chose allowed us to build a dynamic and universal interface that constitutes a single entry point to very heterogeneous electronic administrative services, providing true One-stop Government capabilities.

References

1. Bellin, D., Suchman Simone, S., Booch, G.: The CRC Card Book. Addison-Wesley, Massachusetts (1997)
2. Booch, G., Rumbaugh, J., Jacobson, I.: The Unified Modeling Language User Guide. Addison-Wesley, Massachusetts (1999)
3. Chappelet, J.-L., Le Grand, A.: Towards a Method to Design Web Sites for Public Administrations. In: Galindo, F., Quirchmayr, G. (eds.): Advances in Electronic Government, IFIP WG 8.5 in Zaragoza, Spain, 10-11 February (2000)
4. Fingar, P., Stikeleather, J.: Distributed objects for Business. Sunworld, USA (1996)
5. GCSI: 2e rapport du Groupe de Coordination Société de l'Information à l'intention du Conseil Fédéral. Suisse (2000)
6. Gräslund, K., Krcmar, H., Schwabe, G.: The BTÖEV Method for Needs-driven Design and Implementation of Telecooperation Systems in Public Administrations. Universität Hohenheim, Stuttgart, Germany (1996)
7. Kulak, D., Guiney, E.: Use Cases: Requirements in Contexts. Addison-Wesley, Reading, USA (2000)
8. Ross, R.: The Business Rule Book: Classifying, Defining and Modeling Rules, Version 4.0. Business Rules Solutions, Inc., USA (1997)
9. Suresh Raj, G.: A Detailed Comparison of CORBA, DCOM and Java/RMI. Wisconsin, USA (1998)

Reflections on the Requirements Gathering in an One-Stop Government Project

Johanna Krenner

University of Linz, Institute of Applied Computer Science,
Altenbergerstr. 69, 4040 Linz, Austria
krenner@ifs.uni-linz.ac.at

Abstract. This paper reports on the requirements analysis for one-stop government. It is focused on the work that was done in the analysis phase of the eGOV project - an EU project with the goal to develop an integrated platform for realizing online one-stop government - which is presented as a case study. Different types and sources of requirements are described. Furthermore the user-survey for the eGOV project is dealt. The paper concludes with some of the insights gained from the requirements analysis.

1 Introduction

At the moment e-Government is a highly popular topic. There are many projects running on communal, district, regional, national and European level. Thereby one-stop government is an important factor. The goal of one-stop government is to provide one point of access to the administration, which enables the user to take care of every public services s/he needs from there. Thereby it has to be taken into account, that most one-stop government solutions have three different target user groups: citizens, businesses and the public administrations themselves. As a consequence of this, there is a whole palette of different requirements.

To develop a successful one-stop government solution, careful requirements analysis has to be performed. In the EU-project "An Integrated Platform for Realizing Online One-Stop Government" – short: the eGOV project[1] (cf. [7], [9]) – a lot of effort was put into that phase.

The goal of the eGOV project is to develop a one-stop government platform. This platform includes a portal that can be used by citizens and businesses as the access point to all public administrations of a country. The user can do all of his/her public services there without having to know which one the responsible public authority is. Therefore the portal itself has to have some kind of "intelligence", so that it leads the user to the right public administration in the right region. Furthermore the eGOV platform includes elements through which public authorities can seamlessly connect to their back-office workflows.

The topic of requirements analysis for one-stop government is dealt with in this paper by first discussing different types of requirements. Then diverse sources of requirements will be presented. Based on this a closer look on user surveys will be taken exemplified by the eGOV project surveys. It will be concluded with insights and examples of requirements that were gained within the requirements analysis phase.

---

[1] eGOV Project homepage, http://www.egovproject.org/

R. Traunmüller and K. Lenk (Eds.): EGOV 2002, LNCS 2456, pp. 124–128, 2002.

2 Basic Types of Requirements

When designing an integrated one-stop government system, one focus should be on public services. They are the central elements of the system from the point of view of the public administration as well as from the point of view of the citizens and businesses.

Wimmer [8] has developed the holistic reference framework, which covers three dimensions that have to be considered when dealing with public services: abstraction layers, progress of public services and different views. The dimension of special interest for the requirements analysis is the one of different views. Based on these views, categories of requirements - not only for the public services but for the whole one-stop government system - can be defined: process specific, technical, user, security related, law based, organizational, social and political as well as data and information specific requirements.
By taking a special point of view (the necessity of considering multiple viewpoints is also pointed out in [6]), the requirements of one category can be distinguished. Thereafter these requirements have to be analyzed, rated and reconciled within a category as well as category spanning.

3 Sources of Requirements

The following section presents different ways of gathering user requirements based on the experiences gained in the eGOV project. One advantage of that project is, that the project partners already represent different points of view, since they are representatives of different sectors: science, economy and public administration. Hence many requirements could be gained by exploiting the knowledge and know-how of the different project partners. The representatives of the businesses are from the IT sector and could therefore contribute much to the technical requirements, security related requirements and also data and information specific requirements. The scientific partners (partners from research institutions) have special knowledge concerning processes, organizational requirements and also data and information specific requirements. A decisive role is played by the public administration partners, also referred to as user partners, since they represent the field that the platform should finally be used in. Consequently they added not only to the user requirements, the organizational requirements and the social and political requirements but also made demands in the other categories.

Another source of requirements was the literature research. Besides the traditional way of gaining knowledge from books and papers on e-Government in general (e.g. [2], [4], [5]), as well as on project-task specific topics (e.g. process modeling, meta languages, etc.), the internet was used as a knowledge source too. Many public administrations already have their own web site. These sites differ a lot regarding the level of technical progress, implementation of progress of public services (cf. holistic reference framework [8]), usability, etc. Many requirements could be gained by evaluating the pros and cons of these features. To enable this, a state of the art report [1] was worked out. In this report, a closer look was taken on several existing e-Government web-sites / portals and their connection to the back office.

Special attention has to be paid to the acquisition of user requirements, which are especially crucial, since they also have an effect on requirements of other types. The needs and demands of the users can be gathered and evaluated best by asking them in an ordered and concrete manner. Therefore user surveys were performed in the eGOV project which included questionings of citizens, representatives of businesses and public administrations.

4 Surveys in the eGOV Project

The method utilized for gathering user requirement was based on questionnaires. Since the potential users of the eGOV system can be divided into three user groups - citizens, businesses, public administrations - the demands, expectations and fears of each group had to be figured out. Based on this the user requirements could be concluded.

In order to gather user requirements, three questionnaires have been developed, one for each target group (citizens, businesses, public administrations). The survey was mainly performed by handing/sending out questionnaires. In addition online-versions of the citizen questionnaire were created, which could be filled in via internet. In Austria and Switzerland the survey at the public administrations was done by performing interviews based on the public administration questionnaire. Switzerland also chose this way of gathering input for the businesses.

The persons questioned were divided into focus groups according to different attributes to ensure the coverage of the whole bandwidth of citizens, businesses respectively public administrations.

The focus groups for the citizens were categorized according to:

- Gender: male/female
- Internet usage: not used to use the Internet / used to use the Internet
- Age: <18, 18-34, 35-54, 55+

The businesses were divided into focus groups depending on their number of employees (<50 (Small/Medium Enterprise), >=50 (Large Business Unit)) as well as on the sector the business was acting in.

The public administrations were distinguished by their level (national, federal state, municipality, other).

5 Insights Gained from the Analysis

There is a wide range of detailed requirements that could be drawn from the analysis in the eGOV project. Only an overview on some of them can be presented as examples in this paper.

5.1 Citizens and Businesses

Even though citizens and business need a different content in the portal, the requirements they pose are similar. Most citizens and businesses considered a common one-stop portal for the whole public administration of a country as important and helpful and also explained their willingness to use such a portal for transactions. However certain conditions have to be met to reach a high acceptance.

Users, especially inexperienced ones, will need a good guidance through the portal. A well thought out structure, a comprehensible navigation as well as a consistent look of the web pages are a cornerstone for that. A good guideline to reach this can be found within [3].

It is also necessary to present the content in a way that is understandable for a normal user. This means that the public administrations must not describe things using their terminology, but have to make "translations" into everyday language. This also includes the representation of the structure of public services. Unlike public administrations, which often categorize public services by their affiliation to a public authority , citizens and businesses consider a structure according to life / business events understandable.

When navigating through the portal, it is of importance, that the user is aware at any time where s/he is and where s/he can go next. This makes it less likely that the user finds him/herself doing something s/he did not intend to and it prohibits that s/he gets lost.

Personalization is also very important to improve the acceptance of the portal. Since an authentification of a citizen is necessary for transactions anyway, this can also be used for personalization purpose. Then, when filling in an application form, the citizen would get a pre-filled form where the data known about him is already entered. It should also be possible to reuse former applications. This would be convenient as well as timesaving for the user and would reduce the probability of mistakes.

A business is likely to use the portal more often than a citizen, since they are obliged to do more public services. Therefore the demand of businesses for personalization is even higher than the one of the citizens. Especially the possibility to create shortcuts to often used services and to reuse former applications is of importance for businesses.

Users have concerns about data security especially in connection with personalization since they are aware that their data has to be stored therefore. Transparency can counter these concerns. The user needs to be informed what happens with his data, for whom it is accessible and how it is protected. This creates confidence in the portal.

It is also desired - especially by the citizens - that one can gather information without having to register. Any kind of registration should only be obligatory if the user wants to undertake transactions.

Another requirement for a one-stop government portal is that it needs to be intelligent in some way. If a user wants to use a specific public service, s/he should be automatically connected to the right public administration (e.g.: marriage - registry office) in the right town/municipality/region/... The location information should be drawn from the information known about the user if s/he is registered, or, if that is not the case, from the information given by the user.

Additional help should be available if it is needed. An element that might need further explanation can have a link added, which connects to the according help-file. This way, the clarity of a page is not impaired and experienced users do not have to deal with information they already know.

5.2 Public Administrations

Public administrations often have a strict hierarchical order and the way they work is fixed to a certain degree by laws. Hence, much attention has to be paid to legal aspects during the development of a one-stop government system, but since these aspects are very task specific, it cannot be gone into detail in this paper.

The way public administrations work can be reflected in models of their processes. Process models need to be part of the one-stop government system in order to enable a clear connection to the back office system. In contrast to e-Commerce processes, e-Government processes are not always well structured but often semi-structured and sometimes even unstructured. Therefore the process models representing the processes in the one-stop system must be flexible and allow nonlinear process flows. Also the processes themselves have to be thought over and changed with regard to the new medium to profit most of the new system.

At the moment many different back office systems are in use at the different public authorities. They are the backbone of the back office work and exchanging them would be extremely difficult and expensive if not even impossible. Therefore a new one-stop government system - as the one that is developed in the eGOV project - has to enable the integration of a whole range of different types of back-office systems to have a chance to be successful.

Furthermore, a one-stop government system needs to be easy to administrate and the communication between different public authorities should be facilitated by it. Therefore a standardized language for communication would be useful.

To profit most of a one-stop government system, the back office work at the public administrations must have a certain level of technical advancement. A workflow management system as well as an electronic record system are a good basis for the profitable usage of a one-stop government system.

References

1. eGOV (IST-2000-28471), Deliverable D121: Services and Process models functional specifications, January 2002
2. Lenk, Klaus, Traunmüller, Roland, Öffentliche Verwaltung und Informationstechnik, v. Decker, Heidelberg, 1999
3. Nielsen, Jakob, Usability Engineering, Morgan Kaufmann, Academic Press, San Diego, CA, 1993
4. Reinermann, Heinrich, von Lucke, Jörn (Hrsg.), Portale in der öffentlichen Verwaltung, Speyrer Forschungsberichte, Speyer 2000
5. Snellen, I. Th. M., van de Donk, W.B.H.J. (eds.), Public Administration in an Information Age, IOS Press Ohmsha, Amsterdam, 1998
6. Sommerville, Ian, Sawyer, Peter, Requirements Engineering, A good practice guide, John Wiley & Sons, Chichester, 1997
7. Tambouris, Efthimios, An Integrated Platform for Realising Online One-Stop Government: The eGov Project, in: Proceedings of the DEXA International Workshop "On the Way to Electronic Government", IEEE Computer Socity Press, Los Alamitos, CA, p. 359-363, 2001
8. Wimmer, Maria A., A European Perspective Towards Online One-stop Government: The eGOV Project. Electronic Commerce Research and Applications, Volume 1, 2002 (forthcoming)
9. Wimmer, Maria, Krenner, Johanna, Next Generation One-Stop Government Portale: das Projekt "eGOV", In Bauknecht, Brauer, Mück (eds.), Informatik 2001, Tagungsband der GI/OCG Jahrestagung, Band 1, OCG, Vienna, S. 277 - 284, 2001

Understanding and Modelling Flexibility in Administrative Processes

Ralf Klischewski[1] and Klaus Lenk[2]

[1] Hamburg University, Department for Informatics, Software Engineering
Vogt-Koelln-Strasse 30, 22527 Hamburg, Germany
klischewski@informatik.uni-hamburg.de
http://swt-www.informatik.uni-hamburg.de

[2] Oldenburg University, Department of Economics and Law, Public Administration
26111 Oldenburg, Germany
lenk@uni-oldenburg.de
http://www.uni-oldenburg.de/verwaltungswissenschaft/

Abstract. Aiming to provide a platform for collaboration across agencies and to design appropriate IT support for the variety of administrative processes and decision making, concepts need to go beyond current approaches in business process modelling as well as workflow and record management. Drawing on these approaches, we suggest to focus on the unique tasks and activities of each actor involved and to present the relation of each individual contribution to the overall process as something tangible in order to support flexibility in the execution of administrative processes.

1 The Challenge to Support Administrative Processes

Electronic government is increasingly drawing attention to the need of reorganising many business processes within the public sector. In order to select appropriate forms of IT support for these processes, it is necessary to get a clear picture of their nature and of the purposes which they serve. Business processes in the public sector cover a wide range of tasks and of work arrangements. Whilst some of them can be fully automated, others rely on human agency and professional knowledge and require flexibility to a large extent. Unleashing the full enabling potential of IT for modernising the public sector requires a wider approach which presupposes a thorough familiarity with the "business" of the public sector and the characteristics of non-standardised work processes.

In this paper we re-examine approaches to understanding and modelling administrative processes and try to highlight the unique involvement of actors in administrative processes. The human element stands central in this approach, and we are looking for ways of modelling business processes which draw on the full range of the enabling potential of IT. We will propose a relational actor-oriented approach to modelling administrative processes and decision making across agencies, thus paving the way for a more appropriate IT support for public administration.

R. Traunmüller and K. Lenk (Eds.): EGOV 2002, LNCS 2456, pp. 129–136, 2002.

1.1 Characteristics of Administrative Processes

The characteristics of business processes in the public sector have much to do with the fact that most of the work there requires professional knowledge and experience. Mass-production of a type which can be fully automated does exist, but its scope is limited to simple processes of registering information, accounting and calculating. Of much greater importance are processes in which individual cases are dealt with, in more or less direct contact with the stakeholders. Legal rules and the explicit and implicit knowledge of administrators play an important role in such processes [4].

It is therefore adequate to say that the bulk of administrative processes in fields like assessing claims, granting licenses etc., is situated on a continuum which has on its one end fully standardised "production processes", and unstructured decision processes on the other. For processes of policy making, of legislating and of rendering justice, it is obvious that they depart to a large extent from the assumed model of production processes on which standard software in the private business sector is predicated. The same holds true for many processes occurring at the operative level of administrative agencies. Examples of such weakly structured decision processes include the granting of a license, assessing social benefit claims, issuing building permits, etc. When such processes start it is often not clear how long they will take, how much information is needed, and whether negotiations between the various agencies involved in the processes will take place.

Unfortunately, most computerised information systems in the public sector are still based on an understanding which takes well-structured and fully standardised processes as its starting point. These processes are recurrent in the private sector, e.g. in the field of accounting. Since many standardised processes can also be found in the public sector, e.g. in the fields of financial and personnel management, standard ERP (Enterprise Resource Planning) software such as SAP's R/3 is also making its way into institutions of the public sector. But besides such processes of an auxiliary nature, much of the work of public administration is characterised by primary processes at the operative level in which claims are processed, decisions made, and services rendered.

It is important to state that decision making in public administration occurs not only at the level of organisational management or policy, but is characteristic of its operative work. The officials in charge must remain flexible as to the workflow at stake. They must be able to ask for information, to ask a colleague for help, or to organise a meeting and insert the outcome of this meeting into the sequential work process.

1.2 The Quest for Flexibility

The options for standardising processes involving decision-making on individual cases or negotiations are very limited. Determination of some typical steps, which such processes should follow, may decrease service quality, effectiveness and efficiency. There are at least four reasons (which are not confined to the public sector) why more flexibility should be built into the execution of business processes where human agents collaborate using software of different types:

1. **Support of professional work:** from the perspective of an individual worker, an IT-supported workflow crosses a "workbench" which supports his or her work in all aspects, not confined to the work process to be acted upon. The official as a knowledge worker draws on many resources. He or she is used to invoke office tools, search for additional information, and uses platforms for collaboration with teams or groups, either on a steady basis or ad hoc. The interface between the flow of a process and the resources which this knowledge worker marshals can be construed as a situation where, according to the situated requirements of an ongoing process, he or she formulates demands toward the supportive environment which are met by "satisfiers" either available locally or brought in from elsewhere [5].
2. **Client's concern:** the occurrence of "service encounters" where an agent providing a service (or mediating it e.g. in a one-stop front office) is confronted with individual customer requirements flowing from a wide variety of life situations. These are difficult to specify in advance. Standardised service models should not preclude behaviour which caters to special wishes or needs of the customer. Rather, such models should serve as a resource, providing orientation in processes of service delivery which are adequate for a given situation [2].
3. **Unpredictable decision making processes:** decision-making is an important characteristics of its operative work in public administration. The officials in charge must be able to involve additional actors in decision making and to change the course of the process at stake.
4. **Limitations of cross-organisational feasibility:** actors co-operating across organisational borders have less or no possibility to discuss and commonly decide on details of their case-based collaboration during execution time. They frequently make assumptions or draw on commitments on what other agencies can contribute to the process execution. And it might turn out that actors in charge cannot act as planned and therefore must be able find other ways according to their available means and resources.

2 Modelling for Flexibility in Administrative Processes

Understanding the particular problems of different types of business processes in the public sector is a prerequisite for developing an adequate modelling approach. As pointed out above, many of the business processes in the public sector must not be predefined. We need to understand the details of how the actors involved bring in their expertise and how they collaborate and participate in decision making throughout the processes in order to choose or develop a modelling approach, which allows the design of appropriate IT support without losing sight of human agency and discretion in performing knowledge work, as well as of the collaborative aspect of such work.

Up to now, modelling of administrative processes is based on approaches known from business process reengineering, workflow management and/or record management. Those approaches and the current research in this areas do not focus on administrative processes. However, they do offer some support, but fall short of providing the flexibility required (see table 1).

Table 1. Major modelling approaches used to support processes in the public sector

approach	**business process reengineering (BPR)**	**workflow management (WFM)**	**record management (RM)**
original focus	business processes with the aim of reengineering, often based on event-process-chains (e.g. ARIS), usage of reference models for different domains	automatic management of data objects (e.g. documents) "flowing" through the work organisation while relating work items, work capacity and IT applications during run-time	processing administrative documents, i.e. creating, sharing (managing access authorisation), manipulating, registration/archive, retrieval etc.
current research	e-commerce	inter-organisational WFM	inter-organisational RM, semantic web
support for admin. processes	enables identification and overview of core processes of the organisations at stake in the process of modelling, actors are not taken as human agents working in a situated environment, but as attributes of process elements does not address workplace perspective, collaboration or flexibility (e.g. for officials the daily work is not triggered by "events")	adequate support of well-structured and standardised routine processes inclusion of independent subprocesses (e.g. across organisational boundaries) and other flexibility issues (e.g. exception handling) come increasingly into focus no support of officials or agencies in their way to organise or redirect processes according to situated needs	standardised IT support for record management throughout the organisation – but not beyond creates a more or less flexible collaboration environment, but no support for process management poor support for officials or agencies in their way to organise or redirect processes according to situated needs (e.g. to make adequate annotations)

2.1 Focussing on Actors and Relations within Processes

All of these approaches above can be applied successfully in public administration. But as they are inherently limited in supporting flexibility required for collaboration, we need to look for and/or develop modelling approaches which acknowledge the broad range of human work practices and take into account the understanding of human agency within administrative processes. Combining modelling practices from collaborative computing and process modelling (as in workflow and re-engineering methodologies), we try to identify the *unique involvement of actors* in administrative processes and present the *relation of each individual contribution to the overall process* as something tangible within the situated execution of processes and of related decision making. Focussing on actors and relations during modelling and IT implementation allows (even while a particular administrative process is ongoing) the official in charge to decide whom and what kind of contribution to enrol into the process.

From that point of view, the interconnection of process elements can be regarded as a form of individual contracting, framed by standards providing process patterns and rules for contracting. We radically depart from the assumption of a hierarchical world (in which process re-engineering is still caught). Instead of predefined processes being imposed and implemented from above, we assume contracting relationships – not only between external customers and an agency or an agent, but also within a process in which the results of each step performed should serve the next step of process execution. The actor in charge of this next step is thus considered as an internal customer. Such a contract model closely corresponds to the philosophy of New Public Management. One of its tenets is replacing hierarchical relationships with performance contracts, according to the principles of Management by Objectives. It is also related to a view which assigns tasks to units or agents not in a hierarchical way but by means of contracts, thus allowing for a wider range of institutional arrangements than classical administrative thinking.

2.2 Modelling Admin Points and Process Patterns

To support contracting and collaboration between officials / agencies as well as design of appropriate IT support, we suggest to model relational and actor oriented process patterns based on a repertoire of "admin points". The approach is a domain specific enlargement of serviceflow modelling ([1], [6]) which has been developed to model

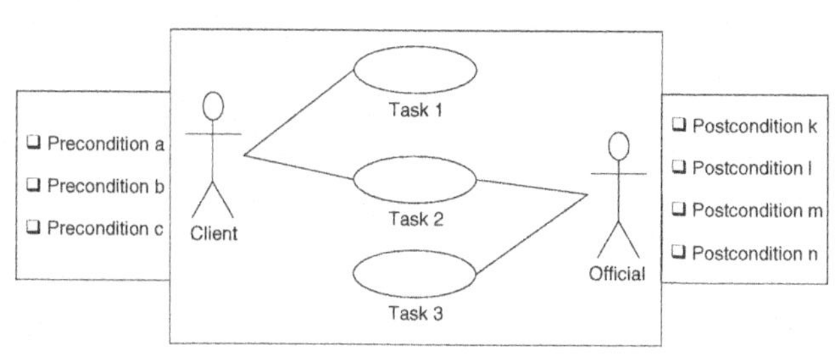

Fig. 1. Abstracted model of an admin point with three tasks (involving an official and a client, e.g. a citizen) and its pre- and postconditions

service processes in the field of tension between given standards and case-based reasoning. Within serviceflow modelling, the series of service points (denoted as a list) serves as the process plan or schedule (looking ahead) as well as the process history (looking back). Each of those points include a (UML-) specification of actors carrying out certain tasks/activities as well as the pre- and postconditions at each point (see abstracted model in figure 1) for "contracting" within process execution.

In principle, each of those points can be defined as needed. However, for communication and co-operation within domain specific processes and across organisational borders, it is most helpful to share a modelling "language", i.e. a common repertoire of premodelled admin points (see figure 2) in which each of these points are specified in terms of tasks/activities to be carried out, based on lists of pre- and postconditions.

Admin Points

Take-in Point: official or agent accepts application of citizen or other client of the administrative unit; if needed: official or agent evaluates situated concern and provides application assistance

Give-out Point: official or agent gives out result of application of citizen or other client ; if needed: evaluation of situated concern and assistance for next steps

Legal Inspection Point: official inspects legal aspect of incoming process/record and documents inspection results for further use within process

Research/Inquiry Point: official determines and/or inquires additional information in relation to an incoming process/record and documents results for further use within process

Record point: official processes incoming record according to assigned tasks and given standards/ rules, and documents results for further use within process

Approval/Decision Point: official approves or decides on a matter in relation to an incoming process/ record and documents results for further use within process

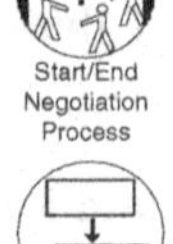

Negotiation Point (or start/end of negotiation process): several actors negotiate a matter in relation to an incoming process/record in order to decide on continuation of process (the responsible actor may also start a negotiation subprocess if necessary); results are documented for further use within main process

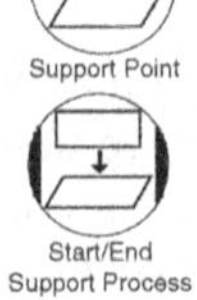

Support Point (or start/end of support process): a responsible actor may call on a contribution not specified a priori to support processing (the actor may also start a support subprocess if necessary); results are documented for further use within main process

Fig. 2. Example of an admin point repertoire

Based on these ideas, administrative processes may be predefined as process patterns, i.e. a series of admin points (e.g. figure 3 denotes an admin flow for the postal vote application through the web portal www.hamburg.de). In practice, actors involved in administrative processes may use these patterns as a general agreement for standardised co-operation and, in each particular case, as a template for the process and for the related individual documentation.

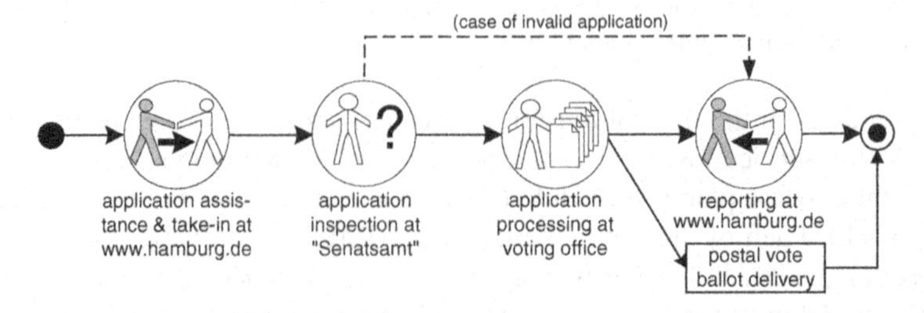

Fig. 3. Process pattern for postal vote (through the web portal www.hamburg.de) based on the admin point repertoire

Here, we can only briefly indicate how modelling of admin points and process patterns improves flexibility. While processing an individual case it is possible that, e.g.,

- the admin point schedule is predefined, but may be changed if necessary
- the task list at each admin point is predefined, but may be changed if necessary
- skilled work is needed to compare preconditions at each admin point with the accumulated postconditions of process history, and to decide about action to take
- skilled work is needed to document results at each point, in particular to compare actual postconditions with those premodelled and/or expected by other points

In addition, there are various ways of integrating negotiations into process execution:

- the point schedule is suggested, but may be suspended for negotiation (point or subprocess) any time
- the task list at each admin point is suggested, but may be suspended for negotiation (point or subprocess) any time
- skilled work is needed to compare preconditions at each admin point with the accumulated postconditions of process history, and to decide about action to take
- skilled work is needed to document results at each point, in particular to compare actual postconditions with those premodelled and/or expected by other points
- skilled work and/or legitimated decision is needed to evaluate negotiation results and to decide about course of process continuation

In practice, processes may be of mixed character. E.g. the postal vote application is well structured except for handling individual cases with unexpected characteristics revealed at inspection or at application processing (in rare cases these might involve negotiations). For many kind of processes (e.g. when an application for a building permit is filed) it is not clear at the outset if they relatively straightforward or not, and how complex they will eventually become.

3 Discussion

Aiming to provide a platform for collaboration across agencies and to design appropriate IT support for the variety of administrative processes and decision making, we need new approaches to understand and model such processes. We suggest to focus on the unique tasks and activities of each actor involved and to present the relation of each individual contribution to the overall process as something tangible in order to support flexibility through case-based contracting of process elements. Prior to modelling it is essential to comprehensively understand the work situations which unfold every time when an individual process is started. In these processes, process patterns based on admin points should serve as a guide, allowing for departures from the patterns at the discretion of human actors.

Reviewing the different types of work encountered in public administration (mostly based on professional knowledge and processing information), we find that not all types can be adequately described as processes, especially those having to do with organisational learning [3]. There, approaches from collaborative computing might be a better choice for modelling. E.g. in the case of multilateral negotiations, such as in the area of house construction and issuance of building permits, the situation can best be rendered by assuming a platform for free collaboration where the architect, the owner and the issuing agency meet to discuss the relevant questions.

However, in most instances the process view is inherent in the nature of the work of public administration and also of the judiciary and of legislative bodies. It is always (except for individual actors trying to improve their knowledge without producing any tangible results) about delivering a product – mostly an informational product such as a legally binding administrative decision – to some actors in their environment, or to society at large. This implies that an input (demands, supports, legal constraints) is transformed through a conversion process into an output. It is therefore not advisable to model collaborative working situations without a process structure.

We do not expect that the modelling approach presented here will soon be adopted widely. It still needs further research and empirical evidence to (1) prove the feasibility of modelling admin points and admin flows in public administration, (2) provide guidelines as to the identification of process elements and the required granularity of modelling admin points and respective flow patterns, and (3) gain experience on the scope of the modelling approach and its (possible) impact.

Finally, we do not claim that the modelling approach exposed here is the only one possible permitting to escape from a view which implies the strict co-ordination of process steps and which treats human actors as simple executing agents. But we do argue that this approach offers more potentials for flexibility and for acknowledging the central role of human agency than any of the concepts applied up to now. A radical departure from standard workflow approaches is now required in order to achieve an integrative understanding of the work in the public sector and to avoid blocking opportunities for improving productivity, performance as well as working conditions.

References

1. Klischewski, R., Wetzel, I., Baharami, A.: Modeling Serviceflow. In: Godlevsky, M., Mayr, H. (ed.): Information Systems Technology and its Applications. Proceedings ISTA 2001. German Informatics Society, Bonn (2001) 261-272
2. Klischewski, R., Wetzel, I., Serviceflow Management: Caring for the Citizen's Concern in Designing E-Government Transaction Processes. In: Proceedings HICSS-35. IEEE (2002)
3. Lenk, K: Notwendige Revisionen des Geschäftsprozessdenkens. In: Maria A. Wimmer (ed.): Impulse für e-Government, Internationale Entwicklungen, Organisaton, Recht, Technik, Best Practices. Österr. Computer Gesellschaft, Wien (2002) 65-74
4. Lenk, K. Traunmüller, R., Wimmer, M.A.: The Significance of Law and Knowledge for Electronic Government. In: Grönlund, A. (ed.): Electronic Government – Design, Applications and Management. Idea Group Publishing, Hershey, London (2002) 61-77
5. Mowshowitz, A: Virtual Organization. Communications of the ACM, 40-9 (1997) 30-37
6. Wetzel, I., Klischewski, R.: Serviceflow beyond Workflow? Concepts and Architectures for Supporting Inter-Organizational Service Processes. In: Proceedings of 14th CAiSE (International Conference on Advanced Information Systems Engineering). Springer, Berlin (2002)

Business Process Management – As a Method of Governance

Margrit Falck

University of Applied Sciences for public administration and legal affairs Berlin, Alt-Friedrichsfelde 60, D-10315 Berlin, Germany
margrit.falck@fhv.verwalt-berlin.de

Abstract. Practical examples of the public administration are discussed, in order to show the fact that business process management does not only serve the purposeful reorganization of administrative expirations but transports at the same time processes of organizational learning in the sense of Governance. Under the signs of e-Government there are above all the capability to cooperate and to act under on-line conditions, which must be learned by coworkers and administrative organizations. It is reported on a Virtual Community, which is used as instrument both for the business process management and for e-Governance.

1 It Does Not Go without Strategy and Examinable Goals

Business process management concerns the modelling of business processes with the goal of its gradual improvement (optimization). Which kind of improvement is aimed and in which measure improvement to be achieved, is to be defined before and specified on the basis examinable criteria.

With orientation on e-Government one goal is already set: the improvement of the process organization by means of the opportunities, which offer information and communication technologies on the state of the art. Decisive for kind and range of the use of technology however are such capability characteristics, those the organization of the process to be sufficient depending upon conditions are or the quality criteria, to which the result of the process correspond.

That has the consequence that the technological design concepts differ as a function of the goals, e.g. whether as improvement the balance of an administrative decision is the center of attention or the speed, with which it is present. In case of a development plan the employment of technology has to concentrate probably rather on transparency argumentation and decision making as well as on support forming of an opinion and bargaining processes, when in the case of a building permit, with which rather the automation of the processing and transportation logistics would have priority, after which the editors are supplied with information and documents for the quick task completion.

Administrative procedures are rarely in their capability characteristics homogeneous or in their quality criteria clearly. Rather many subprocesses are involved in a result, which concern one or more authorities and have in different segments different improvement goals.

R. Traunmüller and K. Lenk (Eds.): EGOV 2002, LNCS 2456, pp. 137–141, 2002.

Example: The management of a building exists in a set of service processes, which include planning, decision, production and negotiation processes and in those at least two authorities are involved: the building-using authority ("tenant ") and the building-managing authority ("landlord"). The improvement goals extend of the quick accessibility of the responsible partners, over the rapid and economical completion from repairs to the flexibility of the demand services. Accordingly possible solutions for improvement will exist in a mosaic of single solutions, which must be developed by a systematic analysis and organization of the processes in the sense of administrative engineering. In the example of building management belonged to:

- the conclusion of a service agreement,
- the list with the telephone numbers of the responsible partners,
- making available forms and document patterns,
- the modernization of different workplaces in the equipment with means of communication as well as with hard- and software,
- authority-internal and authority-spreading solutions for information supply and collection,
- the access to the Intranet,
- the budgeting of financial ressources as well as
- new regulations in the contract design.

2 Organizational-Learning As an Important Goal

The selection of the realizable solutions depends primarily on it which possibilities the surrounding field of the business processes permits. In addition, it should be made dependent on it, in which measure with the modelling and optimization process possibilities of the organizational learning for the involved ones are created, which refer at the request of the future process organization in the sense of the e-Government desired.

What do I mean with it?

Under orientation e-Government the feasibility of an improvement solution essentially depends on the following factors:

- of available resources (finances, technology, personnel including their qualification),
- of the valid right and security requirements,
- of available standards, information and reliable knowledge,
- of the talents of the involved ones in handling new media as well as
- of preferential action standards and learned behavior.

Those are factors, which are to be searched in the surrounding field of the business process, i.e.. in the application environment of the planned process organization.

In addition, many of these factors belong to the success factors for the work on the project, i.e. for the joint work during the modelling and optimization procedure, in whose process the new process organization is agreed upon and established, which is to become after conclusion of the project the lived organization everyday.

What lies more near to already learn and learn as during the common work on the project exactly the qualifications and talents or train such behaviors and set standards, which are in the future necessary for the planned process organization and for e-Government?

In the example of the service processes between tenants and landlords the agreed upon process organization required of the involved ones:

- to deal with information openly and transparency; i.e.. to make and maintain information offers, to give the tenants insight into documents upon achievements agreed or planned measures of the landlord or inform about problems and complaints, in order to be able to react in time.
- to work and communicate service-oriented; i.e.. to consider desires appropriately or to make also demands appropriately, for confirming dates and inquiries, to inform about intermediate treatment states;
- to call up available information regularly or as required (pull-principle); i.e.. to use information offers and to inquire missing information.
- to act team-oriented and cooperatively; i.e.. to regulate deputy, to be able to give information, to use electronic communication efficiently, to share knowledge.

Many of these requirements refer to the fact that with the transition for e-Government competence in electronic communication and in the electronic task completion is needed, which could not be developed so far. Therefore we tried to set during the work on the project standards and promote behaviors, which contribute to the development of the demanded competences. So we have

- transparency and openness promoted, by informing in the Intranet about the course of the project continuously. In addition reports belonged to beginning and conclusion of the individual work packages, over the planned goals, proceedings and over the reached results.
- the communication ability promoted and time standards for reactions and feedbacks set, by answering electronic inquiries rapidly or Reactions to inquiries called consequently, to made calculable our accessibility by absence assistant, as well as, results of common meetings documented and distributed nearly in time, regular Reviews to the fulfilment of agreements and measures.
- authority-spreading co-operation develops, by restoring the common confidence basis with existing conflicts by appropriate presentation and were endeavored around mutual understanding.

Like that the business process management is not only a method administrative engineering but also an instrument to carry the culture change on the way to the e-Government.

3 Co-operation between Authorities As an Element of the Cultural Change

From our view the development of the ability for authority-spreading co-operation is a particularly important behavior for e-Government, because usually several authorities are involved at the production of a product or a decision for citizens, economics or the public. With the development of citizennear on-line services the contributions of the authorities involved must be integrated to a continuous process chain.

It becomes clear by the example of the registration of a trade (Fig. 1.), which will be possible for on-line over the Internet or off-line over an office for citizen or the district office for economy either. The expiration of request integrates office for citizen, call center, office for economy and senate administration to a process organization, whose optimal organization for citizens and administration is dependent on the close co-operation of all involved ones.

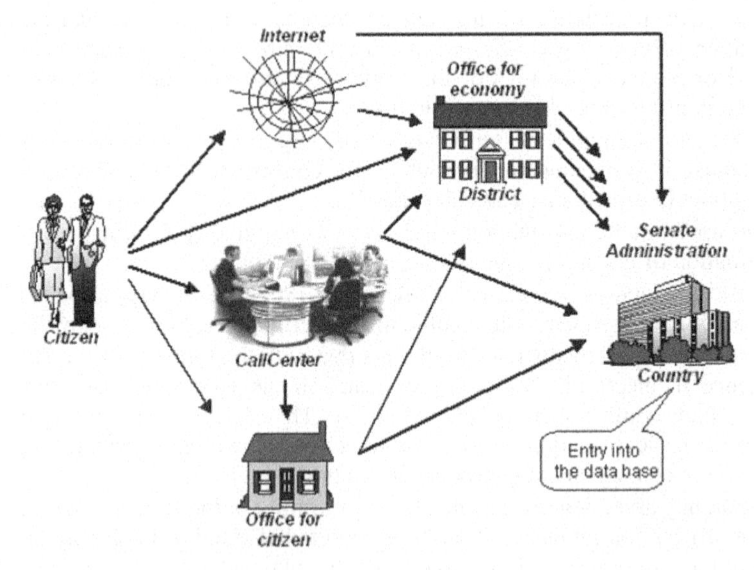

Fig. 1. Process variants „registration of a trade" Which process variant is going through, depends on different factors and decides in different places. For the citizen the choice of the entrance depends on its technical possibilities, on the use comfort, on the costs, on the quality of the consultation or other criteria. In the citizen office the run of the process depends on whether the case can be worked on finally or whether the case must be passed on to the economic office. The decision depends on which technical assistance and which know-how the coworkers have. In the Internet it decides whether the case can be worked on automatically or must be worked on manually. The decision depends on the fact whether the state of affairs is clear or must be interpreted. In the call center the further way decides likewise after the technical possibilities, after the existing qualifications and assistance as well as after the state of affairs of the case. The conditions at the different interfaces change with the time. Technical possibilities can be created, know-how grow, qualifications improve, assistance can be extended, etc.. Therefore the process optimization must be understood and realized as dynamic process, which presupposes the constant co-operation of the involved ones

An additional difficulty is that process organizations are subject to dynamics, with which the original optimization goals and criteria loses at validity. The dynamics can have their causes

- in the change of tasks,
- in the increase in experiences, competences and reliable knowledge,
- in the change of the division of responsibilities and the interfaces between the places involved and/or
- in the technology development.

The capabilities of cooperation of the authority involved, which was necessary for the development of the optimized process organization, is necessary thus also for the maintenance of the process organization.

In the case of the registration of a trade the representatives of the authorities were requested to the discussion in a virtual work space in the Intranet under the presentation of a neutral authority, in order to realize the difficult tuning processes between the numerous involved ones in a justifiable time. In this way the attempt is undertaken to establish the cooperative work forms necessary for the future process organization and to learn the appropriate abilities of the involved ones during the process design.

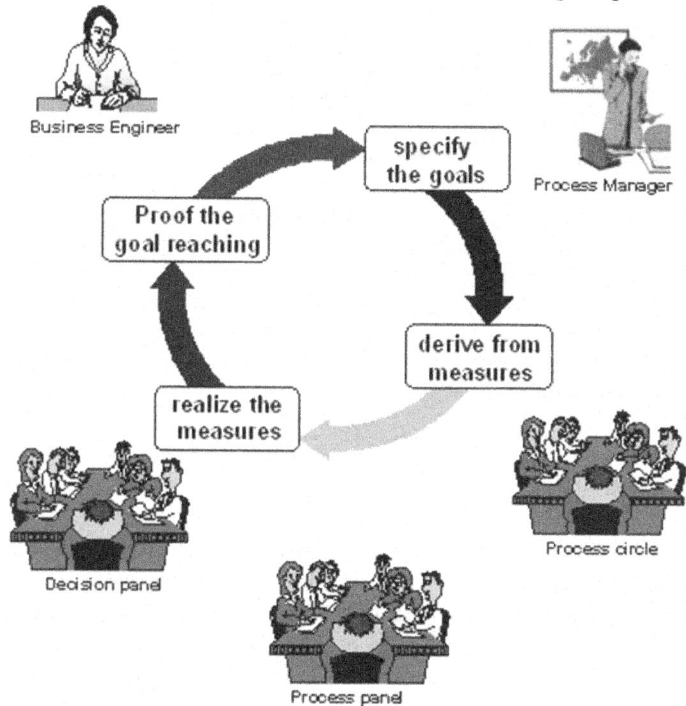

Fig. 2. Sustainability by process responsible person. The process manager observes the process organization, controls the process goals and suggests measures for further improvement. The process circle consists in theparticipants of the prozess, which collect information about the process, judges together, identifies weak points and looks for possibilities for the improvement. (similarly quality circles) The process panel consists of the process managers. They coordinates the suggested measures for process improvement. The Decision panel consists of the line managers of the organizational units taken part in the process. They decide on the realization of the suggested measures for process improvement. The business engineer supports the process responsible person in the process documentation. It knows GPM methodology and creates the connection to the information technology

4 Process Responsibility As Instrument of Sustainability

In order to reach the sustainability of an optimized process organization, it requires the continuous observation, analysis and improvement, in the sense of the process Controlling. As instruments of the process Controlling the function of the responsibility of process is currently tested. (Fig. 2.) About experiences can be reported at the time of the conference.

Proposal for a Dutch Legal XML Standard

Alexander Boer, Rinke Hoekstra, Radboud Winkels[1],
Tom van Engers[2], and Frederik Willaert[3]

[1] Dept. of Computer Science & Law, University of Amsterdam, Netherlands
[2] Dutch Tax and Customs Authority, Utrecht, Netherlands
[3] Application Engineers, Belgium

Abstract. This paper presents a proposal for an XML Standard for legal sources in the Netherlands. The standard intends to provide a generic and easily extensible framework for the XML encoding of the structure and contents of legal and paralegal documents. It differs from other existing metadata schemes for legal documents in two respects; It is language-independent and it aims to accommodate uses of XML beyond search and presentation services.

1 Introduction

This paper presents a proposal for an XML Standard for legal sources in the Netherlands. The research is carried out in the context of the E-POWER and e-COURT[1] projects and the results will be brought into the LeXML initiative[2]. The standard intends to provide a generic and easily extensible framework for the XML encoding of the structure and contents of legal and paralegal documents. This obviously includes legislation and case law, but also written public decisions, internal and external business regulations (for instance ship classification rules as in [7]), and contracts. XML elements and structure are defined in schemas that can be used to validate a document. Since there is a great variety of legal documents that cannot be covered by one normative standard, the standard consists of multiple schemas defining vocabularies that can be mixed in a document. While the standard aims to cover all possible legal sources, the focus of current work is on Dutch legislation: the 2001 Dutch law on income tax in the context of the E-POWER project, and the Dutch penal code of 1881 in the context of the e-COURT project. Later we will cover the structure of (Italian and Polish) court room transcripts (for e- COURT) and case law. The standard differs from other existing metadata schemes for legal documents in two respects; It is language-independent and it aims to accommodate uses of XML beyond search and presentation services.

[1] See Acknowledgements at the end of this paper.

[2] The LeXML initiative is the European equivalent of the Legal XML community in the Unites States.

R. Traunmüller and K. Lenk (Eds.): EGOV 2002, LNCS 2456, pp. 142–149, 2002.

1.1 Why a Legal XML Standard?

Clearly, legal documents serve a plethora of purposes, and for each purpose contain information that can be grouped in various general and specific categories. Many of the most common classification and reference systems even pre-date the storage of legal information on computers. XML promises dramatic improvements in the efficiency of managing and processing information in legal documents. XML and related technologies are essential parts of the more general idea of an integrated *semantic web*: A machine-readable and machine-understandable version of the current web coupled with new smart applications to exploit it. This machine-readable extension is called metadata: information about information.

Current XML schema and metadata definition efforts in the legal domain are concerned with government document locator services for general use like the British Legal and Advice sectors Metadata Scheme (LAMS) for 'Just Ask!'[3] and Australian Justice Sector Metadata Scheme (JSMS)[4]. In addition, there are XML schemes used by legal publishers for document locator services. These schemes focus mostly on classification of legal documents using traditional differentiae like:

- The author and legal status of the document
- Creation, modification, and promulgation dates
- The jurisdiction to which it applies, and the language(s) it is available in

These attributes are rather crude in meaning, and the resulting classification is superficial; It leaves out a lot of relevant detail and its quality is questionable for automated reasoning. Identification of documents by jurisdiction, for instance, assumes that the user of a search service knows what jurisdictions he is in – and that jurisdictions can be meaningfully delineated. Establishing jurisdiction is however a domain of legal research in itself. The meaning of the dates is clearer and can be directly used to establish validity of a document, but only relative to a known jurisdiction.

In addition to these traditional categories documents are usually classified in substantive terms with an attribute identifying a 'domain' of the document in a fixed classification. Viewing these attributes as metadata, or extra information about the document, creates a potential maintenance issue: The values of these attributes may change over the lifetime of a legal document, even if the document itself does not, as the concepts employed in the document change over time and become associated to (disassociated from) other concepts (see e.g. [5]). In reality, the agent of such a change in the web will usually be the creation or modification of a new legal document with certain attribute values. The information about the information is often not extra: It was in the document in the first place, or in another document referring to it. Classification at a 'heuristic level' always places the burden of maintenance on human domain experts.

[3] See e.g. http://www.lcd.gov.uk/consult/meta/metafr.htm

[4] See e.g. http://lawfoundation.net.au/olap/guidelines/metaintro.html

Furthermore, it is hard to conceive of any 'smart application' using the classifications as more than filters on search results. This is not because the meaning of 'jurisdiction' or 'appellate court filing' cannot be made understandable to the computer, it is because the level of the classification presupposes that the user of the classification system can read the document to find out why the classification was attached.

1.2 RDF and Representation of Meaning

The alternative approach to such a domain classification, is the direct identification of statements in the contents of documents. This includes statements about other documents and (fragments of) the document itself. For this purpose the Resource Description Framework[5] (RDF) was designed. In RDF, statements are encoded as as (*subject*, *predicate*, *object*) triples. By describing legal concepts in different jurisdictions in an RDF 'dictionary'[6] as conceptual prototypes, the LexML initiative assumes that it is possible to identify and describe similarities and differences between legal concepts in different languages. Existing schemes (like JSMS and LAMS) rely on compatibility with HTML's META tag only allowing RDF-like statements about documents; In RDF terms that means that the subject of a statement is always a HTML document. The META tag can only be used to make statements about a document in the document itself.

The RDF data model is based on the concept of a statement and the concept of 'quotation' (or reification) of statements. Quotation is used to make statements about (reified) statements – the statement is treated as the subject or object of another statement. In addition to this data model additional RDF schema statements can be defined, whose meaningfulness depends on the extent to which they can be transformed to other schemas and whether there exist any applications that can validate statements against that schema with query or inference languages. Since RDF is serialized in XML documents, it is important to realize that the validity of an RDF statement is relative to an unspecified set of other statements in the memory of the validating application. There is no central authority guarding universal consistency or coherence of the semantic web and anyone can publish any truth, falsehood, opinion, or judgment.

The problem, obviously, is that of trust. Quotation allows one to create trust, because it allows one to express where something was stated, who stated it, who modeled a statement made by someone, and what level of guarantee they dare to associate with it. 'Syndication' of information is based on this notion of trust. You may be willing to pay for advice, for instance, if you trust the party that gives it even if advice on the same subject is also available for free from other parties. It is not merely about good intentions: Heuristic notions like 'domain' always remain open for disagreement. The quotation mechanism in RDF allows others to make qualifications about statements, solving another limitation of META tags.

[5] See http://www.w3.org/RDF/. RDF, like XML is an open standard from the World Wide Web Consortium (W3C) that is well-supported with free software.

[6] http://legalxml.org/Dictionary/

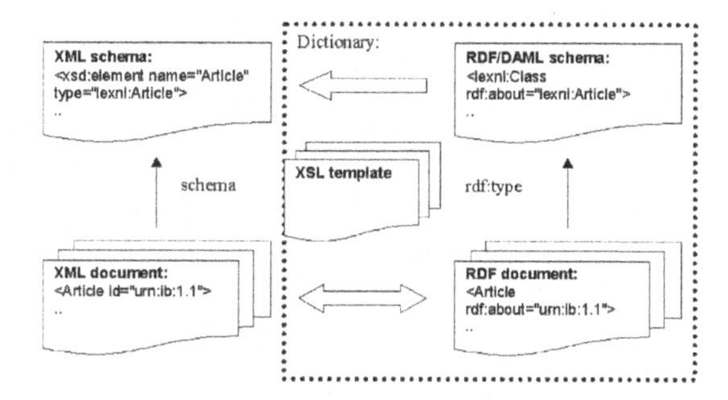

Fig. 1. RDF representation of legal documents.

Figure 1 shows the relationship we propose between the XML encoding of a document and the RDF encoding of the same document. If a document element is encoded as a set of RDF statements, any element of it can be subject, predicate, or object of statements *about* the document.

1.3 Aims of the Proposed Standard

This proposal is more complicated than comparable standards because it is motivated not just by concerns for 'smart' search, but also with document management and maintenance of existing and future decision support software used by public authorities like the fiscal applications developed by the POWER group in the Netherlands (cf. [1]). This proposal aims to standardize legal documents for the purposes of:

- Filtering
- Presentation
- Document Management
- Knowledge Representation
- Search
- Code Generation
- Rule generation
- Classification and Verification

Each meaningful element of the text up to a full sentence can be separately selected or changed by software through an XML API. Each element of the text can be marked with an ID attribute that is unique in the document. An element is easy to transform to (X)HTML (including hyperlinks) with familiar layout by application of XSL transformations. Because each element can have a unique identifier, it is possible to make external statements in RDF refer to the right part of the document. Only RDF, RDF Schema, and HTML support a widely accepted standard for reference anchors embedded in XML. The recent XLink specification released by the W3 Consortium creates a generic facility for referring unambiguously to parts of XML documents, but is not yet supported by most tools.

The standard consist of coupled schemas expressed in RDF Schema and XML Schema that are equivalent in meaning. A a two-way translator between the 'basic' XML standard and a more flexible RDF standard for software engineers is in development. RDF can be used to encode inverted file indices, association rules and, if time permits, self organizing maps (cf. [3]) to search for document elements using a concept index. A great variety of IR techniques requiring special indexing and active learning algorithms have been used for information retrieval on legal documents (see e.g. [6,4] for an overview). RDF is syntactically also sufficient to represent description logics (in DAML+OIL) or UML. The standard thus allows use of existing UML and DAML+OIL editors for legal knowledge representation. DAML+OIL is an RDF Schema[7] that provides a number of standard 'logical' constructs for RDF descriptions (cf. [2]). The content of documents, e.g. norms and validity constraints, are expressed using the DAML+OIL vocabulary, because this vocabulary can be validated by free description logic theorem provers (FaCT and RACER) and translated to an expert system rule engine (JESS).

2 Description of Documents

We roughly distinguish three different viewpoints on how we look at legal documents:

Form In the Netherlands and in general a legal document can be 'recognized' and classifed by certain required phrases and formulas. Formal requirements on structure mostly reflect considerations of consistency of language and ease of access for the reader, but it also provides a context for the interpretation of the content of the document. This latter role is very specific for jurisdiction and timeframe and not part of the basic standard. Generic structural desiderata are defined in XML schemas.

Role Although we may look at the phrases and formulas in a written decision to classify a document as a law, we know that it is not the structure of the document that makes it a law, but the role the document plays in the activities of public bodies - most importantly the activities that produced the document. This is captured in RDF statements 'about' the document.

Content We also classify documents depending on what its content means: It represents a type of decision. If it is just a public decision its meaning is limited to a particular occurrence or case. If it is a norm or policy its meaning extends to general class of occurrences or cases and it postulates a value theory for making and judging decisions. This is captured in RDF statements 'about' this content: acts, norms , agents etc.

Figure 2 shows the relation between the various XML schemas that are part of the standard. An XML document on the bottom leftside that adheres to

[7] See http://www.w3.org/SW/ or http://www.daml.org for DAML+OIL or e.g. http://www.ontoknowledge.org/oil/ for its precursor OIL.

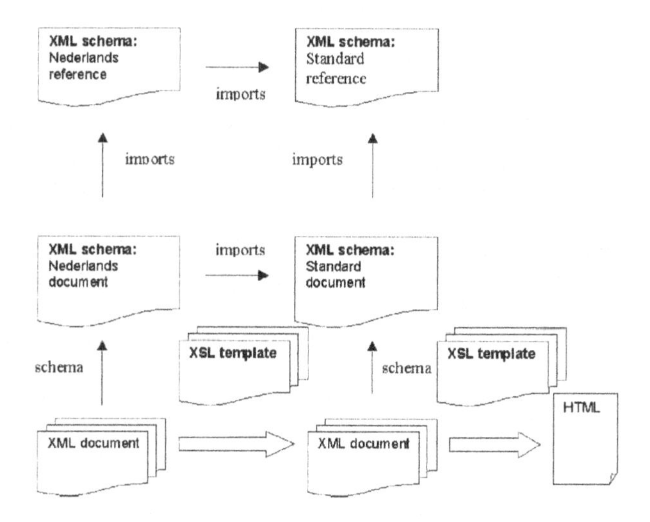

Fig. 2. Relations between components of the standard. The white arrows represent possible XSL transformations.

schema *Nederlands-document* uses the Dutch (Nederlands) XML element vocabulary for document structure and indirectly imports vocabulary *Nederlands-reference* and the structural requirements of *Standard-document* and *Standard-reference* – the 'language-independent' core vocabulary (in simple English). The language-specific schemas contain direct word-for-word translations (with the XML schema substitutionGroup element) to the standard vocabulary. An XML document is first transformed with an XSL transformation sheet to standard vocabulary, and the standard vocabulary can be transformed to HTML or RDF. RDF is used for further description of role and content.

Extensions built on top of the standard consist of relevant language-dependent vocabulary schemas, a standard schema that defines formal requirements, a simple XSL template that translates language-dependent schemas to standard vocabulary, an XSL template that translates the standard schema to standard-document, and optionally XSL templates for specialized presentations in HTML.

3 Terminology

Regardless of whether one sees the written law as an imperfect description of something else (*the* Law) or as a product of an social agreement that requires clear communication and predictable and foreseeable decisionmaking, it is clear that the written law refers to things 'in the world'. To regulate a world it must be described, after all. To be able to describe the meaning of the document the standard specifies a general vocabulary that can be used to describe it in RDF for specific purposes and domains. This work is in progress.

The conceptual core of these ontologies describes the relations between the various public documents published by public bodies in a fairly generic terminology, based on the general administrative act (AWB) of the Netherlands, that positions public documents in a legal system in terms of:

Public body A public body or body created by an act of law to serve a public interest.
Decision A written decision of a public body to perform a public act using a public power assigned by law.
Power A permission to perform certain public acts in a public interest.
Assignment A public act that attributes a power to perform a public act to a public body.
Delegation A public act of a public body that transfers a power assigned to it to another public body.
Mandate A decision of a public body to allow another person or body to use its power, without transfering the power, to perform certain public legal acts.

3.1 Time and Change

The importance of capturing the relations between public legal documents is made apparent by considering the requirements for correct updating of a collection of documents in time. Changes in laws are announced in separate documents and publishers must keep track of all documents from certain publication channels to be able to reconstruct what the form of an organic law is at some time point. Similarly, if you find a written decision on your doormat its validity status changes when a new written decision retracting it follows two days later.

To keep track of versions the standard provides a number of attributes for every structural XML element in the document that can be identified, selected, and thus changed; The *date-publication* of an element is the time the element is officially published or announced. The *date-enacted*, the time the content becomes applicable in decisionmaking, is always later than or the same as date-publication, but before *date-repealed*, the time the content becomes inapplicable in decisionmaking. Between date-enacted and date-repealed the element and its content is *active*, and outside this interval it is *inactive*. The *date-version* attribute represents the date the correctness of the content and other dates of the XML element was last verified. The XML document looses its value as a normative reference as time progresses and the time-interval between date-version and today increases.

If an XML element in a newly published document refers to another XML element the content of the element may repeal, enact, or change the other element. A XML element may also refer to another XML element to invoke its 'power'. This is the case if the official author of the document obtains the power to take the decision(s) communicated in the document from a specific assignment by law, delegation decision, or mandate decision. In such cases the element, or an ancestor element, becomes inactive if the element from which the power is obtained becomes inactive (at least in regulations from public bodies in the Netherlands). The standard aims to provide representational primitives that adequately capture legally relevant acts of this nature on legal documents without commitment to a specific 'model' for updating documents.

4 State of the Proposal

We currently have XML schema definitions of the structure (form) of Dutch legislation, Dutch documents in general and their standard counterparts, i.e. effectively all that is presented in Fig. 2. We have tried it out on the new Dutch income tax law and are in the process of applying it on other legislation. We have also created XSL translations from the XML format of a major dutch legal publisher and an XML format supported by the POWER tools (cf. [1]). We are building a large ontology for the legal domain, concentrating first on fiscal and penal law. By the end of June 2002 a substantial part will be finished.

Acknowledgements

E-POWER is partially funded by the EC as IST Project 2000-28125; partners are the Dutch Tax and Customs Administration, O&I Management Partners, LibRT, the University of Amsterdam (NL); Application Engineers, Fortis Bank Insurance (B); Mega International (F). e-COURT is partially funded by the EC as IST Project 2000-28199; partners are Project Automation, Ministry of Justice, CNR (I); Ministry of Justice (POL); Sema Group S.a.e. (SP); Intrasoft International (L); Universit Paul Sabatier (F); University of Amsterdam (NL).

References

1. Engers, T. van, Gerrits, R., Boekenoogen, M., Glass'ee, E., Kordelaar, P.: POWER: Using UML/OCL for Modeling Legislationn -an application report. In: *Proceedings of the 8th International Conference on Artificial Intelligence and Law (ICAIL 2001)*, pp. 157–167. ACM, New York, 2001.
2. Fensel, D., Horrocks, I., Harmelen, F. van, Decker, S., Erdmann, M. and Klein, M.: OIL in a nutshell. In: R. Dieng et al. (eds.) *Knowledge Acquisition, Modeling, and Management, Proceedings of the European Knowledge Acquisition Conference.* Lecture Notes in Artificial Intelligence, LNAI, Springer-Verlag, October 2000.
3. Kohonen, T.: Self-Organizing Maps. Springer Series in Information Sciences, Vol. 30, 1995. Springer, Berlin. Third edition 2001.
4. M-F. Moens, 'Innovative techniques for legal text retrieval', *Artificial Intelligence and Law*, **9**, 29–57, (2001).
5. E. Rissland and T. Friedman, 'Detecting change in legal concepts', in *Proceedings of the Fifth International Conference on Artificial Intelligence and Law (ICAIL-99)*, pp. 127–136, New York (NY), (1995). ACM.
6. H. Turtle, 'Text retrieval in the legal world', *Artificial Intelligence and Law*, **3**, 5–54, (1995).
7. R.G.F. Winkels, D. Bosscher, A. Boer, and J.A. Breuker, 'Generating Exception Structures for Legal Information Serving', in *Proceedings of the Seventh International Conference on Artificial Intelligence and Law (ICAIL-99)*, ed., Th.F. Gordon, pp. 182–195, New York (NY), (1999). ACM.

Size Matters – Electronic Service Delivery by Municipalities?

Ronald Leenes and Jörgen Svensson

University of Twente, P.O. Box 217, 7500 AE Enschede, Netherlands
{r.e.leenes,j.s.svensson}@bsk.utwente

Abstract. The development of e-government in the Netherlands shows two different worlds. The large national organisations implement Electronic Service Delivery (ESD) fairly successfully, while municipalities are slow to adopt ESD. This is a pity, since municipalities account for over 70% of the public services. They are expected to implement ESD on their own although they lack the necessary resources and distributed development is inefficient. In this paper we address the role of municipalities in the real and virtual world and argue that development of electronic (local) public services may be organized on a larger scale, depending on the type of service in question.

1 Introduction

In the international rat race for e-government, the Dutch government too has set ambitious targets. Twenty-five percent of the public services are to be delivered online by the end of this year. Although twenty-five percent does not seem much, and certainly is a long way from full-blown electronic government, even this goal is difficult to meet. In fact, in order to claim success, the Dutch government is already in the process of massaging the data. By redefining the phrase "twenty-five percent of services" into "twenty-five percent of service transactions", it is giving a disproportionate weight to the few large national service programmes, such as the Internal Revenue Service and the Dutch student bursary programme, which have succeeded in implementing electronic service delivery.

As a result of this redefinition, the twenty-five percent target may be reached. But, what about the next seventy-five percent, or even the next ten percent? With the large central programmes digitised, any future increase in electronic service delivery (ESD) will have to come from the Dutch municipalities. These municipalities deliver the vast majority of public services in the Netherlands, and they are currently considerably less successful in implementing ESD.

In this paper we discuss the reason why further ESD-development by the municipalities is problematic and we suggest some solutions to improve the chances of success.

The structure of the paper is as follows. In section 2 we compare the ESD track records of large national public service organisations on the one hand, and municipalities on the other. Then, in section 3 we provide the simple, but fundamental, problem of ESD-development by the municipalities: scale. Real ESD does not develop well in the context of small scale service delivery by municipalities. Then, building on this

R. Traunmüller and K. Lenk (Eds.): EGOV 2002, LNCS 2456, pp. 150–156, 2002.

conclusion, the central question in the remainder of the paper is whether the scaling up of ESD will be a viable option. To answer this question we first address the background of small scale service delivery in the Netherlands, the various arguments supporting the current arrangements for service delivery and the actual practices (section 4). Understanding this background provides insight in the conditions under which scaling-up will be desirable and possible, and in the methods that can be applied in specific circumstances (section 5). Section 6 provides some concluding remarks.

2 ESD, Fast and Slow

For some decades now, the Dutch government has had an eye on introducing ICT in public service delivery. Experiments with the use of legal expert systems in service delivery, for instance, date back to the late eighties [1]. The spectacular development of the Internet in recent years has boosted the expectations and prompted for even higher ambitions. Not only would ICT help to make service delivery more efficient, it would also improve service quality. Moreover, by giving the right examples, our government aims to propel Dutch society to the forefront of the information age [2–5].

The policy as outlined by the Dutch Government is aimed at both the quality and the quantity of electronic service delivery. All public service providers are to have a web site that not only provides basic information, but also allows for integrated service delivery (based on life events or demand patterns). Public service providers should also consider implementing pro-active service delivery. In 1998 the quantitative ambitions were expressed in measurable criteria. One of them was that the Dutch government aimed at bringing twenty-five percent of services on line by the year 2002 [3].

What progress have we made so far? Will the targets be reached this year? What can we say about quantity and quality of the services provided?

If we look at ESD in the Netherlands, we may distinguish two very different worlds. On the one hand, several organisations are indeed progressing on the road to electronic service delivery. Import duties in the port of Rotterdam are handled electronically. Most people's car license registrations are renewed by means of automatic bank transfers. The IRS offers its clients a computer programme, free of charge, for filing their tax returns. The data produced by the programme can either be sent to the IRS on-line or by means of a floppy disk. The student bursary system has been highly automated for over a decade and uses modern ICTs to communicate with its educated clientele. It not only offers general information to its clients, but also shows them the data that are stored about them and allows them to change certain data on-line.

Organisations such as these benefit immensely from electronic service delivery. The electronic communication with clients improves the speed and efficiency of communication and lowers the error rate in data entry. Also the clients benefit from ESD. They can interact with these organisations whenever and wherever they want.

Contrasted with the world of large organisations is the world of municipalities where progress in implementing electronic services is much slower.

The municipalities were supposed to be the driving forces in the development of electronic public services [3]. This idea was largely based on the fact that they account for some 70% of public services. Municipalities therefore have most citizen contacts, and they are also the biggest beneficiaries of a successful implementation of ESD. The quality of services may improve and also processes may become more efficient.

The strong focus on municipalities manifested itself in the Public Counter 2000 project (Overheidsloket 2000). Municipalities were encouraged to submit plans for funding local experiments. In 1996 the first phase of the Public Counter 2000 project started with 15 subsidized municipal pilots. Others were encouraged to follow these forerunners.

Some 5 years later we may conclude that municipalities have not come very far. Although many do their best, some still do not even have a website. In fact, in May 2001 only 282 of the 504 municipalities had one and the current aim is to have all 504 on-line in May 2002. If we look at the content of the websites, we may conclude that most of them only provide (sparse) basic information. Very few offer forms that can be downloaded. Only a few municipal websites offer the possibility of on-line transactions.

In sum, in 2002, we can conclude that some organisations indeed succeeded in realising advanced ESD, where at the same time others failed. The question is of course: why?

3 ESD, Large and Small

The fact that some organisations innovate and others do not can of course be related to many aspects [6]. Success in innovation depends, for instance, on the (correct) realisation of an actual necessity, on adequate management and on the organization's (work floor) culture. Indeed, with regard to ESD, all these arguments have been expressed, especially in explaining the lack of development in the municipalities. However, these explanations tend to overlook a simple factor. The organisations that have succeeded differ quite a lot from the organisations that did not.

The ESD champions in the Netherlands share a number of characteristics. First of all, each of these organisations is highly centralised, in the sense that the services are coordinated by a central administrative body. Second, each of them is only responsible for a limited number of related services: import duties, licences, taxes or bursaries. Third, the services are typically *high volume* services; they are offered to a larger audience (IRS, car-licences) and/or with a high frequency (import duties, bursaries). Finally, the organisations in question all have vast resources to develop electronic services.

The Dutch IRS (Belastingdienst) can serve as an example. It only administers the national tax programme. Of this programme, the income tax applies to six million citizens, who have to file their tax returns on a yearly basis. The tax office has a yearly ICT budget of roughly 300 million and has an *IT* staff of over 2000 people, for a large part working in the special tax automation centre [7].

The municipalities contrast sharply with these ESD champions. Amsterdam, the largest city in the Netherlands, has some 750.000 inhabitants, followed in size by Rotterdam, The Hague and Utrecht. These large cities house 13% of the Dutch population. The rest of the population lives in the other 500 municipalities, resulting in an average number of inhabitants per city of about 30.000. Many towns of course, have even fewer inhabitants.

The size of the administration in municipalities depends on the size of the population. The same goes for the resources spent on ICT and ESD development. A town, such as Woudrichem (14.000 inhabitants) has an IT staff of 1.4 person and an annual IT budget of 160.000 euro.

In the meantime, every municipality, Woudrichem included, is expected to offer a very large number of services (300 to 400), ranging from garbage collection to education and from building permits to social assistance.

Can we expect these municipalities to develop the same advanced types of electronic service delivery as the Dutch IRS? Of course not!

Most municipalities, due to their size, lack the necessary resources to develop ESD for all of their services. But it is also questionable whether it makes sense to develop ESD locally for services that only target a limited population and generally have a low frequency [8].

The problem of developing ESD is clearly related to scale. For most service delivery the current development scale is that of the individual municipalities. This is not only inefficient, in many cases the scarcity of resources at the local level actually prohibits the development of ESD.

We may turn this argument around. If we really want to get ESD of the ground we have to develop electronic services on a larger scale (for groups of municipalities) thereby increasing the number of 'clients' and bundling the available resources.

Although this makes sense from an economic perspective, it also raises important questions from other perspectives. Economy of scale has always pleaded against local service delivery, so what is the rationale of providing services locally? Is the birth of an electronic government reason to change the existing arrangements?

4 The Rationale of Local Service Delivery and Local ESD Development

Public service delivery in the Netherlands is rooted in historical and legal grounds. Service delivery in most cases relates to a decision taken by some public body pertaining a right or an obligation of a citizen or an enterprise. Government is ultimately bound by law. The constitution and the laws based on the constitution determine the powers of the various government bodies. In the Netherlands most powers are distributed to the local level (art. 124 Dutch Constitution). The municipalities are therefore at the core of the public sector; the Netherlands are a decentralised unity state. In this the Netherlands differs from a country such as France, which is far more centralised.

There are various reasons for this distribution of powers. Among them are simpler democratic control, adaptability to local circumstances, better means for people to have their say and better integration of policy [9].

Besides the necessary powers to govern, municipalities also have the powers to provide public services. This of course makes sense, especially in the pre-internet era. Delivering services on a local level is practical (efficient) from the citizen's perspective. Having a service provider nearby saves time. Although municipalities formally are at the core of the public sector in the Netherlands, the system has become more complicated and obscured over the years. With the coming of the welfare state, central government became a more active player, intervening in the autonomous municipalities. Policy no longer is developed and executed primarily on the local level. The higher levels of government (provincial and state level) nowadays have a much stronger role in policy formation and even in its execution.

In many cases policy is developed at the national level with municipalities only administering the national policy. This form of joint governance has serious consequences for the shaping of the public landscape and for public service delivery. In the current practice the powers and responsibilities are distributed over the various levels of government, sometimes making it unclear which level is responsible for a particular task.

In this blurring of powers and responsibilities, municipalities are observed to provide essentially three very different kinds of services:

- Truly local services: i.e. services which are provided based on local policy and local autonomy, concerning the management of the municipalities' own affairs free from interference by the State. Examples of such services are: street and community care and safety, local taxes, sports, recreation and culture.
- Joint governance services: i.e. services which are rooted in national legislation, but which are administered by the municipalities, with the municipalities having their own (additional) policy responsibilities and discretionary powers. An example in the Netherlands is the municipal social assistance, based on the General Assistance Act.
- Municipal delivery of national services: i.e. the administration of national policy by the municipalities, where the policy is completely defined at the national level and discretion is limited and the administration by the municipalities is simply a convenient means of bringing the service to the citizens. Clear examples of such services are the issuing of driver's licenses and passports.

In sum, municipalities deliver very different services, and there are clearly very different reasons for service delivery on the local level. However, when we consider the strategy in developing ESD in the Netherlands, we see that these differences are not being considered. Initiatives like Public Service 2000 simply place municipalities at the centre of improving public service delivery, regardless of the type of services.
A rethinking of which services really have to be delivered by the individual municipalities and which services might benefit form co-operation has not taken place.

The adage seems to be: services that are currently delivered at the municipal level should be informatized at the municipal level. But is this really the case? As we argue, the differentiation between truly local, joint and national services provides a basis to differentiate in the way ESD for the various services may be developed. While it underlines that some of the truly local services really require local ESD development, it also shows that some electronic services may be developed in co-operation and some may even be taken up by more central, national organizations.

5 Possibilities for Co-operation in, and Centralisation of ESD

Economies of scale provide the key to improve the speed at which ESD can be developed, and as we argue, the type of service determines the possibilities for increasing this scale.

Truly local services, addressing local problems and based on local policies are indeed best dealt with at the level of the municipalities. The municipalities determine the content of these services and therefore should also determine and organize service delivery. When a service is really typical for the municipality in question, there seems to be little choice regarding ESD development: it will require the special development of this ESD for this municipality (either by the municipalities staff or by a commercial organization hired for this task). The possibilities of gaining economies of scale are very limited in this case, which implies that it may be wise not to develop ESD at all. Perhaps, however, solutions may be found in using general tools for developing simple service modules (JAVA applets, ASP code), or by co-operating with other municipalities with similar local services. However, this number of very specific local services is generally limited, and even differences in 'truly local' policies are not always as big as claimed. Many local bylaws are based on standard bylaws as produced by for instance the Association of Dutch Local Governments (logging permit, fire and safety measures), which makes co-operation between similar municipalities a viable option. In some circumstances it may be efficient to look for other, perhaps non-governmental, organizations to actually deliver these truly municipal services, based on specifications from, and tailored to, the needs of the various municipalities [10].

For joint-governance services the core of the service delivery, the possibilities of co-operation and centralization of ESD development are far greater. Joint governance services, such as General Assistance, typically are based on a national core of regulation, which applies to all municipalities. This means that, in some cases, it may also be possible to partly centralize ESD development. Where the practice of municipal service delivery consists of combining pieces of national and local regulation, ESD development may be approached as a question of integrating local and national ESD-modules, thereby limiting the effort needed by each municipality. An important question concerning the feasibility of this approach is the amount of variation in local policy. For general assistance in the Netherlands, the viability of this approach has already been shown in the development of the MR-Expert systems, which contain the national legislation straight out of the box and to be supplemented with local rules [11].

For service based on national policy, the development of service modules on a national scale is an obvious choice. The responsible ministry could develop service modules, such as intelligent forms or expert system modules, and provide them to the municipalities to incorporate them in their websites. But, in this case also another obvious step can be taken: concentration (or centralization) of service delivery. An example where this already is possible, is the housing benefit. People fill in forms provided by the Ministry of Housing, Spatial Planning and the Environment (VROM), and send them to the Ministry, which takes care of the administrative process. The step to electronic service delivery by the ministry in this case is relatively small.

6 Conclusion

E-government dawns slowly in the Netherlands. One of the causes of the slow progress towards realizing true electronic service delivery is the choice for local ESD development. In the traditional service delivery model the central role for local government made sense. In the internet era the same choice hampers progress because it is inefficient to have all towns and cities develop ESD modules on their own with limited resources. An alternative is to move development to a different scale. This is possible for services that are the same for every town and city.

If we take e-government serious, we have to rethink the development process. We also have to rethink the way services are to be provided by the various levels of government. Which services should be provided by local councils, which may be provided by independent service providers (either public or private) and which services may be provided by the national government? ICT offers different means to establish better services: in the virtual world time and space lose importance. This opens the road to economies of scale by offering services on a supra municipal level. However, it is important that these advantages are weighed against other goals and functions of the public sector, such as universal service principles, and democratic control [12, 13]. The typology in types of services we have provided offers a starting point in this discussion.

References

1. Nieuwenhuis, M. A.: Tessec: Een expertsysteem voor de Algemene Bijstandswet. Kluwer, Deventer (1989)
2. Ministry of the Interior and Kingdom Relations (BZK): Terug naar de Toekomst: Over het gebruik van Informatie- en communicatietechnologie in de Openbare Sector, Ministerie van Binnenlandse Zaken en Koninkrijksrelaties. Den Haag (1995)
3. BZK: Actieprogramma Elektronische Overheid. Den Haag (1998)
4. BZK. Voorbij het Loket: Over de mogelijkheden en onmogelijkheden van pro-actieve dienstverlening voor de Nederlandse Overheidsorganisaties. Den Haag (1999)
5. BZK. Contract with the future: A vision on the electronic relationship between government and citizen. Den Haag (2000)
6. Levine, Arthur: Why Innovation Fails. State University of New York Press, Albany (1980)
7. Belastingdienst: Jaarverslag Belastingdienst 1999, (IRS, annual report) (1999)
8. Hoogwout, Marcel: Leuker kunnen we het niet maken, maar willen we het wel makkelijker? Waarom overheden geen haast hebben met het verbeteren van de dienstverlening. In H.P.M. van Duivenboden and M. Lips (eds.): Klantgericht werken in de publieke sector: inrichting van de elektronische overheid, Lemma, Utrecht (2001) 149-66
9. van Wijk, H.D, Konijnenbelt, W, van Male, R.M. et al.: Hoofdstukken van administratief recht. VUGA, 's-Gravenhage (1997)
10. Johnson, Peter: Knowledge management, knowledge based systems and the transformation of government. In: Leading People into 2000. Australian Human Resources Institute and the Public Service & Merit Protection Commission, Australia (1999)
11. Groothuis, M.M., Svensson, J.S.: Expert System Support and Juridical Quality. In: Breuker, Joost, Leenes, Ronald and Winkels, Radboud (eds.): Legal Knowledge and Information Systems (Jurix 2000) IOS-Press, Amsterdam (2000) 1-10
12. Bellamy, Christine, Taylor, John A.: Governing in the Information Age, Public Policy and Management. Open University Press, Buckingham; Bristol, PA (1998)
13. Stedman Jones, Daniel, Crowe, Ben: Transformation Not Automation the E-Government Challenge. Demos, London (2001)

Administration 2000 – Networking Municipal Front and Back Offices for One-Stop Government

Volker Jacumeit

Siemens Business Services GmbH & Co OHG,
Rohrdamm 85, 13629 Berlin, Germany
volker.jacumeit@siemens.com

Abstract. Administration 2000 is the first solution that combines legacy applications of different local administrations in order to offer a One-Stop Shop solution to citizens and private enterprises. The solution uses the Internet technology for a high secure and modern Intranet application and will be extend to a real Internet application later. The solution follows consequently the so called life event approach and also integrates services from private companies that belong to the specific life events. Besides the implementation of new technology the solution requires a paradigm within the public administrations and its business processes. In most cases the solution requires new contracts among the participating administrations and can lead to a change of the respective laws that define responsibility and process of the services.

1 One-Stop Government – New Services for Entrepreneurs, Public Administration and Citizens

Due to the growing competition and limited financial budgets Government, politics and public administration are forced to reform their Business. A new comprehension on public services hence changed and new tasks demand reengineering of business process that are consequently oriented by the needs of citizens, enterprises and other users of the public services.

Administration 2000 offers a complete access to services processed of public administration via the so called Citizen's Office (One-Stop-Government). Offerings for all situations of life in the field of e-Government fosters the connection with citizen, administration and private businesses.

A consortium of three counties and over 5 towns in German's state Schleswig-Holstein has run the project together with Siemens Business Services. To introduce a seamless One-Stop-Government for the citizens with the integration of local businesses and independent from the need of citizens to invest in equipment and technology, was the goal of the project. The services were on focus all the time and technology only used as a tool. Hence the reengineering of the business processes of the public authorities has consumed a big deal of time and money of the project. New ways of cooperation of the different departments as well as cooperation between different local administrations (municipalities and county council) have been identified and lead to organizational changes and proposed legal provisions.

Currently the new services are offered via Citizen's Offices or Walk-In Centers as these kind of public offices and administrations are called. In these offices civil ser-

R. Traunmüller and K. Lenk (Eds.): EGOV 2002, LNCS 2456, pp. 157–162, 2002.

vants can offer the entire bunch of public services related to certain life events. The citizens do not have to walk from office to office in order to have their requests processed. "Instead of people we let move the data" is the concept of Germans Minister of the Interior and with the solution Administration 2000 we let this happen. Civil servants can now change data in databases of other municipalities and even of the county council, if this is necessary for the respective request. Also the very close cooperation with local businesses allow to extend the solution to real life event management.

1.1 Example of the Life Event Move House

People who move house today have to visit different offices of public administrations which are located often in different towns or districts of big cities. So they have to spent plenty of time standing in a queue in order to get to the right civil servant. They have to give redundant data several times and fill in many forms. In Germany the event Move House means that people have to deregister at municipality they live in before and then go to the municipality they have moved into for the entry in the new citizen register (update the ID-card and passport). A third administration must be visit for the change of the car-plate and the according car register. Additionally people have to inform service providers such as the gas- and power supplier, the waste collection company and more. In order to inform all those providers people fill in more forms (sometime via the internet) or send letters by traditional mail. But each time redundant date must be typed. Surely these boring tasks are more or less the same within all countries.

Now with the solution of Administration 2000 people can go to any municipality. The civil servant is the human life event manager for the citizen. The civil servant will first deregister the citizen at the former municipality, register the citizen at the municipality he has moved into, update ID-Card, passport and print necessary documents that certify the new address. Also the civil servant can process the data for the county in terms of changing the car-register and car plate. If the citizen wishes, the civil servant can even inform other service providers such as gas- and power supplier.

Technically spoken we have integrated all existing legacy applications of the public authorities and involved private companies in a new and unique user interface. This user interface is oriented to the respective services and works independent from the legacy applications with its proprietary interfaces.

With the growing number of households connected to the Internet, these kind of services will be offered also via the Web. But at this time we (Siemens Business Services as a solution provider and the public authorities and private businesses as the service providers) have gathered plenty of experiences and enhanced the solution to a mature version which can be used by everybody and that is comprehensive. An other advantage of this approach is, that due to the fact that the civil services are using the same interface as the people via the Internet, each civil servant can assist a citizen via telephone if any problem occurs with the application.

2 Demands of Administrations

Administrational decision maker intend to drive the modernization of the administration by using e-Government solutions.

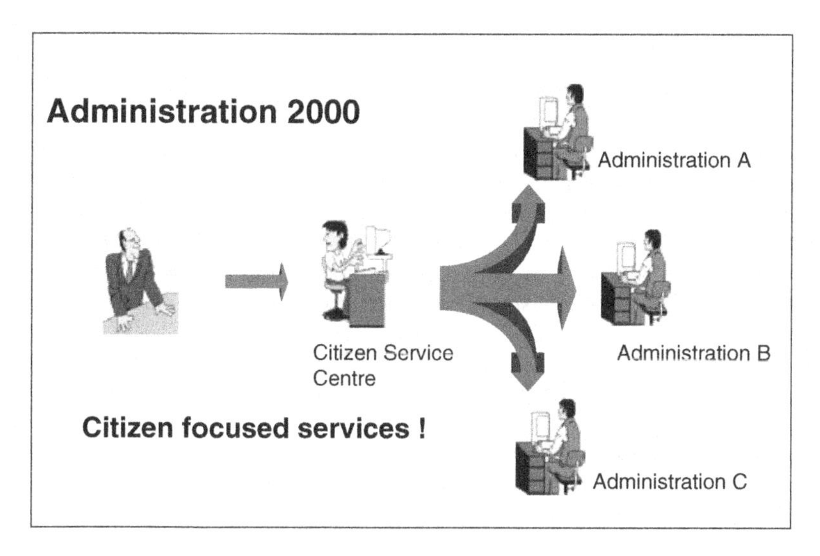

Fig. 1. Administration 2000

Table 1. Access Points for Processing Life Events

Portal	Citizen Office	Life Event Manager
• Information public and private sector • Search until you find the proper information and form if available • Fill in the form • Submit the form • No combination of data to other forms • No direct processing of data	• Any Citizen Office of the region / county • Civil servant collects data and processes these data online to all respective governmental processes and other entities • On-Stop Shop solution with human interface	• Integration of all governmental processes and of the private enterprises concerned by the life event • Intelligent and comprehensive dialog allows as much integration as the user wants • One-Stop Shop via Internet
All users via PC, Kiosk	Application for Citizen Offices and Call Center operated by civil servants	Citizens via PC, Kiosk Citizen Offices Call Center

Administrations like to offer integrated services to citizens, local enterprises and other administrations. These services must be secure, legally binding and optimized.

Using new Online-Services administrations like to improve the contact to its citizens. Services for local enterprises strengthen the competition of the region.

Administrations like to make use of experiences and the solution architecture of best practices.

Administrations like to built the new Online-Services based on standardized data and interfaces and keep already existing interfaces to legacy applications.

3 Advantages

3.1 Advantages for Public Administrations

- Improvement of quality of services for citizens by local and customer oriented solution offerings, independent from the structure of the administration.
- Realization of the political goals on e-Government and modernization of the administration.
- Improvement of transparency of the tasks of the administration
- Shortcut and simplify the business processes, communication and work time.

3.2 Advantages for Citizens

- Flexibility in terms of time and location.
- One access point for all services related to a life event.
- Avoid typing of redundant data.

3.3 Advantages for Enterprises

- Seamless interface to public administrations.
- Exchange of documents via e-mail.
- Online access to important data for planning.
- Online access to up-to-date laws and directions.
- Online Access to information on bid and tender process and monitoring data.
- Online procurement.

4 The Solution Components

Administration 2000 is a standardized and modular solution. Already existing best practice modules can be integrated on demand. Solutions for new business processes can easily be developed due to a tested an proven methodology approach and data descriptions of the new business processes.

4.1 Technical Overview

The technical concept is based on the usability of an internal networks of the administration and the legacy applications of the involved administrations. The local administrations and the county council are connected via a secure County-Backbone-Network. In a demilitarized zone an information server, application server and a communication server are operated. On the application server runs a web based application which is responsible for the processing of the integrated work flow of the administration. Using an unique user interface the different legacy applications of the participating administrations are operated. The communication server connects the user to the respective legacy applications because most of the legacy applications are operated at the local administrations. For this exchange of sensible and personalized data we are using Biz-Talk server from Microsoft. As the communication server. The communication server also takes care for the transfer of data, monitoring and reporting of the transactions via the network. The processing of data is done at the legacy applications on the local administrations. Security and integrity of data is guaranteed by the overall concept of the solution.

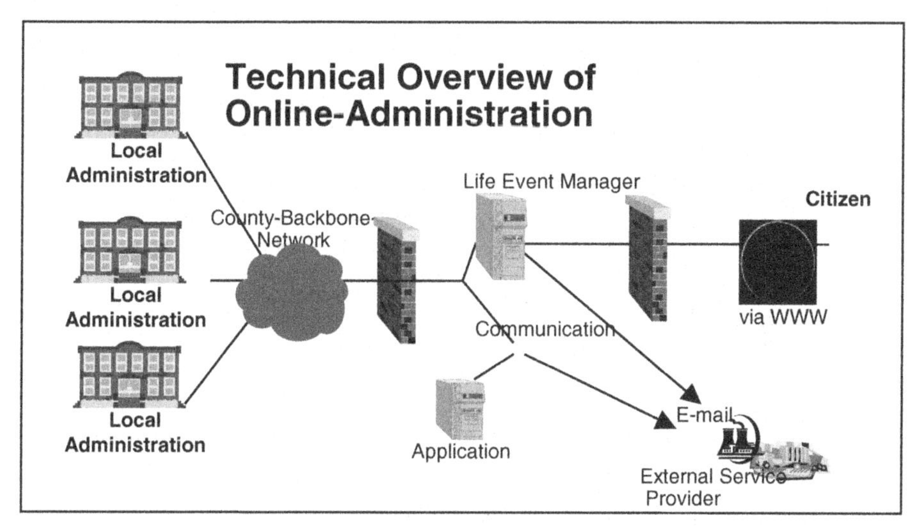

Fig. 2. Technical overview of Online Administration

Citizens will receive information about the services that the local administrations offer via the citizen offices. Hence the citizens can prepare themselves with all relevant documents and information necessary for the applications. They can arrange dates for visits of the citizen offices and thus avoid queuing.

4.2 Best Practice Modules

For the German market we have prepared business process descriptions and according XML-Schema. We have developed interfaces to main applications and a complete solution architecture available for the following processes:

- Solution package Citizen Register
- Car Register
- Construction Permission

Even if the solution of Germany can not be transferred without any changes to other countries, we can adopt these best practice modules.

4.3 The Methodology Approach

A tested methodology approach simplifies and accelerates the modeling of new business processes as well as the development of a solution architecture of new processes.

Existing definitions of interfaces, XML-Schema and a toolbox to customize the user interface are fostering the development and integration of new business processes.

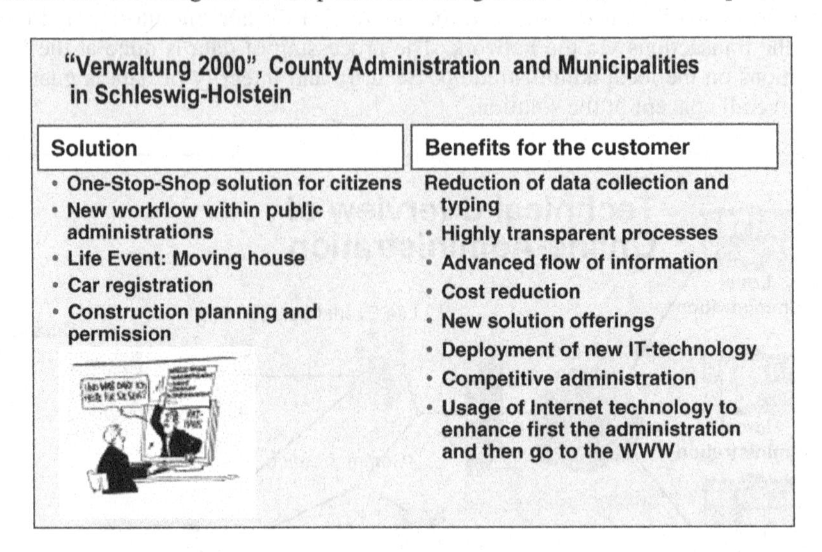

"Verwaltung 2000", County Administration and Municipalities in Schleswig-Holstein
Solution
• One-Stop-Shop solution for citizens
• New workflow within public administrations
• Life Event: Moving house
• Car registration
• Construction planning and permission
Benefits for the customer
Reduction of data collection and typing
• Highly transparent processes
• Advanced flow of information
• Cost reduction
• New solution offerings
• Deployment of new IT-technology
• Competitive administration
• Usage of Internet technology to enhance first the administration and then go to the WWW

Fig. 3. "Verwaltung 2000", Country Administration and Municipalities in Schleswig-Holstein

4.4 Operation Concept and Security Concept

The operation concept defines the organizational tasks and the organizational environment as well as service and operation tasks for the operation of the solution Administration 2000. The security concept describes the threat to the system and possible defense activities. From this basis a step-by-step plan for the implementation of the new demands on security will be derived.

4.5 SBS Services

Siemens Business services fosters administration by the development of en e-Government strategy, Concept and Architecture of new Online-Services, Implementation and if demanded service providing, application support and operational services.

We develop for administration the appropriate e-Government solution and support administration in all phases of the development. Beginning with the vision and strategy up to the operation of the services.

4.6 Our Solution Partners

- The realization of the web application has been done by HEC GmbH, Bremen
- Standardized software products have been provided by Microsoft
- The County-Backbone-Network has been provided and is operate by Datenzenztale Schleswig-Holstein

The Experience of German Local Communities with e-Government – Results of the *MEDIA*@Komm Project

Tina Siegfried

MEDIA@Komm accompanying research , German Institute of Urban Affairs,
Strasse des 17. Juni 110, 10623 Berlin
siegfried@difu.de

Abstract. The largest multimedia project of the national government – *MEDIA*@Komm – has now run for almost two years. Over the course of time, the model municipalities and other German local communities have gained experience with the design of legally binding on-line transactions and the use of electronic signatures, and they have made progress with virtual town halls and the e-government project. The following article shows the results achieved in the *MEDIA*@Komm municipalities and considers the experience, obstacles and unsolved problems which have arisen in designing business processes without discontinuity of media in many German towns and cities. Also, as an important result of the accompanying research on *MEDIA*@Komm, it describes the central factors for success in the creation of virtual town halls.

1 The Multimedia Initiative of the National Government *MEDIA*@Komm

The *MEDIA*@Komm project is the largest multimedia-project of the national government to test legally secure electronic transactions between the administration, citizens and business companies. The aim is to design on-line transactions without discontinuity of media and to test the use of electronic signatures. A municipal competition was announced in 1998, and 3 municipalities were then selected to implement their concepts with the support of the national government. The winning projects were presented by the city of Bremen, the municipal association of Nürnberg, Erlangen, Fürth, Bayreuth and Schwabach and the town of Esslingen in partnership with Ostfildern. These projects have been implemented since 2000 with federal subsidies of 25 million euros. In addition to these funds the municipalities themselves, together with private partners, are raising about another 35 million euros for the implementation of the projects. This was originally scheduled to be completed by the end of 2002, but due to unforeseeable delays (cf. section 3) this is likely to be extended to about the middle of 2003.

The German Institute of Urban Affairs (Difu) was commissioned by the Federal Ministry of the Economy to act as the consortium leader and provide academic support and guidance to the federal government's largest multimedia initiative. Together wirth three other academic insitutions, each of which deals with specific aspects of the subject, Difu deals with economic issues and questions of adminstrative science. In addition, specialist events are held and a coopreation and communication network is being established which also includes the Internet presentation on *MEDIA*@Komm (www.mediakomm.net).

R. Traunmüller and K. Lenk (Eds.): EGOV 2002, LNCS 2456, pp. 163–168, 2002.

2 Present State of Implementation in the Model Municipalities

2.1 Free Hanseatic City of Bremen

The Bremen *MEDIA*@Komm project "Legally binding multimedia services with digital signatures in the Free and Hanseatic City of Bremen" is being implemented by bremen online service GmbH & Co. KG and comprises three large core areas with further sub-projects: Access to secure and legally binding on-line services, platform and designing a standard for municipal online transactions and applications/life situations.

2.1.1 Successfully Implemented Local Community Applications

The first "visit" to a public authority carried out completely on-line, in which a marriage certificate was ordered and the fee was paid directly with a purse card via a secure Internet link, was presented in September 2000. Since then, the citizens of Bremen have been able to call on various on-line services of the administration and private service providers. In the life situation residence and change of address, for example, they include registration and cancellation of gas, electricity and water with the municipal utility companies, changing bank accounts at the Sparkasse and mail forwarding arrangements with the post office. Birth, marriage and death certificates can also be ordered on-line from the registry office in Bremen – and paid for electronically. Since June 2001 the residential registration office of the city Bremen has also offered on-line services. There, legally binding entries and deletions can be made in the residential register via the Internet by using the electronic signature. Since May 2001, applications for students have also been in place. A service for solicitors and business companies is currently in the test phase. They can call up information in electronic form free of charge from the register of companies database at the Local Court (Amtsgericht). A further process, which is also likely to interest small and medium-sized companies, is the on-line debt collection application, which has been developed for the Free Hanseatic City of Bremen. With this software, applicants can file debt collection applications electronically via the Internet, thus saving time, effort and costs. And finally, the prototype of a digital tender platform for the award of public orders in the construction industry is currently being developed in Bremen. The process is being developed in cooperation with Administration Intelligence (AI) and will make it possible in future to read announcements and download tender documents over the Internet. Since December 2001 citizens have been able to issue a direct debit authorisation with an electronic signature. This new possibility to pay any relevant charges can now be used to order a marriage certificate. Previously, marriage certificates ordered on-line were only sent out with an invoice by the registry office. Now the citizens no longer need to fill in the bank transfer form; instead, they simply issue the public authority with an electronically signed direct debit form for the respective transaction directly in the on-line application.

2.2 Nürnberg Municipal Association

The goal of the project "*MEDIA*@Komm in the Nürnberg region", implemented by the project sponsor, Curiavant Internet GmbH, is to offer legally binding multimedia services with digital signatures in the municipal association. A regional communication platform is therefore being created which will support secure communications and offer the citizens various communal and private services. The Nürnberg municipal association consists of five municipalities of different sizes in the region. The special challenge here is to develop on-line services and products which are equally "fitting" for all municipalities.

2.2.1 Successfully Implemented Local Community Applications

The citizens of Nürnberg have been able to use out municipal applications on-line since October 2000. With the multi-functional chip card from Curiavant Internet GmbH and a class 3 card scanner, they can apply and pay for their residents' parking permit from their PC. The provision of information from the register of residents is fully implemented, but it cannot yet be placed on-line due to legal regulations. Moreover, a municipal council information system has been developed which has been in use since April 2002. Another local community sub-project, the building site information system, has been in place since October 2001. The integration of geographical information systems and town maps in the Internet presence of Nürnberg enables citizens to call up information on traffic obstructions in the urban area via the Internet. Similarly, data and maps can be ordered on-line and paid for by invoice. Since the end of 2001, citizens in the Nürnberg urban region have also had access to a comprehensive information and booking system for regional cultural and educational facilities. Adult education centres, the municipal theatre and the municipal libraries already present their services on the Internet. The citizens can even enrol and receive a confirmation with their digital signature. The chip card also makes convenient payment possible – either on-line or by direct debit. On-line enrolment for an adult education course is possible; the payment (for now) is made by direct debit. In the next stage of development, this service will be extended to include access to data, e.g. lists of participants for the teachers. The municipal theatre not only presents its programme of performances, it also enables visitors to select, book and pay for their seat on-line. Tickets are then sent by post or deposited for collection at the evening ticket office. The libraries project, which offers citizens a number of library functions via the Internet, e.g. borrowing, reservations and research, has been in pilot operation in Erlangen since November 2001. Since December, the draft of the zoning plan for the city of Nürnberg has been available on-line in the framework of citizen participation.

2.3 Esslingen/Ostfildern

The central focus of the project in the towns of Esslingen and Ostfildern is on communication between citizens and the administration. With their community-based approach, the two towns aim to enable their citizens to participate actively in the life of the local community and to cooperate in municipal activities. The aim of the Esslingen approach is to create the necessary acceptance in the population for the use of signature cards to implement legally binding transactions and to make on-line "visits" to the municipal administration via the Internet. Six sub-projects make up Mediakomm Esslingen: Communal services, Education, E-commerce and e-business, Culture, Social affairs and Cross section.

The responsibility for each sub-project lies with a different institution or company; the sponsoring body Mediakomm e.V. in Esslingen is responsible for coordination.

2.3.1 Successfully Implemented Local Community Applications

In accordance with the principle of the local community of citizens, projects have now been implemented which provide information for the citizens and involve them in the process of discussion on developments in the town. Thus, citizen forums have been set up on the Internet pages which enable interested persons to discuss topics relevant to Esslingen. The zoning plan provides an even more practical example. The formal participation of the citizens in the development of a new construction area in Esslingen was also implemented via the Internet. To achieve goals such as greater transparency of the administration, customer orientation, better accessibility for the citizen and faster handling of administrative procedures, the Esslingen citizen information service ESSOS ("Esslingen On-line Service") was founded. It offers, or will soon offer, information and on-line services for various life situations. And to counteract the "digital divide", projects such as the supervised citizen PC (Bürger-PC™) have been implemented, which particularly aims to help population groups in Esslingen with little experience of new media to make use of the Internet and digital signatures. At central points in the town (e.g. in schools), PCs are available for the use of the public. Great attention was attracted by the world's first legally binding on-line election of a public body – the Municipal Youth Council in Esslingen – which was (partly) carried out via the Internet with the aid of digital signatures. The on-line election fulfills all legal requirements; the municipal regulations had to be adapted for this purpose. For construction projects, a prototype of the new service for on-line building permit applications is offered from the first quarter of 2002. In this service, all information (plans, drawings, correspondence) is available in digital form on an Internet platform after the user has identified himself with his digital signature. This can be used on-line by the administration, the citizens, contractors and architects. The aim is to make the construction process considerably more efficient because the approval of the digital original is also issued by an electronic signature.

3 Experience with the Use of Digital Signatures

The dependence of this development on the general framework is still underestimated. A number of factors can be identified which are major obstacles to the widespread dissemination and use of electronic signatures.

3.1 Development of the Legal Framework

The development and adjustment of legal requirements has taken a great deal of time, and for a while it caused great uncertainty among the users. The Digital Signature Act and the Digital Signature Ordinance were revised in 2001. And the adjustment of private law to enable the electronic signature to have equality of status with handwritten signatures also did not take place until 2001. The revision of administ-rative procedure law in the spring of 2002 is still to come, although the direction is now apparent and it is anticipated that the Administrative Procedure Act will prescribe the use of qualified signatures, with accredited signatures only required in exceptional cases.

3.2 Technology and Security

Users reported that the software which had to be installed for the use of signatures was often faulty or completely useless, and lay persons could only install it with great difficulty. Reports of complete and irrevocable system crashes during attempts to install it were no exception. These "teething troubles", which are so annoying for the user, are immense obstacles to widespread use. At least the problem of the lack of interoperability between digital signatures from different suppliers has been tackled since October 2001. TeleTrusT e.V. and the trust centre association T7 are working on behalf of the Federal Ministry of the Economy and Technology (BMWi) to develop a standard for electronic signatures which will draw together the present individual solutions.

The technical security requirements for the integration of signatures into the on-line processes between the administration and its clients and the requirements of the Data Protection Act were other stumbling blocks that needed to be overcome. For example, it is desirable that the client only enters his data into an on-line form once and can then assume that the administration knows his data, but the use of such existing data often collides with the provisions of the Data Protection Act.

3.3 Benefits and Acceptance

The high price of the necessary equipment is still an obstacle to the dissemination and use of digital signatures. The chip card and the associated software currently cost about 25 euros from the market suppliers, and there is also an annual fee of 25 euros for the administration of the certificates. The prices for the scanner, which is also needed, fluctuate depending on the level of security. And there is no specific application which leads citizens to feel that they absolutely need a signature card. The hopes of the *MEDIA*@Komm municipalities that the banks would include digital signatures as a standard feature on EC cards, and that a large proportion of the population would then automatically have a digital signature, have unfortunately not yet been fulfilled.

4 Practical Problems of Use in the Local Community Administration

The design of on-line transactions should involve more than simply reflecting the existing processes and "converting" them into electronic form. Instead, consideration should be given to the question of whether processes can be redesigned in the interest of greater efficiency. To date, this point has been neglected because of the difficulties of integrating digital signatures into existing departmental procedures. The creation of public key infrastructures and concepts for key management within the local community are still central issues which remain to be solved. It must be considered whether every member of the administration should have a signature card, or whether there should be a central office to administer and process signatures. The present trend in the discussion among specialists seems to be towards a virtual postal centre which separates the incoming signatures and then passes the transaction itself on to the responsible person. But at the time of writing (spring of 2002) there is still no product which could take over these tasks. Another significant point in this connection is the training and integration of the staff; such projects cannot possibly succeed if the staff are not willing to cooperate.

5 Factors for the Success of e-Government

The *MEDIA*@Komm accompanying research has the task of advising and assisting the projects, but at the same time it must also evaluate them systematically. To ensure the comparability of the various approaches and enable the projects to be evaluated, a total of 9 dimensions were identified which play a central role in the implementation of e-government and virtual town halls. These dimensions in turn can be sub-divided into individual factors; the importance of each individual factor will be explained in greater detail. Criteria to assess the "maturity" of e-government and virtual town halls round off this measuring instrument for the evaluation of the *MEDIA*@Komm projects. In addition to evaluating the *MEDIA*@Komm municipalities, the factors for success can also be used as an instrument to measure the success of other local communities. The critical factors for success in the designing of virtual townhalls in Germany are: Vision and strategy, Organisation, Applications, Benefits, Use of Technology, Qualifications, Marketing, Cooperation and Resources. A more detailed description of the dimensions and sub-factors is expected to be published by the accompanying research in mid of 2002. For more information see www.mediakomm.net.

Electronic Public Service Delivery through Online Kiosks: The User's Perspective

Ruth Ashford[1], Jennifer Rowley[2], and Frances Slack[3]

[1] Department of Retail Marketing, Manchester Metropolitan University, Manchester M15 6BH, UK,
R.Ashford@mmu.ac.uk

[2] School of Management and Social Sciences, Edge Hill University College, Ormskirk, L39 4QP, UK,
rowleyj@edgehill.ac.uk

[3] School of Computing and Management Sciences, Sheffield Hallam University, Sheffield S1 1WB, UK,
F.Slack@shu.ac.uk

Abstract. This paper reports a case study of Knowsley Metropolitan Borough's response to the UK Government's White Paper 'Modernising government' [1]. It provides unique data on user behaviour in relation to electronic public service delivery through public access kiosks and highlights some of the issues relating to the 'digital divide', the reduction of social exclusion. It offers a perspective on the uses for which customers perceive public access kiosks to be valuable and indicates barriers to kiosk use for other functions. Some of the messages reflect issues that have been debated in consumer responses to e-commerce and communication over the Internet. This is important because it suggests some consistency in the public reaction to IT-based service delivery, irrespective of the platform.

1 Introduction

In common with governments across the world, the British government has pledged to use information and communication technologies to increase participation in a greater part of local governance by 2005.

1.1 UK Government IT Initiatives

Both central and local government have been encouraged, through access to funding, to develop a co-ordinated approach to information technology procurement, working in partnership with the private sector. The main aim of this policy is to ensure that public services could be delivered 24 hours a day, seven days a week. This includes the use of interactive kiosks and call centres to enable the local community to access more convenient services. A recent survey of electronic service delivery by English District Councils (EDCs) found that 21% of councils surveyed currently use public access kiosks and 25% plan to use them in the future [2]. However, 54% of EDCs do not use them now and do not plan to use them in the future.

R. Traunmüller and K. Lenk (Eds.): EGOV 2002, LNCS 2456, pp. 169–172, 2002.

1.2 The IT Strategy at Regional Level

The roles of English local government are defined [3] as providing local democracy, local public policy making and local services to the community. The Government's White Paper [4] has had a dynamic effect on the way that UK local government now operates. At a regional level, the Mersey Wide Web (MWW) is a grouping of the public, private and voluntary sectors that aim to establish Merseyside, in the north-west of England, as a global player in the information society. MWW, established in 1997, aims to help the economic regeneration of Merseyside by ensuring that its workforce is geared to the future and raising the ICT presence in the region.

Knowsley Metropolitan Borough Council, an area of significant social deprivation in Merseyside, established the Knowsley Community Information Programme. The aims of the case study reported here were to investigate the behaviour and attitudes of users and non-users of the online kiosk located in the Knowsley Metropolitan Borough.

2 The Context of Online Kiosks

Online kiosks can be viewed as a medium through which it is possible to train, educate, inform, communicate, persuade, and relate. But, as with other public access systems, it is important that the kiosk is designed to support the task, the user profile and the environment in which the task is to be performed. A useful recent article in the context of public access kiosks is Maguire's [5] review of user interface design guidelines for public information kiosks. Rowley and Slack [6] argue that there are four components of public access systems that need to be considered: user characteristics, environment, task and technology, and they point to the limited attention that has been paid to environment or context.

Online kiosks may have both commercial and community functions. Early applications were designed to provide access to community or government information [7]. Integration of both community and commercial services and information is likely to characterise the kiosks of the future. Previous articles have tended to describe specific kiosks [8] or to develop taxonomies of the different types of kiosks [9]. This article offers unique data on user behaviour and attitudes.

3 Research Methodology

An I-plus information point created by Adshel and media technology company, City Space has been designed to equip the residents of Knowsley with access to information and learning systems and to promote personal and community development. This is a deprived area with an unemployment rate of 13.5% (the national average is 4%) and a high level of crime. The kiosk offered: free e-mail, business finder, child-care finder (in partnership with Mothercare), TV licence payment, access to local council services, transport information and entertainment information.

The multiple methods of data collection used a questionnaire survey with 1068 respondents, two focus groups and two in-depth interviews. The questionnaire comprised a mixture of behavioural, attitudinal and classification questions. In both the pilot study and the main study the questionnaire was administered by students who were trained in the process.

The sample for both the questionnaires and the focus groups were convenience samples; their composition was affected by the willingness of potential respondents to participate in the study. The individual interviews were used to supplement data collected in the focus groups and to develop some themes more fully. Usage statistics were also available to provide comparative statistics on usage levels, length of usage times and frequency of access to the local council web page.

Questionnaire data was entered into SPSS and analysed for descriptive statistics and relationships between variables. Content analysis was performed on the focus group and interview transcripts to identify recurrent themes that influence user behaviour and attitudes.

4 Results

Usage statistics revealed that the average number of users per day was 45. The average time per user was 3.05 minutes and one third of usage was to access the local council web page.

The majority (90%) of people questioned had not used the kiosk. This is a significant finding and suggests that for kiosks to fulfil their potential there is a need to understand their optimum role in communities. The main reason for non-use was a lack of awareness of the kiosk's existence. A significant group was aware that the kiosk existed but 'did not know what it did'.

Of kiosk users (slightly more than 100 respondents) half had only used it once, while 20% had used the kiosk five times or more, suggesting that they had found something of interest and value on the kiosk. It appeared that females were more likely to use the kiosk than males. Over 75% of users were under 35 years old and 45% of users were students. Users were generally quite positive about usefulness and ease of use. A correlation was evident between the perceived usefulness of the public access kiosk and the frequency of use.

Focus groups and interview respondents were asked to suggest solutions to the low level of awareness of the kiosk. An advert in a free local newspaper had generated some interest. It was felt that 'a lot of people would walk straight past it, because they don't want to talk to a robot'; and not be aware that there were 'things for them' on it. Suggestions ranged from creating a signpost, to point people in the right direction, to information leaflets, explaining the uses of the kiosk and a poster campaign in the surrounding shops.

Many kiosks are located under cover and where people are 'milling around' or waiting. This kiosk was located out of doors on a street and this provoked a number of negative comments, thus confirming the importance of location.

Internet access was available to 60% of respondents through work, home PC, or the local public library. The library is the only other free method of access available to the general public. On the other hand, 40% of respondents were disenfranchised. In addition, however, many of the kiosk services were available through other channels, such as the telephone or by making personal contact.

Focus group respondents expressed concern about service delivery through a kiosk. There was evident technophobia and a lack of trust in the technology (kiosks, Internet or telephone help lines) and the service exchange that it supported. Personal contact generates a sense of accountability and reassurance. Concern was also expressed about giving credit card details over the Internet and there was a preference for buying tickets in person.

5 Conclusion

Online kiosks have potential to offer Internet access and other information and services to people who might otherwise be excluded from participation in the 'information society'. This study has emphasised that local authorities and businesses need to treat a public access kiosk as a service to be promoted, particularly in contexts and communities where kiosks and other IT applications are unfamiliar. An enhanced understanding of the tasks that users are prepared to perform on a public access kiosk would also contribute to their success. Local government provision of electronic service delivery must be implemented with the needs and participation of the users in mind, not simply to comply with national Government initiatives. Without this level of engagement, online kiosks will remain a technology looking for a problem, rather than becoming a solution to people's problems.

References

1. 'Modernising government' (1999) HMSO, http://www.cabinet-office.gov.uk/moderngov/whtpaper/index.htm, [accessed 23 January 2002].
2. Phythian, M J and Taylor, W G K (2001) Progress in electronic service delivery by English District Councils. *International journal of public sector management*, 14 (7), 569-584.
3. Pratchett, L (1999) New technologies and the modernization of local government: an analysis of biases and constraints. *Public administration*, 77 (4), 731-750.
4. 'Modernising government' (1999) *loc cit.*
5. Maguire, M. C (1999) A review of user-interface design guidelines for public information kiosk systems. *International Journal of Human Computer Studies*, 50 (3), 253-286.
6. Rowley, J E and Slack, F (1998) *Designing public access systems.* Gower.
7. Ellis, B (1993) Kiosks handle employment queries. *Computerworld*, 27 (46), 88.
8. Miller, C (1996) New services for consumers without home page at home. *Marketing News*, 30 (9), 1-2.
9. Tung, L. L. and Tan, J. H. (1998) A model for the classification of information kiosks in Singapore. International Journal of Information Management, 18 (4), 255-265

FASME – From Smartcards to Holistic IT-Architectures for Interstate e-Government

Reinhard Riedl[1] and Nico Maibaum[2,*]

[1] University of Zurich
riedl@ifi.unizh.ch
[2] University of Rostock
maibaum@informatik.uni-rostock.de

Abstract. In this paper we shall present the results of the change management in the European research project FASME *(Facilitating Administrative Services for Mobile Europeans)*. First we shall compare the original objectives as they had been described in the Technical Annex with the achieved results. Second we shall highlight those issues of change, which were responsible for the success of the project. Only after the prototypical development of a product was abandoned and replaced by basic research on holistic solutions for interstate e-government, the already achieved progress became visible to the participants and a true interdisciplinary co-operation emerged. Concluding from this experience, we shall draw some conclusions for future e-government projects.

1 Introduction

A European citizen moving from city *A* in a European member state *X* to city *B* in a European member state *Y* has to deal with a lot of bureaucracy - such as the registration and deregistration of her living place, the change of her car license and of her insurance contracts, or the enrolment of children in school. This usually requires the following: First, the citizen has to investigate on administrative regulations and culture in *Y* and on the municipal services in *B*. Second, she has to collect personal documents from government agencies in *X*, or from the municipal administration in *A*. And third, she has to deliver these personal documents to agencies in *Y*, or to the municipal administration in *B*. In some cases this may fail because the authorities in her original place of living are unable to provide her with the documents needed; in other cases this may fail because the authorities in her new place of living are unable to interpret her foreign documents. As a consequence, even if the citizen manages to understand the procedures, it may become necessary that a tedious individual exception handling has to take place, which is expensive for both the citizen and the authorities, and whose outcome is unsure.

* Supported by the EU Fifth Framework Project FASME, http://www.fasme.org, and by a grant of the Heinz Nixdorf Foundation

R. Traunmüller and K. Lenk (Eds.): EGOV 2002, LNCS 2456, pp. 173–178, 2002.

FASME *(Facilitating Administrative Services for Mobile Europeans)* was an interdisciplinary R&D-project supported by the 5th Framework of the European Union[1], which intended to develop an intelligent Javacard for the authentication of citizens and the digital transport of personal data. That card was supposed to minimize the high, administrative migration efforts depicted above. During the project, we have found out that such an approach is indeed feasible, but it would reduce administrative migration efforts only marginally, and it would not alter the basic problems for the citizen significantly. In order to make life really easier for both citizens and civil servants, a much broader approach is required. In particular, a secure and trustworthy ad hoc procurement of so far not existing documents from remote authorities is needed, whereby it should be possible to tailor these documents to the needs of remote administrative processes, following different laws and guidelines formulated with the use of different administrative ontologies. This requires the following:

1. A **holistic solution for digital identity**, which flexibly realizes *the in effigy principle*, and which supports individual, group, and role identities, as well as anonymous identities, and the delegation and the revocation of rights.
2. A **decoupling of inter-organizational processes** and **the provision of interoperability**, based on the concept of a loose C/S-coupling of local processes (i.e. no inter-organizational workflows, no global service point architectures), interface definitions for document services and hierarchical directory services.
3. The design of a **vertically and horizontally scaling, virtual information transfer space** for inter-organizational – and in particular international – exchange of personal information and other affiliated information, plus guidelines for the use and for the management of the space (e.g. the locality principle for relevance management).
4. Joint provision of guaranteed and enforced **protection of privacy** and of **monitoring facilities** for both civil servants and citizens – i.e. a fully-fledged transparency management in accordance with European law and user requirements.

We have developed an architectural framework, which meets all these requirements, although with different degrees of detail, as we shall discuss below. Hereby the Javacard plays the role of a digital ID-card. After the card has verified the identity of the citizen with a biometric check, it speaks in effigy of her with local e-government services and remote document services. Thus, the Javacard is the enabling token for digital citizen-to-agency communication, while group and anonymous identities, as well as delegations of rights, are realized with soft identities [1]. The Javacard is used as a cache for digital documents, and as a key to a virtual memory for such documents called Secure Card Extension (SCE) [2], but its main function goes beyond simple document transfers, it is the digital representative for the citizen interacting with e-government services. See also [3] for an illustration of the problem of heterogeneity faced in FASME and [5] and [6] for a presentation of our solution.

We have implemented demonstration versions for the core IT-components in a simplified setting with the municipalities of Cologne (Germany), Grosseto (Italy),

[1] Approximate Figures: Total costs = close to 4 million , duration = 18 months plus extensions adding up to 21 months, head count = close to 60.

and Newcastle-upon-Tyne (England). Both the overall solution and the demos were successfully validated with user groups from six European member states.

In this paper, we shall depict how we have changed the main objectives during the project. Notably, this change was initiated by the work on the IT architecture. ***Only after the prototypical development of a product was abandoned and replaced by basic research on solutions for interstate e-government, a true interdisciplinary co-operation emerged.*** As long as the project was understood as a feasibility study based on a prototype close to a product, there was no common view of what the project was all about and the project partners did not realize its achievements. This changed only when the new focus on holistic solutions became accepted and thus leveraged inter-disciplinary work on the multiple dimensions of the solution. The observation that IT engineering activities contributed to the emerging interdisciplinary perspectives somewhat parallels the observation of the role of IT managers as brokers reported in [5].

Our bold statement above is confirmed by the results of the two project reviews carried out by the European Commission. The first took place, when the project was still run as an applied technology project. It severely criticized the progress of the project. Shortly afterwards we changed our goals and we directed activities towards the development of a holistic IT solution (for the price of providing demos which were less mature than the prototypes originally intended). The final review then was quite a success, as we presented the solution depicted above, which integrates the research results from the various disciplines with IT engineering R&D work to achieve a generic, customizable IT architecture for international e-government services. The reviewers' only point of criticism was the missing business plan for making money out of the project results.

2 Objectives versus Results

In the following, we shall compare the objectives (formulated according to the guidelines of the European Commission) with the achieved results.

Overall Plan: The main benefit of the FASME project will be a Javacard, which facilitates the mobility of citizens as well as the entailed administrative services. Resulting from the new possibilities offered by the Javacard, the subordinate benefits are reflected in the facilitation of procedures, concerning migrating citizens, municipalities and enterprises.	**Overall Achievement:** The main achieved benefits of the FASME project are a holistic understanding how to build scaling international e-government services, the identification of a next step of R&D topics in order to proceed systematically towards real world implementations, and a principal change of the engineering perspective for e-government project.
Overall Planned project goals: • The customizing of existing technology … for novel administration procedures and the exploitation of potential benefits of interdisciplinary engineering processes …	**Overall project goals achieved**: • Technological preferences shifted during the project; a new perspective on convergence management with so-called boundary objects for interdisciplinary R&D emerged.

• Simplification of the co-operation of municipalities, creation of public acceptance of innovative technologies (Javacards) ... • Facilitation of the European citizens' mobility ... • The early exploitation of Javacard technology ... to extend the European lead in Smartcard technology. • The reduction of financial and time efforts incurred by the migration of citizens.	• The way to more effective co-operation has been shown, public acceptance for Javacards is still an issue as its legal admissibility. • This goal was too far reaching for a small project like FASME. • The future success of Javacards is still unclear. • No empirical validation was provided (due to a lack of time) in the project, but the evaluation by civil servants has shown that it is indeed reasonable to believe it.
The **main planned project objective** is the development of a Javacard for the easy, safe, and secure transport of personal data needed for administrative procedures.	The **main result achieved** is the development of a holistic architecture for interstate e-government services.
The six **planned operational project objectives** are: • Development of business process models ... • Development of a social context model including social constraints for the application design process. • Development of a design framework (infrastructure, processes, and application software). • Implementation of a prototype. • Evaluation of the prototype by users (citizens, civil servants). • Development of plans for a European wide deployment.	The **achieved project results** are: • Development of citizen process models ... • Research has shown that problems with private service providers are hard to solve. They should be integrated in future R&D perspectives. Further, an analysis of the internal project development processes has suggested various guidelines for future R&D management (see below). • Done, but with a broader focus and a smaller degree of details than originally intended. This was vital for the project survival. • The same as above. • More standards for evaluation processes are needed. A lot of resources were spent on the discussion of the evaluation process. • The only point of criticism in the final review was that no plans for commercial use of project results were developed.

Measures of Success:
There will be two main measures of success (cf. section on 'Innovations')
• The number of transport routes for documents ...implemented... functionality, security, and ease of use.
• The quality of the interplay of all components:
  • ... privacy and information self-determination granted.
  • ... transparency of the usage processes.
	Reasons for Success:
	• Interdisciplinary results (as opposed to plain interdisciplinary work).
	• Feasibility, holistic nature, and scalability of the architecture.
	• Citizen control on what is done with her data.
	• The possibility to achieve proper transparency – although further work is required on this issue.

Milestones and expected results:	**Results achieved:**
Business process models of the administrative activities created by the migration of European citizens from one member state to another.	Change to citizen process models & transfer of service context for citizens from one municipality to another (if the citizen requests this).
Generic HW, SW and process design for Javacard enhanced administrative procedures.	Done on a conceptual level – too little time to compare the pros and cons of different technological solutions.
Prototypical implementations (infrastructure, applications, processes).	Demos instead of more complete prototypes as resources have been shifted from implementation work to basic research work.
Evaluations of the prototype under real world constraints.	Broadening the view and shortage of time resource did not allow enough real world testing (just the six cities).
Plans for a European wide deployment.	Business plans are still missing.

3 Divergence and Change

The positive result of the change management in FASME project may be traced back to the following key issues (compare [4]:

1. It changed from an applied research project to a basic research project. It delivered a validated concept for a solution rather than prototypes.
2. The planned floating multi-functionality was replaced by very few, easy to use, static functions, which can be used for different scenarios.
3. The intended international coverage could be resolved by local adaptation rather than by global standardization (except for the information transfer).
4. The intended main function of the Javacard as an information carrier was replaced by the implementation of digital identity as a main function.
5. The (traditional) transfer of documents with context specifications replaced the intended transfer of data and the idea of data consistency management was rejected.

6. The project team resisted the temptation to integrate payment functions, which are believed to be killer applications, but rather tend to kill projects.

The main difficulties of FASME were

1. It was too short and the efforts for interdisciplinary learning were underestimated.
2. We could not react to the finding that some major problems of migrating citizens arise from the need to consume private services.
3. The project plan relied on interfaces between different work-packages, none of which worked. Successful communication happened mostly during workshops.
4. Qualified human resources for research leadership had to be provided by the technical partners, who themselves faced serious staffing problems.

4 Conclusions

The diversity of public administration, reflecting the richness of socio-cultural differences, is one of the key challenges for a united Europe. This particularly refers to digital identity, data protection, and trust services. In the future, the main challenge for a European-wide implementation of interstate e-government services will be the heterogeneity of Europe with respect to administrative ontology and law, administrative processes and culture, the expectations of citizens, the existing legacy systems, and evolving new e-government legacies. Interdisciplinary R&D is required to further develop vertically and horizontally scaling, holistic solutions. We suggest focusing efforts on architecture in its holistic sense (rather than on prototypical products). Interdisciplinary projects should be provided with enough time, experienced convergence management, change management by scientists (or IT architects), and the possibility to integrate further partners at a later stage of the project.

References

1. C. H. Cap, N. Maibaum. *Digital Identity and it's Implications for Electronic Government.* In Towards the E-Society - E-Commerce, E-Business, and E-Government, Kluwer Academic Publishers, Boston, 2001.
2. C. Cap, N. Maibaum, and L. Heyden. *Extending the Data Storage Capabilities of a Java-based Smartcard.* In Proceedings ISCC 2001, IEEE Computer Society, 2001.
3. *Oostveen A.-M., van Besselaar P.*: Linking Databases and Linking Cultures, In Towards the E-Society - E-Commerce, E-Business, and E-Government, Kluwer Academic Publishers, Boston, 2001.
4. A.-M. Oostveen, P. v. Besselaar, and I. Hooijen. *Innovation as learning: three case studies of multi-functional chipcards.* In Proceedings EASST 2002, York 2002.
5. S.D. Pawlowski, D. Robey, and A. Raven. *Supporting Shared Information Systems: Boundary Objects, Communities, and Brokering.* In Proceedings ICIS 2001
6. R. Riedl R. *Interdisciplinary Engineering of Interstate E-Government Solutions,* Proceedings 4th International Conference on Cognition Technology: Instruments of Mind, Warwick 2001.
7. R. Riedl. Document-Based Inter-Organisational Information Exchange. In Proceeedings ACM SIGDOC 2001, Santa Fe, 2001.

The Local e-Government Best Practice in Italian Country: The Case of the Centralised Desk of "Area Berica"

Lara Gadda and Alberto Savoldelli

Politecnico di Milano, 20133 Milano, Piazza Leonardo da Vinci 32, Italy
Tel. +39 02 23992796; Fax. +39 02 23992720;
lara.gadda@polimi.it

Abstract. The reform impulse, which in the last years characterised the Public Administration of all the word, underlines the necessity to join legislative changes with process and change management. In more recent years, the attention to process oriented change management techniques has also emerged in the public sector, through attempts to draw from private sector, searching for new methodologies and managerial approaches that could satisfy the need of organisational innovation. The article aims to present a successful case related to the application of an e-government experience in Italy in a local context: the development of a centralised desk for issuing building permits, grouping twenty-two villages in north-eastern. This experience represents a best practice which could be transferred in many other context.

1 Introduction: The Context

In the last few years, industrialised Countries faced up the necessity of *reforming Public Administration*, a crucial problem since that context is quickly evolving. The change in progress is moving along two directions: on one hand, the users require a Public Sector's "product" risen in value and, on the other hand, there's the need to provide better services using the same resources (more trend towards efficiency) [1]. The progressive shift from a slow and inefficient bureaucracy toward a lean and dynamic organisation is one of the topic joining a number of interventions of the last decade, within the public sector administration all over the world. Though this deep change has been faced in each country in different ways [2], the fundamental characteristic is the presence, finally contemporary, of two separate trends: *the decentralisation and the modernisation*, intended as procedures and process simplification based on importing management models typical of the private sector and on the regulation of the relationship between politic and administrative responsibility in management of the public organisation.

After years of unsuccessful attempts also in the Italian context, the two fundamental reforming pressures are carried out and their most significant expression is the so called *Bassanini laws*[1], which on one hand determine the conferring of new functions and tasks to regions and local authorities (Bassanini 1) [3], on the other hand they

[1]The laws n°59/1997 and n°127/1997.

R. Traunmüller and K. Lenk (Eds.): EGOV 2002, LNCS 2456, pp. 179–186, 2002.

enact the simplification of control and decision procedures (Bassanini BIS) [4]. In particular, the decentralisation laws transfer to regions, provinces and municipalities strategic tasks which have been performed up to now, by the central level, so including in the reforms the *sussidiarity principle* which states that "it's not due to the superior government all for which the inferior level is enough" [5].

The decentralisation represents a leverage for obtaining a more effective and efficient administrative management, nevertheless the functions delegation, up to now centralised, can be only a necessary condition, not certainly enough, for approaching of the public administration to the citizens. The enlargement of functions the regions should perform risks to undermine consolidated and, on the whole, efficient organisation, that could reveal inadequate to the "new decentralised system".

In order to shift the change from the legislative level to the organisational-functional one many models of change management have been experimented, as the *Total Quality Management* and the *Business Process Reengineering* [6] [7] [8]. Among the most important experiences, it's possible to remember:

- the necessity of PA to consider not only the services users, but also all the actors who, even if they aren't "clients", are interested or have influence on their allocation;
- the necessity to control equity, clearness and legitimacy principles of public processes;
- the presence of legislative constraints, the political commitment and many stakeholders

In this context the simple application of the business models from private to public sector creates several problems. So it's important the identification and evaluation of the actual processes and the definition of performance levels which should be reached by the change.

The article aims to present a successful case related to the application of an e-government experience in Italy in a local context: the development of a Centralised Desk for issuing building permits, grouping twenty-two villages in north-eastern ("Area Berica"). These administrations are bound by a Territorial Treaty which is finalised to reorganise the process of local development and to create new economic opportunities. Analysing the specific characteristics of the reality, this case is a best practice which could be transferred in many other contexts.

The next paragraph explains what Centralised Desk is, which are its aims and the involved actors. The third paragraph presents the definition of the case. At the beginning the "as is" analysis was realised on the base of an appropriate check list in order to identify the changes which allow the definition of the "to be" process. The fourth one is based on the chosen methodology, the change management approach and the introduction of IT solutions, and on the results of the implementation. Finally the article presents the conclusions: the relevant local changes and the possibility of extend the experience in other different local contexts.

2 The Centralised Desk

In the modernisation context of the local, regional and national Public Sector the Centralised Desk (CD) plays a very important role since it's a hinge between the local administration level and the economic-productive system.

The normative evolution of the CD is bounded to the DPR 447/98 and to its modification, the DPR 440/2000 [9]. Its subject is related to the localisation of productive system of goods and services, their realisation, restructuring, extension, closing-down, reaction and re-conversion of the productive activity, as well as the realisation of the internal works in building used as an enterprise.

The CD is the only one administrative enterprise interlocutor for all is related to the localisation, construction and restructuring of productive systems. It is the unique responsible of the whole administrative procedure and it's able to assure to everybody is interested the access to all information about the authorisation procedures and to the service and assistant activities to the enterprises. The creation of the CD transfers the collect of all needed authorisations from the enterprise to the Public Administration. The CD receives a schedule and its task consists in the production of all the documents useful to authorise the realisation of the required activity [10].

The Law establishes that the CD does either the administrative activities or the informative ones in order to help the contractors interested into the starting of an activity on a specific lot. So, on the administrative hand, the CD requires:

- the definition of the timing of the issue of an authorisation by PA;
- an internal organisation of the different offices;
- a change in the relationship among local PA which are involved in the process of the administrative authorisation related to different kind of productive activities;
- the necessity to develop forms of inter-institution co-operation which are based on the use of new technologies;
- a continuos process finalised to the administrative simplification

On the other hand, the informative activities could be related to:

- public information about who have already presented the requirement to the administration;
- the access to the knowledge of the state of art of each procedure by applicant;
- the facilities which the administration, manager of the CD, is able to assure to everything who wants to invest on its territory.

In conclusion it's possible to affirm that the CD is born in order to:

- simplify and speed up the administrative procedures;
- make administrative action clear and available to the citizens' participation;
- promote the local economic development also by the diffusion of information related to promotional and technical assistance activities.

The introduction of the CD finds its *problems in the change of the process*. it was difficult to convince the involved people to change their habits and to think that the new technologies could make the process better. This problem is due to the fact that in the little local administrations *the process is hindered by politic and parochial reasons*. Each administration is convinced that its working method is the best and they are worried to loose power and contact with the citizens/SMEs if the CD will take place[11].

3 The Definition of the Case

The realisation of the Centralised Desk started from a social/economic analysis. It was the result of several interviews made to the representatives of local administrations in order to identify the main characteristics of twenty-two villages which constitute the "Area Berica". The identified characteristics could be divided in two categories:

- specific elements of local administrations:

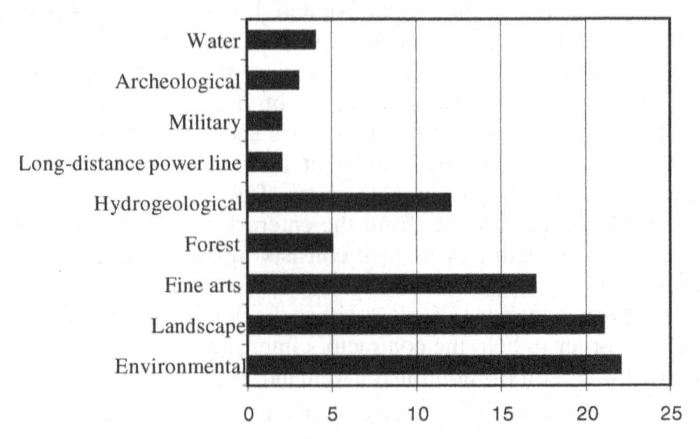

Fig. 1. The bonds of the protection of the territory

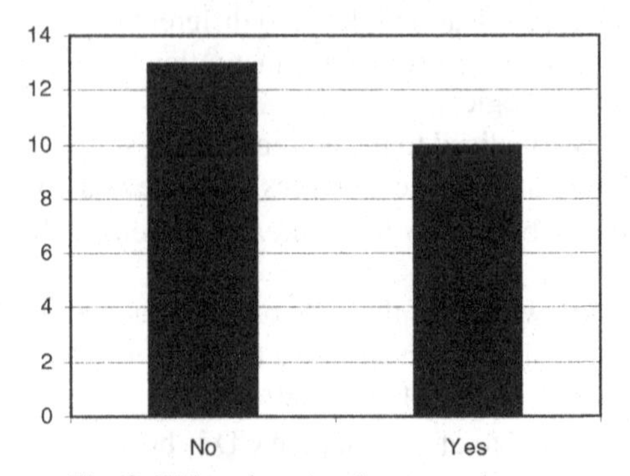

Fig. 2. IT in urban development plan

- specific elements of process of building permits:

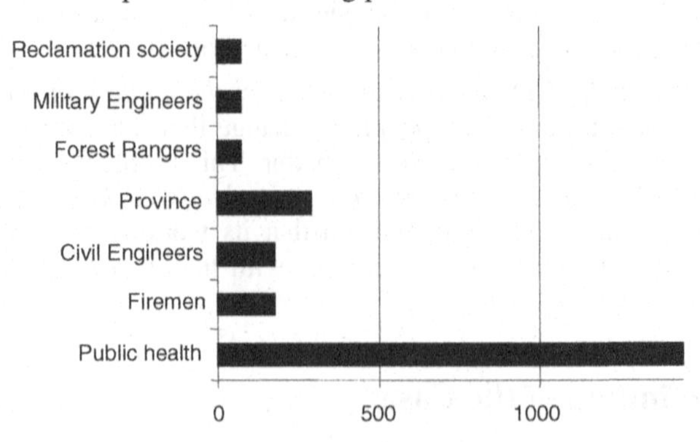

Fig. 3. The involved Third Administrations in procedures

4 The Identification of "As Is" Analysis to the Definition of "To Be" Process

The collected data through the social-economic analysis were the starting point to build the "to be process". First of all, in order to reach the best solution for the specific case, *weakness points & expected performances* were defined:

Table 1. Analysis "as is" – weakness points and expected performances

Weakness points	Expected performances
• Lack of operative and organisational structure; • Lack of accurate information for the citizens: all the procedures need to be integrated; • Lack of opportune technological tools; • Need to have a wide knowledge, but lack of available time; • Lack of standard procedures	• To standardise the procedures; • Major investments in technologic infrastructure; • Employment of high qualified people who carry out consulting activity; • Major information to users; • Simplification of procedures and reduction of timing of building permits; • Change of information between local and Third Administrations

This working approach let the drawing up of the *Centralised Desk Income Statement*: it was realised on the based of the collected data, the number of procedures and the involved people in "Area Berica".

Table 2. Analysis of the Income Statement

	"As is" analysis	"To be" process
Number of building permits	1440	1500
Number of involved people	25	30
Time dedicated to procedures by each person	94 days/year	75days/year
Costs (personnel, redemption and managerial expenditure) for each local administration	about 5.000	about 20.000
Covering of costs for each local administration	• a tax per head = 0,4	• a tax per head = 0,95 or • a payment for the management of the relationship with the Third Administrations = 7.000

According with the introduction of Centralised Desk, the higher number of involved people is due to the presence of qualified personnel who requires an additional cost. On the other hand the Centralised Desk takes advantages to the final beneficiaries such as the possibility to have information in quick time, a major transparency in information exchange and the saving related with consulting costs.

5 Implementation Plan

Once the analysis and search phase was finished, the action plan was implemented:

- the *collected results were submitted to the twenty-two villages* in order to focus the attention on the common and the different characteristics of the "Area Berica";
- during the meetings with people involved in each local administration, some *proposals for the realisation of the Centralised Desk* were collected;
- in order to implement the CD it was important to open the dialogue towards the majors. It was possible to achieve the *signature of the treaty for the activation of the CD*;
- according with the characteristics of the villages, the project consortium decided *to divide the "Area Berica" in three parts called sub-areas*. In each sub-area, the villages referred to the biggest local administration which was in charge of co-ordinating activity;
- to realise the effective introduction of the CD *work teams* were organised. These teams convened directly people of different local administrations. The *training areas* dealt with:
  - *technological area*. Its programme was divided in three levels: presentation of CD web-site (www.sportellounico.net); guidelines to fill in the web-site the local administration rules and forms; simulation of procedure management by CD;
  - *administrative/managerial area*. Its programme dealt with the process analysis and the forms analysis. These training courses were firstly organised in three groups and then in only one group: each one expressed his/her opinion in order to define the standard form which satisfied all interests.

The main problems of the implementation plan dealt with:

- the *different opinions expressed by the involved representatives of local administrations*. Each of them was convinced his working methodology was better than anyone else and they were not in favour of the change;
- *the necessity to interact with the majors* who followed their political interests. In the twenty-two villages the majors belonged to different political currents and so it was very difficult to make them agreed;
- the importance to *find an agreement between the technicians and the majors* since their interests were different.

6 The Actualisation Status

According with the implementation plan and considering the emerged problems, the following decisions were taken:

- *An associated managerial model was created*. The three sub-areas were joined in a unique centralisation at the administration which heads the line of the Territorial Treaty. This model foresees the presence of two different administrations: ones which is responsible at local level and a co-ordinating ones. The activities of the local administration are:
  - to inform and help the citizens by information instruments;
  - to provide the standard and defined form;

- to receive the filled in form and the attached documentation;
- to send the form to the Centralised Desk and to sign building permits.

The activities of the co-ordinating administration are
- to record the form and to manage the unique procedure;
- to establish the relationships with the Third Administrations and to check it;
- to issue the authorisations for starting the productive installation;
- to transfer to the local administration the form which has completed the iter;
- to assure all the communications to the involved people.

- *The standard unique procedure was define. It refers to different documents.* Also these forms were standardised and they deal with building permits, starting activity declaration, habitability and feasibility certificate, and commerce;
  - *The standard form was filled in the web-desk (www.sportellounico.net).* The realisation of the on-line sharing of data simplifies the technicians' work, reduces the time of authorisation permits and makes the procedure monitoring and the data computerisation easier.

The project started on March 2001 and finished on October 2002. Now the administrations of the "Area Berica" carry out their building permits through the CD: all the local administrations take part in the project and they are enthusiastic for the achieved results.

7 Conclusions

The key elements of the CD project consist of the implementation method at local administrations. The outcomes are the result of acquiring relevant information from people involved in conducted analysis: that was allowed by the widely sharing of knowledge and strong co-operation of Majors. Specifically, public Authorities contributed to the work groups in a decisive way through their strong commitment to build widely accepted procedures and models, and on the basis of a sound experience in the field of knowledge and document management.

Moreover, the obtained model can easily be applied to complex information flows in local, but widely diffused, contexts. In fact, notwithstanding the regional characters of such procedures, the developed model provides a wide structure and every necessary feature for modelling CD. For that reason, the project constitutes a best practice to be applied in the European context.

The last element which should be considered is the possible future developments. The economic development area will become in the next future one of the most important and prestigious scope of local administration participation. Obviously, this is possible if the old competence "administrative" conception is overcome by politics characterised by local economic development. This passage requires an integrated vision of the different areas where it's possible to work.

References

1. AIPA, "La reingegnerizzazione dei processi nella Pubblica Amministrazione", pp. 1-24, 1999).
2. Klages, H., Loffler, E., "Administrative modernisation in Germany – a big qualitative jump in small steps", International Review of Administrative Sciences, pp. 373-384, (1995).

3. Legge 15 marzo 1997, n.59, "Delega al Governo per il conferimento di funzioni e compiti alle regioni ed enti locali, per la riforma della Pubblica Amministrazione e per la semplificazione amministrativa".
4. Legge 15 maggio 1997, n.127, "Misure urgenti per lo snellimento dell'attività amministrativa e dei procedimenti di decisione e di controllo".
5. Balboni, E., "I principi di innovazione del decreto Bassanini", Impresa&Stato, Aprile/ Giugno, pp.44-45, (1998).
6. Thompson, J.R., Jones V.D., "Reinventing the federal Government: the role of theory in reform implementation", American Review of Public Administration, June, (1995).
7. Kock, N.F., McQueen, R.J., Baker, M., "Re-engineering a public organisation: an analysis of a case of successful failure", The International Journal of Public Sector Management, Vol. 9, n. 4, (1996).
8. Pfiffner, J.P., "The National Performance Review in perspective", International Journal of Public Administration, 20(1), pp. 41-70, (1997).
9. D.P.R. 20 ottobre 1998, n.447 come modificato dal D.P.R. 7 dicembre 2000, n. 440,
10. Aa. Vv, "Realtà locali – periodico di informazione e cultura", n. 19, (2001).

The Immanent Fields of Tension Associated with e-Government

Otto Petrovic

Karl Franzens University Graz and evolaris foundation,
Universitätsstrasse 15/E4, A-8010 Graz, Austria
otto.petrovic@uni-graz.at

Abstract. In the past few years the establishment of e-government has been given numerous new impulses. Totally new horizons have been opened both on the level of information and communication technologies, especially by the Internet, and on the level of administrative processes by new methods and tools applied in process design. But nevertheless especially e-government is faced with much more difficulties and opposition in its implementation than its counterpart in business, namely e-business, which in turn meets difficulties that are anything but small. E-government looks back upon more than 40 years of history [1,2,3,4], and the recipes for success that have been presented are often older than the people who today are in charge of implementing them. The present article tries to show that in approaches to e-government that are merely limited to the technological and the administrative process level many immanent fields of tensions remain unconsidered.

1 Many Unsolved Problems Despite Long Traditions

Many arguments such as the necessity to consider organization and technology in an integrated way (i.e. organizational reform taking place simultaneously to a technological reform) have similarly long traditions as e-government itself, date back even to the 1950ies and are thus anything but new [5]. This also applies in a very similar way to the demand to coordinate internal administrative changes and external services as well as to coordinate various decisions on technical investments within public administration. It seems rather unlikely that the two core innovation potentials of the Internet for e-government, namely the creation of uniform standards for data transmission and data representation and the penetration of the non-public sector with information and communication technologies, will be sufficient to make up for deficits in the implementation of e-government.

2 There Is Something That Gives a Meaning to Administrative Processes

As soon as in the late 1980ies and early 1990ies both business and public administration realized that the organization of enterprises and public administration should not be oriented merely to individual functions but to a sequence of related activities – namely the business and administrative processes [6]. E-business – i.e. the counterpart

R. Traunmüller and K. Lenk (Eds.): EGOV 2002, LNCS 2456, pp. 187–190, 2002.

of e-government in business – however does not just concern the transformation of essential business processes but also a redesign of business models [7] with the aid of Internet technologies. Business models lend a more profound meaning to business processes. They determine which players are involved, which products and services these players offer and which revenues they are able to earn from which particular source of revenues. So business processes are always an implementation of a certain business model. Analogously there is a kind of 'administrative logic' that gives an intention-driven meaning to the underlying administrative processes. Merely considering the implementation of existing law as the only meaning in administrative processes would definitely be a drawback to a level that would roughly correspond to the sense of reality of the homo oeconomicus and would thus be of little pragmatic relevance for e-government. Activities in public administration and politics are based on limited rationality due to incomplete information, unsolved conflicts of interests and decisions that are characterized by emotions.

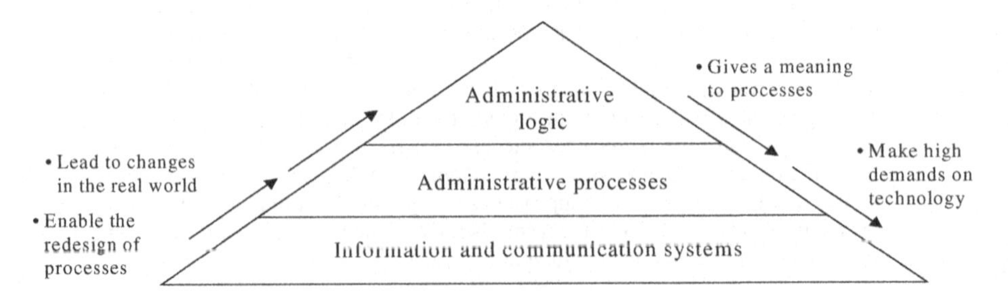

Fig. 1. The process levels of the e-government

As long as e-government primarily deals with technological matters on the level of the information and communication systems and with organizational matters on the level of administrative processes, citizens, public administration and political players will substantially object to changes and show their resistance to these changes either in an open or in a more subtle way. As a result the success of these changes will remain rather modest or at best just perceivable.

3 Understanding Public Administration As a Complex System

Public administration can be considered as an open system that internally is characterized by an extraordinarily high degree of functional differentiation, or to be more precise, of specialization. If such a complex system is faced with a highly dynamic change such as the development of the Internet, a field of tension is created because it is necessary to consider as many conflicting interests of the various parties involved in the process as possible (citizens, public administration, politics), as well as the costs incurred in terms of funds, time and contents.

While in the past ten years many highly efficient methods and tools for optimizing business processes were developed by means of information and communication technologies, the development of business models and administrative logics is still in its infancy. When developing business models, intuition, a 'funny feeling' or some knowledge gained through experience often play an important role. But intuition is very difficult to communicate and thus difficult to share with others so that decisions made by intuition very soon meet the limits of usefulness in a society that is increasingly based on division of work. An essential task of the e-government therefore should be to help develop methods and tools, show complex correlations within an administrative logic, make it communicable and analyze it from various points of view, in order to finally be able to make substantial use of the potential factors of the Internet when redesigning administrative processes.

4 True Administrative Logics Do Exist – Even Numerous Different Ones

Administrative processes could be managed rather easily in an efficient and effective way by means of the e-government if there was a consistent administrative logic underlying them. This administrative logic eventually determines which form of administrative process is considered as meaningful by the various players involved and is thus fostered by them either openly or in a more subtle way.

An example for such different administrative logics are e-government information systems for every day matters such as jobs, education, housing or health. The *citizen* expects that such services are provided by the public sector free of charge – and of course in a high quality and always up-to-date. The citizens' willingness to pay for such services is very low. From the point of view of *public administration* information systems for every day life represent a totally different logic. Due to the continuously growing cost-push, public administration is increasingly interested in e-government systems that will immediately contribute to cutting costs. Information systems for every day life, however, may well increase the quality of the service provided by public administration but also cause additional costs. *Political players* consider such information systems as highly valuable, additional tools of communication which can be used for self-expression, for communicating achieved goals, legitimizing public expenditure and above all for a very targeted formation of opinions.

5 But Who Is Entrepreneur, Bureaucrat and Politician at the Same Time?

The last field of tension associated with e-government is created by divergent requirements for planners, developers and operators of the e-government. Every-day applications such as tourist information systems, regional marketplaces for enterprises or information systems for businesses and culture require entrepreneurial thinking and actions. The ability to detect – or to generate – substantial gaps between demand and supply and bridge them with big economic success is in the center of interest. Thus searching for imbalances and balancing such divergences between demand and supply in a targeted way, as well as the variable 'generation of revenues' determine entrepreneurial thinking and actions.

Just the contrary applies to the demands made on e-government, especially when it comes to contacts with authorities, because in this field it is particularly important to fulfill specific requirements as efficiently as possible. Minimizing expenditure and optimally using existing funds as well as trying to establish a balance are in the center of interest. This means that the most important goal is to establish a balance between clearly defined demands made on public administration and the degree of fulfillment of these services.

Applications in the field of political participation in turn require an in-depth understanding of the impacts on the political structure of power, the functions of political processes and in many cases even political legitimization.

Acknowledgements

The author would like to acknowledge the funding of the evolaris foundation by contributions from the federal ministry of commerce and employment, the styrian business support association, the styrian provincial government's departments of commerce, finance and telecommunications and of innovation, infrastructure and energy, and the city of graz.

References

1. Kubicek, H., & Hagen, M., et al. (1999): Internet und Multimedia in der öffentlichen Verwaltung (Gutachten). Medien- und Technologiepolitik, Bonn: Friedrich-Eber-Stiftung, http://www.fes.de, retrieved on 6th August 2001.
2. Prins, J.E.J.(ed.) (2001): Designing E-Government: On the Crossroads of Technological Innovation and Institutional Change, Kluwer Law International.
3. Kraus, H. (1983): The Impact Of New Technologies On Information Systems In Public Administration In The 80, Verlag-North-Holland, Amsterdam - New York.
4. Traunmüller, R.; & Reichl, E.R. (1980): Die Entwicklung von Verwaltungsinformationssystemen, Proceedings of the 2nd Joint Technical Meeting of ÖGI/GI: Information Systems for the 1980ies.
5. Leavitt, H.; & Whisler T. (1958): Management in the 1980's, Harvard Business Review, Nov./Dec., S. 41-48..
6. Davenport, Th. H. (1993).: Process innovation: reengineering work through information technology.. Boston, Mass.
7. Petrovic, O.; Kittl, C.; & Teksten, R. D. (2001): Developing Business Models for eBusiness, Proceedings of the International Conference on Electronic Commerce, 31st October 2001 to 4th November 2001, Vienna, Austria.

VCRM – Vienna Citizen Request Management

Josef Wustinger[1], Gerhard Jakisch[2], Rolf Wohlmannstetter[3], and Rainer Riedel[2]

[1] Vienna City Administration, MKS, Rathausstraße 1, 1082 Vienna, Austria
wus@mks.magwien.gv.at
[2] Vienna City Administration, MD 14-EDP, Rathausstraße 1, 1082 Vienna, Austria
{jag,rie}@adv.magwien.gv.at; gerhard.jakisch@aon.at
[3] Vienna City Administration, MD 55, Niederhofstraße 23/304, 1120, Vienna, Austria
woh@bue.magwien.gv.at

Abstract. The vCRM is one of the eGovernment applications, which was awarded by the European Commission's eGovernance Competition 2001. [1 eGovernment] It is the only workflow application within the Vienna City Administration, which handles tasks administration wide and not like the standard procedures within a department. It deals with all comments, complaints, requests, which are not routine eGovernment processes and have a specific legally defined procedure. It deals not with requests that are designated to a specific department. One main advantage of this procedure is, that the officer in charge may easily adjust the workflow according to the specific need of the case. It is one of the finalists of the Global Awards for Excellence in Workflow [2 WARIA 2002]

1 Introduction

Vienna's Citizen Request Management (vCRM) is one of the most frequented applications of the implementation of the eEurope initiative [3 Liikanen] by the Vienna City Administration. The first pilots of the vCRM were the "street damage information system", which started 1989. It included the co-operation of 12 departments, the central citizens' matters system, (Zentrale Bürgeranliegen) which is in operation since 1985.

The vCRM formerly named Central Complaint management (CCM) was initiated 1998 by the CEO of Rapid Relieve to reduce the workload of those departments, citizen service centers and administrative units in the 23 districts dealing with Citizens' complaints. [4 Jakisch], to realize suggestions made during the organizational assessment and evaluation of the Vienna City Administration in 1998, to implement proposals made by administrative units with intensive client interaction and to meet citizen's expectations as described by the VCA's CEO in his inauguration speech 1997 [5 Theimer].

The vCRM may be described as one important action to build up CRM – Citizen Relationship Management – the adoption of Consumer Relationship Management concept in Vienna. It will play a very important role in the future of the interaction between Public Administrations and Citizens/Inhabitants to meet more successfully the citizens' expectations, requirements and needs.

R. Traunmüller and K. Lenk (Eds.): EGOV 2002, LNCS 2456, pp. 191–194, 2002.

vCRM deals with those complaints, comments, requests, which are not well structured, need workflows that cannot be designed in advance, handle multiple complaints about one subject and which every inhabitant /citizen may ask.

It deals not with routine governmental activities as legally based requests, petitions, etc. that are based on a specific law and have a defined workflow and those have legally stated participants like building permissions etc.

2 Goal and Objectives

The goal of the vCRM is to improve substantially the administration's citizen orientation – the relationship between the citizens on the one hand and the administration on the other –. This will be achieved by implementing a new administration-wide tool – the vCRM – to record and process all citizens' requests, comments complaints and suggestions concerning all spheres of work of the Viennese administration.

The Vienna Citizens' Request Management has five objectives:

1. Coordinated workflow of comments, complaints and suggestions,
2. Recognition of parallel cases
3. Avoidance of different solutions for similar cases
4. 24/7 availability of information
5. Improvement of citizen's information

All incoming complaints, comments, informal request and suggestions are recorded administration-wide in one application the Vienna Central Request Management. It uses one specific basic workflow and predefined search tree of keywords including colloquial notions and synonyms.

The recognition of parallel and similar cases is performed via the keywords. New incoming cases are recorded by recognizing location (address and Geo-Reference System -code) and predefined keywords. Thus in future new cases concerning same subject and location will be added automatically to the file of already recorded and processed cases.

This will be achieved twofold by filing all cases concerning one matter in on file. such already processed cases are the basis for a new one and free access to stored data for all administrative officers.

The information stored in vCRM is available for every user, the registered administrative officers. In future also the citizens will have a specific access to their cases. Only those cases dealing with citizen's privacy, other privacy or public security issues are secured via a special structured user ID and password.

The situation of the person sending a comment, informal request, complaint or suggestion will be improved by changing actual web form into a more interactive version, direct feed back of filing key and password and possibility for the citizen to access vCRM and read the actions performed and running.

3 The Procedures

The cases recorded in the vCRM [6 Resel] arrive via several modes of access: Email htttp://www.wien.gv.at/m55/b-bue.htm , fax or letter, Call Center or Face to face: in the Citizen Service Centers or all other participating administrative units.

At the moment all cases have to be recorded by a human intermediate – a civil servant –. The next version will offer an interactive web client too.

Every administrative unit participating accepts all interventions. This eases the procedure for the citizen because he needs no knowledge about the internal organization and responsibilities of the administration.

The procedures are divided into several steps, the recording, the take down, the different processes to finalize the complaint and the response to the requester.

4 vCRM Web Client

To overcome the obstacles caused by hard and software requirements of Fabasoft components running on each PC used to deal with vCRM on the one hand and to implement an interactive application offering advantages to administration by reducing the workload of the vCRM officers and the citizens by easier access and the overall goal to reduce costs led to the development of the Web Client solution.

It will be available by on all public terminals and may be accessed via Internet without having implemented Fabasoft Components. This application enables the VCA to spread the usage of vCRM faster to 120 Administrative Departments and 500 units because no additional hard and software is required and thus reduces their costs in future.

The main differences to the Version described in paragraph 3 are: the recording of the request is performed by the requester, the confirmation is accompanied by the creation of a specific User ID and PIN-Code presented to the requester, each requester may trace the current status of the processing of his request at any time by typing the ID and Pin in the Authentification form. This will reduce time of the vCRM officers spent on answering requesters' calls on status of current affairs.

5 Vision

Currently the Vienna City Administration is participating in a European Union Framework 5- IST project – EDEN [7 FP5-IST] – which deals with the citizen participation in administrative processes and eDemocracy. It develops several tools to improve interaction between citizens and administration.

These tools are based on NLP technologies (Natural Language Processing). One of them, the automatic routing system, is designed to analyze unstructured texts, find the relevant keywords and phrases, and pass the document automatically to relevant expert available. In case of vCRM such a tool may assist the vCRM officer recording requests by automatically filling in the record. It will be able to handle requests received by web client, email, fax and partially letters.

The necessary information about the experts may be taken from the advanced version of the VCA's electronic organization expert system "Überblick Wien" which may include not only the tasks and responsibilities of administrative units, job descriptions but also the expert knowledge of the administrative officers.

Thus matching the vCRM search results and expert information it could suggest next steps. Taking into account the developments of speech text conversion the authors will make no predictions how avatars may change administrative acting.

6 Conclusion

vCRM is the first city-wide document handling system introduced by the VCA. It is based on Faba Components workflow technologies allowing the Citizen Service department to document all incoming interventions.

It benefits as well Citizens as administration because it offers all involved parties actual information about the state of affairs. It reduces the processing time. The citizen may participate via email and public terminals. The incoming information is analyzed according the addressee, the location and content, thus allowing linking the information to the GIS maps.

The combination of this application with the EDEN Tools as described above will enable the administration to analyze the citizens complaints and comments multi modal. By giving all involved parties the possibility to interact, improvements will be achieved, which haven't been possible in the past.

Thus the (inter)-active Citizen will become an equal partner of the administration in the decision making process.

References

1. eGovernment, from policy to practice, 2001 Brussel http://europa.eu.int/information_society/eeurope/egovconf/projects_selected/austria/index_en.htm#CCM – Central Complaint Management
2. http://www.waria.com/awards/awards.htm
3. Liikanen, Making eEurope a reality, European IT Forum 2000 Monaco
4. G. Jakisch, eVienna living situation based eGovernment and eDemocracy, DEXA, Munich, 2001, 397 ff
5. E.Theimer "Electronic Services" http://wwws14.advge.magwien.gv.at/w4base/mdi/default.asp Vienna 1997
6. Resel , Eschner, ELAK ZBM, VCA MD 55 internal training document, Vienna 1999
7. Framework Programme 5 Contract IST-1999-20230, Brussels 2000

Public-Private Partnerships to Manage Local Taxes: Information Models and Software Tools

Mario A. Bochicchio and Antonella Longo

SET-Lab (Software Engineering & Telemedia Lab)
Department of Innovation Engineering, University of Lecce, Italy
{mario.bochicchio,antonella.longo }@unile.it

Abstract. The present work is about the first results of SOSECO, a public-private company operating in the southeast of Italy, created by Servizi Locali SpA and the Municipality of Castrignano, in cooperation with the University of Lecce (Engineering School and Law School). The goal of SOSECO is to improve the management of local taxes and to enhance the relationship between Citizens and Public Administration.

1 Introduction

From surveys [West] made on US federal and local agencies, e-government is still in its infancy: even if many governmental units are putting on-line a wide range of information and services for citizen and employees, a coherent and mature design of the underlying information infrastructure (software applications, standards, protocols, ...) is still at the down, far from the evolution already reached in the private world. Some argue that private-sector management techniques can be applied to government, to produce more efficient, effective and responsive public agencies. At the same time, new ideas about governance have also emerged, stressing collaborative relationships, network-like arrangements and hybrid public-private partnerships to enable a more effective problem solving and a greater citizen participation in public affairs [Mechling], [O'Toole], [Koppel].

In Italy, as in many other countries, we have interesting experiences about online services to citizens (e.g. in the "ICI On-Line Project" [Prato], or in "National Online Fiscal Service Project" [Finanze]), but the situation is not homogeneous. From an informal survey in the Southeast of Italy, for example, we found a scenario similar to the one depicted in [Vintar]. In this context we decided to support SOSECO in putting up a "Citizen Relationship Management" (CRM in the following) strategy, with the aim to create trust between Local Public Administrations (LPA in the following) and citizens, and to provide it with a modern, multichannel, personalized set of services.

The CRM strategy has 3 main parts:

- **Interaction CRM**, i.e. the ability to create new multichannel services (one-stop counters, call center, Web portal, ...) based on an integrated information resource.
- **Insight CRM**, i.e. the ability to continuously acquire updated info and to relate it to the existing ones. This is used to construct new, more effective user profiles.

R. Traunmüller and K. Lenk (Eds.): EGOV 2002, LNCS 2456, pp. 195–198, 2002.

- **Satisfaction CRM**: i.e. the ability to evaluate/correct the citizen satisfaction level, with respect to new services.

SOSECO information system has been designed and developed to fulfill all the main aspects of our CRM strategy.

2 The Proposed Model

Up to some years ago, the main model in management of local taxes has been its concession to private partners: so the concessionary (invested of public powers) replaced the Municipality in collecting local tributes and in checking for tax evasions. Moreover, municipalities were few interested in managing tributes efficiently, because the main source of incomes came from the Central Government.

After the last constitutional referendum about local autonomies (Oct. 2001), a new model of local government management came out, asking for management savings and for a more active role in the territorial development. So, many municipalities has been forced to find new founding ways and/or to better use all existing sources of incomes. Thus the management of tributes and entrances is becoming a core business for local administrations, and tax evasion is now a big issue at local government level.

As a consequence, a lot of municipalities start to manage entrances and local taxes and to double check evasion by themselves. The problem is that most LPAs, especially in the South of Italy, lack of the resources to design and to manage good fiscal services. To skip this problem they can adopt two possible solutions: acquiring resources (ICT infrastructure, technologies, skills, people, etc.) to directly manage fiscal entrances inside the municipality, or outsourcing the services necessary to support the core management of tributes.

A detailed comparison between the two solutions is out of the scope of this paper, but we can observe that, in general, the first approach is ineffective because of the lack of standards (in software, interoperability, data formats etc.) and because of the rapid evolution of technologies and skills. The second approach, instead, can be very effective if performed with the following constraints:

- it must be "transparent": data must be provided to the LPA with well documented formats, on open architectures (like standard DBMSs), ready to be reused for different purposes;
- the outsourcing must be on-line: each municipality must hold every time the actual ownership of its own data, on-line with all other municipal data bases;
- a sharp separation must be enforced, between the integrated data resource of the LPA and the services built over the data, so several service providers can concurrently offer services over the same data resources.

Thus public-private companies are the right component to create and to manage all the online LPA's data resources: the public part guarantees the respect of the previous constraints, while the private part provides the technological know-how. At the same time the role of the private partner changes: from substitute of the Administration (as in the case of the concessionary) to coach of innovation.

Basing on this model, in our project 3 partners cooperate to provide innovative fiscal services to citizens: the Municipality of Castrignano, the SOSECO (i.e. the data provider) and Servizi Locali (SL in the following), (i.e. the service provider); other private providers can cooperate with the Municipality and SOSECO to concurrently offer further services to citizens, business, and employees. At the same time with this approach the private partner can provide the same service to several Municipalities, without the need to manage the overall LPA's data resources.

3 The SO.SE.CO Project

SOSECO (Società Servizi Comunali – Society for Municipal Services) was born in Spring 2000 and is made up for 51% from SL as a private partner and for 49% from public partner (The Municipality of Castrignano del Capo, in the district of Lecce in the southeast part of Italy). The mission of SOSECO is:

- To adopt innovative tools and services to manage local tributes, like data quality assessment and citizen relationship management (CRM);
- To train Municipality's employees about the recent administrative reform;
- To technology support Municipality with data digitalization and integration;
- To support territory's research and development;

The relationship between the SOSECO and SL is based on a service agreement to supply innovative services and to manage fiscal entrances. SL provides innovative fiscal services but the whole management is still owned by the local Administration (through SOSECO).

The agreement between SL and SOSECO is about what they call "Introductory activities to tribute and entrances' collection", including:

- Census of real estates in the Municipal territory
- Estates data cleansing
- Cross checks among several data flows in order to check out evasions
- Services to citizens who can access their fiscal position and communicate with Administration and manage pre contentious, contentious and post contentious phase through different integrated channels

The Municipality has 15.000 inhabitants and last year, with this approach, it found 30% evasion level on a 2.5 M approximate global entrance. The services provided by SL are paid by SOSECO with a percentage on plus evasion cashing assessed.

Major's Point of View is: "We are not outsourcing our fiscal services, but we are outsourcing innovation on fiscal services", with the following advantages:

- learning by doing for employees involved in the project, and knowledge transfer from the project team to the rest of employees;
- better aptitude to innovation;
- forcing the application provider to keep high the quality of service and innovate it not to be substituted by another provider.

Currently, SOSECO employs 3 people and 10 freelance consultants, but today we foresee the society growth like double over the next 12 months.

4 Implementation Issues

SETLab has supported both SL and SOSECO to define the model previously described, to build the innovative information infrastructures to the whole fiscal process, to design the multichannel applications for the CRM strategy and the communication with citizens.

To build, design and develop SOSECO's information system we used UWA (Ubiquitous Web Application) [UWA] and W2000 methodology [Garzotto]. Moreover, we adopted a fast prototyping approach (based on MS Access and MS SQL Server) to finely tune algorithms and filters for data cleansing and data management for the SL side.

UWA helps us to reduce overtime costs in development, moving most of the issues at the design phase, to improve effectiveness and ergonomics in functionality and features, and to to dominate multichannel design and multichannel integration through a proven methodology.

Four main software prototypes have been created in cooperation with SL to support its core processes (census taking, information cleansing and matching, tax assessement and injunction of payment, citizen reception and advise).
The main research effort made by the SET-Lab has been to produce reusable tools and standardized procedures for data cleansing, data integration and integrated multichannel services to citizens. In the first year of operation the information system prototype we designed allowed SOSECO and SL to correctly manage the whole fiscal process over 15,000 citizens with 3,5M of incomes, that is 30% more than the previous year. This result was achieved without contentious.

References

[Vintar] M. Vintar, "*Reengineering Administrative Districts in Slovenia*", Discussion Paper, No.11, Local Government and Public Service Reform Initiative, Open Society Institute, ISSN 1417 – 4855, 1999

[West] D.M.West, "*Assessing E-Government: The Internet, Democracy, and Service Delivery by State and Federal Governments, 2000*" Brown University, Sept., 2000

[Prato] http://www.comune.po.it/servcom/tributi/ici/htm/icionline.htm

[Finanze] http://www.finanze.it

[Mechling] J. Mechling, A customer service manifesto: using IT to improve government services. Government Technology, January, S27-S33, 1994

[Koppel] Koppel, J. G. S. (1999). The challenge of administration by regulation: preliminary findings regarding the U.S. government's venture capital funds. Journal of Public Administration Resarch and Theory, 9, 641-666.

[O'Toole] O'Toole, L. J. (1997). Treating networks seriously: practical and research-based agendas in public administration. Public administration review, 57, 45-52.

[UWA] UWA (Ubiquitous Web Applications) Project: Deliverables D2 (General def. of the UWA framework), D7 (Hypermedia and Operation Design: Model, Notation, and Tools).

[Garzotto] F. Garzotto, "*Ubiquitous Web Applications*" - Invited Talk, In **Proceedings of ADBIS 2001**-5th East European Conference Advances in Databases and Information Systems, A. Caplinskas, J. Eder (Eds.): Springer LNCS 5121, 2001

E-MuniS – Electronic Municipal Information Services – Best Practice Transfer and Improvement Project: Project Approach and Intermediary Results

Bojil Dobrev[1], Mechthild Stoewer[1], Lambros Makris[2], and Eleonora Getsova[3]

[1] Fraunhofer Institute for Secure Telecooperation, Schloss Birlinghoven
D-53754 Sankt Augustin, Germany
www.sit.fraunhofer.de
[2] Centre for Research and Technology – Hellas, 6th km Charilaou – Thermi Road
57001 Thermi – Thessaloniki
www.iti.gr
[3] Elisa Consult, 93, James Boucher Blvd., 1407 Sofia
Elisa@solo.bg

Abstract. The E-MuniS (Electronic Municipal Information Services – Best Practice Transfer and Improvement) Project aims to improve the best practices of the European Union municipalities regarding the use of information technology in municipal administration working processes and services to citizens and to transfer those results to South-Eastern European municipalities in particular from the Balkan region thus integrating it to the EU municipal network. The project consortium involves as participants couples of local municipality – IT-company partnerships from EU countries and from South East European countries. Within the project solutions for an e-municipality office (as back-office system) and prototypes of e-services to citizens and business (as front-office system) will be developed and implemented.

1 Project Main Goals and Expected Benefits

1.1 Project Background

Municipal services to citizens and municipal administration working processes have always posed challenges to both citizens and municipal employees. They were mostly associated with enormous paper work, procedural formalities, long queues in front of always busy administration offices, complete waste of time and a lot of nerves – again both for the citizens and administrative staff.

Although cities in the European Union have made rapid advancements in the field of implementation of innovative Information Technology for improving and facilitating the citizens' life, there are still deficits especially in providing e-services to citizens and integrating these services to internal applications. In the South-East European (SEE) region there are limited IT applications (mostly off-line) and a lack of use of internet technologies for municipal applications.

The E-MuniS (Electronic Municipal Information Services – Best Practice Transfer and Improvement) Project as a result of the successfully completed SEEmunIS pre-

R. Traunmüller and K. Lenk (Eds.): EGOV 2002, LNCS 2456, pp. 199–206, 2002.

project (South East European Municipal Information Infrastructures, 1999/2000) provides an original contribution to the solution of the problem bringing together the efforts of technology providers and local authorities from EU and SEE based on public-private partnership relations.

Its results will be disseminated firstly to the project municipalities and following the dissemination strategy - to about 50 other municipalities from EU and the Balkan region, thus integrating the latter to the EU Community municipal network.

The project is realised with the financial support of the IST Programme of the European Union. It started in November 2001 and will end in October 2003.

1.2 Description of the Project Objectives

The E-MuniS ultimate goal is to provide opportunities for user-friendly implementation of the information technologies achievements in municipal administration working processes and services to citizens. The E-MuniS project main objective is to bridge the gap between EUMs and SEEMs regarding use of Information Technology in administration working procedures and services to citizens, thus facilitating the work of the municipal employees and making the life of he citizens easier.

E-MuniS project contributes to improving of the quality of life by:

- Better and more accessible services provided by local administrations, following the identification of the citizens' needs.
- Improving the working conditions of the municipal administration employees by reducing the time and efforts needed for information processing.
- Creation of new IT based jobs related to the latest achievements of the information technology usage in the municipal administration processes.
- Improving citizen's participation in all levels of local government life through a facilitated participation and interaction between citizens and local government.

2 Consortium

In compliance with the specifics of the priorities VIII.1.6-VIII-1.5 of the EU IST Programme, the project Consortium involves as participants couples of local municipality-IT company partnerships (municipalities and technology providers) from EU countries (Germany, Spain, Italy, Greece) and from South East European countries (Bulgaria, Former Yugoslav Republic of Macedonia, Croatia).

The E-MuniS tends to act as a driver for cooperation between public administrations of a number of cities in different countries, to contribute to public-private partnership development between the local administration and the IT industry, and to establish links and joint activities between IT companies from SEE and EU.

3 Key Issues

3.1 Project Aims

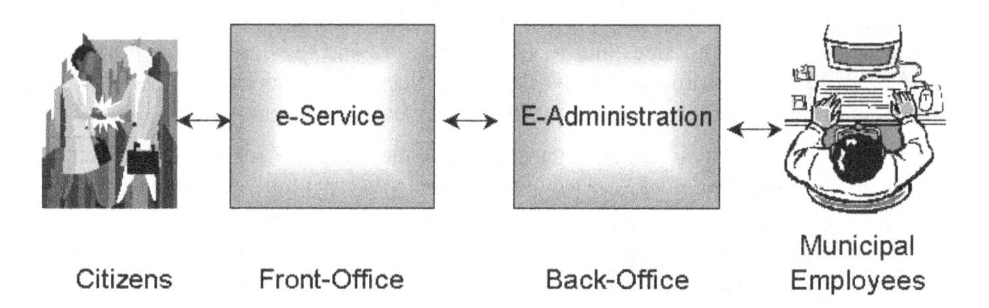

Fig. 1. Components of system development within the project E-MuniS

The project aims at development and implementation:

1. prototypes realizing an E-Municipality office (Back-office solution).
2. on-line services to citizens as front office solution with municipal and city websites and sets of interface tools allowing citizens access to municipal administration applications to ensure transparent information services to citizens. These solutions will be ported to an information kiosk system, too.

3.2 Project Approach

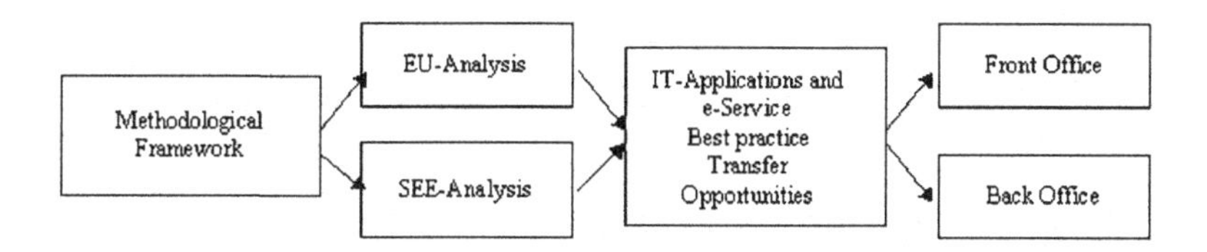

Fig. 2. Project approach

Basis of the project activities is the development of a methodological framework to analyze the situation in cities of the EU and South East European cities. The results will be used to identify best practice transfer opportunities for municipal IT-applications and e-services for front office and back office solutions.

The overall project methodology will follow the top-down approach. All the tools for study, analysis, development and dissemination will be centrally elaborated. The responsibility for this will be delegated to the Work package leaders. The tools will be implemented in the project municipalities following a decentralized manner. All adapted or developed software tools will be completed to a prototype level with the relevant accompanying documentation. Pilot implementation will take place at two or three sites. All mid-term and final results will be immediately disseminated to all the project participants.

4 Results from the 1st Stage of the Project

4.1 Methodological Remarks

The first stage of the E-MuniS work plan focused on the study and analysis of EU municipalities' best practices and ICT infrastructure and the demand for improvement and ability for e-performance of working practices and services to citizens in 4 SEE municipalities. Therefore a set of questionnaires have been realized to:

- identify the current situation regarding implementation of Information Technology in EU municipalities
- identify best practice applications and elaborate on their improvement
- identify those applications that are better suited for transfer to the SEE municipalities based on certain criteria
- describe the most important services currently provided by SEE municipalities as well as information sources, IT applications and IT infrastructure.

For selecting best practice applications which are better suitable for transfer to SEE-municipalities a set of criteria has been used:

- is it relevant?
- is it feasible?
- what are the risks involved?

Is it relevant? Considering the scarcity of resources and the need to make an impact the E-MuniS project needs to carefully choose applications, which address a real need and whose results make a definite impact. The relevance of a given application depends on two considerations:

- is it relevant to the core activities and services of the municipality?
- would the solution make a measurable difference?

Is it feasible? The question about feasibility is more difficult to answer than the one about relevance, because there are a number of aspects that have to be considered. These aspects can be grouped into mainly three categories:

- organisational aspects - this concerns the current structures, legal restraints, communication mechanisms and procedures that are in place and which may be amenable to the introduction of IT systems or alternatively which require a complete reorganisation for their effective use.
- technological aspects - looking at the current state of technological development, especially concerning infrastructure availability, security technology, telecommunications and the Internet.
- human aspects - people are a significant factor in any organisation, and considering the need for change, there is a need for looking at the interest in new technology, trust in reliability, the need for training and potential cultural barriers.

What are the risks involved? In most cases the introduction or expansion of new technology like IT systems has a potentially dramatic effect on the organisation, especially when it is meant to lead to significant improvements in efficiency. There is a need for change in three different areas: in the organisational structure, in the organisational culture, in the information technology.

Fig. 3. Approach of selecting best practice applications which are suited for transfer

4.2 Intermediate Results

4.2.1 IT in EU Municipalities

The questionnaire, designed to gather information from EU municipalities, was disseminated to municipalities in Greece, Italy, Germany and Spain. In total, twenty-seven questionnaires from EU municipalities were selected and evaluated.

Some results:

- All municipalities use MS Windows (98, 2000 or NT) as operating system while a large number (67%) have UNIX-based servers. Only 17% have Linux installed although a large number (83%) of municipalities consider the use of Open Software.
- All municipalities taking part in the survey have a LAN, 67% have PCs connected to a WAN. The majority of the municipalities have access to Internet (83%).
- Administrative municipality tasks (as citizens registration and financial administration) are in general supported by IT-systems. The use of various database systems is common.
- Geographical Information Systems are already introduced or will be implemented in near future.
- Beside this the prevailing trends within EU municipalities regarding administrative tasks are the use of modern technologies as electronic documents management systems, workflow components and the use of Internet technologies.
- Our survey shows that the municipalities already offer several e-services but there are still deficits concerning interactive systems using security technology and the integration of e-services to internal applications. The following applications seem to be proved:

- Municipal information system realising the life event concept with quick access using keywords.
- Forms server for downloading of forms (pdf files).
- Services referring to citizens' registration such as extracts of citizens registration, requests of delivery of an income tax card - changes of income tax cards, indications of changes of address. With changes to registration laws it is expected that such Internet based procedures will expand.
- Presentation of geographical information as land-use plans, city development plans, cadastral-plans, digital maps and cultural maps.
- Information about cultural events and ticket reservation
- City council information system
- Services of the municipal waste management
- Online library services
- Online registration for courses at adult evening classes

- A few cities have started to realise challenging services such as:
  - Virtual market places
  - E-democracy applications that support citizens' participation in municipal planning procedures

4.2.2 IT in South East European Municipalities

Twenty-five questionnaires from SEE municipalities, districts and communities were answered. All of them were evaluated and summarized.

The situation and demand for Information Technology in SEE municipalities have been analyzed during the SEEmunis pre-project. According to the results of this project some problems are common to almost all municipalities:

- Lack of IT strategy as part of the strategy of the municipality
- Insufficient integration of IT applications (often using different tools and platforms)
- Insufficient computer equipment

This situation has not generally be changed.
Some results from the survey:

- The cities use personal computers which are connected to a LAN. All have Internet access.
- Microsoft technology is generally used (Windows as operating system, MS Office as personal productivity software). UNIX and Linux are not installed.
- The cities have websites which are used for provision of basic information.
- Administrative tasks as financial management, tax related administration, citizens registration, social services are supported by IT-applications, but the application related service to citizens and business is unsatisfying. The most important deficit is the lack of transparency of administrative procedures.
- The most demanded procedures are services to business and to issue permits and certificates for citizens.

4.2.3 Further Proceeding

Using the results of the survey and applying the selection criteria the following best practice applications will be realized within the E-MuniS project:

1. Information and E-Services Portal providing:

- Municipal information system – Information about all administrative issues of the municipality structured according to the life event concept
- Integration of forms (pdf-files) (one – way communication)
- Provision of services
  - For citizens: request of certificates (two-way interaction) (interactive forms are provided via portal).
  - For citizens: announcement of change of address (as all municipalities taking part in the project have already introduced a citizens' registration system, it will be possible to connect the e-service to this internal application).
  - For business: permits and certificates (using the example of starting or changing a trade). The e-service solution must be integrated in an internal application.

2. Electronic document management system with workflow components. This system will be connected to the administrative application and to the e-service application. This will ensure transparency and may be an example of multi channel approach, as there would be several ways of access to the municipal service: through internet, by post with the necessity to handle paper documents or via telephone.

5 Outlook

One of the most important issues of the project E-MuniS is to ensure on-going dissemination and exploitation of the project outcomes. The target is that about 50 municipalities of the Balkan Region should participate in the results.

At the Project level of dissemination workshops will be organized in each project municipality to demonstrate the applications, discuss the opportunities for their implementation in the particular municipality and conduct training for a designated group of staff to be working with them. The Project web site will contain a data base with information about all results achieved during the project stages and tools for analysis, modification and visualization. It is installed on a server and accessible via Internet.

The dissemination will be through professional societies and associations, NGOs, government bodies, supported by IT companies, based on public- private partnership.

At the International level of dissemination municipalities from other countries that have expressed interest in the E-MuniS outcomes will be potential end-users. The dissemination will take place through European associations and networks, such as Global Cities Dialogue, Tele-cities Network, Regional Innovation and Technology Strategies Networks, Association of the Balkan Municipalities, GISIG Network.

The project consortium will take great efforts to distribute the project outcomes aiming the integration of SEE-municipalities to the EU municipal network.

References

E-MuniS-Project Team:

1. Electronic Municipal Information Services – Best Practice Transfer and Improvement Project – a short project Presentation

2. Project Presentation and State-of-the-art of the ICT Infrastructure, IT Applications and Demand for new IT Projects in the SEE Municipalities
3. Methodological framework for study and analysis of the municipal administration working processes and services to citizens
4. IT applications in the municipal administration working processes and provision of e-services to citizens in the EUMs and SEEMs with best practice transfer opportunities

All documents are available on E-MuniS-Website: www.emunis-ist.org

Some Specific e-Government Management Problems in a Transforming Country

Nicolae Costake

Consultant Bucharest,
ncke@starnets.ro

Abstract. e-Government (e-G) raises specific technical, and also managerial problems. The managerial ones are particularly important for the transforming socio-economic systems. Recent experience of Romania in this field is shortly presented . Key managerial requirements are derived from the e-G business requirements. The proposal of orienting the back-office reengineering implied by e-G also on enforcing virtuous societal closed circuits and minimizing vicious ones is formulated.

1 Introduction. Statement of the Problem

E-Government (e-G) makes full use of ICT in the managerial subsystem of socio-economic systems (SES), implying the reengineering of its processes and links with its clients: citizens / organizations and internal users (managers, MPs, magistrates and civil servants). Last year, an EU Ministerial Declaration stated that e-G could improve the services, strengthen the European societies, raise the productivity and the well-being and also strengthen the democracy [1]. Technical aspects of e-G are studied in a vast literature.(e.g. [2]). Managerial aspects are, certainly, also important. The problem was studied by a number of authors. to name a few: Lenk and Traunmuller (e.g. split back-end from front-end activities, [tele-]cooperative work, organizational knowledge management.[3]).. Traunmuller and Wimmer (*i. a.* decisions and knowledge, including organizational memories [4], Lockenhoff (a general system theory for guiding the societal change.[5]). The managerial aspects of e-G are particularly important in the transforming countries, for assuring, and accelerating their trajectories towards the advanced Western SESs, as suggested e.g. in [2] and [6]. The present paper tries to answer the question what managerial e-G requirements (e-GMR) could result from its business requirements (e-GBR). Case study Romania is considered. The methodological approach is:e-G BR => e-GMR => comparison with the actual evolution =>consequences. The structure of the paper is:a) the challenge; b) e-G management requirements; c) recent developments for the information society in Romania: .d) conclusions.

2 The Challenge

The higher the decision level, the more adverse effects a mismanagement can induce. For the transforming countries, this conclusion is more evident. In the case of Romania, the difference in the interval 1990…2001 between the average evolution of the

R. Traunmüller and K. Lenk (Eds.): EGOV 2002, LNCS 2456, pp. 207–210, 2002.

GDP of the former European communist countries and the actual evolution, represents a total value not far from that of the GDP in 1990. The 2001 Report of the European Commission on Romania's Accession [7] also identifies managerial problems, such as: the reform of the Public Administration, the reduced administrative capacity and the corruption.

3 Specific e-G Managerial Requirements. Case Study Romania

The experience in the field of e-G suggests a number of BR, which imply specific information system (IS) requirements [2]. The e-GBR can be structured in some categories: a) better services supplied to citizens, organizations and internal users: e.g. friendly interactive access to public information, minimization of the time and effort consumed in solving problems with the public institutions; b) better services for the users within the state institutions: e.g. adding flexibility and scalability; c) operational support: e.g. electronic documents and archives, centrally updated standard procedures, state and trend of resources, costs and revenues; d) better support for the executive management: feedback form the public, decision support based on data warehousing; e) general requirements: e.g. 24x7 availability, data security, and protection; protection against corruption and terrorism; possibility to build the coherent e-G information system in parallel by various state institutions. The general resulting key e-GMR is the need to assure the synergy within the SES. Following directions of managerial action can be enumerated: a) initial prerequisites support for the development of the communications infrastructure; political will to move towards the Information Society; readiness to fight bureaucratic resistance and corruption; .precise delegation of authority between central and local levels and. between / within state institutions; creation of the necessary legislative, regulatory and institutional infrastructure); b) support of the integrated informatisation of the public sector (including legislative and judiciary); c) creation of an empowered organizational entity for e-G; d) production of a national e-G strategy, support for : cooperative IS, e) identification of possible societal closed circuits, to be enforced if virtual, else minimized; f) training in basic ICT and e-G oriented applications, starting with high-level decision-makers. It is advisable to start with the creation of the informational coherence information system (as described in [2]), a shared informational resource for practically all information systems (IS) of the SESs.

Developments in Romania in 2001 and 2002 show the following: a) the mentioned managerial pre-requisites for e-G are now practically achieved.. In particular:, the telecommunications sector is growing,, the political will regarding the information society is shown by a specific chapter in the 2001..2004 government's action plan [8] and its update [9], the start for introducing networked PC's in all schools, growth of the national ICT industry; the adoption of a national anti- corruption strategy;. b) the essential specific legislation (copyright protection, electronic document and signature, data protection etc.) is in place. c) empowered authority for e-G exists. (A "Group for Promotion of the Information Technology", headed by the Prime Minister approves all major ICT projects. *I. a.,* it approved a Strategy for the Informatisation of the Public Administration [10]). The Ministry for Communications and Information Technology, quite dynamic, launched a bid for 19 IT projects, which became demonstrable prototypes produced by consortia of companies in less than one year.(end 2000). One of them, e- procurement, is now successfully experienced on a moderate scale and started to demonstrate the economic benefits for public acquisitions.. The scope and quality of the on-line access to public information was improved. The 2001-2004 action plan contains a number of provisions for e-G, many of them inspired by the eEurope+ plan of actions.. Its 2002-2003 update [9] foresees also the creation of unitary nomenclatures and informatised registers vital for achieving the coherence of the societal information. The national server for general interest nomenclatures (first defined in the strategic planning of 1992) started functioning.

However, from the point of view of the management science, some problems still exist:

a) As a strategy based on the e-GBR (covering all the three Authorities of the state) is still under development, the action plans are not necessarily complete and do not necessarily represent the synergic trajectory towards the goal..
b) The direct implementation of actions defined in the context of the advanced countries is not necessarily without difficulties. E.g., front-office oriented applications in general imply a back-office applications support. The version of the strategy [10] based on the Italian one [11] may encounter contextual problems. However, other experiences (e.g. [12]) could be directly implemented.
c) Some of the action plans do not specify responsibilities and resources. and their financing source. Full use of advanced project planning and management methods and instruments can certainly be useful, also contributing to synergy.
d) A general problem in societal management is the selection of the governance methods. Apart identifying invariant models to rely on, a rational choice is necessary between open- circuits commands (e.g. Ordinances)-or "hard governance (HG)" and regulations which create / enforce virtuous societal closed circuits, to automatically generate most of the actions provided otherwise in detailed action plans or "soft governance (SG)". This question is closer examined below.

A study [13] suggested the still great importance of the HG and also the need of closed informational circuits in the judicial field. A preliminary research suggested the existence of some tenths of societal closed circuits, some of them being switchable vicious / virtuous ones. [6]. One example, concerning the Legislative Authority follows:

(i) Examples of possible actions of the state are: a) precise delegation of the authority between the Parliament, the Government and the other state institutions able to generate regulations; b) minimization of modifications of the laws and other regulations, e.g. via the creation of instruments for simulating the likely consequences of their drafts and the standard structuring of their content; c) adoption of an efficient electoral law and friendly electoral interface; d) informatisation of the legislative processes, providing also transparency; e) public free access legislative database.

(ii) Example of a virtuous closed circuit: performant electoral law & implementation of the above exemplified actions of the state => elections => best choice of persons => improved legislation and better people in state managerial political positions =>good results => new better alternatives => elections.

(iii) Example of a vicious closed circuit: poor technological support => incomplete or incoherent legislation => possibilities to by-pass the law => growth of the underground economy => encouragement of incomplete or incoherent legislation.

4 Conclusions

Following conclusions are proposed: a) e-GMR derived from e-GBR suggest the importance and feasibility of creating synergy; b) for transitional SESs e-G presents a high economic potential; c) a rational governance mix must be found between HG and SG, with the accent on the latter.

References

1. EU Ministerial Conference on e-Government November 2001 http:// europa. eu. int / iinfoorma- tion_society
2. Costake, N., Jensen, F., H..:Towards an Architectural Framework of e-Government information systems Paper prepared for KMGov 2002 Copenhagen, 2002
3. Lenk, K., Traunmuller , R.: Perspectives on Electronic Government IFIP WG8.5 Information Systems in Public Administration. Working Conference on Advances in Electronic Government Zaragoza, 2000
4. Traunmuller, R, Wimmer, M: Daten- Informtion – Wissen – Handeln: Management des Wissens. *In: Reinermann, H (Hrsg): Regieren und Verwalten im Informationsalter v.Decker Heidelberg, 2000 482-498*
5. Lockenhoff, H.: Simulation for Decision Support in Societal Systems: Modelling for Guided Change to meet Complexity and the Future *In: Hofer, C., Chroust, G. (eds): Proceedings of the IDIMT- 2001 Interdisciplinary Information Management Talks Trauner Linz, 2001*
6. Costake, N., Dragomirescu, H., Zahan, E: E- Governance: a Mandatory Reengineering for the Transforming Economies *In: Wimmer, M. (Ed): Proceedings KMGov Trauner Linz, 2001* 30-38
7. European Commission : 2001 Regular report on Romania's accession SEC (2001) 1753 Brussels, 2001
8. Romanian Government's Plan of Actions 2002-2004 (in Romanian) http:// www. gov. ro
9. Plan of Action for the Social Democratic Party Governance in 2002 and 2003 (in Romanian) http:// www. gov. ro
10. National Strategy for the Informatization of the Public Administration (in Romanian) http:// www. gov. org
11. Presidenza del Consiglio dei Ministri, Dipartimento Funzione Pubblica: E- Government Action Plan 22 june 2000:// www. funzionepubblica. it/ download/ action plan. pdf (2000)
12. IDA Architectural Guidelines Technical Handbook Version 5.3 http:// www. ispo. cec. be (link to IDA)
13. Costake, N.:E- Governance and the Judicial System. A point of view. In:Tjoa, A., M., Wagner, R., R.: 12th International Workshop on database and Expert Systems Applications IEEE Computer Society Los Alamitos, 2001

Towards a Trustful and Flexible Environment for Secure Communications with Public Administrations

J. Lopez, A. Maña, J. Montenegro, J. Ortega, and J. Troya

Computer Science Department, E.T.S. Ingenieria Informatica
Universidad de Malaga, 29071 Malaga, Spain
{jlm,amg,monte,juanjose,troya}@lcc.uma.es

Abstract. Interaction of citizens and private organizations with Public Administrations can produce meaningful benefits in the accessibility, efficiency and availability of documents, regardless of time, location and quantity. Although there are some experiences in the field of e-government there are still some technological and legal difficulties that avoid a higher rate of communications with Public Administrations through Internet, not only from citizens, but also from private companies. We have studied two of the technological problems, the need to work in a trustful environment and the creation of tools to manage electronic versions of the paper-based forms.

Keywords: Public Administrations, Secure Communications, Electronic Forms, Certification Authorities, Public Key Infrastructure

1 Introduction

Approaches to electronic versions of many of the paper-based administrative procedures between Public Administrations and citizens can bring meaningful benefits. These benefits concern accessibility and availability of documents, regardless of time, location and quantity.

Although there are some experiences in the field of e-government there are still some technological and legal difficulties that avoid a higher rate of communications with Public Administrations through Internet, not only from citizens, but also from private companies. Any type of digital transaction is influenced by typical open networks risks. Agents involved (public organizations, private companies and citizens) need to work in a trustful environment. This environment must satisfy the required security levels in such a way that privacy and authentication of digital information is guaranteed to senders and receivers OGIT96 [4]. Also there is a lack of software tools that help to create, distribute and manage in an easy and flexible way the electronic versions of paper-based forms, which is the usual way of interaction with Public Administrations. Clearly, these tools must incorporate authenticity and integrity mechanisms that mimic those ones existing in the traditional paper-based documents [5].

R. Traunmüller and K. Lenk (Eds.): EGOV 2002, LNCS 2456, pp. 211–214, 2002.

In this paper we present the results of a research project that has focussed on the problems we have mentioned. We also show how the integration of the approaches produce a solution that enhance many of actual developments. Thus, the structure of the paper shows the two main works done in the project. Section 2 presents the design and development of a real hierarchical Public Key Infrastructure (PKI), which we consider the most convenient type of infrastructure for operation of any administrative procedure that involve a digital signature. Section 3 presents the design of a language for the description of electronic forms that allows the utilization of signed forms in all communications with Public Administrations. The paper finishes with conclusions in Section 4.

2 Development of a PKI Based on New Design Goals

Digital signatures schemes are based on the use of public-key cryptosystems [3]. The reasons are that these schemes offer the same functionality than hand-written signatures, and also a high protection against fraud. However, the global use of any of those cryptosystems needs a reliable and efficient mean to manage and distribute public keys, by using digital certificates. Such functionality is provided by a Public Key Infrastructure, which is formed by a diversity of Certification Authorities. A PKI becomes essential because without its use public key cryptography is marginally more useful than traditional symmetric one [2].

Although addition of certification capabilities in commercial electronic mail programs is a very helpful feature, a detailed analysis shows that these schemes result not satisfactory for e-government applications. Some design features that may compromise the security of the systems have been detected. We summarize some of the most important ones:

- It is common in most of Public Administrators that users share the same computer system. Therefore, private keys belonging to different users are not completely "isolated". This drawback does not allow the appropriate use of a very important security service for e-government applications, the non-repudiation service [7].
- Certificates needed for a verification of documents that have been digitally signed must be obtained from sources that are external to the electronic mail programs. Therefore, it is very possible that users do not verify them properly (as they are not forced to). Moreover, use of Certificate Revocation Lists (CRLs) is constrained.

These considerations has taken us to develop our own PKI [11], that has the following features:

- Adapted to the multi-hierarchical Internet structure because this is the operational environment.
- Provides secure means to identify users and distribute their public keys.
- Uses a CAs architecture that satisfies the needs of near-certification so the trust can be based on whatever criteria is used in real life.
- Eliminates problems of revocation procedures, particularly those associated with the use of Certificate Revocation Lists.

The main element in the hierarchy is the Keys Service Unit (KSU), which integrates certification and management functions. We use a scheme with various KSUs operating over disjoint groups of users, conforming a predefined hierarchy.

KSU hierarchy is parallel to the hierarchy of Internet domains. KSUs are associated to the corresponding e-mail offices. Every KSU is managed by a CA.

Additionally, it contains a portion of the certificates database to store the certified keys of its users. The third component is the key server, which receives requests and delivers the certificates. The key server manages a certificate proxy that keeps some of the recently received external certificates. The certified keys are managed solely by the corresponding CA; therefore, key updating and revocation are local operations that do not affect the rest of the system.

3 Description Language for Eforms: A New Design

Structured forms has been the traditional method of interaction with Public Administrations [6]. Moreover, the use of hand-written signatures in this type of documents has provided the necessary legal bindings for most of scenarios. Our previous study of common e-government applications has showed us that if paper-based forms have to be substituted by electronic forms, then these ones must have the following characteristics: integrity, or non-modification by external entities; non-repudiation, or non-deniability of agreements; and auditability.

Taking these features as a staring point, we have tried to design an appropriate language for the description of forms.

These reasons recommended us to try to design and develop our own Formal Description Language. Its name is FDL, and XML-like. To be more precise, it is based on XFDL [1].

The use of our own specific language, with its own tools, and completely adapted to XML [8,9], introduces many advantages in comparison with traditional use of HTML [10].

The most important advantages are briefly summarized next:

- *Regarding forms status*: It is easy to add useful components not included in standard HTML, and it provided automatic data validation without programming specific code. Also, the definition of the structure of the fields where signatures are contained simplify the (automatic) process of signature verification, and finally, the particular design of our language, together with the standard where it relies on, facilitate creation of parsers that automatically translate forms to any other language.
- *Regarding forms management:* The signer can store a copy of a partially filled document and one or several persons can sign these forms.
  FDL has two fundamentals concepts oppositely to HTML, the first notion is there are some extensions defined to distinguish different parts and formats in the same document, and the second one is the status of the form is preserved, thus our solution has been designed to organize any form in several pages while having data in memory continuously.

- *Regarding communication:* A proprietary format facilitates that the context of the signature is not lost.Moreover, the document is audited on its own. Oppositely to HTML, FDL provides a data structure and separates application, presentation and logic levels.

4 Conclusions

In this paper we have presented the results of a research project that studies the need of using security for communications over open networks, and the use of electronic versions of the paper-based forms to interact with Public Administrations.

We have shown the main features of the PKI specifically developed and the reasons for its design. Regarding the electronic forms we have designed a language for their description.

Modular design and development of those tools facilitates that the outcome of the work is integrated into e-government broader systems, and can be immediately applied to the social environment. These new solutions also help in establishing the basis for future design and development of schemes oriented to electronic forms signature in the communications with Public Administrations.

References

1. B. Blair, J. Boyer, "XFDL: Creating Electronic Commerce Transaction Records using XML", Computer Networks, n. 31, pp. 1611-1622, 1999.
2. W. Burr, "Public Key Infrastructure Technical Specifications (version 2.3). Part C: Concepts of Operations", Public Key Infrastructure Working Group, National Institute of Standards and Technology, November 1996.
3. W. Diffie, M. Hellman, "New Directions in Cryptography", IEEE Transactions on Information Theory. IT-22, n. 6. 1976, pp. 644-654.
4. European Commission, "Directive 1999/93 of the European Parliament and the Council on a Community Framework for Electronic Signatures", December 1999.
5. European Commission, "Directive on Certain Legal Aspects of Electronic Commerce in the Internal Market", February 2000.
6. "Improving Electronic Document Management", Guidelines For Australian Government Agencies, Australian Office of Government Information Technology, 1996.
7. "Non-Repudiation in Electronic Commerce", Artech House, 2001.
8. Extensible Markup Language (XML)" http://www.w3.org/XML
9. Canonical XML, Version 1.0, W3C Working Draft, September 2000. http://www.w3.org/TR/2000/WD-xml-c14n-20000907
10. HTML 4.01 Specification, W3C Recommendation, December 1999. http://www.w3.org/TR/html4/
11. J. Lopez, A. Mana, J. Ortega, J. Troya "Distributed Storage and Revocation in Digital Certificate Databases", Dexa 2000.

Supporting Efficient Multinational Disaster Response through a Web-Based System

Ignacio Aedo[1], Paloma Díaz[1], Camino Fernández[1], and Jorge de Castro[2]

[1] Laboratorio DEI. Departamento de Informática.Universidad Carlos III de Madrid
Avda de la Universidad 30, 28911 Leganés (Spain)
aedo@ia.uc3m.es, {pdp,camino}@inf.uc3m.es
[2] Dirección General de Protección Civil. Ministerio del Interior
Quintiliano 21, 28002 Madrid (Spain)
jcastro@procivil.mir.es

Abstract. The current process to deal with disaster mitigation has a number of drawbacks that can be solved using web technology. The basic problem is that there is a unidirectional and asynchronous flow of information among the different agents involved in a disaster mitigation procedure. This situation often results in a lack of coordination in the resources provision and in a useless assistance. In this paper we introduce ARCE, a web based system envisaged to cope with the lack of synchronism among assistance requests and responses in a multinational environment as the Latin-American Association of Governmental Organisms of Civil Defence and Protection is. ARCE makes uses of role-based access policies (RBAC) and information flow mechanisms to offer an efficient and reliable communication channel.

1 Introduction

In the last years, Internet is emerging as an essential platform to support cooperative tasks, as demonstrated in projects like the parametric earthquake catalogue [1], Arakne [2] or WebSplitter [3], particularly useful when the users of a system, who can be geographically scattered, do not access it in a synchronous way. This is the case of multinational disaster response, a process which involves different countries and organisms giving assistance to mitigate an emergency situation affecting to one or more countries. The use of a web system to coordinate and synchronize the efforts of each assistance supplier would make the emergency management quite more efficient avoiding duplicates and misunderstandings and providing reliable information on the actual needs. One of the main concerns of the Latin-American Association of Governmental Organisms of Civil Defence and Protection[1], involving 21 Latin-American countries, is to promote cooperation and mutual assistance in emergency situations. Nowadays, when one or more countries of this association are affected by a disaster they ask for cooperation to other countries or organisms in an unilateral way. In turn, each requested country or organism supplies assistance according to its possibilities but without taking into account how the others are contributing. As a consequence, an affected country can receive lots of perishable food, even more that what can be con-

[1] http://www.proteccioncivil.org/asociacion/aigo0.htm

R. Traunmüller and K. Lenk (Eds.): EGOV 2002, LNCS 2456, pp. 215–222, 2002.

sumed by the victims, while the need for clothing is not been addressed by anyone. Moreover, most communication among countries is made through obsolete means like the phone, fax or telex that introduce delays and problems of flexibility, accessibility and reliability that can be easily overcome using the web. In this context, the Association in its IV Conference held in Azores in September 2000, committed to design a mechanism to make easier and more efficient the cooperation and mutual assistance tasks in emergency situations. As a consequence, the Spanish Civil Protection Department along with the DEI research group at Carlos III University of Madrid started to work on the development of ARCE, a web based system oriented towards enhancing the management of multinational disaster response within the scope of the Latin-American Association. ARCE is aimed at becoming a platform to share updated and reliable information among the associates in order to orchestrate an integrated and efficient response, respecting at the same time the peculiarities and autonomies of each member. In order to deal with different access rights (such as the ability to create an emergency, to ask for resources, to offer assistance and so on), the basic principles of role based access control (RBAC) [4] have been adopted and, therefore, a hierarchy of kinds of users makes up the basis to define how the system will be accessed. Moreover, not only the role assumed by a specific user will determine her capabilities to modify or browse the information but also the country she belongs to, since ARCE is implemented in a multinational environment. Finally, roles are also the basis to establish an information flow policy is used to push valuable information to each ARCE user.

2 Design Principles to Support Efficient Disaster Response

As mentioned before, the current process to ask for multinational assistance to mitigate the effects of a specific disaster has a number of drawbacks that can be solved thanks to the web technology. ARCE is a web application that provides mechanisms to notify an emergency, to ask for resources and to offer assistance. In a few words, when an emergency happens the country or countries affected, that we will call throughout this paper the emergency owner, can use the system to keep informed the other associates, called the assistance suppliers. If assistance is required, the emergency owner can ask for resources which are classified according to a multilingual glossary included in the system. Whenever an assistance request is received in ARCE the rest of the associates are notified by e-mail, so that they can access the application to see what the emergency owner is asking for and how they can help. For each emergency, the application will provide updated information about which resources are requested, which quantity was originally needed and how many items have been already supplied. Thus each associate can decide what to contribute taking into account what the others are providing. In order to develop a useful application a number of requirements were taken into account, including: accessibility; multi-user support; inter-state support; multi-purpose application; reliability; efficiency and maintainability.

Accessibility. The application has to be accessed anytime and anywhere to improve the communication among associates. With this purpose ARCE has been designed as a web system which can be used with any Internet browser (e.g. Internet Explorer, Netscape and Konqueror) and e-mail client.

Multi-user support. ARCE is accessed by different users with different responsibilities. This implies that each user has to be allowed to perform or not an operation (e.g. create an emergency, offer assistance, accept assistance) according to the role she plays in a specific organism. To deal with this issue ARCE assumes a role-based access policy (RBAC) described in [5]. Users are classified into roles identified in the Civil Defence and Protection departments as well as those of the invited organizations (e.g. EU, UN). Roles are structured in a hierarchy, so that general roles can be specialized into more concrete roles which inherit the access rules of their parents. Moreover roles can be aggregated into work teams. In this case, access rules are not propagated from the parent to the children. Figure 1 shows the roles hierarchy supported in the current version of ARCE. In the figure, aggregations denote the composition of teams, whereas generalizations represent the specializations of a role.

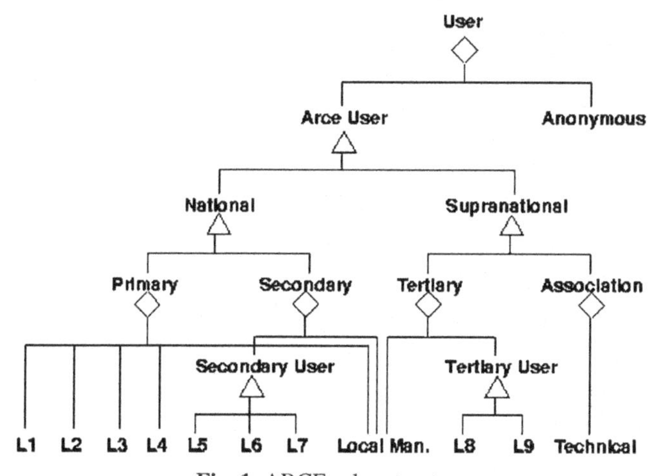

Fig. 1. ARCE roles structure

Each user of the application will be assigned one or more roles. This role along with the country or organism the user belongs to determines the information each user receives as well as the tasks she can perform on the system. For example, the Local Manager can add new users to her country structure while the Technical can perform maintenance operations such as create new Local Managers or new entities. In addition, different users have different views of the same information as shown in Figure 2 and Figure 3 where a request and assistance offer are shown respectively.

Inter-state support. ARCE has to provide each associate a certain degree of autonomy and control. For example, it is allowed to invite organisms or countries that do not belong to the Association to take part in a disaster mitigation. The access policy is then carefully designed in order to support enough roles as to provide control to the different associates while maintaining reliability and efficiency. This diversity of countries also affects to accessibility in terms of multilingual support. In the current version the two languages spoken in the Latin American community (Spanish and Portuguese) are supported both in the user interface and in the glossary of terms. The glossary was introduced in order to unify terminology concerning emergencies with a view to improving communication in a multicultural community. An on-line multilingual description of the technical terms is also provided.

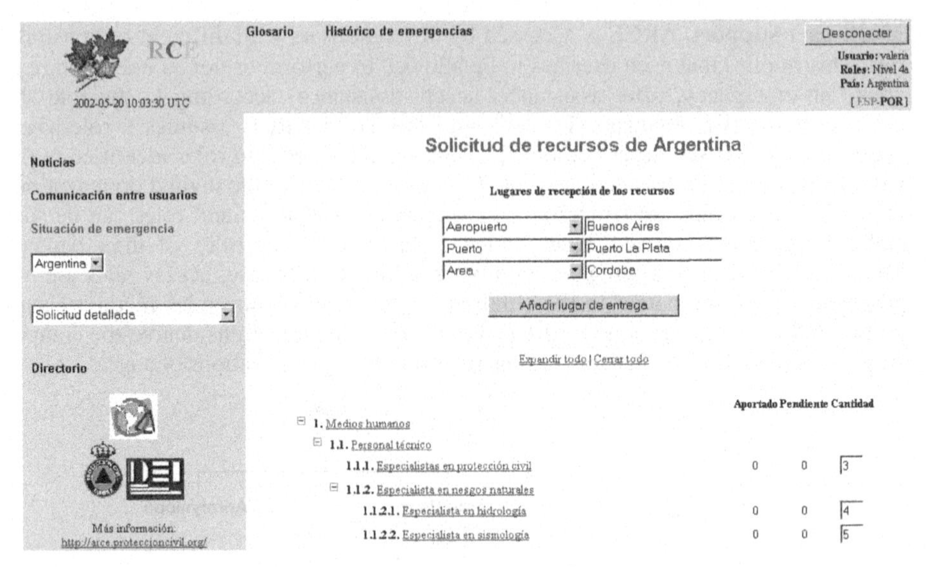

Fig. 2. Assistance from the point of view of an emergency owner

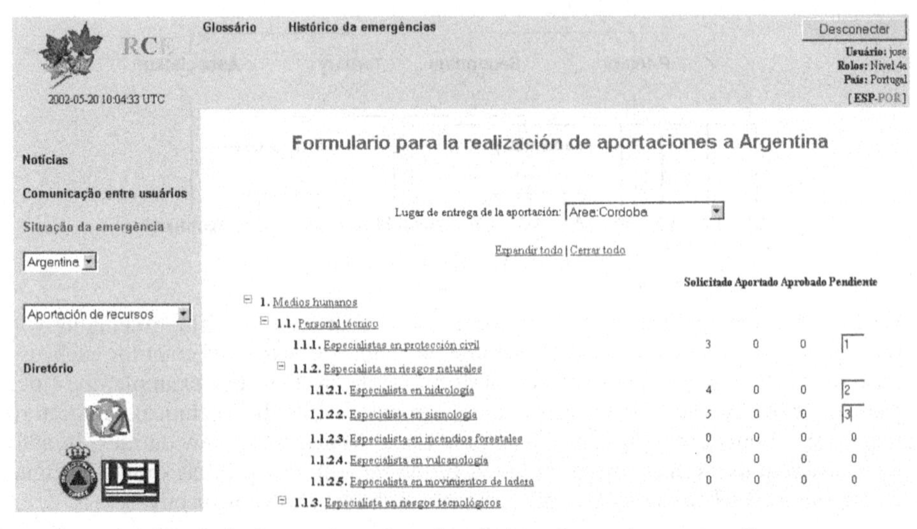

Fig. 3. Assistance from the point of view of an assistance supplier

Multi-purpose application. Even though the main goal of ARCE is to efficiently support emergency responses in a multinational context, it was considered that in order to improve its utility it was necessary to introduce the use of the application in other situations so that users can get used to it before an emergency occurs. Thus, ARCE has two operation modes: routine and emergency. In the routine mode, the application is used as a communication channel and to post news. There is a communication module that uses information flow policies to distribute messages among the different roles. These policies are aimed at ensuring that users only access the information for which they are authorized [6]. With this purpose, information is categorized into strategic, operational, technical, general and public. Information flows from one role to another according to the rules represented in Figure 4. Thus, when a user wants to send a message, first she has to decide which kind of information is going to send and then she will be able to select who is going to receive it from a list of potential targets. For instance, an L2 role will not be able to send strategic information although she could receive it from the upper level, and she can send Operational information to users holding an L2, L3 or L4 role.

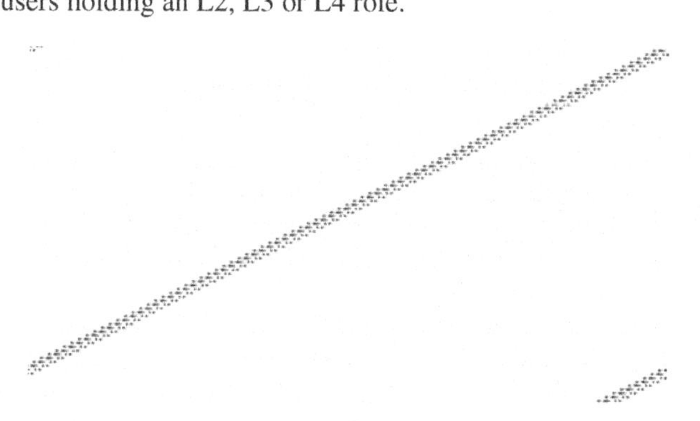

Fig. 4. Information flow policy in ARCE

In emergency mode, countries affected by a disaster can manage the emergency informing to the other associates, preparing a preliminary request for urgent resources, elaborating a more detailed request and coordinating the assistance offered by other countries. Since all the entities involved in an emergency, whether the owners or the assistance suppliers, have access to updated and reliable information about the real needs and how they are being solved, there are no problems of overlapping help. Indeed, before a supplier initiates the protocol to physically send any help, its assistance has to be approved by the emergency owner.

Reliability. This is a basic requirement in ARCE, since users have to trust the information and services offered by an application supporting a critical task. Indeed, users need to sure that information is updated, precise and accurate. All pages include the time of the last update in the system, so that users can have an idea of their "freshness". To avoid improper modification of data ARCE relies upon the use RBAC policies and authentication mechanisms, so that only authorized users can modify the information provided by the system. Moreover, the information flow policy ensures that messages received from the system are trustworthy, as far as only authorized users can send messages to the users who require or need that information.

Efficiency and maintainability. Since ARCE will be used in critical situations users have to receive enough information to plan their response and have mechanisms to provide an efficient response. All associates, through the appropriate role, will be able to create emergencies, ask for help and coordinate the multinational response to avoid redundancies. Moreover, they can invite external organisms or countries to take part in the mitigation of a specific disaster. They will also be able to access to the information concerning to a disaster, for which historical archives are maintained, and offer an assistance that is validated by the emergency owner. In order to increase the system usability, the user interface has been designed applying HCI principles paying special attention to usability and consistency issues [7]. Moreover, in order to improve this collaborative environment an iterative and user-centered development methodology has been adopted as suggested in [1]. Therefore, there is continuous evaluation process involving user representatives in order to empirically test the utility of the application.

3 ARCE Implementation and Evaluation

ARCE architecture relies upon an Apache server running over a number of modules developed on Zope, a platform to generate dynamic HTML pages. A number of modules have been built on Zope to deal with ARCE functions including: emergencies, messages, news, directory and RBAC. Information is held on a PostgreSQL relational database.

In order to assess the utility of ARCE, an empirical evaluation was carried out in the last meeting of the Latin-American Association held in Cartagena de Indias (Colombia) in February 2002. In this meeting representatives of 13 countries took part in a disaster simulation exercise and they used ARCE to coordinate a multinational response. In order to collect information on the system utility and usability, evaluators were asked to fill a questionnaire which was organized in three sections.

The first one ("Personal Information") was oriented towards gathering information on the evaluator command on technical aspects required to use ARCE, such us web browsers and e-mail clients.

The second one ("ARCE utility") gathered information concerning performance and usability, such as utility, speed, appropriateness of the roles structure implemented, reliability, legibility, quality of the user interface, and their degree of satisfaction after having used ARCE. They were asked to rate these features using a Likert-scale with five values, ranging from very good to really bad.

The last one ("ARCE operation") included questions about the different tasks they had performed during the exercise. In particular, evaluators expressed their opinion concerning five operation scenarios:

1. Creating a new emergency, that implies to create a preliminary report to inform the associates, to create a preliminary ask for assistance and to create a detailed ask for assistance. This task is performed by an emergency owner using ARCE in emergency mode.
2. Managing an emergency, that consists of modifying the ask for assistance; managing the proposals of assistance (accepting, denying or modifying them); accessing to the ask for assistance status and to the historical archive of an emergency. This task is performed by an emergency owner using ARCE in emergency mode.
3. Contributing to the disaster mitigation, that implies accessing to the ask for assistance (resources required and supplied), proposing a specific assistance and modifying a proposal of assistance. This task is performed by an assistance provider using ARCE in emergency mode.
4. Using the directory, that consists of accessing and modifying it. This task is performed by any user using ARCE in routine or emergency mode.

5. Using the communication through messages. This task is performed by any user using ARCE in routine or emergency mode. For each scenario, evaluators assessed the difficulty, utility and correctness of the different tasks they had to do to achieve their goals. They could also propose modifications to improve the system.

Evaluation results showed that users considered ARCE a quite useful tool. Thus, as it can be seen in Figure 5, most users were satisfied (80\%) or quite satisfied (20\%) with the application and all of them considered it as a useful tool to coordinate a multinational response. The most important conclusion of this experiment was that ARCE gave place to a positive attitude from the different organisms represented in the Association and reinforced their commitment to make use of web technology to orchestrate their efforts in an efficient way. Moreover, this simulation exercise was considered as a key activity in the iterative design of ARCE, since it has provided quite useful information to improve the application and, particularly, to take into account the specific needs of each entity of the Association.

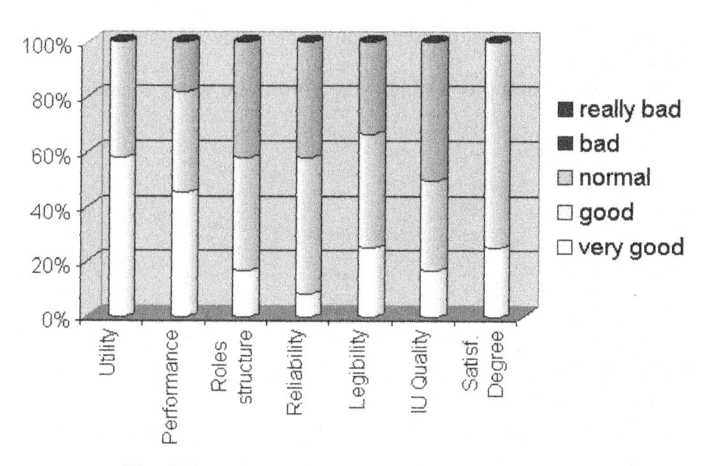

Fig. 5. Results for the part B of the questionnaire

4 Conclusions

In this paper, we have introduced ARCE a system using the web and technologies such as RBAC or information flow control policies in order to make more efficient the multinational response to a specific disaster. ARCE tears down distance and time barriers to make easier the management and coordination of mutual assistance in a multinational community as it is the Latin-American Association of Governmental Organisms of Civil Defence and Protection.

Acknowledgements. This work is supported by an agreement between Universidad Carlos III and "Dirección General de Protección Civil del Ministerio del Interior". RBAC module is based on the MARAH project funded by the "Dirección General de Investigación de la Comunidad Autónoma de Madrid y FSE" (07T/0012/2001). We'd like to thank Juan García and Miguel Àngel Hernández for their cooperation in the implementation of ARCE.

References

1. Padula, M, and Rinaldi, G.R. Mission-critical web applications. Interactions July 1999 Volume 6 Issue 4, 52 - 66
2. Bouvin , N.O., Designing user interfaces for collaborative web-based open hypermedia Proceedings of the eleventh ACM on Hypertext and hypermedia May 2000, 230 - 231
3. Han, R. Perret , V. and Naghshineh, M.. WebSplitter: a unified XML framework for multi-device collaborative Web browsing. Proceeding of the ACM 2000 Conference on Computer supported cooperative work. December 2000, 221 - 230
4. Ferraiolo, D.F., Barkley, J.F. and Kuhn, D.R.: A Role-Based Access Control Model and Reference Implementation within a Corporate Intranet. ACM Trans. on Information and Systems Security, 2(1), February (1999), 34-64.
5. Díaz, P., Aedo, I. and Panetsos, F.: Definition of integrity policies for web-based applications. Integrity and Internal Control in Information Systems: Strategic Views on the Need for Control. Ed.s. van Biene-Hershey, M.E. and Strous L. Kluwer Academic Publishers, USA 2000. 85-98.
6. Denning, D.E.. A lattice model of secure information flow. Comm. of the ACM, 19(5):236-243, 1976.
7. Rubin, J. Handbook of usability testing. New York: John Wiley & Sons. 1994.

KIWI: Building Innovative Knowledge Management Infrastructure within European Public Administration

Lara Gadda[1], Emilio Bugli Innocenti[2], and Alberto Savoldelli[1]

[1] Politecnico di Milano, 20133 Milano, Piazza Leonardo da Vinci 32, Italy
Tel. +39 02 23992796; Fax. +39 02 23992720;
lara.gadda@polimi.it
[2] Netxcalibur, 50123 Firenze, via Alamanni 25, Italy
Tel. +39 055 285859; Fax. +39 055 285760;
ebi@acm.org

Abstract. The paper is composed by two parts. The objective of the first part is to define a new approach to the innovation process in the Public Administration. In fact, in more recent years, the attention to process oriented change management techniques has also emerged in the public sector, through attempts to draw from private sector, searching for new methodologies and managerial approaches that can satisfy the need of organisational innovation. KIWI project analyses these techniques in one of the public management processes, the Knowledge Management. The second part of the paper is more related to the new IST tools that should be used to improve the efficiency and the effectiveness of non-profit organisations. In particular, the project aims at developing innovative, user-relevant, wireless technologies which make the relationship between PA and citizens easier.

1 Introduction

In the last few years, industrialised Countries faced up the necessity of reforming Public Administration, a crucial problem since that context is quickly evolving. The change in progress is moving along two directions: on one hand, the users require a Public Sector's "product" risen in value and, on the other hand, there's the need to provide better services using the same resources (more trend towards efficiency) [1].

Public Sector reform started with the adoption of a new set of rules. The progressive shift from a slow and inefficient bureaucracy toward a lean and dynamic organisation is one of the topic joining a number of interventions of the last decade, within the public sector administration all over the world. Though this deep change has been faced in each country in different ways [2], the fundamental characteristic is the presence, finally contemporary, of two separate trends: *the decentralisation and the modernisation.*

The mere political rules' transformation is not enough and it is necessary to develop a specific method in order to enhance organisations' performance, efficacy and efficiency [3]. The technological innovation and web oriented technology are the necessary starting point for improving PA performance. Therefore, there is the necessity to use a "change management" which should combine with information technology, change of organisation and human resources management [4].

R. Traunmüller and K. Lenk (Eds.): EGOV 2002, LNCS 2456, pp. 223–229, 2002.

2 The Innovation Process in the Public Sector: The e-Government

The change management is guided by four factors: the crisis of bureaucratic model, the monitoring of public expenditures, the pushes coming from the dissatisfaction of the citizen-user and the European integration. The advantages lead to a greater flexibility in activities management, to the possibility to eliminate the no added value activities and to an higher knowledge of the managed activity [5].

The main aim of PA is to be able to answer to the needs of citizens and enterprises. In the previous time the citizen was the receiver of services and public activities, now he is the client of a modern and efficient system which offers certain services [6].

In this context KIWI project is aimed at developing innovative knowledge management (KM) infrastructures able to transform *PA* at any level inside Europe into knowledge driven and dynamically *adaptive learning organisations* and empower *public employees* to be *fully knowledge workers.*

The expected results from KIWI project are:

- *Three re-designed Knowledge Management processes* related to large public administrations in Italy, Finland and France;
- the *KIWI Toolkit* designed to facilitate the implementation of Knowledge Management in large, multi-site, public administrations, based on the *Knowledge Warehousing tools* and the *Mobile Collaboration tools*.

Therefore, the project outcomes are to make a decision making faster and better informed, to improve the customer services, to increase the returns on investment as productivity, to save time and staff resources, to push to the innovation which is further stimulated by capitalising on knowledge and expertise, to reduce the costs.

3 The Key Process: The Knowledge Management

In order to be effective, the innovative process has to be funded on the existence, the development and the integration of the following four leavers related with the innovation key: the *policy*, the *culture*, the *organisation* and the *technology*. These four elements are so linked among them, that it's not possible to act on one without acting on all the others otherwise the effectiveness of the innovative actions is reduced. Along with the leavers, it's possible to identify four engines: *benchmarking*, *project management*, *total quality management* and *digital signature*. Combining the leavers with the engines, it's possible to identify the three main process which characterise the PA innovation: the *Change Management* (CM), the K*nowledge Management* (KM) and the C*itizen Relationship Management* (CRM) [8]. KIWI project is focused on the Knowledge Management (KM).

The knowledge is a combination of experience, values, information and specific competence which provides a framework for the assimilation of a new experience and new information. The knowledge comes from the information and, at the same time, the information comes from the data. So, the knowledge could be obtained by information through the instruments of the comparison, the consequences, the connections and the conversation [10]. But the knowledge is not an information enriched with content, it implies a judgement, involves the values, the emotions and the people perceptions, which generate a relevant impact on the available knowledge in the organisations [11]. The knowledge has to be correlated to the action because it is always a knowledge towards an "objective".

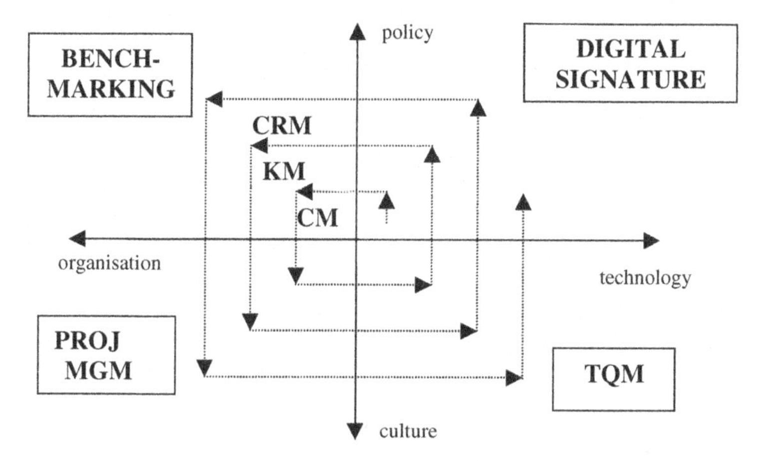

Fig. 1. The innovation sources for implementing e-government processes in public organisations

The knowledge could be classify in three main macro-typologies [12]:

- *declarative knowledge* (knowledge "about" something);
- *procedural knowledge* (knowledge of "how" something occurs);
- *causal knowledge* (knowledge of "why" something occurs).

There's a more general classification of the knowledge. The *explicit knowledge* is codified, expressed by formal and linguistic modalities, easily transmissible and conservable which can be expressed through words and algorithms. For Nonaka and Takeuchi (1995) this kind of knowledge is formalised, easily communicated and shared. The concept of *tacit knowledge* was introduced by Polanyi, who evidences the importance of a "personal" method to build the knowledge, influenced by the emotions and obtained at the end of a process of active creation and of organisation of the individual experiences [13]. It is difficult to define in a formalised way, it is linked to the reference context and it is personal. To be able to spread the tacit knowledge inside the organisation, it needs to convert it in words and numbers which could be understood by everybody. During this conversion, from tacit to explicit, the organisation knowledge is created.

Each organisation has a unique asset of knowledge and its internal problems. Each action of *knowledge management* has characteristics which are specific of the body for which it has been projected. The organisations could assure KM procedures oriented to the results and the strategic needs of the context where they operate, focusing the attention on the planning and the carrying out of the following areas:

- *process*: it assure the *KM* is aligned with the specific managerial processes.
- *organisational dynamics*: they over the barriers which obstacle the sharing of the knowledge and promote the innovation behaviour.
- *technology*: it allow people to share the activities using known instruments

The knowledge has also to be created. The mechanism of the knowledge creation is the condition and the "engine" of the innovation, in the two dimensions of the business innovation and the social innovation. The key point is the mobilisation and the migration of the knowledge across the organisation: it's necessary to rethink the capability of the organisation to encourage the relationships among people, or at least not to put obstacles in their way. The KM process is classify in:

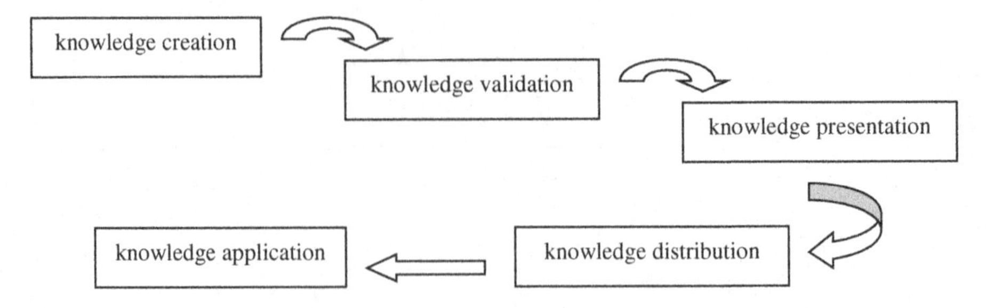

Fig. 2. The five phases of the KM process

The main aim of the organisation is to transfer the knowledge from tacit to explicit, from individual to collective, from collective to organisational:

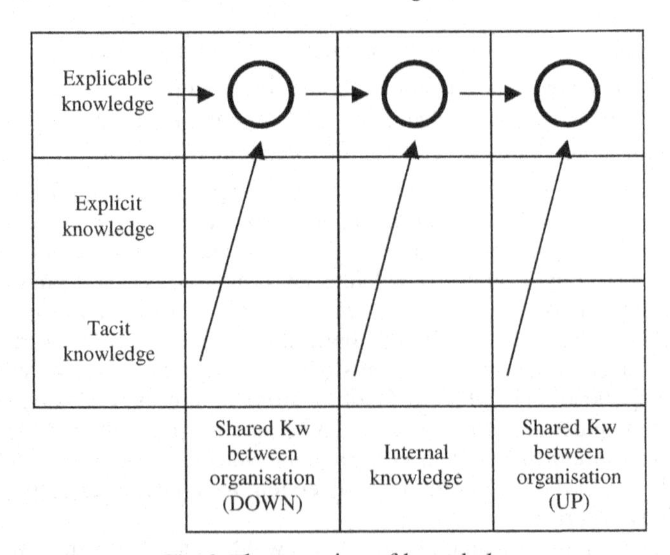

Fig. 3. The mapping of knowledge

4 The System Architecture

The KIWI platform will be structured around the following items:

- A web-based *Intranet Knowledge Warehousing* Toolset that will allow to build a wirelessly accessible knowledge warehouse. The knowledge warehousing will amass *internal* and *external knowledge*;

- *A Mobile Collaborative Environment* (built on Web-based Groupware tools), to support a realistic collaboration and knowledge sharing and transferring also among geographically distributed workforces, within and between public administrations. It will represent the convergence of technologies such as multimedia document/image management, videoconferencing, and mobile 3G technologies helping public administrations transcend all sorts of boundaries (geography, time and organisational structure) by *making available the right information to the right employee at the right place and at the right time.*

4.1 The Components

The KIWI platform is based on a *Intranet Knowledge Warehousing* Toolset. These tools will allow to build a wirelessly accessible knowledge warehouse (knowledge resources will includes manuals, letters, responses from citizens/companies, news, technological, organisational, legal and other relevant information from administrations, as well as knowledge derived from work processes) applications that support inter-organisational learning process.

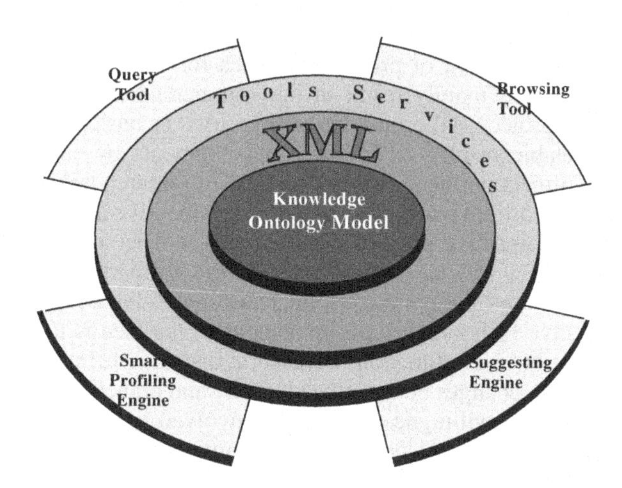

Fig. 4. The KIWI Knowledge Warehouse

The second Tool which composes the KIWI platform is the *KIWI Mobile Collaborative Environment.* The Groupware can help Public Administrations transcend all sorts of boundaries. Geography, time and organisational structure fade in importance in the groupware-enabled process. At the same time, mobile technologies are increasingly penetrating businesses, offering anytime/anyplace access to enterprise information. The novelty of the KIWI approach lies in the convergence of these two technologies.

The challenge of Mobile Groupware Tool development within KIWI is twofold:

- firstly, to allow for web-based, mobile groupware by supporting both *synchronous* (ie, real-time interaction such videoconferencing, chatting, electronic whiteboard, etc.) and *asynchronous* (ie, email, group discussions, etc.) communication, including live video/audio/text communications;

- secondly, to take full advantage of the Knowledge Warehousing tools, by integrating mobile groupware facilities within the KIWI Toolset. *This will lead to set up of a full Mobile Collaborative Environment.*

The project will look at the recent advancements in the mobile groupware standardisation such as the GroupDX (www.groupdx.org) initiative (linked to the WAP Forum), proposing an industrial grade XML Document Type Definitions (DTDs) and Object Schemata for Internet groupware applications. This new standard, called Groupware Mark-up Language (GML), is being established in order to facilitate data exchange among the various groupware applications, and facilitate data synchronisation among groupware applications and their individual counterparts.

5 Conclusions

The KM will result into a decisive improvement in inserting an information database. This allows the public employees to access easier to the needed data, independently from the place where they are. The aim is "to bring every citizens, home and school, every business and administration into the digital age and on line, creating a digitally literate Europe, build consumer trust and strengthens social cohesion".

Concerning the exploitation of project outcomes for the industrial component of the Consortium, this will mainly result in the commercialisation of the prototypes produced within the project. All prototypes will be used as basic elements to develop and produce marketable results: as in-house developments by each partner and in collaboration with project partners. A quick process of research transfer in production will assure to the Consortium partners an essential competitive advantage for a further consolidation of the respective positions on the market. The specific techniques implemented in the project will be used by most of the Partners to enhance the techniques already in use, contributing to consolidate a competitive advantage.

Once the innovative technologies are implemented, the idea is to realise a mobile groupware which let in-house functions be used at distance. It involves an organisation and a management changes between headquarter and branches.

Within this structure, the willingness of the PA involved in the project to provide a common exploitation of the project results, constitutes the cornerstone of the KIWI Exploitation Strategy.

References

1. AIPA, "La reingegnerizzazione dei processi nella pubblica amministrazione", (1999).
2. Klages, H., Loffler, E., "Administrative modernisation in Germany – a big qualitative jump in small steps", International Review of Administrative Sciences, pp. 373-384, (1995).
3. Hammer, M., Champy, J., "Reengineering The Corporation: A Manifesto for Business Revolution", HarperCollins, New York, (1993).
4. Osborne, D., Gaebler, T., "Reinventing Government. How the entrepreneurial spirit is transforming the public sector", New York, Plume, (1993).

5. D'Albergo, E., "Le sfide delle amministrazioni pubbliche al Change Management: una prospettiva socio-istituzionale", Rivista trimestrale di scienza dell'Amministrazione, n.1, (1998).
6. Halachmi, A., "Re-engineering and public management: some issues and conditions", International Review of Administrative Sciences, vol. 61, (1995), pp. 329-341.
7. Kettinger, W.J., Teng, J.T.C., Guha, S., "Business process change: A Study of Methodologies, Techniques, and Tools", MIS Quarterly, March, (1997).
8. Sbrana M., Torre T., "Conoscenza e gestione del capitale umano: la learning organization", FrancoAngeli, (1996).
9. Davenport T. H., Prusak L., "Working knowledge: how organizations manage what they know", Harvard Business School Press, (1998).
10. Takeuchi H., "Beyond Knowledge Management: lessons from Japan", Giugno, (1998), dal sito internet www.sveiby.com.au/LessonsJapan_it.html.
11. Zack M., "Managing Codified Knowledge", Sloan Management Review, Summer, (1999).
12. Polanyi M., "The tacit dimension", Routledge and Keagan, (1966).

Elektronische Steuer Erlass Dokumentation a Documentation on Official Tax Guide Lines

Viktorija Kocman, Angela Stöger-Frank, and Simone Ulreich

Bundesrechenzentrum GmbH, Department A-VA-DX,
Hintere Zollamtsstr. 4, A-1030 Wien
{viktorija.kocman,angela.stoeger-frank,
simone.ulreich}@brz.gv.at

Abstract. E-Government can be defined as "carrying out government business transactions electronically". One aspect of e-government are online law documentation systems. In the Austrian Federal Dataprocessing Center (Bundesrechenzentrum), we have two online law documentation applications under development and support: Electronic custom-law documentation system (EZD) and electronic tax-law documentation system (ESED). EZD has been running since 1995, ESED is currently in a prototype state. The customer of both systems is the IT-section of the ministry of finance. These two applications have been developed especially for the use in the administration and are available for the staff in the Austrian government intranet. In the concrete paper, we will discuss technical and organisational aspects of the online tax-law documentation system ESED which in the first step will be used in the tax department of the ministry. The ministry of finance is responsible for ESED project management. Bundesrechenzentrum is responsible for the technical realisation of ESED.

1 History

The Federal Ministry of Finance, Austria's highest administrative office in tax affairs, is authorized in executing tax legislation. The applying tax rules are decided by the ministry. Further more it is necessary for all subordinated offices to know all those legal decisions. But various forms of information transfer are existing: paper, telefax and e-mail. So, no standardizised form exists to ensure that the information flow is done correctly and just in time. Another important fact of handling information is how to store them. Each office has to handle that by itself. Some of them collect the tax rules as paper sheets, others catalogue them or run an insufficiant electronic database in the office.

2 Declared Aims

Under these circumstances it was ordered to improve the information transfer to standardize the running out and, in addition, create a central archive for searching. Less paper, less time and certainly lower costs were demanded to be used as before. The usability of the new system should be extremly comfortable and trained easily. Additonally, knowledge management should be established within the complete finance administration throughout the whole country.

R. Traunmüller and K. Lenk (Eds.): EGOV 2002, LNCS 2456, pp. 230–233, 2002.

A working group was installed constisting of members of the ministry and tax officers of the subordinated offices. Just in case, BRZ supplied technical support. As a result a paper of demands was provided.

BRZ started the technical realization.

3 Technical and Organisational Aspects of ESED

Basically, there are 2 ESED-user groups: document writers and document readers. These two user groups can overlap. ESED writers work with a sophisticated Winword client (enriched with ESED Macros and ESED ActiveX components). ESED readers can query the documents with the ESED web client.

In general, ESED is a document database that is currently being updated with enactments. The amendments of enactments are also updated in the database, in order to acquire the relevance of the information.

ESED consists of a central database which stores the documents. We use a document database TRIP, which differs from relational databases. The emphasis in document databases lies on accurate (fragment) letter indexing to enable fast text search. The disadvantage of TRIP is that it cannot map relations that well as relational databases can.

ESED documents are organised in document collections which are referenced by a unique ID (Stammnummer). The documents are further divided into segments in order to give long text a structure. On each of these three levels (document collections, documents and segments) the ESED writer categorizes the document.

One of the used standards in the austrian administration is Winword. This was the guide value in designing the application. The input application had to be a Winword client. All of the ESED documents have to be written in the import ESED Winword client, and an easy import/export functionality has to be designed in order to raise the user acceptance and enable law experts to work on their private PCs outside of the ressort and the ministry of finance intranet.

The documents have a complex categorisation scheme. The categorisation scheme is specifical for the austrian tax law. The unique registration number (Geschäftszahl) is a unique ID for the file. Since ESED consists not only of tax enactments which have a registration number, but also of other document types like "announcements" which don't have a case number, we use another unique identifier for the database (Stammnummer). Another important attribute is the category "quota" (Norm), which is used to refer to citated statute. Each document can have more quotas. These quotes are later provided with hyperlinks which link to extern legal databases or to other ESED documents.

The TRIP database is located on UNIX, so there is a complex processing between the local PC and the database. After a document has been released for the ESED database, it is tranfered to a NT Server. NT Server has some conversion programmes running. The main assignement of the NT Server is to convert RTF documents to XML and to establish a communication to the unix database in order to fill the database with ESED text and categories.

This solution enables ESED documents to be processed immediately after the user has decided to release the document for the database. Seconds after the document release the text can be quieried with the web client.

4 Lessons Learned

E-government establishes special requirements on the information technology, especially when the application is designed to be used at the administration internally. There are more aspects to be considered as the mere technical concept.

Our experience has been that in order to gain acceptance by the administration staff, special emphasis has to be given to the *usability* of the application. It has to be considered that the administration staff is a special user group, less flexible in giving up their working habits than an "ordinary computer user". Much more, in the prototype stage it has to allow inaccuracy and individual user behaviour and find concept to cope with them. Once the user group got used to the process improvement enabled through the application and has lost their fears, it is also possible to require compromises from the users in order to win on processing efficiency.

The prototype stage of an e-government project designed for internal administration is crucial for the project success.

5 Prototype

Now a prototype already is running and it seems that the new system meets definetly the expectations: Each tax officer is registrated by using ESED and gets therefore only the documents that are due for his field of work. In addition it is possible to search in the database in various ways. A catalogue of subject, law and catch phrases make it easy finding the right documents. Besides, a query mask with different searching fields offers a comfortable and efficient use. Novels are specially marked, long documents are splitted but could in case be joined together.

A lot of ESED-documents reference to other ESED-documents or to documents of another legal database (for instance RIS-Rechtsinformationssystem). That is why the documents are connected as hyperlinks to get the further information.

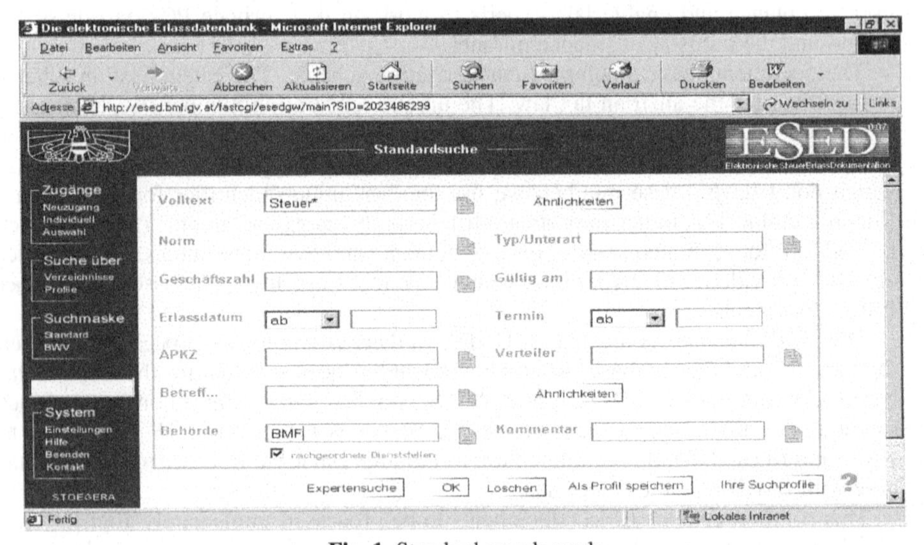

Fig. 1. Standard search mask

6 Prospects

Tax rules as well as important messages by the ministry will be recorded. To establish as a platform of communication the subordinated offices will offer their messages as well. Finally information-transfer will have no longer a one way direction.

Voting in the New Millennium: eVoting Holds the Promise to Expand Citizen Choice

Anthony Watson and Vincent Cordonnier

Professor, Edith Cowan University – Perth, Australia , Professor, Université des Sciences et Technologies de Lille – France and Adjunct Professor at Edith Cowan University.
a.watson@ecu.edu.au, vincent.cordonnier@univ-lille1.fr

Abstract. E-voting is not the same as e-democracy it is however a tool with the potential to help distribute rights in the voting process. The process may well prove to be a new 'killer application' for the Internet and a suitable authentication tool for voting security. A previous model outlined by the authors based on the use of voting smart cards and the Internet is expanded detailing concerns at the operational level and providing alternative solutions for security and the rigors required for voting scrutiny.

1 Introduction

The potential to invigorate the democratic process on a large scale through inclusivity and improved access based on adapted Internet technology may well change the way some social structures operate for the next generation and in particular the voting process. Accenture [1] suggests eDemocracy embodies these stages : 'Citizens access information …, Decision making and influencing politicians, …Voting electronically'. This paper focus on the latter aspect using a smart card based option.

Democracy exists when people who are concerned by a decision may participate in the making of that decision with many systems offering a mixture between direct and indirect voting alternatives (Watson and Cordonnier, [2]). In most cases the direct option would be preferred as the voter has their individual choice recorded as opposed to a delegated indirect vote. The use of information technology has the capacity to overcome some of the aspects of cost, time and distance in the voting process.

In Australia voting rates rose from 57.9% in the non compulsory era to 96.2% at the last election in the compulsory period [3]. In the US voting is by self direction and Kantor [4] suggests that the use of e-voting could have an impact and "help overcome the low voter turnout rate 44.9% in 1998". On the other hand it would appear that a postal voting option in Switzerland did little to improve the voting participation rate at a lowly 42.3% [3]. Voting from the comfort and security of ones home may have appeal in hot or cold climates or those areas with street violence or geographic issues.

Alexander [5] says "Casting a secret ballot in a fair and democratic election is, in fact, unlike any other kind of transaction" and it is true that the failure or abuse of the process can be more significant than a bank fraud. Recognising this it is suggested e-voting is feasible using a combination of smart card, cryptographic techniques and Internet technologies with a focus on authentication and security of data transport and storage.

R. Traunmüller and K. Lenk (Eds.): EGOV 2002, LNCS 2456, pp. 234–239, 2002.

The architecture of the e-voting system proposed is composed of five parts: i - Smart card for each citizen who may vote; ii - Terminal with card reading capacity and communications; iii - An acces to the Internet in some form; iv - A server to collect and manage the votes; v - A component which is the organisation process itself responsible for the voting delivery technology (pencil and paper or smart cards and the Internet).

2 Using a Smart Card As the Key to the Voting Process

Information technology designers recommend smart cards as a component of a system if the required service addresses a large and distributed population where each individual may require a personal treatment according to a specific profile and particularly so, if a high level of security is identified as a major issue of the application. Voting is an ideal domain of application for smarts cards in a technical sense but it is realised that any changes to an established voting practice is likely to encounter political and possibly social resistance.

Watson and Cordonnier [2] presented a scheme for what a digital voting process must offer and in [6] described security index options. The present goal it to address a limited number of issues including: authentication of the voter; verification of the right to vote; security of vote transmission; anonymity of the voter; guarantee of the vote counting process; possibility of reverse verification of some stages.

3 The Role of the Card

To operate effectively the smart card must authenticate itself and must verify in some way that it is used by the legitimate owner. The latter role is difficult .

Many applications already operational in the public area using a smart card use these important features of secure card technology: capacity to lead a mutual authentication with the server; identify the person who uses it (Authentication of the voter); sign an electronic document verifying the transaction; hide the message (encryption).

The way a card leads mutual authentication with the server is well known and not that different from what is used in many other applications.

It is more difficult to implement a satisfying solution to identify the voter. Many applications use a PIN code. This is possible if the card is used frequently however a voting card may be used only once or twice a year as an average. It is then likely that some voters will forget this code with, more abstentions as a consequence.

The most logical conclusion is to use biometrics where the user can be identified by a biological profile such as fingerprints, shape of the hand, characteristics of the eye. Most possibilities require an expensive special device to capture and analyse these biometric data. The argument is important if people are authorized to vote from their home. There is research suggesting Iris scanning technology, protected software and a Web camera could represent an effective and cheap solution.

A voice signature option is suggested because: voice identification techniques offer a reasonable level of security. Voice identification may be dynamic by asking the voter to repeat a word or a phrase he cannot prepare or copy on a recorder; most of the possible terminals as PC have a microphone and a loud speaker; it is even possible to implement these devices in the thickness of the card itself; voice is the simplest and most spontaneous means of expression of an individual; people who cannot easily read or write are not excluded by any other sophisticated identification process.

Another approach is to verify through the Internet that a voter is able to correctly answer a set of questions that relate to their life, family, job, holidays, etc. The method requires the data must be registered prior to the vote and stored in a secure environment, possibly the card itself. Some authors consider this solution as a good balance between security goals and access facilities, even for persons who do not have a high educational level. The system accepts a certain level of mistakes, can be personalised to any particular profile and is fast enough. As the proper responses are stored within the card, there is a limited risk for privacy, especially as any access performed is an internal comparison and then validated by a voting session.

4 Extended Roles of the Card

Microprocessor smart cards provide functionality beyond a static paper document, allowing for the overall architecture of a voting system which may offer new facilities or options. The card is responsible for driving the voting protocol. Voting at a distance implies the roles attributed to a voting officer must be delegated to another and only the card is secure enough to ensure this role. Furthermore as it holds most of the information to be protected, it is better to prevent this information leaving the card by giving the card itself a major role in the global management of the process.

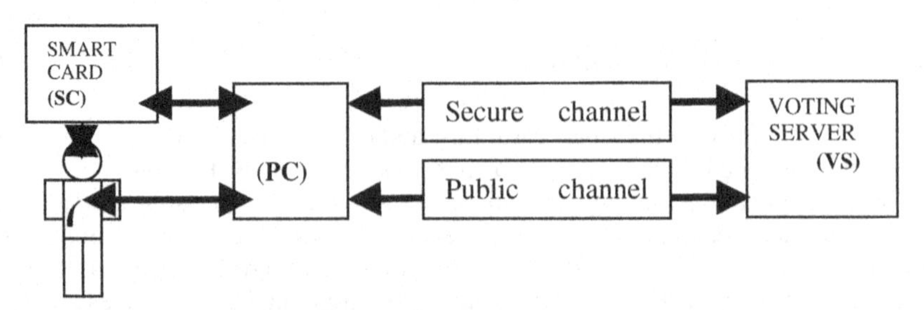

Fig. 1. representing both the voter and the voting bureau

Stage 1: The voter has to login to the server through the usual public channel importing a program that comprises interfaces with the voter, the card; and various electronic forms and documents to be used. The end of that stage is a request to the voter to present his card to the reader.

Stage 2: Is the identification process of the voter; it will use links from the PC that are not secure but the final decision to proceed will not allow any of the references (PIN code, prints of any profile) to be moved out of the card. The card responds through the PC in encrypted form allowing the voting process to continue. This is the point at which a particular token is identified.

Stage 3: May be considered as the informative time to propose to the voter the set of possible choices. It will use links from the voting server via the PC in a public environment. This stage may be personalised to the voter.

Stage 4: The vote is performed on the PC and data immediately moved to the card for encryption.

Stage 5: Is the transmission of the vote details to the voting server via a secure channel (meaning encrypted data).

There are two assumptions, which are crucial, to ensuring this model works:

1. each card uses a unique internal identification number determining that no 2 cards are the same;
2. the card itself is secure. This is an assumption we make with financial and health cards already and is usually a matter of degree.

5 The Voting Server (Verification)

The voting server may be local or global and it is responsible for managing, in real time, all the aspects of the voting session including server security, time stamping, dissemination of voting information and any voting variations.

The first role of the server is to replace the voting bureau with functionally to: verify the identity of the voter; provide voting material and information; receive the vote; date and time stamp votes and ensure the effective closure of the voting process for a given session; collect and aggregate the votes; and ensure annonymity.

The model proposed here should be seen as a framework and may be modified for national regulations There have been concerns expressed regarding the capacity to store individual voting transactions and 're-engineer' the vote to relate a particular vote to a particular individual. The model proposed here would prevent this. The Report from the National Workshop on Internet Voting (2001, [7]), suggests some specific design options for servers using protective firewalls and an archival system (P.7).

6 The User

Authentication of the user is a critical aspect and must be balance: prevention of another person using the card; simplicity of the associated device that allows multiple voting sessions; and capacity for any voter to easily provide the expected proof for identification.

Some countries authorise voters to use proxies. If a person A has to vote on behalf of a person B, the simplest solution is for B to temporarily give his card to A. Then the identification process would require B to give A the whole set of identification data. If A and B may attend together a trusted office with their cards, the transfer could be done prior to the vote. B will authenticate his right to vote by an appropriate identification and will transfer his voting capability to A. Then, at the voting time, A with his own identification process will be permitted to vote twice.

7 The Card and the Voting Session (Transaction)

It is possible to limit the role of the card to an authentication function but it could do more and it is these options which could distinguish the differences in e-voting systems. Firstly, the server will verify that the owner of the card is authorized to vote for that session; this is the result of a comparison of what is described in the card and the profile of the vote situation. For example, for a city council vote, the server will work on behalf of the city hall and verify that the owner is authorized to vote for that city for this particular voting instance.

Most of the voting models need to verify that a citizen has voted but nobody may know the content of that citizens vote. Manual systems are open to abuse with multiple votes recorded against the same name in different venues. This proposed model incorporates alternatives to ensure annonymity and a single vote check.

In addition to the authentication function the next requirement to be implememted in the card is a session token. Actually, as the card will be delivered for more than one session, it will be initially loaded with as many tokens as feasible and appropriate, thus ensuring a single card may last for several years. Each session will be given a number and the corresponding token will be used just once. This prevents an attempt to use the card more than one time for a session. The token will also provide a unique link between a card and a voting session.

The voting process will produce a voting array for that session and will be cyphered and transmitted to the voting server. It is not necessary to communicate the name or any id of the voter. However some countries, such as Australia, apply penalties on people who do not vote. It is preferable, then, to transmit some identifier to another server to ensure a token has been used and to make sure than there is no possible link between them. The single vote check is determined at the time of the sessiona and the token for that voting process erased preventing repettitive voting by a single card/voter.

8 Issues for Consideration

There are some technical matters for consideration in the actual moment of voting and although extreme need to be considered. The first must address the issue as to whether the person selecting a vote on a screen-based system is actually seeing the correct information. For example ensuring that the candidates are in the correct order on the screen. There are audit check mechanisms and the use of pre printed voting material, which should make it extremely difficult to fraudulently represent the information.

Another issue is to prevent 'dead people' voting and events of this nature have been recorded regularly in Italy, USA, France and Australia. The voting process is defrauded when a person pretends to be someone else and presents a vote in that name. It would be possible although difficult to obtain and use a deceased persons smartcard to register a vote. The smartcard option also allows for a defined life of a card. For example the card life could be set at 2 years which would limit the period of the fraud. There is also the argument of proxy vote by smartcard including voluntary and involuntary. The voluntary case could occur when an aged or infirmed person gave their voting card and authorization to a third party to vote on their behalf. The involuntary case occurs when a person is forced to hand over their voting smartcard and identification under duress to another person who uses that vote. This is a criminal activity and should be dealt with by law. In Australia there has been cases of electoral abuse through duress on individual voters and the law was changed to require a secret and scrutinized voting process in unions for example.

There must be a process to deal with lost or damaged voting cards. This is already the case with financial and health cards and is a systems management issue rather than a technical model matter.

9 Conclusion

One of the difficulties in accepting an e-voting approach may arise from social rather than technical issues including possible suspicion of the process and owners. There is still a test of time yet to take place which will look at the various brands of cards for security but the role of a smart card could lead a significant step forward in creating a remote eVoting secure system and the possible benefits that may bring.

References

1. Accenture, (2001), eDemocracy, http://www.accenture.com/xd/xd.asp.
2. Watson, A.C., Cordonnier, V., (2001), *Information Technology Improves Most of the Democratic Voting Processes*, Twelfth International Workshop on Database and Expert Systems Applications, IEEE Computer Society, Los Alamitos, California, p388-393.
3. Department of The Parliamentary Library, Research Note 29 1996-97, Parliament of Australia, http://www.aph.gov.au/library/pubs/rn/1996-97/97rn29.htm
4. Kantor, J., (1999), Obstacles to E-Voting, http://slate.msn.com//netelection/entries/99-11-02_44394.asp
5. Alexander, K. (2001), *Ten Things I Want People to Know About Voting Technology*, http://www.calvoter.org/publications/tenthings.html
6. Cordonnier, V., Watson, A.C.,(1997) *Access Control Determination of Multi-Application Smartcards Using a Security Index*, Third Australasian Security Research Symposium, July 1997, Brisbane.
7. Internet Policy Institute (2001), *Report of the National Workshop on Internet Voting : Issues and Research Agenda*, March 2001, National Science Foundation, University of Maryland.

e-Democracy Goes Ahead. The Internet As a Tool for Improving Deliberative Policies?

Hilmar Westholm

Center for Computing Technology, University of Bremen,
Postfach 330440, 28334 Bremen, Germany
westholm@tzi.de

Abstract. At first, important features of the current situation of citizens' participation in political procedures such as reasons for political apathy, political reactions with new ways of direct participation and the increase of communication-based means of participation are summarized. In the second part, "information", "consultation" and "active participation" are described as main components of non-organized citizen communication which must be embedded into users' and administrations' environment not only technically, but also economically, legally, organisationally, motivationally, and politically. Two case studies illustrate opportunities given by online-participation and underline the requirements to use online participation as a supplement, not as a replacement of traditional participation and to combine the advantages of various online and offline ways of citizens' involvement in a multi-channel approach.

1 Introduction

The situation of e-democracy usage has to be described against the background of formal as well as informal means of (offline-) participation. The current situation can be described with the following features:

Less involvement of citizens in traditional politics but increase in "instant participation". Factors like globalisation and individualisation cause more distance of citizens to the state and its institutions (e.g. administration, political parties). In many democracies, not just participation rates in elections are decreasing, but there is also a decline of membership in civil networks resulting in a precipitous drop in political engagement in general [1] with lower commitment to the political process and less trust in government. [2] Nowadays citizens prefer selective, focussed and limited involvement in political processes (mainly on the local level, often with NIMBY-(not in my backyard) character (i.e. individuals only participate politically if they feel a threat to their existing situation – e.g. if an industrial plant is to be built in their neighborhood) and transparency of political-administrative procedures. Traditional forms of political participation (e.g. elections, lobbying of interest-groups) mostly ignore these demands.

Thoughtful governments are now looking at the Internet not as a threat, but as a positive potential tool to re-engage the citizens in the business of governing. As the Internet has helped to empower a new generation of well informed and demanding consumers, some people argue that it will challenge the essentially passive relationship that the majority of people have with their government and politics.

R. Traunmüller and K. Lenk (Eds.): EGOV 2002, LNCS 2456, pp. 240–247, 2002.

Increasing institutional possibilities for and use of direct participation. In some representative democracies in Europe direct-democratic tools are used more and more – sometimes combined with deliberative elements: Citizens' initiatives, referenda and ballots become part of the constitutional law and, mostly on the local level, normal democratic practice. "Legislatures are increasingly squeezed between the general public and the executive; the new technologies make plebiscite democracy more feasible and this possibility is putting pressure on representative democracy". [3] But these forms of votes are often criticised because they reduce complex political problems to simplifying yes/no-alternatives. [4] Besides, experts agree that ICT-supported voting by "pressing the button" has less importance than innovating the procedures of decisions. [5]

Part of such procedures (as well as indicators) are methods of gauging public opinion like polls, e-mails, or, more sophisticated, focus groups. The increasing use (and number) of these applied methods seem to be relieving the pressure on governments (and their representatives) to develop new and more effective means of public involvement. [6] But it is not finally decided if the main tool will be ICT; as representative experiences from the EU-supported DALI (Delivery and Access to Local Information and Services) project in Sweden showed: "Some politicians expressed doubts about participating in political debate via the Internet and preferred to use the phone when discussing issues with citizens. They felt that a written answer forced them to make a commitment on particular issues too hard, which thus made them more cautious. Also, some politicians felt that some questions from citizens could not be answered in a simple or public way. Such questions typically related to complex or politically sensitive issues, or involved a straightforward misunderstanding on the part of the citizen". [7]

Increase of deliberative involvement methods at various state levels. Most of the countries in Northern and Western Europe created a lot of elements of public involvement, especially in the area of public planning since the 1990s, like "public fora", "working groups", "citizen advisory groups", "caucuses", "visioning workshops" or "round tables", which go beyond legal minimum requirements. The Agenda 21, a document most countries of the world agreed on at the United Nations Conference on Environment and Development (UNCED) in 1992, gave a new input. A main emphasis of this document is to give citizens the opportunity to participate in issues dealing with transformation of societies (and local communities) to ensure sustainable development.

The challenge administrations are now facing is to establish appropriate approaches for engaging citizens as well as for (new) different interest groups in those new policy areas where IT permits the use of flexible and easily accessible means – not only for voting: For example, information can flow fast, it enables citizens to inform others about their opinion in an easy way or makes communication between representatives and voters easier. In regional planning processes, it becomes possible to involve interested citizens by providing shared workspaces with hyperlinks to relevant documents (e.g. maps) and to enable users to make their comments after using a structured filing via an Issue Based Information System (IBIS). Meanwhile interest groups are also better informed, better linked through networks, and, according to a survey in eight OECD-countries, better able to bring pressure to bear, especially on the middle level of bureaucracy. [8]

2 Steps of Participation in the Internet

There is a great hope that IT will improve political communication among citizens (or their interest groups), administrations, politicians and government. "E-democracy" is the keyword to bundle information and public participation and includes activities that follow these steps within the policy-making cycle from policy design through implementation to evaluation[1]:

2.1 Information

Most processes of political participation (should) start with intensive information. This "one-way relation" where government produces and delivers information for use by citizens covers "passive" access to information upon demand by citizens as well as "active" measures by government to disseminate information to citizens. On the state level it is now more or less self-evident that governments provide texts of laws and regulations and structures of their agencies via the Internet like political parties or NGO's do it with campaign-information. It is still a problem on the local level (often due to financial and human resource problems).

2.2 Consultation

This implies a "two-way relation in which citizens provide feedback to government - based on the prior definition by government of the issue on which citizens' views are being sought and requires the provision of information". [9] This happens in the form of opinion polling within city-marketing or as management of complaints. Managed electronically, these measures are cheaper than traditional means (surveys, household-questionnaires, see case study I below). Not representative, but more often used are online speaking hours of political party representatives or chats of politicians (who want to be re-elected). In Denmark, during a campaign in connection with the euro-referendum, cabinet ministers held online chat sessions with people. Resonance is normally rare – in rankings about usage of various online services, chats with politicians are in the last place beside the virtual visit of a museum.

Also for preparation of proposals of national acts or local planning procedures, governments are testing online applications. In Germany for instance, the federal government used the Web to publish the proposals of a Freedom of Information Act as well as another one of a Data Security Law and invited the citizens to comment on them in online forums. In these cases, government received high-quality contributions of the citizens who otherwise maybe would not have influenced the procedure.

A problem of Internet activities in traditional procedures is that they normally do not reach new citizen-groups. In an empirical research in the German town of Osnabrueck, researchers concluded that the function of organising or articulation via the Internet plays a secondary role and those parts of the population who normally do not participate in planning procedures will also not do so in Internet supported procedures. The Internet is a means for the communication-strong elite.

---

[1] Besides, the term e-democracy also includes e-voting, internal organisation among political parties and association, NGOs, etc. as well as new ways of participation like email-bombing.

Case Study I: Online Polling "Traffic Calming"
To gather the public opinion in a controversially discussed issue of traffic calming, both an online and offline opinion polling was launched with a questionnaire (mainly multiple choice-questions, one open question). Citizens were asked about quality of information by the district, if they support or reject the trial to reduce the traffic by installing barriers against non-resident traffic and what they think about the information and involvement policy of the district. Main results were:

- Politicians and administration had a quick feed-back of a large number of citizens' opinions to a controversial issue.
- Also those citizens participated who normally would not do that; the vocabulary was partially very rude.
- Users did not only make "clicks" to the multiple-choice questionnaire but every third user took the open question as an opportunity to express his own opinion.
- The combination of online and offline polling increased the representativity of participants compared to normal Internet-activities which are characterized by social exclusion (in this case, a large number of citizens elder than 60 years replied, only young citizens were represented worse than in reality; compared to other issues related to traffic, participation of women was high with 39%).
- Similar to offline ballot campaigns, interest groups mobilized their followers.
- The polling illustrated that an easy registration procedure to authenticate users is necessary because anonymous clicking enables participating several times.
- The polling process must be logically embedded in the political process (in this case, a citizens' interest group that fought for this issue for two years and had persuaded all the politicians, did not understand that political institutions launched an - even non-representative - polling when the trial started.
- Media were integrated into the polling process; the local newspapers reported about the (new) way of involvement many times thus increasing the number of participating citizens. [10]

2.3 "Active Participation"

A C2G-relation based on partnership with government, in which citizens actively engage in the policy-making process, is called "active participation" by the OECD. It acknowledges a role for citizens in proposing policy options and shaping the policy dialogue – although the responsibility for the final decision or policy formulation rests with the government. [11] Good experiences were made in the arctic Swedish town of Kalix where 7% of the population participated in an online discussion concerning the future development of the city centre. Other good experiences were gathered by the author with an online forum organized for five weeks during November and December which was especially related to the quality of citizens' contributions to the discussion (see case study II below). Also new forms of governance, e.g. the *delphi method*, a long-established approach for developing a consensual position statement through iterations of anonymous comments by a small invited panel, or alternative dispute resolution techniques like mediation can be very well supported by IT (esp. groupware applications) because the possible time-lag (between the contribution and the reading) enables the users to understand contributions and to react after collecting further information from other sources before contribute in a more qualified manner.

Case Study II: Online Forum Horn-Lehe

Within the EU-supported project E.D.E.N. (Electronic Democracy European Network) tools for better relationships between citizens and administrations are developed.[2] Two of them, a tool for online discussions and another one for visualizing maps, were applied first in a discussion about future development of the core district Horn-Lehe in Bremen. The discussion tool enabled anonymous and registered contributions with the right name as well as with a nickname to various issues from urban (district) planning to youth leisure activities. The forum combined information (masterplan and other map, minutes of district board meetings) with discussion, the problem of high expenses for moderation was met with the innovation of moderation by citizens (of the district); besides, it was a multi channel access while online discussion and visualizing maps was accompanied by physical committee meetings (where citizens could express their opinion), household-questionnaire, round-table. Some results were:

- About 6.100 visits to various discussions (not only hits to the homepage) were counted, 67 registered users and a unknown number of anonymous users (some 25-50) participated with 224 postings.
- A large amount of material was "produced", it was a big great effort for decision-makers to read them all (problem of information-handling capacity), an additional effect was that administrative experts became afraid to use the forum-tool for normal procedures.
- Surprisingly two third of the contributions were constructive (good and better) according to a five-point rating list, nevertheless most of the contributions were only expressive, some listening or responding, a small number emphatical and none of them persuasive (in both meanings: being persuaded as well as persuading). The quality of contributions made by registered authors was better.
- The quality of discussions profited by contributions of the head of the district administration and other experts.
- "Citizen-moderators" need to be trained intensively (technically but also "socio-technically" to summarize contributions, to find sensible titles for them (for the index-listing and better use) and to forward contributions to responsible experts for reply.
- A small number of users read the information provided additionally (such as the municipal masterplan, an explanation text to this plan, a zoning-code, minutes of the district committee, etc.).
- The project had a good press feed-back, importance of other PR-means (folder, poster) was small.

3 Adaptability of Applications

The few experiences existing demonstrate that the only technical ability to realize a means of citizen participation through the Internet has no effect on involvement from

[2] Within the E.D.E.N. project (duration until July 2003) tools are developed to improve communication between public administrations and citizens to increase knowledge and therefore support decision-making with urban planning as application domain through the use of natural language processing (NLP) and community tools (like polling and e-consultations) (www.edentool.org).

a content view. The great challenge is to embed these technical constructs into a specific technical, organisational, cultural, motivational, juridical and economical context. In other words: the electronic applications in the so-called virtual world need to become "adaptable" to processes and structures in the real world. [12] Adaptability must be related to the following factors:

From a *technical* point of view, online participation supplies must be integrated in the technical procedures in the back-offices. This is also true for planning procedures as well as for voter-registrations in online-elections. Besides, technical applications must not overstretch the users' equipment (e.g. in the Horn-Lehe forum end-users with modems needed too much time for downloading large files, e.g. for opening the maps).

Another issue are the *economic* framework conditions in an extended meaning of cost-benefit relations both of the public authorities and the users. How can governments both become "open" through practising more and better public involvement in the context of a financial crisis of public agencies and considering the necessity to decrease the role of governments trough outsourcing or privatisation of services? Public consultation is not for free but can be expensive when further requirements are considered, e.g. printing and distribution of popular brochures or the rent of venues.

Political participation (via the Internet or elsewhere) has only a chance of implementation if the target groups are motivated. *The readiness of citizens to participate* has often been over-estimated. According to the standard-model of political participation [13] it's no wonder – only citizens with a good financial as well as educational background and a good position in society believe that they can move something with their own political engagement. There is no reason to believe in a change due to the existence of new technologies. Young citizens will only be involved in planning issues that are related to their interests (such as planning of a youth centre) but not because there is a new attractive technology they use in everyday life. In other words: technically supported supplies must look for adaptability to existing discussions and motivations.

Motivation is also an issue from the point of view *of the internal users* in the administrations: If the personnel-situation is bad, employees will not be motivated to get more work by addressing planning issues via an Internet-forum.

Another precondition is the *legal admissibility* of existing or to be revised regulations. It is not only a question of implementation of the signature law but also one of procedure-related regulations in the circumstances described in this paper.

A further issue is the *political adaptability*: Participation supplies are less credible if a promise of implementation is missing. Online participation supplies must be integrated into the political process. Issues like legitimacy (elected committee, publicness of an online forum) must be weighed against the new input (knowledge of citizens) into the political process and the social demand for participation means other than elections.

4 Multi-channel Strategy

Under the precondition of a restricted budget, new public management models and budgeting within public administrations, cost-benefit relation becomes a decisive factor whether and how single experiments develop to widely used and sustainable applications. Positively influencing this relationship, resources have to be used not only for a single procedure but for many simultaneously and could be used from a larger number of operators. Therefore an economically driven online-participation platform should be operated that offers various functionalities (e.g. opinion polling, discussion fora, maybe also online-decision making – which can combine these ways and which is adaptable do the tasks of the clients) according to the example of commercially driven application service providers which could be rented by interested municipalities.

But nevertheless Internet-based participation will always be used as a supplement and support of offline tools but not alternatively. The social aspect of a meeting cannot be replaced, otherwise the possible anonymous use of Internet applications can be used by citizens who are not very trained in verbal communication. Similar to bank services which are provided physically in the branch as well as via telephone or Internet, we can speak of a multi channel strategy in which all channels relate to the same information sources: In the case of political participation, this includes the combination of public meetings and hearings, press releases, Internet-use and a citizen-hotline.

5 Conclusions

The Internet can be used to improve but not to replace traditional ways of citizens' involvement to political procedures. In a multi-channel application, the advantages of offline and online ways of involvement should be combined. eDemocracy does not implicate a jump ahead of democracy but can improve deliberation in many ways:

- It enables asynchronous communication: participants at virtual discussions have more time to think about the arguments of political opponents before they react.
- Discussions can be lead more rationally than emotionally.
- Internet-involvement-methods are more flexible related to time and location: users do not have to join evening meetings or visit the agencies at public opening hours.
- Planning can be visualized more easily.
- Citizens can prepare themselves before visiting an officer by looking at the planning via the Internet; communication can work on a qualitatively higher level.
- The anonymity of the Internet empowers people to participate who normally would not do that and enables discussion among citizens who usually don't have contacts (nevertheless, it seems to be a medium that favours literate and educated citizens).

If deliberative policies should be innovated by the Internet, it is necessary to take into account that the processes are embedded in an environment that is determined by the innovation not only technically but also politically, legally and economically.

References

1. Coleman, S, Goetze, J.: Bowling Together: Online Public Engagement in Policy Deliberation. Hansard Society, London (2001)
2. National Audit Office, 1999, Government on the web: a report by the comptroller and audit general, commissioned by: Commons, H. o., London (1999)
3. OECD, Impact of the Emerging Information Society on the Policy Development Process and Democratic Quality, PUMA Vol. 15. OECD, Paris (1998)

4. Bellamy, C., Taylor, J. A., 1998, Governing in the Information Age, Buckingham Philadelphia: Open Univ. Press (1998)
5. Lenk, K., Dienstleistungssysteme und elektronische Demokratie. Multimediale Anwendungen im Verhältnis von Bürger und Verwaltung. In: Schneidewind, U., Truscheit, A. Steingräber, G. (Eds): Nachhaltige Informationsgesellschaft. Analyse und Gestaltungsempfehlungen aus Management- und institutioneller Sicht, Metropolis, Münster, 135-153 (2000)
6. OECD 1998, ibid.
7. Ranerup, A., Internet-enabled applications from local government democratisation. Contradictions of the Swedish experience., in: Heeks, R. (Ed.): Reinventing Government in the Information Age. International practice in IT-enabled public sector reform. Routledge, London, New York (1999), 177-193
8. OECD 1998, ibid.
9. OECD, Engaging Citizens in Policy-making: Information, Consultation and Public Participation. f, PUMA Policy Brief Vol. 10, OECD, Paris (2001)
10. Westholm, H., (Mehr) Partizipation über Internet? Fallbeispiel einer Online-Meinungsumfrage. In: Stiftung Mitarbeit (eds.): Rundbrief Bürgerbeteiligung, Vol.. I (2002)
11. OECD 2001, ibid.
12. Kubicek, H., Westholm, H., E-democracy: Quo vadis? Stand und Perspektiven elektronischer Demokratie per Internet. In: Behörden Spiegel, Vol. 12 (2001) 44
13. Dalton, Russel J., Citizen Politics in Western Democracies. Public Opinion and Political Parties in the United States, Great Britain, West Germany, and France, Chatham House Publishers, Chatham N.J. (1988)

Discourse Support Systems for Deliberative Democracy

Thomas F. Gordon[1] and Gernot Richter[2]

[1] Fraunhofer FOKUS, Berlin, Germany
gordon@fokus.fhg.de
[2] Fraunhofer AIS, Sankt Augustin, Germany
gernot.richter@ais.fhg.de

Abstract. The idea of deliberative democracy is to facilitate broad and deep public participation in systematic, constructive discourses about legislation and policy issues, so as to enhance the legitimacy, efficiency, quality, acceptability and accountability of the political process. By **discourse support systems** we mean groupware designed to support structured, goal-directed discourses. The paper discusses the importance of discourse support systems for deliberative democracy, provides a brief overview of the Open Source Zeno system and mentions several e-democracy pilot applications of Zeno, including the DEMOS project of the European Union.

1 Introduction

E-government is about redesigning or **reengineering** the processes of government, taking into consideration the opportunities and risks of modern information and communications technology (ICT). **E-democracy** is a special case of e-government: using ICT to support the core political processes of government, sometimes called **governance**: policy debates, legislation, executive decisions, the resolution of legal and political conflicts, and the election of representatives.

There are various conceptions of how to best make use of ICT to "reinvent" democracy. (See [3] and [6] for an overview and case studies.) Some emphasize the potential of **e-voting** to facilitate processes of **direct democracy**, via referenda, where public interest groups can propose legislation which is put to a popular vote and decided by citizens directly, bypassing elected representatives. Direct democracy is controversial and we will not address its many issues here, except to point out that even proponents of direct democracy emphasize the importance of adequate information and deliberation, before putting issues up to a vote [4]. For us, the main potential of e-democracy is to enable greater citizen participation in political discourses, whether or not the citizens or elected representatives make the final decisions.

There is much talk about overcoming the problem of **digital divide**, to assure that all stakeholder groups have effective access to e-democracy processes. While this is important, we should not forget that only powerful special interest groups have access to the traditional print and broadcast media, creating an even greater **analog divide** which already has been severely detrimental to democratic ideals. What has greater influence on the outcome of an election: a substantial donation to a political party's "war chest", so as to be able to afford media events, or casting a vote at the polls? The

R. Traunmüller and K. Lenk (Eds.): EGOV 2002, LNCS 2456, pp. 248–255, 2002.

new media of networked computers, especially the Internet, has given far more ordinary citizens an effective voice than any other technology in history. Consider, for example, the recent **weblog** phenomena, where thousands of ordinary citizens have begun publishing journals on the web [2,1].

In addition to weblogs, other more established Internet and web technologies can and have been used to facilitate political participation, including email, instant messaging, mailing lists, newsgroups, and web-based bulletin boards and discussion forums [3]. Each of these technologies has its advantages and appropriate uses, but due to a lack of space we cannot compare them here. Rather, the focus of the rest of this paper is on presenting a new kind of system, called **discourse support systems**, which unlike these other technologies are designed specially to support deliberation and other consensus building and conflict resolution processes, and discussing some experiences in applying such systems in e-democracy pilot projects.

2 Conceptual Model of Discourse Systems

Examples of discourse systems in politics and public administration are not difficult to find: 1) If a city plans to build a new airport, the applicable building codes may require the plan to be subjected to a public discussion with affected citizens and interest groups; 2) The cities of a region may work together to revise the zoning laws and plans to find a balance between growth and environmental protection; 3) A political party will need to discuss its political program and strategy for the next federal election; 4) Last but not least, parliaments, city councils and other law-making bodies deliberate about legislation in party factions, subcommittees and in plenary sessions, with input from various experts, lobbyists, professional staff and the public.

According to Walton in [13], a **dialog** is a goal-directed conventional framework in which two or more participants or parties "reason together in an orderly way, according to the rules of politeness or normal expectations of cooperative argumentation for the type of exchange they are engaged in." We define **discourse** as dialog, in Walton's sense, about some language artifact, such as draft legislation, project proposals, or city plans. We use the term **discourse system** to mean a "sociotechnical" system, consisting of human and technical "components", for performing particular discourse tasks within an organization, or between collaborating organizations. Finally, inspired among others by the work of Sumner and Shum [11], by **discourse support system** we mean the system of information and communications technology used as part of the infrastructure of a discourse system.

Conceptually, the main components of discourse systems are: the **actors** participating in the discourse, in their various roles; the **document** being discussed, including the history of changes made to the document; the **dialog** about the document, the subject matter of the document, or the dialog itself; and the **norms** which guide or regulate the dialog and modifications to the document.

Notice that the dialog can consist of many different kinds of speech acts: questions, motions, claims or assertions, arguments, offers, votes, and so on. In a more elaborate model, one might want to define separate components for different classes of speech acts. For example, there could be a component for managing claims and arguments and another for handling procedural issues.

Norms are of various kinds. They can provide mere guidance without imposing any obligations on the participants. Sources of norms are plentiful and varied, some general and some specific to the application. Example sources include social and linguistic conventions, rules of order, laws, regulations, administrative procedures, cases, principles, values, professional standards, and best practices. Norms may be conflicting and substantial reasoning may be necessary to decide which norms apply and how to resolve conflicts among them. Finally, norms are subject to change over time and may even change during and as a result of a particular discourse. For example, a participant might make an issue out of some rule of order.

3 Generic Use Cases

Having defined the relationship between discourse systems and discourse support systems, our next job is to consider what kinds of discourse tasks can be sensibly and usefully supported by modern information and communications technology. While a detailed requirements analysis would be possible only in the context of a specific application, we have been able to adduce some general requirements from our experience in several e-democracy projects. So-called "use cases" are a good starting point for identifying requirements. The use cases describe, at a very high level, the tasks and responsibilities of each role in the discourse.

We distinguish three main roles: **readers** browse the document and follow the dialog; **authors** write parts of the document or actively participate in the dialog; and **moderators** edit the document or moderate the discussion. Notice that we have used the same three roles for actors who interact with either the document or the dialog. This is because the protocol of a discussion can be conveniently viewed as a kind of complex, structured document.

Also, we have not distinguished between rights and obligations in these role definitions. For example, we leave it open whether authors have an obligation to make contributions to the document or only the right to do so if they want to. Here we are interested only in understanding the tasks which could be performed by each role, whether or not there is an obligation or even a right to perform such tasks in particular circumstances. This will ultimately depend on the norms appropriate for the particular application.

As usual, individuals may have several roles at once and several individuals may share the same role. For example, authors are typically also readers and several moderators may be responsible for some document or discussion. In a discourse, the moderators (i.e. editors) of the document being discussed need not be the same persons as the moderators of the discussion about the document.

Reader Tasks. Readers are interested in timely, relevant and accurate information about the participants and their roles, interests, background and activities; about the document and its subject matter; about the discussion; about the state of the proceedings in light of applicable procedures, about any other relevant norms; and about any background information helpful for understanding the issues. In particular, readers would like to be able to find information about similar past cases; to search for information in documents using metadata and the full text of the documents; to browse documents conveniently, using tables of contents, indexes and references (links); to find documents which are similar, in some sense, to a selected document; to cluster a set of documents and to categorize such clusters. Finally, readers would like to be able to keep informed about activity in the discourse, without having to regularly take the initiative (notification services).

Author Tasks. Authors are responsible for adding information to the system, to share them with readers. They need to be able add messages, articles and other kinds of documents to the system or to insert bibliographic information, abstracts and other data about these documents into catalogs and other databases. Authors need ways to refer to other documents or, ideally, parts of documents and make relationships between documents explicit. Finally, authors need ways to keep informed about tasks for which they are responsible and the status of these tasks, such as due dates, whether or not they have been completed, priorities and task dependencies.

Moderator Tasks. Moderators have final responsibility for the quality of the document or the discussion. They oversee and guide the entire process, helping other users to be aware of the applicable norms and thus their roles, tasks, rights and obligations. Their task is to assure that the discourse proceeds according to its purposes, so as to maximize the chance of achieving its goals. Moderators are responsible for applying appropriate moderation techniques to focus and guide the dialog. These include methods for broadening the dialog by gathering information about the problem and the interests of the parties, and brainstorming about possible solutions, and then narrowing the debate by clustering, categorizing and prioritizing options, and arguing about their relative merits. Relevant here is also the moderator's responsibility for opening and closing topics for discussion. Finally, moderators need tools for expressing, visualizing, presenting and analyzing relationships between parts of the dialog. Moderators need support in applying relevant norms to guide and regulate the process. Moderators require resources and methods for motivating other users to perform their tasks well in a timely manner. Moderators need to be able to conveniently monitor the activity in the discourse for which they are responsible. They need to be informed about new additions or changes to the document and new contributions to the discussion, without having to manually search for this information.

4 Overview of Zeno

The Zeno system, an Open Source groupware application for the Web written in Java, has been designed specifically for use as a discourse support system. This includes managing both the communities of users and groups who participate in the discourses and the content which is created and used in the discourses. A simple but powerful role-based access control scheme connects the two functional parts of Zeno: users and groups managed by the directory service are assigned access roles in journals where the content of discourses is stored. Discourse management functions for session management and event monitoring (logging, notification, discourse awareness, etc.) as well as communication services (messages to users and journals) provide the necessary support during a discourse.

Zeno's features are implemented in an extensible, object-oriented system architecture with easily customizable user interfaces, using the Velocity template engine and Cascaded Style Sheets.

4.1 Data Model

The design goal of Zeno led to a simple but general data model with a rich set of data structures and operations. The core of this data model is a persistent content store for **hyperthreads** of journals, articles, topics and links. **Journals** are container-like objects that can be used for many purposes, including shared workspaces, discussion forums, calendars, task management, and as a collaborative editing environment for complex, structured documents and content management. **Articles** are similar to email messages and support multiple MIME attachments. The contributions to a discourse are stored as articles. **Topics** are thematic collections of articles, that is, sets of articles which deal with the same subject. Topics and articles are contained in journals. When used as discussion forums, journals support both the threaded and the linear (topic-oriented) style of discussions. Journals, articles and topics, collectively known as Zeno **resources**, form a hierarchy or tree called the **compositional structure** of the content.

Typed links allow resources to be connected, which results in an arbitrary graph structure with labeled directed arcs called the **referential structure** of the content. A **link** connects a **source** resource with a **target** resource. A resource can be the source or target of any number of links. Thus, links can be used to create arbitrary directed graphs of resources. Links are typed with **labels** chosen from the set of **link labels** in the journal containing the source. The referential structure models non-compositional relationships between resources. Such graphs are much more general than the essentially hierarchical data structures typically used by file systems, shared workspaces, outliners or threaded discussion forums. We call the connected subgraphs of the referential structure **hyperthreads**, since they can be viewed as a generalization of the threads of discussion forums, replacing the reply relation by Zeno links.

Operations of the data model include full text search and powerful support tools for moderators: moving, copying, deleting, publishing and unpublishing articles, opening and closing topics, ranking or ordering articles and journals, and labeling articles and links to build conceptual graphs and visualize relationships. Automatic link management helps moderators to preserve the referential structure when they restructure the content of a discourse.

Journals are composed of a partially ordered sequence of any number of resources. Topics are composed of a partially ordered sequence of any number of articles. Articles are composed of any number of **attachments**. An attachment can be a file of any MIME type, such as word processing documents, spreadsheets, or image files.

Attributes describe the properties of resources, attachments and links which are relevant to the system or to the users. **System attributes**, such as the **creator**, **creation date** and **modification date** of a resource, are not modifiable by users, but rather set by the Zeno system. **Primary attributes** are standard attributes which may be modified by users, such as the **title**, **rank** and **note** of the resource. Finally, **secondary attributes** are ad hoc attributes defined by users for application-specific purposes.

All resources have the following system attributes: **id**, **creator**, **creation date**, **modifier** and **modification date**. All resources also have these primary attributes: **title**, **rank**, and **note**. The rank, an integer value, can be used for many purposes, such as prioritizing tasks or ordering the sections of a document. The note of a resource is a plain text string. Depending on the application, it can be used as the main part or body of a document, for example when journals are used as discussion forums, or as an abstract or description of files attached to the article, for example in journals used as content stores or shared workspaces.

Additional primary attributes of journals include, among others, **article labels** and **link labels**. They define the set of labels which can be used to tag articles and links. This feature enables journals to be used for **concept mapping**, **mind mapping**, **idea processing** and other approaches to modeling knowledge using labeled, directed graphs. For example, to model argumentation as in Issue-based Information Systems [7], one could define **issue**, **position** and **argument** labels for articles and **pro** and **con** labels for links.

Articles also have additional primary attributes, e.g., **label**, **keywords**, **begin date** and **end date**. The **label** is chosen from the set of the **article labels** of the journal containing the article. The **begin date** and **end date** attributes allow articles to be used to describe tasks, appointments or events, which can be used to generate reminders or displayed appropriately in calendar views.

4.2 User and Group Management

Zeno includes a directory service for managing users and groups of users. The directory maintains passwords, contact information, in particular email addresses, and user preferences. The directory can also be used for mailing lists.

For security and administration purposes, the directory is partitioned into a set of subdirectories, called **community directories**. A community directory can be configured so as to allow new users to register themselves in the community directory, without the assistance of an administrator. To allow self-registration, an administrator of the community directory gives **guest** users permission to register as new users if they meet the admission criteria stored in the community directory. The right to register new users is limited, and doesn't imply the right to view or modify existing records.

4.3 Role-Based Security Model

Access rights are controlled in Zeno by assigning the roles of **reader**, **author** or **moderator** to users and groups for each journal. That is, these roles are assigned for journals, but not directly for articles, topics or links. The access rights for an article or topic are those of the journal which immediately contains the article or topic. The access rights for a link depend on the access rights for the source of the link. Anyone with the right to view an article may also view the links from this article. Similarly, anyone with the right to modify an article may also modify the links from this article.

The rights of each role are fixed by the Zeno system. They cannot be redefined by users. Moderators have the most rights; with few exceptions they may do anything which can be done with a journal and its contents. Only moderators of a journal may create subjournals.

The readers of a journal may access and view every article in the journal which is published. Like moderators, readers may also access and view the identifiers and titles of subjournals. Further rights to a subjournal are controlled by the roles defined in the subjournal. The authors of a journal have the right to create new articles and topics in the journal. Participants in a discourse will often be both readers and authors.

4.4 Moderation and Editing Facilities

Based on feedback from users of prior versions of the system, Zeno provides a significantly richer set of features for moderating a discourse and editing its web of contributions. Only a few can be mentioned here.

Articles can be modified by editors at any time. The modification date and user id of the editor who made the changes are recorded in system attributes, to make it transparent to readers that the article has been modified, but Zeno does not currently keep a copy of the original version or provide any other version management services. Several articles, topics and journals can be selected and then, preserving their links, moved in a single transaction from one location to another in the compositional structure. Several articles and topics can be selected and then copied in a single transaction, in which case any links between the original articles are also copied. Resources can be deleted, recovered (undeleted) and permanently removed from the system.

Other features allow editors to close and re-open topics or journals, to publish and unpublish articles, to (partially) order direct components of journals (articles, topics, subjournals), and to define labels and qualifiers for articles.

5 e-Democracy Applications of Zeno

The first version of Zeno was developed as part of the European GeoMed project, which integrated Zeno with a Geographical Information Systems so as to enable citizens to discuss city plans on the Web [5,10]. This tradition has been continued; the current version of Zeno has been integrated with the CommonGIS system [12]. Zeno was recently used in an extensive e-democracy pilot application at the City of Esslingen, as part of the German Media@Komm project [9]. Finally, Zeno is being used as a part of the foundation of the DEMOS system [8]. DEMOS stands for Delphi Mediation Online System and is an e-democracy research and development project funded by the European Commission (IST-1999-20530). DEMOS offers innovative Internet services facilitating democratic discussions and participative public opinion formation. The goal is to reduce the distance between citizens and political institutions by providing a socio-technical system for moderated discourses involving thousands of participants about political issues at the local, national and European level. The vision and long-term goal of DEMOS is to motivate and enable all citizens, whatever their interests, technical skills or income, to participate effectively and actively in political processess which are both more democratic and more efficient than current practice. The DEMOS system is being validated in pilot applications in the cities of Bologna and Hamburg.

6 Conclusion

Many of the use cases we have identified for discourse support systems are not (yet) implemented by Zeno. There is a great deal of work remaining to be done. If there is one point we would like readers of this paper to remember, it would be that current tools only begin to scratch the surface of what is conceivable in the way of support for consensus building, conflict resolution and other core processes of democracy.

References

1. Rebecca Blood. The Weblog Handbook: Practical Advice on Creating and Maintaining your Blog. Perseus Pub., Cambridge, MA, 2002.
2. chromatic, Brian Aker, and Dave Krieger. Running Weblogs with Slash. O'Reilly, 2002.
3. Stephen Coleman and John Gøtze. Bowling together – online public engagement in policy deliberation, http://www.hansardsociety.org.uk, 2001.
4. UNI Unternehmerinstitute e.V. Für Effizienzstaat und Direktdemokratie. ASU Arbeitgemeinschaft Selbständiger Unternehmer e.V., Berlin, 2001.
5. Thomas F. Gordon. Zeno: A WWW system for geographical mediation. In P. J. Densham, Marc P. Armstrong, and Karen K. Kemp, editors, Collaborative Spatial Decision-Making, Scientific Report of the Intiative 17 Specialist Meeting, Technical Report, page 77–89. Santa Barbara, California, 1995.
6. Richard Heeks. Reinventing government in the information age: international practice in IT-enabled public sector reform. Routledge research in information technology and society. Routledge, London ; New York, 1999.
7. Werner Kunz and Horst W.J. Rittel. Issues as elements of information systems. Technical report, Institut für Grundlagen der Planung, Universität Stuttgart, 1970. also: Center for Planning and Development Research, Institute of Urban and Regional Development Research. Working Paper 131, University of California, Berkeley.
8. Rolf Lührs, Thomas Malsch, and Klaus Voss. Internet, discourses and democracy. In T. Terano et al., editors, New Frontiers in Artificial Intelligence. Joint JSAI 2001 Workshop Post-Proceedings. Springer, 2001.
9. Oliver Märker, Hans Hagedorn, Matthias Trénel, and Thomas F. Gordon. Internet-based citizen participation in the City of Esslingen. Relevance – Moderation – Software. In Manfred Schrenk, editor, CORP 2002 – "Who plans Europe's future?". Technical University of Vienna, 2002. 7th international symposion on information technology in urban and regional planning and impacts of ICT on physical space.
10. Barbara Schmidt-Belz, Claus Rinner, and Thomas F. Gordon. GeoMed for urban planning – first user experiences. In R. Laurini, K. Makki, and N. Pissinou, editors, Proceedings of 6th International Symposium on Advances in Geographic Information Systems, pages 82–87. 1998.
11. Tamara Sumner and Simon Buckingham Shum. From documents to discourse: Shifting conceptions of scholarly publishing. In Proceedings of CHI 98: Human Factors and Computing Systems, pages 95–102. ACM Press, Los Angeles, California, 1998.
12. Angi Voss, Stefanie Röder, Stefan Salz, and S. Hoppe. Group decision support for spatial planning and e-government. In Global Spatial Data Infrastructure (GSDI), Budapest, Hungary, 2002.
13. Douglas N. Walton. The New Dialectic: Conversational Contexts of Argument. University of Toronto Press, Toronto; Buffalo, 1998.

Citizen Participation in Public Affairs

Ann Macintosh and Ella Smith

International Teledemocracy Centre, University of Napier,
10 Colinton Rd, Edinburgh, EH10 5DT, UK
{A.Macintosh,E.Smith}@napier.ac.uk

Abstract. Reflecting on the European Commissions stated aim to broaden democracy this paper examines the nature of e-participation and considers concepts of democracy and issues surrounding citizen participation in pubic affairs. The paper describes how citizens are engaging with government and with each other about policy related issues that concern them, using technology specially designed for the purpose. The paper describes a case study of electronic participation developed for the Environment Group of the Scottish Executive in Summer 2001. Using the empirical data from this study the paper explores best practice guidelines for governments who wish to engage citizens in policy-making. The difficult task of addressing the requirements of all stakeholders, i.e. government, civil society organizations (CSOs) and citizens in designing the technology is discussed. The use and moderation of the electronic tools over the engagement period is assessed. Finally, the paper considers how the use of electronic tools can be monitored and their impact on citizen participation and the decision-making of government be assessed.

1 Introduction

Governments around the world are attempting to broaden democracy by providing an effective conduit between themselves and civil society organisations and between citizens themselves using innovative ICT to deliver more open and transparent democratic decision-making processes. It is argued that democratic political participation must involve both the means to be informed and the mechanisms to take part in the decision-making. ICTs have the potential to deliver e-democracy which addresses these joint perspectives of informing and participating. Over the last decade there has been a gradual awareness of the need to consider new tools for public engagement that enable a wider audience to contribute to the policy debate and where contributions themselves are both broader and deeper. A number of commentators have addressed this issue and at the same time highlighted the possible dangers of a technology-driven approach.

Barber [1] highlights the concept of strong democracy, creating active citizen participation where none had existed before. However he goes on to warn of the use of technology in that it could diminish the sense of face-to-face confrontation and increase the dangers of elite manipulation. Held [2] distinguishes nine different models of democracy. His participatory model reflects the need to engage both citizens and civil society organisations (CSOs) in the policy process. However, in order to engage citizens in policy-making, he and others recognise the need for informed and active

R. Traunmüller and K. Lenk (Eds.): EGOV 2002, LNCS 2456, pp. 256–263, 2002.

citizens. Fishkin [3] argues the need for 'mass' deliberation by citizens instead of 'elite' deliberation by elected representatives. Instant reactions to telephone surveys and television call-ins do not allow time to think through issues and hear the competing arguments. Van Dijk [4] addresses the role of information and communication technology with such participatory models of democracy in order to inform and activate the citizenry. However he also warns of the consequences of bad designs of technology.

The arrival of more sophisticated ICTs and the emergence of the Internet during the early 1990s coincided with widespread concern that politics and politicians had become increasingly irrelevant to ordinary people. It is clear from the increasingly low turnout at elections in the UK that traditional democratic processes do not effectively engage people. Other liberal democratic countries have also noted low voter turn-out. To start to address this problem, governments are beginning to consider the concept of e-voting and over the last year there has been a growing interest in internet voting. However, there remains a big question mark over whether this switch in methods of voting will actually change people's attitudes and address their growing disengagement from established political processes.

Several commentators discuss the use of technology to support the broader democratic process. Coleman and Gøtze [5] outline four possible scenarios for technology supporting democracy. The first e-democracy model is where the technology supports direct democracy. For example, Becker and Slaton [6] explore the current state and future of e-democracy initiatives that are designed specifically to move towards direct democracy. The second model is based on on-line communities, where technology is concerned with supporting civic communities. The work of Rheingold [7] on virtual communities assesses the potential impact of civic networks, questioning the relationship between virtual communities and the revitalization of democracy. Tsagarousianou *et al.,* [8] give descriptions of a number of projects involved with e-democracy and civic networking. These authors suggest centrally designed government-led initiatives will clearly differ from grassroots civic developments, but argue also that "civic networking will not realize its objective unless it becomes more realistic in its goals and methods" (page 13). Coleman and Gøtze's third e-democracy model concerns the use of on-line techniques to gauge public opinion through surveys and opinion polls. However, Fishkin [3] questions whether opinion polls contribute to the complex issues of public policy. He argues that, as far as American citizens are concerned, they have the opportunity to be consulted on several occasions by opinion polls without prior warning or preparation, in order to find their views, even when the individual may have had no reason to develop any opinion on the subject being asked. He concludes that all that was gathered was an "attitude" that was created on the spot by the very process of participating in the survey (page 81). Finally, their fourth model focuses on the use of technology to engage citizens in policy deliberation, emphasizing the deliberative element within democracy.

The recent OECD report [9] emphasizes the need for: "a relation based on partnership with government, in which citizens actively engage in defining the process and content of policy-making." The OECD is addressing issues such as how to provide easier and wider access to government and parliamentary information, how to ensure that citizens have the ability to give their views on a range of policy related matters and also how to allow citizens to influence and participate in policy formulation. The OECD [9] report defines three types of interaction, namely one-way information provision, a two-way relationship where citizens are given the opportunity to give feedback on issues and, lastly, a relationship based on partnership where citizens are actively engaged in the policy-making process. Looking at these three types of interaction, the OECD reports that the scope and quality of government provided information has increased greatly over the past decade. With regard to the two-way relationship, consultation is also on the increase, albeit with large differences between countries. For the final type of interaction, the OECD states that active participation and efforts to engage citizens in policy-making on a partnership basis are rare, undertaken on a pilot basis only and confined to a very few OECD countries. With regard to the application of ICT, the OECD reports that, while an increasing amount of government information is obtainable on-line, the use of ICT for consultation is still in its infancy. The case study described in this paper addresses both the two-way relationship through consultation and the notion of active participation where citizens are helping to formulate policy.

2 Objectives of e-Consultation

One can consider three over-arching reasons for better engagement of citizens in the policy-making process:

- to produce better quality policy at the national level
- to build trust and gain acceptance of policy
- to share responsibility for policy-making.

All with the long term objective of strengthening representative democracy.

Given these overarching reasons, we argue that the objective of e-consultation is to improve the policy-making process through a range of devices to facilitate:

1. reaching a wider audience to enable broader consultation - in which case the participants themselves can be profiled to the extent that they have provided any identifying details and they have given their informed consent to use these;
2. supporting participation through a range of technologies to cater for the diverse technical and communicative skills of citizens – in which case the 'ease of use' and 'appropriate design' of the e-consultation can be addressed;
3. providing relevant information in a format that is both more accessible and more understandable to the target audience to enable more informed consultation - in which case participants use of background information that is made available on-line can be analyzed, to give an indication of how relevant it has been;
4. engaging with a wider audience to enable deeper consultation and support deliberative debate – in which content analysis and thread analysis of consultation discussion forums can be considered;
5. analyzing contributions to support the policy-makers and to improve the policy – in which case analysis of what people have said in response to the consultation can be carried out more cost effectively since the responses are received in an electronic form (i.e. they do not need to be transcribed) and responses to closed questions can also be subjected to survey analysis techniques, again with cost-efficiency savings since they do not have to be transcribed from questionnaires;
6. providing relevant and appropriate feedback to citizens to ensure openness and transparency in the policy-making process.

Different forms of technology are appropriate for each of the above issues, but for all, the technology must be designed in ways that generate and support public trust in the process. However, Macintosh et al [10] describe how the very nature of governance means that the design of e-democracy systems becomes complex. For example, democratic needs for openness and transparency may conflict with needs for ease of use and simplicity of access. Whyte and Macintosh [11] discuss technology designed to enhance the transparency of public consultation practices, highlighting that many other problems also need to be addressed. For example, issues of unequal access to technology and the unequal technical and communicative capabilities of citizens demand systems that are simple to use and understand.

The web-based tool, e-consultant[1], which was used in the case study of electronic participation discussed in this paper begins to addresses each of the above issues. The tool is the Center's focus for e-participation research.

3 e-Consultation on Sustainable Development

The case study addresses the 'pre-policy document' stage of civic engagement i.e. where there are a number of pre-identified issues that government wishes to gather opinions on before drafting a policy document. The e-consultation allows government to invite discussion on issues - the aim is to get initial input from a wide audience so as to draft a more comprehensive policy document. The e-consultation was on behalf of the Environment Group of the Scottish Executive and was based around sustainable development issues facing Scotland. The aim was to equip Ministers with views to develop a policy document as input to the World Summit in South Africa in 2002. The study ran from 6th June to 8th October 2001. It aimed to inform people about the key issues facing a future Scotland and asked them to give their views on a range of issues from efficient use of resources to lifestyle and transport. The web site address for the study is: http://e-consultant.org.uk/sustainability/. It received a total of 392 contributions. These were made by 172 individuals and on behalf of 19 groups or organizations.

e-consultant is a dynamic website implemented in Active Server Pages. Scripts written in VBScript generate the HTML, and access and update the consultation data. Data is maintained in a SQL Server relational database, with the exception of the consultation's background information, which is held as static HTML. The e-consultant system resides on the ITC server, which in turn is on Napier University's network. In summary, the main sections of e-consultant for the sustainable development consultation were designed as follows:-

- ***Overview:*** A welcoming page outlining the purpose, target audience and timescales of the consultation, with links to the main websites of stakeholders.
- ***Background Ideas:*** Structured around the 7 key issues of sustainable development for Scotland; also included were links from the consultation issues to a comment page for each issue.
- ***Have your say:*** Here uses could enter their comment on an issue or respond to a previously made comment; also included were links from these comment pages back to the background ideas page for each issue.

---

[1] Developed with support from British Telecom, see internet site: www.e-consultant.org.uk

- ***Tell a friend:*** Here users could email people they felt would be interested in participating in the consultation.
- ***Events and Links:*** Details of the off-line seminars and links to other sites.
- ***Feedback:*** Space for a statement from the Scottish Executive as their response to the consultation once the consultation was complete and responses analyzed.
- ***Review site:*** An online questionnaire for users to complete to help the ITC evaluate the e-consultation.

As well as the above end-user sections, the management of the e-consultation process was facilitated by additional password-protected administrative services. These included functions to

- monitor comments on a 24hour post-moderation basis
- remove from view comments that breached the "conditions of use" statement
- view the entries to the on-line questionnaire
- view the most frequently read comments
- view the comments added in the last 24 hours
- view the comments received from postcode areas.

The ITC had access to all the above services to support managing the e-consultation. The Scottish Executive had access to all but the first 3 of the services. This helped them assess the consultation as it progressed.

In order to give some structure to the debate and encourage discussion topics to form, questions and ideas were divided into 7 issue based sections. The issues were also used to focus discussion on the less obvious aspects of the debate (economic and social) and help to explain how they fitted in. To encourage informed and deliberative debate the "Background Ideas" section was included in the site. This was divided into pages according to the various issues.

4 Analysis of Comments

All comments were read and categorised according to their content. The way this was done was inspired by Anthony Wilhelm's [12] analysis of 10 political newsgroups in October 1996 during the presidential election campaign in the United States. He used relatively simple content analysis categories to evaluate how far the participants provide and seek reasoned argument with evidence to support their contributions.

Our comments were analysed to judge the success of the consultation in terms of the way the e-consultation website was organised and the information given within it. This was done by first looking at the extent to which the comments answered the main questions set out in the e-consultation and second by an investigation of the relationship between the content initially provided on the website (the issues and background information) and the content of the comments.

The questions set by the e-consultation and the relevant categorises were:

- **Q1.** What sort of Scotland do we want to live in?
  - **Analysis categories:** Suggestion; no suggestion
- **Q2.** What could or should be done to achieve this by governing bodies, individuals and business?
  - **Analysis categories:** action-government, action-business, action-individuals, action-all

- **Q3.** What has been done well, so far, and what could be done better?
  - **Analysis categories:** Doing well; Improve

The results of the analysis showed that contributors engaged with the question "What sort of Scotland do we want to live in?". While most suggestions for improvement were directed towards government, there was an extensive realisation that everyone needed to be involved at some level. 97% of comments contained or agreed with suggestions about what could be done to improve our status in terms of sustainable development. Only 3% of comments made no suggestions and merely pointed out a current problem. Over half the comments (57%) included suggestions for government action. A reasonable proportion specified that businesses should take action (19%). A smaller number of comments pointed out where individuals should take action themselves (7%). A third of comments (33%) requested action from all parties (government, business and individuals).

The second part of the analysis was to discover whether the issues and background information provided in the e-consultation were clear and helpful. Here we assessed whether each comment appeared under the most appropriate issue and whether it reacted to or reflected the information provided as well as addressing the question. Almost all the contributors used the issues the way they were intended to be used (96%). Of those which could have perhaps more suitably appeared under a different issue (4%), most had some relevance to where they were placed. Comments that did not seem to fit any category tended to be about how policy was implemented. Some contributors did not divide their opinions into separate comments under specific issues, but grouped them all together under one issue. Most comments directly answered the questions that accompanied the issues or those posed in the "Background Information" section (91%), This is a good indication that the questions were successful in focussing the discussion. A further 7% of comments reflected themes from the "Background Ideas" section, without answering any of the questions posed within it.

5 Evaluation of the Process

The evaluation was based on past experience of conducting e-consultations, the meetings of a consultation Steering Group and the users' opinions of the site as given in response to the on-line "Review Site" questionnaire. Every page of the e-consultation had a link to the "Review Site" questionnaire. This aimed to gather information about the circumstances in which people used the site, how easy or difficult they found it to use and what they thought about using the Internet as part of a consultation process.

The Scottish Executive set up a Steering Committee drawn from key stakeholders across Scottish Society to guide the consultation. This Committee agreed the original design specification and the content of the consultation website through an iterative process. Best practice guidelines for "traditional" consultations were followed as closely as possible. As the target audience was anyone with an interest in sustainable development in Scotland, it was important that people using any Internet enabled computer at minimum connection rates and any web browsers could use the e-consultation. Both the structure and the content of the website were designed to cater for this diverse audience. It contained a clear statement on the conditions of use of the site and also a clear statement on privacy.

It was important to make it as easy as possible for any user to be able to add comments on the consultation, therefore the commenting process had to be as easy and attractive as possible. The ITC had found from previous e-consultations that users were put off by too intrusive a registration process. Therefore, it was agreed not to include a registration process. Instead, the users were asked to provide a minimum of personal information with each comment. Although this simplicity worked in the users' favour, it did lead to some difficulties during the evaluation process.

e-consultations provide a new mechanism to gather public opinion and as such they require new methods to promote them. Traditional promotional routes were augmented with more interactive "on-line" style promotion, "tell a friend" e-postcards and clickable logos advertising the consultation on related websites were used. The number of sites displaying the e-consultation link was disappointing, but they did have some success in involving people in the consultation. 40% of respondents to the on-line questionnaire stated they heard about the e-consultation through an electronic link. Almost half the respondents contributed to the e-consultation from their home, 30% from work, 11% from school or college, the remainder from a friend's house or community centre. Most felt that the site had been easy to use, although 7 people admitted to being confused in places and 4 found it quite difficult.

When one of the objectives of an e-consultation is to reach a wide target audience, the natural concern about the digital divide and hence the bias of an Internet based consultation has to be addressed. An e-consultation should be seen as just one route for participation and be supported by other opinion gathering events. All these participation routes should clearly link to one another. Importantly, an e-consultation provides the opportunity for deliberative participation. Well-structured background information on the website can serve to inform the debate. Allowing users to make their own comments and also to read and respond to comments from others provides a more transparent consultation process.

6 Conclusions

Although examples of e-democracy systems are often cited in the press and by politicians, there are few academic reports of these experiments. Few "real-world" examples of e-participation exist that provide sufficient empirical data on which to base sound research studies. It was because of this that we embarked on a programmed of active research.

We have attempted to show that while the evaluation issues are easy to state at this general level, their assessment has to deal with interdependencies between systems design, policy implementation, and the everyday politics and practice of communications between citizens, in all their variety, and public administrations in all their complexity. We suggest that evaluation of e-consultations has to take place within an analytical framework that takes into account the political, technical and social perspectives [13].

References

1. Barber, Benjamin. (1984). *Strong democracy: Participatory politics for a new age.* Berkeley: University of California Press.
2. Held, Anthony. (1996). *Models of democracy.* Cambridge: Blackwell Publishers.

3. Fishkin, James S. (1995) *The voice of the People. Public Opinion and Democracy.* Yale University Press
4. Van Dijk, Jan 'Models of Democracy and Concepts of Communication'. In Hacker, K.L and Jan van Dijk., (2000). (eds*) Digital Democracy issues of theory and practice.* Sage Publications
5. Coleman, Stephen. and Gøtze, John (2001) (2001). *Bowling Together: Online public engagement in policy deliberation.* Hansard Society and BT.
6. Becker, T. & Slaton, C. (2000). *The Future of Teledemocracy.* Westport, Conn. 2000. LC
7. Rheingold, H. (1993) *The Virtual Community.* Reading M.A. Addison and Wesley. Also see http://www.well.com/user/hlr/vcbook/vcbookintro.html (consulted February, 2002)
8. Tsagarousianou, Rosa. Tambini, Damian. & Bryan Cathy. (1998). (eds). *Cyberdemocracy: Technology, cities and civic networks.* London & New York: Routledge.
9. OECD (2001). *Citizens as Partners: Information, consultation and public participation in policy-making*: OECD.
10. Macintosh, A., Davenport, E., Malina, A.; and Whyte A.; Technology to Support Participatory Democracy*; in Electronic Government: Design, Applications, And Management*; edited by Åke Grönlund, Umeå University, Sweden; published by Idea Group Publishing January 2002; pp223-245.
11. Whyte A. and Macintosh A. Transparency and Teledemocracy: Issues from an 'E-Consultation.' In *Journal of Information Science.* July 2001.
12. Wilhelm, A.G. (1999) Virtual Sounding Boards: how deliberative is online political discussion? In Hague, B.N. and Loader, B. (Eds) (1999) *Digital Democracy: Discourse and Decision Making in the Information Age.* Routledge
13. Whyte, A. and Macintosh, A. 'Analysis and Evaluation of e-consultations; to appear in the e-Service Journal; Indiana University Press, 2002.

An Approach to Offering One-Stop e-Government Services – Available Technologies and Architectural Issues

Dimitris Gouscos[1], Giorgos Laskaridis[1], Dimitris Lioulias[2], Gregoris Mentzas[2], and Panagiotis Georgiadis[1]

[1] eGovernment Laboratory, Dept. of Informatics and Telecommunications, University of Athens
{d.gouscos,p.georgiadis}@e-gov.gr
glask@di.uoa.gr
[2] Dept. of Electrical and Computer Engineering, National Technical University of Athens
{dlioulias,gmentzas}@softlab.ntua.gr

Abstract. The right of citizens to high-quality e-Government services makes one-stop service offerings an essential feature for e-Government. Offering one-stop services presents many operational implications; an one-stop service provision (OSP) architecture is needed that, by means of a layered approach, provides facilities to refer to, invoke and combine e-Government services in a uniform way, in the context of cross-organisational workflows. Although enabling technologies for all the layers of such an architecture are quickly evolving (XML, WSDL, UDDI, WFMS et al) two major issues that need to be solved are (a) abstracting the heterogeneity of the e-Government services that need to be integrated and (b) identifying an appropriate style for cross-organisational workflow control, somewhere in between the fully centralised and peer-to-peer extremes. This paper presents an abstract layered OSP architecture, identifies some major enabling technologies and briefly discusses those two issues.

1 Some Requirements for One-Stop e-Government

The delivery of e-Government services poses certain requirements inherent in the mission of government itself, that differentiate e-Government from other e-service application domains such as e-Business; customers have on *option* to choose the services of a specific business, whereas citizens have a *right* to enjoy the services of their own government. Therefore, e-Government service delivery must achieve (a) maximal benefits (quality and performance of the service, added value of the content), (b) maximal accessibility (simplicity of front-end logic, multi-linguality of user interfaces, anywhere/anytime availability, provision of alternative channels, no demands for end-user IT skills) and (c) minimal costs (transportation, communication, document management); see, e.g. [1], [6], [10], [12], [15].

The idea of one-stop e-Government presents substantial promise for contributing to all those optimality goals, whereas at the same time it is a natural fit to the concept of web portals that by definition can bring together any number of arbitrarily heterogeneous web resources. It is well established, however, that the mere collection of links to e-Government sites is a very superficial implementation of the one-stop concept;

R. Traunmüller and K. Lenk (Eds.): EGOV 2002, LNCS 2456, pp. 264–271, 2002.

true one-stop e-Government calls for *compilation, presentation and delivery* of e-Government services in an one-stop fashion. This means that (a) on a conceptual level, services need to be compiled into some sort of *service bundles* around single real-world situations where they apply; these service bundles then must be (b) logically presented at the user interface level as responses to single real-world problems (the life-event approach) and (c) actually enacted and delivered at the operational level as if they were individual (in contrast to bundled), atomic services.

2 Operational Implications of One-Stop Service Offerings

Conceptual compilation of e-Government services in bundles corresponding to real-world situations must face problems like (a) how to categorize services by means of "ontological indexes" on their content (service ontologies), (b) how to refer to services (service naming schemes) and (c) how to store invocation pointers to service implementations (service repositories).

The issue of actually enacting and delivering these bundles as virtually atomic services sets outs some further requirements of its own. To achieve this sort of virtual atomicity, an one-stop service provider (OSP) must deliver a sense of seamlessness to service requestors. Consequently, an OSP needs to provide not only *bundling transparency*, i.e. hide that a "virtually atomic" service is actually enacted and delivered by invoking multiple lower-level services that may exchange data, synchronize their work/control flows and produce results finally composed into an atomic response, but also *bundling management*, i.e. the actual coordination mechanisms that take care of invocation, interaction and synchronization of the bundled services.

The issue of service bundling management, which is central to the provision of one-stop services, poses some hard problems on the operational level. Talking in an e-Government context, virtual one-stop e-Government services such as the issue of a business permit or the declaration of a change of address correspond, due to public sector functional disintegration, to bundles of multiple atomic services which, in the general case, are delivered by multiple providers. Therefore, it should be noted that (a) one-stop provision of bundled e-Government services entails multiple service providers and, what is more, (b) the technology that is used by individual providers for providing their share of bundled services cannot be assumed to be interoperable (in extreme cases, technology may not be used at all or only in an old-fashioned, merely esoteric, way). What is more, these providers being public agencies that are governed by complex or fuzzy regulatory frameworks in the general case, (c) their internal workflows may contain complex, multi-conditional portions, portions that cannot be modelled deterministically and many cases of exceptions.

One-stop e-Government services are required not only in an intra-border (i.e. local, regional or national), but also in a cross-border context; consider, for example, issuing work permits for immigrants or paying in country X a freelance worker based in country Y. Such as situation is additionally complicated since (d) public service providers from different countries will normally operate in different languages and be governed by non-harmonised regulatory frameworks and, apart from that, (e) it is quite risky to assume a given service level (e.g. response time, exactitude and completeness of response) from a public service provider in a country other than one's own; sometimes, a response cannot even be assumed at all within a reasonable amount of time.

3 Building Blocks for an One-Stop Service Provision Architecture

All these operational level problems are posed as hard design constraints, when one tries to compile one-stop e-Government service bundles, and on the other hand as real-time quality and performance threats, during enactment of the service bundles and delivery of the bundled services. Therefore, work is needed in an architectural level, i.e. a level of systemic analysis and synthesis that considers *both* operational *and* technical issues, along the following directions:

- Service ontologies, service naming schemes and service repositories that offer capabilities to characterize, name, store and invoke a service by means of a well-defined set of "essential service characteristics" (e.g. application domain, content, inputs, outputs, behaviour, exceptions, invocation channel, delivery channel, information exchange format, quality levels, performance levels). Deployment of such ontologies, naming schemes and repositories would result in an *abstract service model*, on top of which it is possible to develop the service invocation and service coordination schemes mentioned below.
- Service invocation schemes that offer capabilities to uniformly invoke services of arbitrary information format and communication channel heterogeneity. Such schemes should be able to abstract from the caller any service-specific invocation and delivery details so that (a) information format heterogeneity could be abstracted by transparent transliteration from caller-native to service-native formats and vice versa (possibly to some standard format so that a star-like transliteration scheme is established), whereas (b) communication channel heterogeneity could be abstracted by having generic service invocation call that accept preferred channels as caller-defined parameters, consider channel capabilities retrieved from the service repository and then call channel-specific communication gateways.
- Service coordination schemes that offer capabilities to implement the atomic enactment of a service bundle and delivery of bundled results. Such a scheme should be able to handle increasing levels of co-ordination complexity, such as (a) invoking (serially or in parallel) *non-cooperating services* (i.e. services able to deliver results without any interim exchange of information or other synchronisation) and passing (possibly transilterated) information delivered from a service to the next invoked one, (b) invoking *co-operating services* with interim synchronization points (so that there are pending states where one service is waiting for another to reach some interim point of progress), (c) handling services (possibly co-operating) which are delivered through *fuzzy*, rather than deterministic, workflows and (d) handling *indeterminate situations* where a service actually does not (due to some service provider failure) or is not perceived to (due to some communication failure) reach expected progress within a reasonable amount of time.
- A service co-ordination scheme should also offer co-ordination management capabilities of increasing complexity, such as (a) enacting deterministic and *fuzzy* co-ordination scenaria, (b) enforcing *atomicity requirements* (i.e. try to complete the scenario or have it rollback as a whole, in a way reminiscent of database transactions), (c) reporting the current progress and status of an enacted co-ordination scenario and (d) mapping one-stop service requests to *pre-compiled* co-ordination scenaria, or *customising* pre-compiled scenaria according to request-specific characteristics, or *deducing optimal performance* co-ordination scenaria according to request-specific characteristics and current workload.

According to the above analysis, an one-stop e-Government service provision scheme should have a layered architecture like one depicted in Fig. 1.

abstract service co-ordination model	formulate co-ordination scenaria	manage co-ordination scenaria
	enact co-ordination scenaria	
abstract service invocation model	information transliteration	channel parameterization
	service invocation calls	
abstract service reference model	service naming scheme	service repository
	service ontology	

Fig. 1. Layered architecture for an one-stop service provision scheme

4 Technologies That Can Serve One-Stop Services

As a complement to the layered OPS architecture a communication platform needs to be assumed that interconnects the OSP with individual service providers in a star-like topology. This platform incorporates all the service invocation and delivery channels that have been mentioned above; in a real-world setting, such a platform would include Internet (HTTP, email, WAP), telephony (voice, SMS, facsimile) or even paper-based connectivity between the OSP and individual service providers.

The current trends for e-Government certainly justify considering the Internet, and in particular the HTTP protocol, as the premium communication platform between an OSP and individual service providers; turning to HTTP, together with its accompanying technologies for deploying and accessing e-service content on the web, already offers a number of very promising, if not definitive, technical choices for implementing parts of this architecture.

Web service description languages, such as WSDL [5], as well as universal naming and access protocols, such as UDDI [13] and SOAP [2], are expected to solve many of the issues mentioned above for deploying service ontologies and naming schemes. The development of service repositories, on the other hand, is an active research issue (cf. the ideas of some IST e-Government cluster projects, such as EGOV and SMARTGOV), possibly looking at techniques from domains such as knowledge management. Not all technical problems are yet solved, but the fact that much research is active in many issues related to e-services, such as e-service composition and integration in heterogeneous environments (see, for example, [3], [4], [8], [9]) gives rise to a certain amount of optimism.

The need for abstraction of technical heterogeneity in information exchange between individual service providers can be effectively addressed by semantic and dynamic content exchange languages such as XML and DHTML; note that XML can also be used for dynamic control exchange in cross-organisational workflow architectures, as in [7], [11]. It should be noted, though, that adoption of a language such as XML just provides us with a syntax for formulating our messages, which is less than half of the solution. The remaining and most important part of the problem would be to establish XML interfaces and a well-defined XML vocabulary with individual service providers, which may well prove to be more of an operational than of a technical issue.

On the level of the abstract service co-ordination model, much of what has been mentioned above strongly reminds of similar issues and requirements in the context of workflow management systems. Indeed, workflow management technology already provides working results in modelling and enacting scenaria of cross-organisational work flows, where individual flows may or may not co-operate, exceptions may occur, etc; see, for instance, [7], [11], [14], [16]. It should be noted, however, that current workflow management research is still active in issues that have to do with management of non-deterministic workflows at different levels of complexity (fuzzy, weak workflows), and there is still much work to be done with respect to workflow recovery from unanticipated exceptions (cf. the CB-BUSINESS project). Fig. 2 depicts the layered OSP architecture annotated with the enabling technologies mentioned above.

As can be seen from this figure, many technological building blocks are there, but technologies still evolve and mature (WSDL, UDDI) and research is still active important open issues (WFMS). Apart from that, there is also a number of operational issues that have to be solved between OSPs and individual service providers in order for such architectures to operate. As a last, but not least, open-ended issue in the situation of Fig. 2, it is worth bringing into attention the underlying architectural paradigm on which this design is based.

abstract service co-ordination model	formulate co-ordination scenaria	manage co-ordination scenaria	WFMS technologies
	enact co-ordination scenaria		
Abstract service Invocation model	information transliteration	channel parameterization	XML, DHTML
	service invocation calls		
Abstract service Reference model	service naming scheme	service repository	WSDL, UDDI
	service ontology		
Communication platform	service invocation messaging	result delivery messaging	Internet/HTTP

Fig. 2. Layered OSP architecture and some enabling technologies

5 A Discussion of Principles

In very general and abstract terms, this OSP architecture may be thought of as an architectural design that attempts to co-ordinate dynamic entities from multiple and independent information systems towards some common goal. Whether these entities are client-server *processes* in a distributed database, blue- or white-collar *workflows* in a manufacturing or office system or individual e-Government *services* to be bundled in one-stop offerings, from an architectural point of view their co-ordination towards a common goal entails some inevitable issues of integration. At some level these dynamic entities must communicate in order to exchange data and/or control. This need for communication and data or control exchange necessitates some degree of homogeneity, in the sense that a minimum amount of common conventions must be applied by all entities involved with respect to the communication channels that they use and the syntax and vocabulary of their messages.

As the technical conventions employed in an information system have mainly evolved out of business choices during the systems procurement or implementation, the technical homogeneity of different information systems is most often organisationally bounded. Systems within the same organisation follow (or are taken care of, in order to follow) compatible technologies, so that communication, information exchange and finally application integration is possible. In such intra-organisational contexts, therefore, one is usually (and justifiably) targeted at applying some standard conventions so that information systems are interconnected, at the networking and database levels, and interoperable, at the processing level.

The feasibility of this approach becomes quite questionable, though, in an inter-organisational setting like the one formed by multiple individual public service providers, whose e-Government services are to be bundled in true one-stop value propositions. In such a context, the diversity of existent IT platforms, as well as various business and political resistances to change leave very little room for assuming homogeneities at a technical or operational level (and this is even more true in the cross-border variant of this situation). Therefore, beyond the tight integration demanded by the interconnectivity or interoperability approaches, a new and more flexible architectural paradigm is sought to make dynamic co-ordination possible.

In the approach that we have presented in this paper, this is the paradigm of intermediation, namely the paradigm of establishing a third intermediator entity amidst all the dynamic entities co-ordinated, with a mission to manage the overall co-ordination scenario. This intermediator entity then (in our case the central OSP scheme), should first of all employ ontologies and naming schemes to acquaint with the dynamic entities that it co-ordinates, memorize their names and features in some sort of repository and isolate heterogeneities by means of specialised communication modules pluggable into an abstract co-ordination model. This, in more abstract terms, is the underlying approach that we have taken in designing the OSP scheme above.

The modelling part of this approach, namely the concept of applying some ontological modelling and some naming scheme in order to gain capabilities of acquainting and invoking multiple heterogeneous resources is already well-established in many computer science fields; this general principle can be found behind agent-based approaches in artificial intelligence, object-based approaches in software engineering and distributed architectures, let alone in the design of the Web itself. Unique Resource Identifiers (URIs) of web resources and Unique Resource Locators (URLs) of web pages are instances of this general concept, whereas description languages like WSDL and naming protocols like UDDI are explicitly based on similar principles to envision "web services networks", an idea which has some common grounds with the OSP architecture.

A substantial difference between these technologies and our own context, however, is that web technologies are completely open-ended as to the architectural location (probably absence) of control. In the true "uncontrolled" spirit of the Internet, these technologies are offered as building blocks that allow constructions of peer-to-peer structures, like the World-Wide Grid (WWG). Nevertheless, it is still architecturally possible to enforce, on top of these structures, higher-level layers that implement some sort of more or less strict control (federated, arbitrated, moderated, centralised styles) which is exactly the option that we exploit in our own OSP approach.

One of the major criticisms against this approach (and against architectural centralisation, in general) is that the center of control becomes at the same time a performance bottleneck and a single point of failure. This is indeed one of the open issues of our work and a major point for further investigation. The final choices on this are going to be made on architectural grounds, as well as on the operational requirements, constraints and capabilities which arise from particular application domain, that of offering one-stop e-Government services.

Acknowledgements

A major part of the work reported in this paper is performed in the context of the CB-BUSINESS (Cross-Border BUSiness INtermediation for Electronic Seamless Services, IST-2001-33147) project, which officially started in April 2002. The CB-BUSINESS consortium is led by Planet Ernst & Young S.A., and includes partners such as the eGovernment Laboratory of the University of Athens, SEMA Group sae, ComNet Media AG, public service providers and professional chambers.

References

1. Atkinson, R., and J. Ulevich (2000) Digital Government: The Next Step to Reengineering the Federal Government, Progressive Policy Institute Technology and New Economy Project, March 2000.
2. Box, D., D. Ehnebuske, G. Kakivaya, A. Layman, N. Mendelsohn, H.F. Nielsen, S. Thatte, and D. Winer (2000) Simple Object Access Protocol (SOAP) 1.1 Official Specification, W3C, May 8, 2000, http://www.w3.org/TR/SOAP/.
3. Boyer, K., R. Hallowell, and A. V. Roth (2002) E-services: operating strategy: a case study and a method for analyzing operational benefits, Journal of Operations Management, Vol. 20, No. 2, April 2002, pp. 175-188.
4. Casati, F. and M.-C. Shan (2001) Dynamic and adaptive composition of e-services, Information Systems, Vol. 26, No. 3, May 2001, pp. 143-163.
5. Christensen, E., F. Curbera, G. Meredith, and S. Weerawarana (2001) Web Services Description Language (WSDL) 1.1 Official Specification, W3C, March 15, 2001, http://www.w3.org/TR/wsdl.
6. Dawes, S., P. Bloniarz, K. Kelly, and P. Fletcher (1999) Some Assembly Required: Building a Digital Government for the 21st Century, Report of a Multidisciplinary Workshop, Center for Technology in Government, March 1999.

7. Lenz, K., and A. Oberweis (2001) Modeling Interorganizational Workflows with XML Nets, Proceedings Hawaian International Conference on System Sciences (HICSS), January 3-6, 2001, Maui, Hawaii.
8. Mecella, M. and B. (2001) Designing wrapper components for e-services in integrating heterogeneous systems, The VLDB Journal, Vol. 10, No. 1, 2001, pp. 2-15.
9. Sahai and V. Machiraju (2001) Enabling of the Ubiquitous e-Services Vision on the Internet, e-Service Journal, Vol. 1, No. 1, 2001, pp. 5-20.
10. Scheppach, R., and F. Shafroth (2000) Governance in the New Economy, Report of the National Governors' Association, Washington D.C., 2000.
11. Shegalov, G., M. Gillmann, and G. Weikum (2001) XML-enabled workflow management for e-services across heterogeneous platforms, The VLDB Journal, Vol. 10, No. 1, 2001, pp. 91-103.
12. Sprecher, M. (2000) Racing to e-government: Using the Internet for Citizen Service Delivery, Government Finance Review, October 2000, pp. 21-22.
13. uddi.org (2000) UDDI Technical White Paper, September 6, 2000, http://www.uddi.org/pubs/Iru_UDDI_Technical_White_Paper.pdf.
14. Vonk, J., W. Derks, P. Grefen and M. Koetsier (2000) Cross-organizational transaction support for virtual enterprises, O. Etzion, P. Scheuermann (Eds.), Proceedings of Cooperative Information Systems (COOPIS 2000), Springer, 2000.
15. West, D. (2000) Assessing E-government: The Internet, Democracy and Service Delivery by State and Federal Governments, Brown University, September 2000.
16. Workflow Management Coalition (1999) Interoperability Proving Framework 1.0, Technical Report WfMC-TC-1021, April 1999.

e-Governance for Local System: A Plan and Implementation Experience

Cesare Maioli

Cirsfid, University of Bologna, Via Galliera 3, 40122 Bologna, Italy
maioli@cirfid.unibo.it

Abstract. In November 1999, the Regione Emilia-Romagna administration (RER) launched the Regional Telematic Plan (RTP), an initiative to increase the awareness of ICT at the different levels of the public sector, to foster their utilization and to allow better services for the small and medium sized enterprises and the citizens. The lead purpose of RTP is to increase the economic competitiveness of the regional enterprises having better served and ICT conscious citizens. We present the main findings after two years of experience.

1 Introduction

Regione Emilia-Romagna (RER) has 4 millions inhabitants, about 400.000 enterprises and, in year 2000, ranked 13th - on 190 regions in Europe - for gross per capita product.

At the end of 1999, RER launched the Regional Plan for Telematics [1], an initiative to foster the ICT insight and utilization and to allow better services for the enterprises and the citizens. The lead purpose of RTP was to increase the economic competitiveness of regional enterprise having better served and ICT conscious citizens, trying to balance and to integrate the efforts and investments in the ICT of more than 500 autonomous Local Administrations; nine are the Provinces, which actually are managing wide area networks and information services for other bodies.

RER has been a leader in several cases with regard to the ITC, at times even on a European level [2]. They were, however, almost always isolated cases: therefore we involved a "teamwork" of politicians, managers and technicians to support the identification of regional problems and interests in a regional process, integrating them in an inter-regional and European context so that the whole regional Public Administration, might improve itself and its ways of operating[3].

Cheaper and faster internet access, promotion of digital literacy, more advantage for the public sector from the digital technologies were the initiatives to bring on-line every citizen, every company, every administration [4].

RER planned to avoid the risks [5] that:

- its main cities or a few of its industrial districts would become increasingly estranged from their hinterlands leaving these areas excluded from global currents and unable, on their own, to chart a viable path towards participation in the information society
- cities, especially the university ones, would specialize in knowledge-based activities making wider the economic gulf between them and their hinterland.

R. Traunmüller and K. Lenk (Eds.): EGOV 2002, LNCS 2456, pp. 272–275, 2002.

The origin of the Plan dates back to 1997 when the regional Government identified computer networks and information services as the bases to leverage in order to increase the competitiveness of the whole regional system instead of considering them just as organizational services [6].

2 The Telematic Plan

The points of intervention proposed in the RTP to pursue the telematic development of the region in the perspective to reach an Agreement with the Local System, were the following:

1 - Innovation of services for citizens and enterprises

The objective is the development on the whole regional territory of high quality and efficient public services, integrated as much as possible, with easy and convenient access taking up the numerous opportunities offered by ICT.

2 – Strengthen and completion of the Unitary Network of Emilia-Romagna

To simplify and improve the public services for citizens and enterprises it is necessary to achieve a greater integration and a better internal communication in the whole regional public administration.

3 - Modernisation of the regional government

Innovation in the operational methods of the regional public administration fully concerns the Region administration itself.

4 - Diffusion of the "fourth knowledge" and public access to information society

Without a rapid and generalized development of competence on ICT the abovementioned aims risk being unachievable..

5 - Promotion of electronic commerce and new media industry

A more intensive use of ICT by small and medium sized enterprises, in order to take advantage of the electronic commerce opportunities, and a reorganization of operative and managing processes represent a crucial challenge for the economic development of Emilia-Romagna

6 - Promotion of a competitive regional market of telecommunications and development of internet services

We considered points 2, 3, 4 as prerequisites for point 1 and points 1, 4, 6 as prerequisites for point 5, which constitutes the rationale of the entire RTP.

3 The Implementation of the Plan

During year 2000, RER made relevant investments to strengthen the regional telematic network: the co-operation with the Provinces in the design and financing made the connection between the offices of the different institutions more widespread.
Later the local telematic plans, compliant with RTP, were collected on the basis of a call issued by RER.

Since we wanted the political boards of the local authorities to consider the ICT not just a technical matter, we accepted the plans only in presence of a clear commitment of the political the boards. *We think that without a strong growth of interest and in-*

sight by the political decision makers in the ICT politics, the public sector risks a governance deficit about the processes reshaping everybody's life.

Also if we recognize the need for a constructive understanding between politics, industry and the public, we consider the ICT industry as a solution provider but we stress the need for politics to take the lead interpreting and anticipating the public requests and wishes.

In particular we asked the local authorities to contribute for about 50% of the cost in case of plan approval.

We financed the planning effort; often the plans were prepared by consortia of small bodies who never before had the chance to plan their information service development in a framework of political objectives.

We received 51 different plans, clustering about 300 different *projects* from more than 250 different authorities.

Finally, at the end of year 2000, RER refined its original PTR, and identified the fields of intervention to fund. They are of three different kinds:

A - Provision of a broad band telecommunication infrastructure for the regional territory

The solution will follow the final choice between the involvement of local utility companies, big ICT players, RER autonomous implementation through a project financing approach.

B - Design and implementation of ICT tools and services

They are developed together with Municipalities, utility companies and Provinces' administrations. The actual efforts are:

- integration of large archives dealing with health data, enterprise data, geographical information systems ,
- one-stop shop services in the fields of health services booking, managing of the tourist sector, monitoring of the natural environment data, information and procedures for the enterprises
- a unified regional Web portal on top of the numerous initiatives of Local Authorities' civic networks and internet information sites
- provision of tools for e-procurement and support to e-commerce initiatives via a regional electronic card.

C - Promotion and balance of the local administrations' information systems according to the RTP

The activities concern geographical information systems, statistical systems, electronic document management, interface usability and citizen-friendship, upgrading of technical skills of the personnel.

The financial investments are in the order of 200 million EUROs in five years for B and C, and of 70 million EUROs for item A.

4 Conclusions

We can summarize the main findings of our experience as follows.

From the organizational point of view, they are:

- the rate of involvement and insight of the political and regulatory bodies are of paramount importance; many times the decision making about apparently technical aspects given to information officers and technologists is a short minded and ineffective measure

- the appointment of the project ownership to local authorities make the inner energy grow.

From the technical assessment point of view, they are:

- the network and the internet paradigms are more and more accepted and assimilated and virtual communities of users of existing services spread the perception of the technologies as an enabler for sustainable communities
- in particular the one-stop approach and the availability of a bundle of transactional services behind a single query is felt as a target "at reach" by the wide public
- while great concern is detected about privacy and data transparency, some basic technical issues, such as data integrity, information redundancy and system consistence, are given for granted

From the management point of view, they are:

- the PTR experience provided a valid laboratory for experimenting new ideas, methods and new regulatory arrangements (e.g. area plans, consortia of bodies at various institutional levels)
- new and more effective partnership emerged among the local authorities
- the involvement of politicians has permitted a wider dissemination of best practice at local level, an increase of available resources and has encouraged the creation of the critical mass required for many projects.

References

1. Guidelines for the telematic development of Emilia-Romagna: a proposal to the local system, http://www.regione.emilia-romagna.it/pianotelematico/ Nov. 1999
2. E-Italia, Forum Società dell'informazione, Il Sole24Ore, June 2000
3. Sandri A., InRete: first action plan of the Regional Telematic Plan, RER, February 2000
4. Liikanen E.; Europe in the information age – accelerating the transition, Telecom Conference, Helsinki, October 1999
5. Gillespie A., Information technologies and integrated city-region development, TeleRegions Conference, Tanum, June 1999
6. La Forgia A., The Global Region, RER, July 1997

Transactional e-Government Services: An Integrated Approach

C. Vassilakis, G. Laskaridis, G. Lepouras, S. Rouvas, and P. Georgiadis

e-Gov Lab, Department of Informatics, University of Athens, 15784, Greece
{costas,glask,gl,rouvas,p.georgiadis}@e-gov.gr

Abstract. Although form-based services are fundamental to e-government activities, their widespread does neither meet the citizen's expectations, nor the offered technological potential. The main reason for this lag is that traditional software engineering approaches cannot satisfactorily handle all of electronic services lifecycle aspects. In this paper we present experiences from the Greek Ministry of Finance's e-services lifecycle, and propose a new approach for handling e-service projects. The proposed approach has been used successfully for extending existing services, as well as developing new ones.

1 Introduction

e-Government development may be quantified through a set of indicators to measure comparative progress. *eEurope* has published a list of 20 basic public services [1] which should be considered as the *first steps* towards "Electronic Government", along with a methodology for measuring government online services [2]. It is worth noting that among the basic public services listed in [1], 75% of them include e-forms filling and submission. Moreover, the citizen interface must be connected to the back-office in order to provide the transactional capability, which will allow it to offer the rich mix of services that customers want and governments have promised [3].

However, transactional services widespread currently lags behind the expected level, taking into account potential offered by technology. Besides structural reforms, and the adaptation of a customer-centric model [4], the development and maintenance processes for such services are quite complex: firstly, service requirements must be analysed; secondly, the service has to be designed, considering functional requirements, user interface aspects, and administrative issues. Implementation and deployment should then commence, and the e-service platform should be linked to some installed IT system, for exchanging data. Finally, when changes to the service are required, the whole process must be carried out, resulting in costs and delays.

2 Experiences from an Electronic Service's Life Cycle

The development of the electronic tax return form service for year 2001 followed a classic software engineering paradigm, starting off with the user requirements analysis. Four requirement dimensions were identified in this procedure (a) appropriate input forms (b) *input validation checks*, which verify that user input is conformant to tax legislation rules (c) forwarding of collected data to a back-end system for the tax computation process and (d) collection of the results of the tax computation

R. Traunmüller and K. Lenk (Eds.): EGOV 2002, LNCS 2456, pp. 276–279, 2002.

process and user notification about the final result. These four requirement dimensions are in fact the "computerised" counterparts of the paper-based tax return form and tax computation process; some additional issues had also to be faced due to e-service operation (a) issuing of *identification credentials* service users (b) personal data management for authenticated users and (c) provision of a full service administration framework for the GmoF's administration team, including account management, database backup and recovery, statistics reports etc.

The design phase followed, producing detailed specifications of the various components and procedures, along with specification of the interfaces between the different stages of the information flow. Subsequently, each portion of the work was implemented "autonomously", and an integration step consolidated the different work dimensions into a single, operational platform. After the service became operational, maintenance tasks were performed, mainly for the purpose of correcting appearance problems and modifying or enhancing input validation checks.

During the life cycle of the project, a number of shortcomings of the followed approach were identified, which led to increased product delivery times and the need to involve more staff. These shortcomings are discussed in the following paragraphs.

1. Knowledge existed *implicitly* within the organisation, usually under the possession of experienced individuals, rather than stored in some publicly accessible repository in an explicit form. This affected all subtasks dealing with the organisation's *business logic*. During interviews, the analysis team often collected partial, or contradicting descriptions of the rules that applied to different cases.
2. User interface design did not include adequate domain expertise into forms. For instance, semantically related fields should be placed close together on the form for easier access. The user interface should also compensate the lack of expert assistance offered by tax officers in local tax offices.
3. Code reusability remained low, since code incorporated business rules, presentational elements, administrative issues and data repository accesses in an environment of increased cohesion. Thus changes to one of these issues triggered modifications to the other dimensions.
4. *Communication with back-end systems*. A full transaction processing cycle usually involves data exchange between the service platform and some organisational information system, either in an on-line or an off-line fashion. In both cases, building interfaces tailored for each case is not a good practice, since it requires substantial programming effort to address the peculiarities of the back-end system.

Regarding these issues, a number of technological solutions are available. Recording knowledge in an explicit and reusable format is addressed by knowledge management tools, while a number of research projects introduce novel methods to knowledge management ([5], [6]). For the service development and deploying phases both commercial products (e.g. XMLForms™ and Oracle E-Business Suite™) and Open Source solutions. In the standards area, the W3 consortium has published the *XForms* specification, an XML-based standard for specifying Web forms, while user-centric approaches for interfaces are discussed in the ISO 13407 standard.

3 Proposed Approach

Although the state-of-the-art provides sufficient tools for tackling the various phases of e-service lifecycle, these phases are still handled in isolation. To provide an alternative solution for the above-described problems one has to adopt a layered approach that introduces higher levels of abstraction (enhancing maintainability and re-usability), isolate knowledge from code, allow asynchronous development of modules and enable people with domain expertise to implement new services.

This approach shifts implementation focus from programmers to domain experts, who possess the knowledge needed to process the data gathered through the e-service. In this view, an e-service consists of the e-form and the processing of the submitted data. Processing has two phases: validation and information extraction. In the first phase submitted data is checked against a set of rules and, depending on the result, the user may be asked to change and re-submit the form. Although some simple error checks (e.g. date and number checking) are usually coded by programmers, for more complex dependency checks domain expertise is of the essence. For example, a check "if user selects field 170 and 180 and has less than two children, then field 190 has to be completed according to the law 1256/98" requires deep domain knowledge. So far, either the programmer worked closely with domain experts to program these checks, or the e-service gathered data that were processed manually, resulting in late error detections. The proposed approach enables experts to play a much more active role and carry out a much greater percentage of the implementation effort.

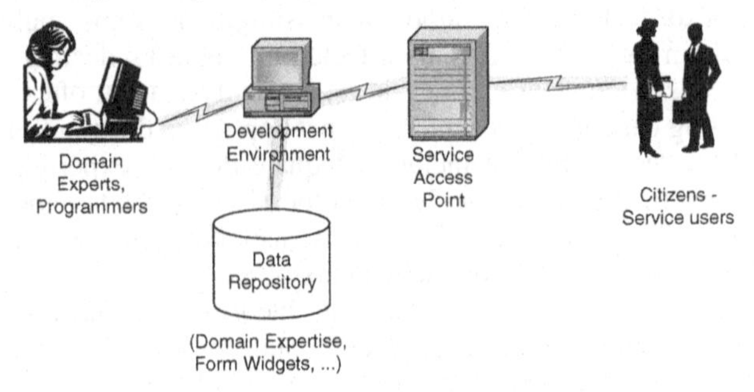

Fig. 1. System Architecture

As illustrated in Fig. 1., the development environment employs a data repository, where the ingredients of an e-service can be stored. These can be the widgets that compose the e-form along with their properties (type, multilingual labels, etc.) and the domain expertise needed to implement the service. Domain experts and/or programmers can use the development environment to implement and activate new services. The development environment has been designed to be friendly and intuitive, since a significant portion of its expected users (i.e. the domain experts) is not usually too familiar with the programming issues concerning the implementation of an e-service. To this end, an appropriate set of tools is provided.

Service development begins with the definition of the *forms* that comprise the service, and each form is then populated with the respective fields and labels. Fields and field groups may be directly drawn from the data repository, promoting thus uniformity between various services and enhancing reusability. For instance, the area displaying the user's personal data may be defined once, and then be used in all taxation e-forms. If no suitable component exists, the developer may create a new one, by entering information such as the description, its type (string, arithmetic, etc.), whether it is editable or not, its appearance on the screen etc. Additionally, a high-level specification of data interchange with the organisation's IT systems is provided, while *validation rules* may be attached to fields. A second level of validation rules may be attached at *form level* and *service level*, to cater for checks involving multiple components. Validation rules are defined using *semantically rich elements*, e.g. "the total husband's income from salaries", rather than implementation-oriented terminology, such as *table4_field22*. This enhances readability and facilitates maintenance, since the level of abstraction remains high and semantic information in retained in this representation. Finally, domain experts would also provide the test cases to verify the validation process.

The adoption of the proposed approach offers significant advantages: Firstly, development is faster since previously developed and tested e-service components can be directly used in new services; this allows for cost reduction as well. Secondly, maintenance is facilitated , since it is easy to locate the components and services affected by some change either in legislation or policy. Moreover, changes are now mainly conducted by domain experts. Thirdly, consistency is increased, through the introduction of a central repository for knowledge storage. Uniformity in the look and feel may be also achieved through usage of service templates. Finally, *multiple dissemination platforms* may be supported, through appropriate content generators that create content suitable for a variety of platforms (Web, WAP, etc).

4 Conclusions

In this paper we presented experiences from developing and maintaining a set of e-services for the Greek Ministry of Finance. The traditional software engineering approaches employed in the first development phases proved to be inadequate in handling all aspects related to the lifecycle for electronic services. In the second phase we used a new approach, together with appropriate software tools, which allowed for using higher levels of abstraction, enhancing thus the maintainability, portability and reusability of the project's results, and reducing overall development time.

References

1. eEurope, "Common list of basic public services"
2. eEurope, "eGovernment indicators for benchmarking eEurope".
3. Vivienne Jupp, "eGovernment – Lessons Learned, Challenges Ahead", eGovernment Conference: From Policy to Practice, 29-30 November 2001, Charlemagne, Brussels.
4. Frank Robben "(Re)-organising for better services", eGovernment Conference: From Policy to Practice, 29-30 November 2001, Charlemagne, Brussels.
5. Know-Net Consortium, "Manage Knowledge for Business Value"
6. DECOR Project, "Delivery of context-sensitive organisational knowledge"

Electronic Vote and Internet Campaining: State of the Art in Europe and Remaining Questions

Laurence Monnoyer-Smith and Eric Maigret

University of Technology of Compiègne, CNRS, France
University of Paris 3- Sorbonne, CNRS, France
EVE partner

Abstract. Recent experiments shows that internet voting is not the political blessing that lots of politicians had hoped it would be to solve the non-ending legitimation crisis modern societies are going through. Our article focuses on four questions about internet voting to stress some experimental results and important remaining questions. We would like to insist on the lack of studies trying to apprehend how people trust these new electronic voting systems and how do they cope with the end of the voting rituals.

1 Introduction

The internet is quite commonly viewed as being able to serve as a new linkage institution enhancing voter information about candidates and elections in general, and as a consequence mobilizing an apathetic population and even, increasing voter turnout[1].

This is far from being a simple equation. Recent studies are showing interesting and mixed results about the ability of the internet to enhance political involvement: it is not quite obvious that internet is a political blessing. Rather, it seems to evolve toward a managerial model which concern "the 'efficient' delivery of government information to citizens and other groups of 'users' ; the use of ICT's to improve flows of information within and around the organs of government; a recognition of the importance of 'service delivery' to 'customers' ; the view that speeding up information provision is by itself 'opening up' government[2]..."

Our current EVE project, financed by the European Commission (DG IST), allows us to compare 11 different e-democracy experiments that permits us to stress a few points on which our knowledge is more or less salient and some remaining very important questions that have to be answered by those who are developping new processes aiming at voting, delibarating and participating in political life.

---

[1] See for exemple, the motivations developped by the UK government on their new on-line voting program: www.sheffieldvote.com,

[2] Chadwick A., "*Interaction between citizens in the age of internet: "e-goverment" in the United States, Britain and the European Union*", Paper given to the APSA's Annual Meeting, San Francisco, August-September 2001, p. 2.

R. Traunmüller and K. Lenk (Eds.): EGOV 2002, LNCS 2456, pp. 280–283, 2002.

2 Do People Seek Political Information on the Internet during Election Campaign?

It is here very interesting to note how far from the expectations european and canadians users stand, especially compared to the USA. In the UK for exemple, James Crabtree[3] shows that during the june elections, although the UK is catching America's acess levels, british citizens lag far behind in understanding the real potential of internet.

Although we have to be cautious with extending the UK case to the other european countries, it seems that some important barriers prevent people for opting-in the internet contrary to the US citizens. The available information seems not to be interactive, attractive, and different enough from the one available off-line. During the last French presidential elections for exemple, none of the candidates has made specific efforts to present interactive informations.

The first contact with the information is crucial if the user is to come back for more: until more efforts are made, internet users won't pay enough attention to it.

3 Do the Use of Internet Enhance Voting Turnout?

Here again, we can't give straight answers considering the available datas. A few remarks can be made, concerning some specific populations studied during the past elections and the famous Arizona case.

There is no evident proof of an existing link between the act of voting and the use of internet whether it is to vote or to get political information. If there is a link, it might be very different from one election to another (from a presidential to a parlementary election): this might explain the differencies shown between the US and the UK and between the 2000 studies and the 1998 studies.

One has to consider as well that situations mights vary a lot from one country to another: in the UK last local elections for exemple (may 2002) electronic voting has attracted more people than usual: the publicity given to the Sheffield and Liverpoll experiments might be nevertheless a bias. Country in which elections are taking place on a week day might be more keen to vote via internet than those, like France, which organise their election on sundays.

Except from the Arizona study, which is very different from a national election, they all show that internet users are no more keen on voting than non users. It does not answer the question that if they could do so via the internet, people would vote more or less.

The Digital divide remains a fundamental issue for internet voting. The Arizona studies are congruent with each other underpinning that particular demographic groups already noted for being democratically disadvandaged are excluded even more than others with the introduction of internet voting.

---

[3] Crabtree J., « *Whatever happen to the e-lection* », a survey done by isociety, juin 2001. available on: www.ipf.co.uk/egovernment/posts/post99.htm

4 What Do We Know about the Young People and Politics on the Internet?

One of the most stable findings is that the young, are more likely to use the Internet than the elderly. The internet thus may represent an important venue for mobilizing young voters who have historically been underrepresented in the electorate, and traditionnaly less voting. Correlated with what have seen before, ie that internet users are not more likely to vote than non users, we can assume than this is partly caused by the importance of young people amoung users. Their political cynicism is spreading on the internet.

According to Henn and Weinstein (2001), the internet voting is not considered by young people to be a substancial reform within the political system and therefore will not more than marginnaly modify their behavior. –Which is better than nothing but will not in itself solve the legitimacy crisis that the political system suffers from- Neverthless, some of the "substancial" problems could be partly solved by an interactive use of the internet (provide information, interact with voters...)[4].

This has been confirmed by the european project E-Poll which stressed that in Avellino where has been organised the first biometric recognition voting system for the Italian constitutional referendum on october 7th 2001, elderly people were more keen on trying this system that young people and they do trust the system more than under 30 years old people.

5 Trust, Rituals and Electronic Voting

No sociological research has been done on those specifics and yet fundamental aspects of the question of e-voting.

If one of the advantages of internet voting is certainly that it eliminates counting mistakes and saves a lot of time when giving the results, lots of security aspects remain uncertain for citizens. Security is not only a technical problem that might be solved by ingenieurs and improved as the technology is getting more complex and close to the 100% identification certitude (with fingerprint for ex), it is also a psychological issue. Then it is more appropriate to talk about "trust" than security from the citizen's point of view. If some research has been done on the e-business, none has tried to understand how to build a trust relationship via the internet with the organisation that will run the election on line. Time will probably be necessary to implement the system and to bring people to trust it as they do when they buy on line (Filser[5]). All e-business studies on that matter show as well the role played by intermediary actors in reinforcing the faith people have in a system (hot line, etc.): who could they be in the case of electronic voting and what kind of intervention could they have? These questions need to be studied or we might be disappointed by the low use of online voting systems.

---

[4] Henn and Weinstein, « Youth and Voting behavior in Britain » ", Paper given to the APSA's Annual Meeting , San Francisco, August-September 2001.

[5] Filser M, "Etat des recherches sur les canaux de distribution", in *Revue Française de Gestion*, Sept.-Oct 1998, pp. 66-76.

Another issue never analysed but stressed by some researchers[6] is whether the withdrawal of such an solemn event as voting, might loose his signification as a symbol which unite people in a common commitment toward democracy. In other words, electronic voting might remove from its pedestal such an important ritual from which generations have fought for. Does the move toward electronic voting correspond to this trend that tend to bring people closer to the elite or does it dilute democracy into a more vast consumer society were voting becomes a consumer act as buying a train ticket on line? One might underline that by introducing information technology at the heart of the political decision process, the society emphasises the passage from a culture of effort to a cultur of service in political matters. This has a lot of implications that should be studied and taken into account in future technological developments.

6 Conclusion

The potential for electronic governance is almost certain only in appropriate conditions. The lack of information that citizen often stress could be fullfiled by the new media. In concertation with political actors who have to be involved in the implementation of new technology, and in consideration of all the citizen's claim –not only access and practical questions[7]-, the development of e-governance processes could have then a positive influence on political participation. On its own, internet is not a blessing.

One recent hypothesis (Bucy et Gregson[8]) is that new media formats might satisfy this need for popular involvement by delivering a continuous stream of opportunities for civic engagement without overextending the government ability to respond. The multimedia thereby increases the likelyhood that citizen concerns will be heard. By making allowance for continuing mass involvement, new media formats serve the socially valuable purpose of bringing closer to reality to classical goal of full participation without over extending already burdened political institutions.

---

[6] Maigret E., Monnoyer-Smith L., « Le vote en ligne: Usages émergents et symboles républicains », Colloque ICUST, Juin 2001 ; Shnapper D., *Qu'est-ce que la citoyenneté*, Paris, Gallimard Folio, 2000.

[7] As C. Aterton already stresses it in 1987, the renewal of political life has very little to do with the technical aspects of voting.*Can technology protect democracy*, London, Sage.

[8] Bucy E.P., Gregson K.S., " Media participation: a legitimizing mechanism of mass democracy", *New Media and Society*, Vol.3(3)., pp. 357-380.

A Citizen Digital Assistant for e-Government

Nico Maibaum*, Igor Sedov**, and Clemens H. Cap

University of Rostock, Department of Computer Science,
Chair*** for Information and Communication Services
{maibaum,igor,cap}@informatik.uni-rostock.de

Abstract. In this short paper we describe the architectural concept of a Citizen Digital Assistant (CDA) and preliminary results of our implementation. A CDA is a mobile user device, similar to a Personal Digital Assistant (PDA). It supports the citizen when dealing with public authorities and proves his rights - if desired, even without revealing his identity. Requirements for secure and trusted interactions in e-Government solutions are presented and shortcomings of state of the art digital ID cards are considered. The Citizen Digital Assistant eliminates these shortcomings and enables a citizen-controlled communication providing the secure management of digital documents, identities, and credentials.

1 Introduction

"The quality of life is also depending on how good and how fast a government is doing its services to the public" said Germany's minister Otto Schily (Federal Ministry of the Interior) at the congress 'Efficient Government' in Berlin 2001 [1]. Today, every public authority is talking about a modern, efficient and citizen-friendly administration. The vision is to map the possibilities obtained through the new IT technologies, and demonstrated within the e-business sector, to the e-government sector. At the end, the citizen will communicate (via Internet) with a modern, efficient, cost-effective and transparent administration.

In order to achieve the above-mentioned goals research projects are dealing with different digital signature, national ID or citizen cards for usage within e-Government applications. Examples are the research project FASME[1], the FineID[2] and the DISTINCTID[3] project. The process of registering at a new place of living within Europe is facilitated by FASME. Administrative data, documents and profile infor-

* Supported by the EU Fifth Framework Project FASME, http://www.fasme.org

** Supported by the National Research Foundation (DFG)

*** Supported by a grant of the Heinz Nixdorf Foundation

[1] Facilitating Administrative Services for Mobile Europeans (FASME), http://www.fasme.org

[2] Finnish Electronic Identification (FineID), http://www.sahkoinenhenkilokortti.fi

[3] Deployment and Integration of Smartcard Technology and Information Networks for Cross-Sector Telematics (DISTINCTID), http://distinct.org.uk

R. Traunmüller and K. Lenk (Eds.): EGOV 2002, LNCS 2456, pp. 284–287, 2002.

mation of a citizen are stored on a JavaCard. One main problem within this project was the low memory of the JavaCard. The solution is a secure card extension using a kind of virtual memory for smart cards [2].

2 Requirements for Secure and Trusted Interactions

The following four requirements are necessary for a secure and trusted interaction between a citizen and public authorities. Presently, they are not satisfied by any consumer device (for example smart cards as digital ID cards).

The **Subscriber Identity Framework** is responsible for access control and controlled disclosure and release of identity (*real identity*, *pseudonymity* and *anonymity*). The on-card functionality of smart cards (e.g. digital signature, payment etc.) is normally restricted through a personal identification number (PIN) with a length of four up to eight digits, sometimes changeable through the cardholder. Today, the increasing number of different passwords and PIN codes is a serious problem for many people, because they are not able to remember all these codes. Furthermore, the verification of a PIN realises no real *proof of identity*, just *a proof of knowledge*.

The **Monitoring and Confidentiality Framework** is responsible for the management, supervision, and logging (security audit) of trustworthy transactions. Smart cards are the carrier devices for digital signatures. A resulting problem is obvious: *"How can it be guaranteed that the document to be signed by the smart card is indeed the document delivered to the smart card for signature?"* The smart card does not include a monitoring module. The cardholder must trust the card accepting device and the system behind it.

The **Secure Document Management Framework** is responsible for the secure storage, retrieval and administration of digital counterparts of paper-based documents (e.g. XML documents as e-docs), certificates, citizen profile information, and credentials [6]. At the moment everyone is talking about digitally signed documents but no one is talking about the secure storage of these e-docs. Smart cards are not utilisable because of their limited storage capacities.

The **Communication and Cryptographic Framework** is responsible for the communication between the citizen and public authorities. This includes unilateral, bilateral and multilateral security, secure key management, en- and decoding and the definition of different security levels. A smart card must rely on the card accepting device or the connected personal computer. The smart card is not able to connect to another device on its own.

3 Implementation Aspects and Application Scenarios

To fulfil the e-Government requirements described in Section 2, smart cards, PDAs, or mobile phones, are just partial solutions not fulfilling all these requirements. A Citizen Digital Assistant can be realised as a combination of a PDA and a mobile phone. We propose an architecture integrating a biometric component (*data storage*, *comparison process*, and *integrated biometric system*) for a real citizen authentication / identification. Biometric information never leaves the device (and therefore cannot be misused), comparable to private keys, and there is no need to rely on other external devices.

Every CDA is identical. Only after a personalization the CDA gets its own "citizen" identity. Therefore at least one card slot for a digital signature card or an electronic, national identity card is required in order to personalise the CDA. Problems, which still exist by using smart cards, can be solved with the CDA. The user does not need to insert her plastic card in any "unknown" device. She inserts the card in her trusted CDA and there is no danger that regular operations are in the sphere of influence of another instance or card-accepting device [4] with not checkable hard- and software.

The CDA is a device which *"talks"* and *"interacts"* vicariously for the citizen (see Figure 1). The user device is a combination of existing and widespread components. The costs continuously going down and the unbroken trend for miniaturisation of the components make the appropriate realisation not only thinkable but also possible.
The CDA can be used for e-voting and can be the carrier device for temporary credentials (see Fig. 1), any kind of digital documents or profile data.

Fig. 1. Citizen remains anonymous via CDA

The citizen will no longer have to carry personal documents in paper form with them in order to obtain certain administrative services. Asynchronous data transfers (e.g. document verifications and document requests) with long latency can be supported, too. The citizen will not have to wait with her card inserted into a kiosk terminal until the complete administrative process is finished.

Such a user device can be used not only within the e-government sector but also within the e-business and internet sector, too. Everywhere the protection of privacy and identity is required the CDA can act.

4 Conclusion and Future Work

Our preliminary CDA implementation is based on the Compaq IPAQ 3660 running Linux 2.4.14[4]. To illustrate spontaneous networking in an e-government scenario the simple, low cost and low energy consuming Bluetooth technology (Ericson Bluetooth Application & Training Tool Kit) is utilised.

Enabling a personal identification without PINs, fingerprint technology based on the Siemens TopSec ID Module was selected. The advantage of this module is the

[4] Some parts are developed within the DFG research project "Security architecture and reference scenario for spontaneous networked mobile devices" at the University of Rostock [5].

secure enrolment, storage and verification inside the exchangeable module. The Service Discovery Protocol provides mechanisms for discovering and associating services within an ad-hoc community. To support multilateral security and the integration of security interest of communication partners an advanced security manager concept was developed, that is partly implemented [3] and is now evaluated.

Work is in progress to implement all the required interfaces. On top of the architecture will be a small Java-based e-government application, that was implemented within the European Research Project FASME including the usage of the Java-based FASME smart card.

The realization of such a CDA is the core module within an e-(government) system. Fixed IPs and unique processor serial numbers for the identification are avoided. Networks without user observability would facilitate a lot of problems [7].
In the future it will be possible that every citizen will have her own CDA including the desired management system in her pocket, possibly distributed by the authorities for a low fee like other official documents or different cards (e.g. passports and health insurance cards).

Smart cards are ideal for many application areas, but the missing monitoring functionality, the limited memory and the insufficient possibilities for biometric authentication requires new solutions.

References

1. O. Schily. Federal Ministry of the Interior, Germany. *Auf dem Weg zu einer modernen Verwaltung - BundOnline 2005*, 2001. Talk within the congress "Efficient Government".
2. C. Cap, N. Maibaum, and L. Heyden. *Extending the Data Storage Capabilities of a Java-based Smartcard.* In Proceedings of the Sixth IEEE Symposium on Computers and Communications ISCC 2001, pages 680 - 685, IEEE Computer Society, July 2001.
3. H. Buchholz. *Sicherheit in Ad-hoc Netzwerken.* Diploma Thesis, 2002.
4. H. Federrath and A. Pfitzmann. *Bausteine zur Realisierung mehrseitiger Sicherheit*, in G. Müller and A. Pfitzmann (Hrsg.): Mehrseitige Sicherheit in der Kommunikationstechnik, Addison-Wesley-Longmann, 1997.
5. Sedov, M. Haase, C. Cap, and D. Timmermann. *Hardware Security Concept for Spontaneous Network Integration of Mobile Devices.* In Proceedings of the Innovative Internet Computing Systems Conference, Ilmenau. Springer, June 2001.
6. Cap, Clemens H. and Maibaum, Nico. *Digital Identity and it's Implications for Electronic Government.* In Towards the E-Society - E-Commerce, E-Business, and E-Government, Kluwer Academic Publishers, Boston, 2001.
7. A.Pfitzmann and M.Waidner. *Network without User Observability.* Computers & Security, vol.2, no.6, pages 158-166, 1987.

A System to Support e-Democracy

Jan Paralic, Tomas Sabol, and Marian Mach

Technical University of Kosice, Letna 9, 041 20 Kosice, Slovakia
{Jan.Paralic,Tomas.Sabol,Marian.Mach}@tuke.sk
http://esprit.ekf.tuke.sk/webocracy/

Abstract. This paper briefly describes functionality and customisation support of the system called WEBOCRAT, which is being developed within the EU-funded project Webocracy (IST-1999-20364 "Web Technologies Supporting Direct Participation in democratic Processes"). The WEBOCRAT system represents a rich set of communication supporting tools that will bring public administration closer to citizens, making it more accessible and more accountable.

1 Introduction

The Webocracy project responds to an urgent need for establishment of efficient systems providing effective and secure user-friendly tools, working methods, and support mechanisms to ensure the efficient exchange of information between citizens and public administration (PA) [1]. This project addresses the problem of providing new types of communication flows and services from public PA institutions to citizens, and improves the access of citizens to PA services and information.

In [6] a three-phase strategy, for implementing e-democracy consisting of initiation, infusion and customisation phases, has been proposed. The first, *initiation phase*, starts with establishment of a portal that conveniently links citizens to PA. Next, *infusion phase*, means restructuring the organisation in order to accommodate innovation. And, finally, *customisation phase* of e-democracy system implements a one-to-one relationship between citizen and PA. The *WEBOCRAT* system focuses on support of infusion and mainly customisation phases by various specialised modules.

One of the main novelties of our approach is the knowledge-based support [4]. Information of all kinds produced by various modules or segments of these documents is linked to a shared ontology representing a domain of PA. The advantage of clear structuring and organising of information is more powerful search and retrieval engine, and more user-friendly content presentation [2].

2 WEBOCRAT System Functional Overview

From the point of view of functionality of the *WEBOCRAT* system it is possible to break down the system into several parts and/or modules [5]. They can be represented in a layered sandwich-like structure, which is depicted in Fig. 1.

R. Traunmüller and K. Lenk (Eds.): EGOV 2002, LNCS 2456, pp. 288–291, 2002.

Fig. 1. *WEBOCRAT* system structure from the system's functionality point of view

A Knowledge Model module occupies the central part of this structure. This system component contains one or more ontological domain models providing a conceptual model of a domain. The purpose of this component is to index all information stored in the system in order to describe the context of this information (in terms of domain specific concepts).

Information stored within the system has the form of documents of different types. Since three main document types will be processed by the system, a document space can be divided into three subspaces – publishing space, discussion space, and opinion polling space. These areas contain published documents to be read by users, users' contributions to discussions on different topics of interest, and records of users' opinions about different issues, respectively.

Since each document subspace expects different way of manipulating with documents, three system's modules are dedicated to them. Web Content Management module (WCM) offers means to manage the publishing space. It enables to prepare documents in order to be published (e.g. to link them to elements of a domain model), to publish them, and to access them after they are published. Discussion space is managed by Discussion Forum module (DF). The module enables users to contribute to discussions they are interested in and/or to read contributions submitted by other users. Opinion Polling Room module (OPR) represents a tool for performing opinion polling on different topics. Users can express their opinions in the form of polling – selecting those alternatives they prefer.

In order to navigate among information stored in the system in an easy and effective way, one more layer has been added to the system. This layer is focused on retrieving relevant information from the system in various ways. Citizens' Information

Helpdesk module (CIH) is dedicated to search. It represents a search engine providing three different types of searches – keyword-, attribute- and concept-based.

Reporter module (REP) is dedicated to providing information of two types. The first type represents information in an aggregated form. It enables to define and generate different reports concerning information stored in the system. The other type is focused on providing particular documents – but unlike the CIH module it is oriented on off-line mode of operation. It monitors content of the document space on behalf of the user and if information the user may be interested in appears in the system, it sends an alert to him/her.

The upper layer of the presented functional structure of the system is represented by a user interface. It integrates functionality of all the modules accessible to a particular user into one coherent portal to the system and provides access to all functions of the system in a uniform way.

In order for the system to be able to provide required functionality in a real setting, several security issues must be solved [3]. This is the aim of the Communication, Security, Authentication and Privacy module (CSAP).

3 Customization Support in the WEBOCRAT System

Since the system can contain a lot of information in different formats (published information, discussion contributions, polling results), it may not be easy to find exactly the information user is looking for. Therefore he/she has the possibility to create his/her profile in which he/she can define his/her interests and/or preferred way of interacting with the system.

When defining an area of interest, user can select elements from a domain model (or subparts of this model). In this way user declares that he/she is interested in topics defined by the selected part of the domain model.

The definition of user's area of interest enables alerting – user can be alerted, e.g. on changes of the domain model, when a new discussion or opinion polling has been opened, new documents published etc. User has the possibility to set alerting policy in detail on which kind of information he/she wants to be alerted in what way (including extreme settings for no alerting or alerting on each event taking place in the system). The system compares each event (e.g. submission of a discussion contribution, publishing a document, etc.) to users' profiles. If result of this comparison is positive, i.e. the user may be interested in the event, then the user is alerted.

Alerting can have two basic forms. The first alternative is represented with notification using e-mail services. User can be notified on event-per-event basis, i.e. he/she receives an e-mail message for each event he/she is alerted on. Alternatively, it is possible to use an e-mail digest format – user receives e-mail message, which informs him/her about several events. The way of packaging several alerts into one e-mail message depends on user's setting. Basically, it can be based on time intervals and/or the size of e-mail messages.

The other alternative is a 'personal newsletter'. This does not disturb user at unpredictable time – user simply can access his/her newsletter when he/she desires to be informed what is on in the system. Moreover, he/she can access it from arbitrary gadget connected to the Internet. The personal newsletter has the form of a document published in the publishing space. This document is generated by the system and contains links to all those documents, which may be of interest for the user. Since the document is generated when user logs in, it can cover all information submitted and/or published since the last user's visit.

User registered in the system as an individual entity (i.e. not anonymous user) is provided with a personal access page ensuring him/her an individual access to the system. This page is built in an automatic way and can consist of several parts. Some of them can be general and the others are user-specific.

The personal access page hides division of the system into modules. Terms 'publishing space', 'discussion space', and opinion polling space' do not confuse users. The personal access page enables user to access all functionality of the system, which he/she is allowed to access in a uniform and coherent way.

Acknowledgements

This work is done within the Webocracy project, which is supported by European Commission DG INFSO under the IST program, contract No. IST-1999-20364 and within the VEGA project 1/8131/01 "Knowledge Technologies for Information Acquisition and Retrieval" of Scientific Grant Agency of Ministry of Education of the Slovak Republic. The content of this publication is the sole responsibility of the authors, and in no way represents the view of the European Commission or its services.

References

1. Becker, T. (2000). Rating the Impact of New Technologies on Democracy. *Communications of the ACM*, 44 (1), 39-43.
2. Dourish, P., et al (1999) Presto: an experimental architecture for fluid interactive document spaces, *ACM Transactions on Computer-Human Interaction,* 6 (2), 133–161.
3. Dridi, F., Pernul, G. and Unger, V: Security for the Electronic Government, *Proc. of the 14th Bled Electronic Commerce Conference*, "e-Everything: e-Commerce, e-Government, e-Household, e-Democracy", Bled, Slovenia, June 2001.
4. Dzbor, M., Paralic, J., and Paralic, M. (2000) Knowledge management in a distributed organization. In *Proceedings of the 4th IEEE/IFIP International Conference BASYS'2000*, Kluwer Academic Publishers, London, pp. 339-348.
5. Mach, M. and Sabol, T. (2001). Knowledge-based System for Support of e-Democracy. In *Proceedings of European Conference on e-Government ECEG'2001*, Trinity College, Dublin, pp. 269-278.
6. Watson, R.T., and Mundy, B. (2001). A Strategic Perspective of Electronic Democracy. *Communications of the ACM*, 44 (1), 27-30.

IST-Project: AIDA – A Platform for Digital Administration

Anton Edl

anton.edl@infonova.com, http://aida.infonova.at/

1 Project Overview

1.1 The Consortium

The AIDA Consortium has a clear structure based on the objectives of the project. It includes:

- one leader in internet security: INFONOVA
- one big hardware equipment manufacturer: HP
- two universities: Politecnico di Torino working on the application side and Technical University of Graz, working in cryptography and software research
- An innovative telecommunication-applications company with great know how in market access: I&T
- Four organizations willing to support the design process and to implement and validate the services and technologies. Euro Info Correspondence Center Ljubliana, Ministry of economic affairs of Slovenia, Mestna Obcina Celje,

1.1.1 INFONOVA

INFONOVA is one of Austria's largest companies for the development of Internet-based telecommunication services and networks. Founded in 1989 by a small group of engineers, it has grown considerably and is now one of the biggest companies in telecommunication research and development in Austria and technology-leader for Public-Key-Infrastructure, E-Business and network integration.

Role in Project
Beside the project management, INFONOVA developed the WYSIWYS-software and deployed AIDA's core-platform.

1.2 General Objectives

AIDA's objective is to implement **A**dvanced **I**nteractive **D**igital **A**dministrations by providing them with an infrastructure which supports administrations and other public bodies, to improve businesses' and citizens access to information and regulation and facilitate contacts, exchanges and feedback between administrations and between administrations and third parties, i.e. citizens, institutions and business.

The European Signature Guidelines as well as digital signature laws in different countries will lay the foundation for the issuing of electronic documents by public

R. Traunmüller and K. Lenk (Eds.): EGOV 2002, LNCS 2456, pp. 292–297, 2002.

institutions such as administrative bodies, professional associations or universities, which – digitally signed – will be held equal to conventional paper-documents. This will enable these institutions e.g. to issue innovative forms of conventional documents, such as electronic certificates of birth, electronic trade licences, electronic diplomas and many more. Electronic documents like these can then be used instead of paper-documents.

1.2.1 Advantages of Electronic Documents

Using electronic documents has a lot of ádvantages - for the purpose of this proposal, documents providing certain qualities will be called e-documents :

- Electronic documents can be sent by means of the internet, thus enabling future applications for situations in which one would have to show up in person today – just for presenting paper-documents. Examples are: car-registration, registration of birth etc.
- Quite a lot of e-commerce activities require proof of identity, legitimization or authorization – which can easily be realized by presenting e-documents
- In combination with ITU-T X.509 digital certificates, the application and distribution of which are currently encouraged all over Europe, e-documents provide significant additional value to e-administration or, more general, to e-business environments.

1.2.2 Aims of the Project

Such electronic documents will only be valid if the digital signature they contain has been created by using a special environment. The European Signature Guidelines as well as national legislations recommend or even require the use of a trustworthy environment for the generation as well as for the verification of advanced digital signatures generally.

Now, this project has all these aims:

- to define and implement a trustworthy environment by means of a signature terminal, which does not only provide a secure solution but is also easy to handle by users,
- to define and develop machine-readable datastructures for electronic documents which can be used for national **and** international purposes,
- to define and develop electronic documents to replace conventional documents used at demonstrator sites and make proof of the usability of such documents in their electronic form for electronic administration environments, and, furthermore, to implement an environment and an infrastructure where such electronic documents can be used.

Summarising one can say, trustworthy digital signatures will be fundamental for various next generation services in the digital era. One of these services shall be an "e-Administration Service Provider" – and to achieve this, all components, environments and techniques required shall be developed resp. integrated.

1.3 Technical Objectives

To enable Advanced Interactive Digital Administrations an infrastructure has to be developed, that consists of the following entities:

1.3.1 Signature Terminal

To create a trustworthy environment, all components have to be trustworthy and thus have to contribute to an overall secure solution.

- The signature creation device must make forging of signatures as difficult as possible by requiring an active user confirmation before signing anything. Entering a pin or fingerprint or any other authenticating information should ideally be done directly on the signature creation device.
- The signature creation device must be embedded into a secure environment, that also makes forging signatures difficult. This environment is responsible for preparing the data to be signed and to be sent to the signature creation device for hashing and signing. As the signature creation device might not be capable of displaying the data to be signed, the security of this environment and the binding to the signature creation device is crucial for the overall functionality of the system. Ideally, the signature creation device and the signature environment are integrated into one component.
- Displaying the data correctly is also very important. Correctly in this environment means that the user sees exactly the data he is going to sign and nothing else. A secure signature environment must therefore be able to parse and understand data and reject signing data that contain illegal parts which are unknown to the display unit. Examples for illegal contents are white text on white background or unknown HTML-tags etc. We call this module the WYSIWYS-module: **W**hat **Y**ou **S**ee **I**s **W**hat **Y**ou **S**ign.
- The software modules used must be authentic and therefore can only be run in an environment where such authenticated modules are supported and missing or wrong signatures would be detected reliably.

Theses project's objectives are to create and to incorporate all existing components needed to provide secure signature environment. It will provide a software-solution and an appropriate soft- and hardware solution, for example:

- Software-only solution for different platforms. To a large extent, Such a solution must rely on the environment, as it contributes largely to the security of the system. However, even if such a solution cannot provide perfect security, sometimes perfect security is not absolutely required. Features like the ability to display the data to be signed accurately, will still be very useful and appropriate for mass-market.
- Solution using small personal devices like subnotebooks or PDA's Integrating WYSiWYS-features into small portable computers like subnotebooks or PDAs equipped with a smart-card reader seem to be the ideal solution for secure signing equipment: the devices are small and can be carried around easily, while the display is usually still capable of displaying different forms of data to be signed.
- Solution using mobile phones. Mobile phones are also an ideal platform for a signing terminal: the devices are small and are carried around anyway and they already use smartcards. The main problem with traditional mobile phones is the limited display. For certain applications – like payment applications - this limitation causes no problem. Newer generation equipment integrates PDA-features with phones and should be the basis of an excellent future solution.

1.3.2 e-Documents

For e-documents, the first step will be to develop a general framework for such mutually acceptable documents for administrative purposes. This definition stage will concentrate on local administrations, e.g. cities and regions, but will have the European perspective in mind from the beginning. Documents like these need to be defined in a structured language, like XML, that easily allows translation into different languages and thus enables international deployment of this concept.

To provide e-documents for European citizens, the following techniques have to be integrated as well:

- Integration: signed XML. XML is of great importance for all data transactions in all domains of e-administration and/or e-business. In order to be able to warrant authenticity and integrity of transmitted data it is essential that digital signing is added to the features of XML.
- Specification: XML-data definitions. Appropriate XML-data definitions for of medical services have to be developed, especially taking into account the transnational network of correlations.

1.3.3 Management Platform for e-Documents

Around these document definitions, a basic universal infrastructure for e-administration services shall be implemented and/or integrated. An environment for generation and issuance of e-documents including a secure signing-infrastructure, a platform for citizens to enable them to display electronic documents and to verify their contents as well as the signatures of the authority, and a suitable directory infrastructure are required.

Special security devices in order to ensure authenticity and reliability as well as non-repudiation and retraceability have to be integrated into the fundamental platform for e-administration procedures.

Without any intention to anticipate the results of the corresponding evaluation task, well suitable security techniques applied in the field of public key infrastructures such as time stamping service, OCSP, chip cards (e.g. Java cards) etc. should be mentioned in this respect.

The management platform shall be a fundamental part of the "e-Administration Service Provider" strived for.

2 Case Studies

This is an overview of some Services, implemented during the AIDI Service Validation. More details and practical demonstrations will be shown at the conference live!

2.1 Exam Admission Service

The *exam admission service* is a widely and frequently used service inside any education institute and is a good candidate for e-administration due to the large spectrum of users and to the large number of paper documents usually involved.

Exam sessions at POLITO are programmed three times a year. Two exam sessions (in February and July) are ordinary sessions and one exam session (in September) is organized in order to help students pass the exams failed during the precedent sessions. Usually a student would have to sustain 6 exams/year. Starting with the academic year 2000/2001 the examinations for some courses will be split in 2 exams/course, so a student will have to sustain about 10-12 exams/year. Thus, he would usually need 12 paper EAC/year. Considering that a student could fail an exam, so he would need to obtain additionally another EAC. Thus, a student approximately needs $12 \times 1.5 = 18$ EACs per year.

POLITO has about 25.000 students, thus the total number of paper EACs released in a year would be approximately 25.000 × 18 = **450.000.**

2.2 Request for Extract from Birth-Book and Birth-Certificate

Extract from Birth-book (Birth-certificate) can be acquired by all physical persons for different purposes. It is a “personal birth-identification” of a person. There are different situations when an extract form Birth-book is needed:

- inscription of children to school,
- acquiring health card,
- marriage…

To get a Birth-certificate a request has to be put in to the competent public service. It can be written by hand or a pre-printed form can be used.

For passing the Birth-certificate the Administrative unit in the place of birth is in charge. Procedures for passing the Birth-certificate are regulated by different national laws, Paris and Vienna convention.

The Birth-certificate is handed on basis of correct fulfilled request. The request can be put in by:

- citizen for himself
- citizen for another person (in that case an authorisation is necessary)
- parents for children younger than 18 years.

The form of request is not prescribed. It can be also written by hand (e.g. MS Word). In that case the form has to contain all data needed for passing the certificate. Administrative unit Celje offers a standardised pre-printed form, which needs to be filled with required data. The pre-printed form is tax-free and available in the central office of Administrative unit, it can be also taken from Internet.

2.3 Permission for Lowering Accommodation Costs for Kindergarten Care

Duties for forfeiting the permission for the height of the payment for kindergarten care are according to several laws and directives valid in Slovenia in the working field of the local community. The process takes place in the municipality of Celje in the Department for social affairs, which is a part of the Ministry of work, family and social affairs of the Republic of Slovenia.

The right for lowering accommodation costs for kindergarten care is granted to any parent, on the basis of formulary, which is put in to the Department for social affairs.

In the next step the form is classified and a map with data is created. The application form is forwarded to the referent responsible for pre-school education in the Department for social affairs.

The referent is then in charge of the whole process and for informing parents about the result of the application. One copy is sent as information also to the kindergarten. Another copy is added to the application map and to the documentary files of the department.

City municipality of Celje has 49.000 citizens. The right for lowering accommodation costs for kindergarten care is taken in focus by 2000 parents. On a month basis we receive around 100 to 150 applications.

Main Achievements

The public demo platform at http://aida.infonova.at/demo/ proved to be a very stable one. All participants had numerous live-presentations via the public internet, which gave a lot of feedback for future enhancements of the system.

By the end of the project there are two **commercial products** - out of AIDA - on the market.

1. The Security Software Package – the so called "Crypto Library" is available on the market and a very stable and innovative product. See http://jcewww.iaik.tu-graz.ac.at/products/index.php for more information.
2. The WYSIWYS-Viewer was delivered to a customer in Austria by December 2001.The first certified Austrian Trust Center "A-Sign" http://www.a-sign.at/ recommends the SW-based version of the viewer for their strongest certificate (premium class certificate = equal to handwritten signature) to show signed XML-content.

So commercial success already started even before the project ended! This is actually one of **the** preferred outcomes of an EC-funded project.

For more information please contact: anton.edl@infonova.com

e-Government Strategies: Best Practice Reports from the European Front Line

Jeremy Millard

Danish Technological Institute, Kongsvang Allé 29, 8000 Aarhus C, Denmark
Tel: +45 72 20 14 17; Fax: +45 72 20 14 14;
jeremy.millard@teknologisk.dk

Abstract. This paper reports on some of the recently completed work of the EU-supported Prisma project examining the best of e-government experience across Europe in relation to technology, organisational change and meeting the needs of the user (citizens and business). Future work of Prisma involves developing scenarios of change over the next ten years, building future-oriented best practice models and providing comprehensible and useful tools for practitioners and researchers to guide their decision making and research priorities respectively. Apart from examining e-government and e-governance generally, Prisma is also examining six service areas in detail: administrations, health, persons with special needs (the disabled and elderly), environment, transport and tourism.

1 Context and Drivers of Change in Government and Governance

The importance of government is clear. Not only are we all dependent upon its services and the framework of law, peace and stability it provides, but in Europe it also contributes 40% of GDP. Over the past few years, however, the concepts of government and governance have been dramatically transformed. Not only is this due to increasing pressures and expectations that the way we are governed should reflect modern methods of efficiency and effectiveness (like the best of business) but also that governments should be more open to democratic accountability. This cauldron of change is now finding itself once again brought to the boil by the impact of new digital technologies on government. In many ways e-government enables the dual goals of efficiency and democracy to be met more cheaply and easily than previously envisaged, but the new technologies go much further than this. They are starting to redefine the landscape of government by changing the relationships (power and responsibility) between players – between service providers and industry, between the public and private sectors, and between government and citizen – by forging new organisational and economic structures, by introducing new processes at work and in the community, and above all by opening new opportunities as well as posing new challenges, not least the threat of new digital divides.

Despite the power of the new technologies in providing global reach and interactivity, models of e-government are cultural and political rather than technical. This

R. Traunmüller and K. Lenk (Eds.): EGOV 2002, LNCS 2456, pp. 298–306, 2002.

can be seen in Europe by comparing the very diverse levels of take up and of policy approaches, ranging from the Scandinavian and Anglo-Irish models of northern Europe, through the more statist, public sector driven responses of central Europe, to southern Europe's strong tradition for family, community and city-region driven approaches.

Across Europe, however, similar forces are acting out despite the variety of policy responses on the ground. The citizen is being treated more and more like a customer, the ICT literate are demanding the option of self-service through the increasing availability of on-line services, the roles of supplier and user of government services are blurring, as indeed are the distinctions between public and private sectors, for example as seen in the increasingly important role of the third sector and intermediaries and the establishment of many PPPs (public-private partnerships) to administer and deliver what have traditionally been seen as purely public services. Networks rather than hierarchies of interests, alliances and actors are increasingly determining how government, and especially e-government, is mapping out.

2 Lessons and Experiences from Europe

E-government is not just about a government portal with services offered electronically. It is also very much about the need to use ICT to support better quality "warm" human services, so that government 'on-line' complements rather than substitutes for government 'in-person'. These goals are simultaneously driven by and the result of:

1. both intra-and inter- governmental reengineering
2. the needs to strengthen efficiency, the public service ethic and democratic and open government.

E-government shares basic notions of business and work requirements with e-business, while at the same time it has its own distinctive features and problems manifest at European, national, regional, and local government levels. Thus, government, like business, requires greater efficiency, productivity, cost reductions, and treating citizens like customers. In this sense it shares the need for business process re-engineering. On the other hand, government, unlike business cannot choose its customers and, indeed, people are more than just customers, they relate to government as legal entities, e.g. as taxpayers, information users, hospital service customers and, generally, citizens who want to be aware, considered, recognised as participants in the democratic process, and free to express their opinions (e.g. through e-voting). Government also has stringent requirements such as:

- exemplary public service ethic with a focus on issues such as the welfare and health of the citizen and the equal treatment of all
- access for all
- caring for a sustainable environment, affordable public transport, etc.
- supply and demand for e-services, without a profit incentive, but subject to accountability and benchmarking
- provision of an institutional and a service framework for the wider economy.

In this sense, it is appropriate to talk of "government process re-engineering" (GPR), including not only services and relations with customers but also transparent, open and accountable government processes and relations with citizens in the digital democracy. This deep modernisation of government implies the development of a new culture for governance in public services, including new mindsets and sets of values, and new behaviours, as well as a re-engineering of structures.

2.1 Inter-governmental Process Re-engineering

A simple model of how the internal structures of government, typically when adopting e-government solutions, are being re-engineered in Europe, is provided in the accompanying four diagrams.

Diagram 1 shows the traditional structure of government before re-engineering. The first structural re-engineering tends to take place by the creation of a front-office (in diagram 2). This creates a one-stop shop enabling citizens and business to access government services through one point, regardless of their purpose, rather than through a multitude of points depending upon the organisational structure of government. Diagram 2 represents the state of art in many local and regional government agencies in Europe today.

In diagram 3, as the front-office logic (customer interface and service) takes hold and starts to determine the development of internal structures and processes, it is necessary for the back office to re-organise in order to reflect the needs of the front office and the customer. Traditional back-office departments, relationships and processes thus give way to ones determined by customer service through a front office.

Finally., in diagram 4, as demonstrated by a few European leaders, there is a decisive shift from "cold" administration (i.e. the back office) to "warm" ICT-supported human services (i.e. front office) in terms of personnel, and increasingly also in terms of resources. Small, ICT-automated back offices can serve and support very large front offices with more frontline ICT-supported human services based upon the improved cost-effectiveness and increased quality of administrative back office procedures.

In other words, the best of new e-governance in Europe is certainly not just about a government portal with services offered electronically. Rather, it is about this, plus internal reengineering of government, greater transparency, citizen access and involvement, and, above all, ICT support for better quality "warm" human services. For example, better direct health care of patients wherever they are, care of the elderly and disabled in their homes by government staff using interactive digital terminals to increase the timeliness and quality of information and support systems, improved responsiveness and efficiency of one-stop shop staff when dealing with direct enquiries, more efficient planning and control systems for delivering frontline services, etc. E-government could, and maybe should, lead to more, not less, human-orientated services.

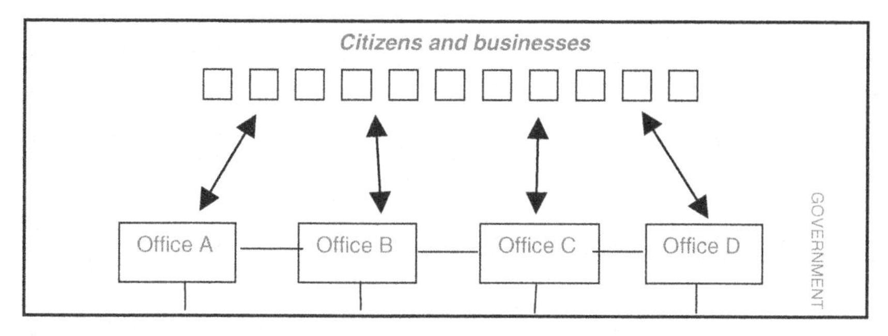

Fig. 1. Before re-engineering

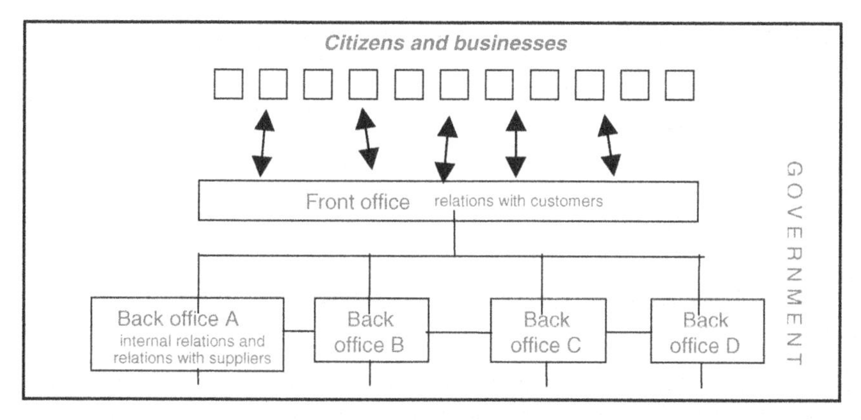

Fig. 2. Front office re-engineering

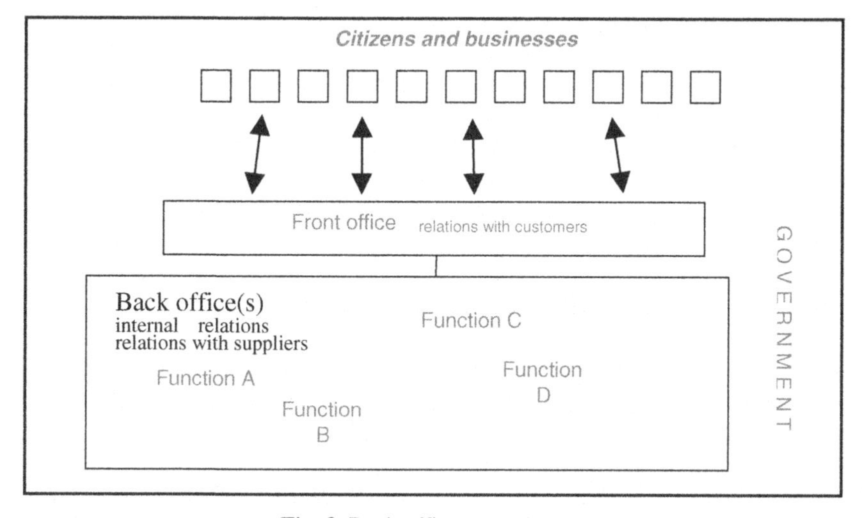

Fig. 3. Back office re-engineering

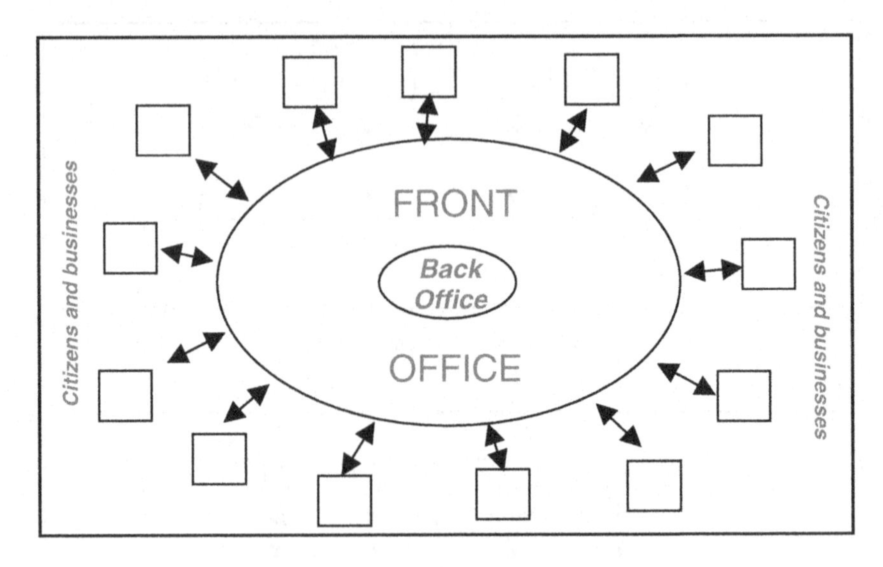

Fig. 4. Total re-engineering

2.2 Inter-governmental Process Re-engineering

Government in Europe is not only confronting massive internal changes but is also grappling with the need to re-examine the whole structure of government. Apart from the macro issues of EU enlargement and the need to specify more clearly the respective roles of European, national and regional/local institutions, is the need to look again at how government at the meso, as well as the micro, level is organised. E-government cannot reap the potentially large benefits it promises without sharing information and responsibility amongst its various parts – and this demands also a reconsideration of respective tasks. and powers. Already the best European examples of e-government require only one minimal data input or request by citizens or business who do not wish to know where, in the often vast apparatus of government, his or her case is being processed. The citizen only wants an effective, rapid and high quality service and is generally not interested in who provides it or how. The citizens' information is often already held somewhere, and as long as data protection, security and anonymity issues are safeguarded where warranted, Europeans generally trust government sufficiently on this point.

In order to achieve these objectives, electronic transactions thus imply that government agencies need to co-operate and even, where necessary, integrate and re-engineer their structures and processes. They certainly need to be *'joined-up'*. Such re-engineering should be considered both horizontally and vertically, as shown in the following diagram. The terms '*integration*' and '*joined-up*', as used here, should not be confused with '*centralised*'. Indeed, government can be both highly integrated, joined-up and de-centralised. In fact, a third '*centralised-decentralised*' dimension could usefully be added to the diagram.

It is, of course, often the case that both vertical and horizontal integration, driven by ICT as well as other political and financial imperatives, are often part of the same policy, which is itself normally driven by a central government, top-down agenda (whether or not part of a centralising or de-centralising initiative).

2.3 Services Delivery

E-government enables many services to be delivered around the clock, throughout the year and ubiquitously across space. As shown above, it can put substantial power in the hands of the ICT-literate citizen so that the whole purpose, structure and mode of operation of government becomes re-engineered to the needs of those it serves. Many European government agencies and administrations are re-organising service access to take account of ICT, so that, in addition to face-to-face, new service channels are available which can either completely replace traditional channels or, as is more usual and desirable, supplement them, for example, phone and fax, Internet, kiosks, WAP, mobile, digital TV, etc. Not only are the number and types of channels proliferating, so too is the organisation of the service delivery itself. Especially in the case of Internet portals, the best of e-government in Europe now organises services:

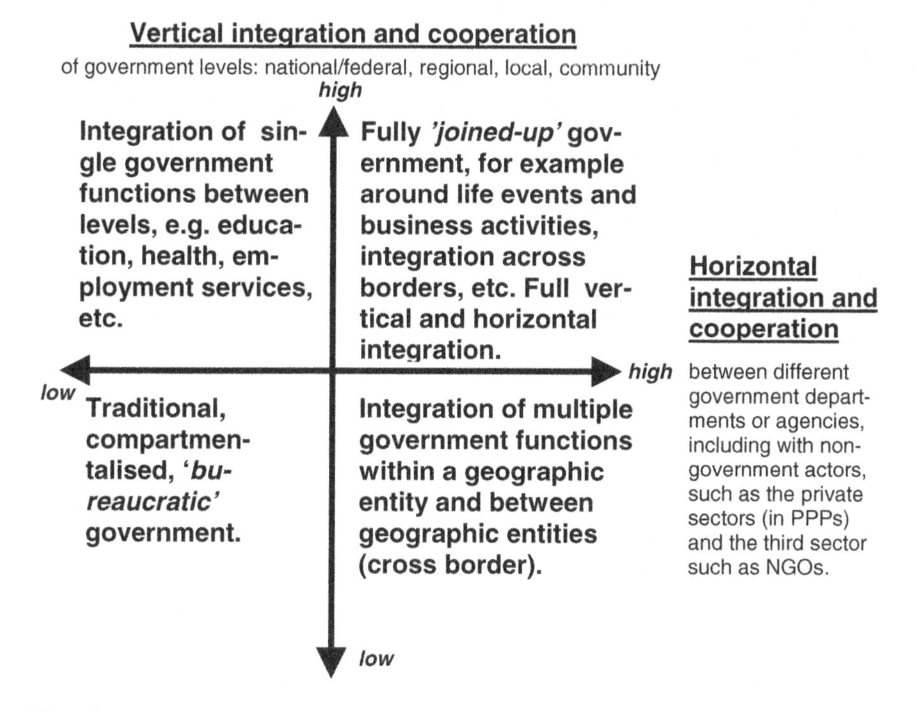

Fig. 5. Vertical and horizontal integration of e-government - inter-government process re-engineering

- for citizens, around life events, life episodes or the life cycle, such as birth, marriage, death, unemployment, education, living, home, working, sport and leisure, etc., etc.
- for business, around discrete business activities, such as VAT, tax, finance, employment, etc., typically supported by the business case approach where examples of typical situations are given and the service needs of businesses exemplified.

In this way, government re-organises to reflect citizen and business concerns rather than its own internal concerns and structure, and provides multiple interface channels.

2.4 Access for All

In many ways, the new technology, as mentioned above, does provide greater opportunities for access, for example by people in remote locations, those who work unsocial hours, or those who are immobile and thus cannot attend offices in person. However, there are important problems in determining precisely what citizens want and need, and how to provide e-government services in a user-friendly and effective way. User needs are always conditioned by what they already get, or imagine they can get. The best examples in Europe break down users into different categories, not just citizen and business, but also, for example, the elderly, the disabled, the visitor, the voter, professional users, etc., rather than treat users en bloc. The presentation of e-services through life events or business activities often supports such segmentation, although there can sometimes be problems of critical mass, i.e. a sufficient mass of a user segment to make the approach financially viable especially where special technologies like smart cards are needed for access and/or transactions like payment by or to the user.

The principle of access for all is also an important policy objective of the EU as well as of national and regional government authorities. This particularly means tackling the digital divide between citizens who do have access to ICT and are ICT-literate, on the one hand, and those who do not have access or the necessary skills, on the other. This is the biggest challenge facing the European Information Society in general, and e-government implementation in particular. Good examples to tackle digital exclusion include special training, the provision of PIAPs (public internet access points) and an acceptance that e-government does not only mean government-on-line but also a continuation of the public service ethic through government-in-person, though supported in terms of effectiveness and quality by ICT, both at the delivery point as used by government staff and in the back office.

2.5 Legal Issues, Security and Trust

Given both that governments are expected to be trustworthy and responsible guardians of the rights of citizens, and that they tend to hold vast amounts of data covering all aspects of the lives of citizens and businesses, systems and data security and trust in using them is vital. As a result, many European countries now have strong data protection laws, but these need constantly updating as new technologies come on stream and as new political and financial requirements manifest themselves. For example, much stress is now being placed upon trust, security and confidence, also in relation to fighting crime and terrorism against electronic, physical and personal property.

Security of transactions, for example using smart cards, is one of the biggest issues, although the barriers to widespread easy to use systems are often as much political and commercial than technical. Above all, there is a need for trust in interoperable systems, including data protection and privacy, as well as the need for anonymity where this is required, but balancing this with the legitimate needs of the authorities is not always easy. For the individual citizen or business, conducting electronic transactions requires cheap, easy to use and secure solutions in which information is easily authenticated and is reliable. The biggest challenge to interoperability across Europe is that legal systems between countries are highly incompatible. However, many trials and examples of roll-out have been successful, so there is confidence that solutions to these challenges are in sight.

2.6 Technology

Technical issues have been raised elsewhere in this paper, but all these need to be seen in the context of the speed and complexity of technological advance. This often makes it difficult for governments to keep pace and adapt their organisational structures, the skills of their staff and the actual e-services offered, in a timely, effective and financially responsible manner. European best practice in relation to technology shows that there is a need to adopt open standards, multiple access platforms, scalability, striking a balance between customised large scale turnkey solutions versus outsourcing and standardised solutions, and new technology models for eGovernment (e.g. mobile solutions, ambient intelligence, etc.)

2.7 Finance

Finance is, understandably, often the crunch factor. Like government the world over, European authorities tend to see their budgets squeezed in all directions, yet, at the same time, they are expected to deliver "more for less".

Experience shows that there is a trade-off between electronic and traditional services, in which e-services typically require heavy capital investments but can save substantial amounts in the longer term, both through improved efficiency and rationalisation, as well as because of greater service outreach and quality. Finance is needed not just for the technology but also for organisational change, as well as skills and competence development costs. Ultimately, government budgets are a political matter, especially as demand for government services tends to be infinite, particularly when offered free at the point of delivery. Good examples in Europe have shown how increased sources of investment and revenue can be generated, for example through greater efficiency of tax collection and the elimination of waste using ICT, well constructed agreements with private investors, the introduction of e-procurement particularly on a national or regional basis, etc. Getting the finance right, but not being driven only by a focus on costs whilst ignoring benefits, is a necessary condition for e-government.

References

1. OECD (2001) *Engaging Citizens in Policy-making: Information, Consultation and Public Participation*. PUMA Policy Brief No. 10 http://www.oecd.org/pdf/M00007000/M00007815.pdf
2. PRISMA (2002) *Pan-European changes and trends in service delivery*, deliverable D2.2 of Prisma, a research action supported by the Information Society Technologies Programme of the European Union, 2000-2003, contact jeremy.millard@teknologisk.dk
3. PRISMA (2002) *Pan-European best practice in service delivery*, deliverable D3.2 of Prisma, a research action supported by the Information Society Technologies Programme of the European Union, 2000-2003, contact jeremy.millard@teknologisk.dk.

CITATION
Citizen Information Tool in Smart Administration

A. Anagnostakis, G.C. Sakellaris, M. Tzima, D.I. Fotiadis, and A. Likas

Unit of Medical Technology & Intelligent Information Systems, Dept of Computer Science, University of Ioannina & Biomedical Research Institute - FORTH, PO BOX 1186, 451 10 Ioannina, Greece

Abstract. CITATION is an innovative software platform designed to facilitate access to administrative information sources by providing effective information structure, indexing and retrieval. CITATION improves electronic government services, ensuring that citizens have easy and direct access to essential public data and promoting online interaction between citizens and government. It targets the improvement of administration services offered by governments and therefore the creation of "smart" and flexible government structures that will be able to provide their citizens with precise and personalised information.

1 Introduction

With the explosion of the Internet and communication technologies, the concept of "information" transforms into a value per se. However, the overwhelming amount of administrative information remains scattered and unstructured -due to an evident lack of modeling and standardization-, stored in multi-site distributed repositories that cannot be accessed, fused and delivered in a seamless manner [1-2].

CITATION is an innovative software brokering technological platform that helps and supports information suppliers and consumers in their info-transactions. It provides groundwork for the creation of "Smart" governmental structures that support the provision of flexible and customised information services to the citizens.

CITATION focuses on the following:

- Effective access to a variety of information sources on the information global net.
- Transformation and representation of the retrieved information on a meta-level
- Matching operations between users' profiles and the information extracted from the various information sources
- Personalised information delivery.

1.1 Background

The first step in building a seamless information fusion environment such as CITATION is the establishment of the "domain of discourse". Ontology is a formal explicit description of concepts in a domain of discourse. Therefore, utilization of domain-specific ontologies is the medium towards seamless information fusion [3-4]. Ontologies provide a compact, formal and conceptually adequate way of describing

R. Traunmüller and K. Lenk (Eds.): EGOV 2002, LNCS 2456, pp. 307–312, 2002.

the semantics of XML documents. By deriving DTDs from ontology the document structure is grounded on a true semantic basis and thus, XML documents become adequate input for semantics-based processing. The ontology provides a shared vocabulary that integrates the different XML sources, making the information uniformly acceptable and thus mediated between the conceptual terms and the actual mark-up used in XML documents [5].

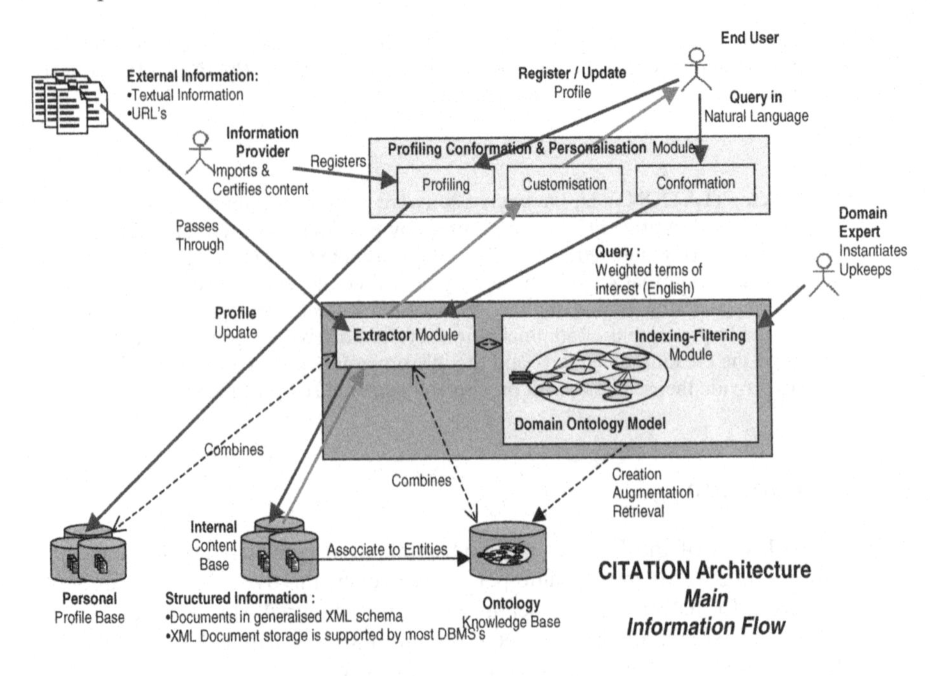

Fig. 1. CITATION architecture

2 Architecture

The CITATION architecture incorporates several technological advances, to deliver the desired innovative functional features.

CITATION internal Content Base holds assimilated external information, which is matched against the Internal Domain Ontology during the phase of import. The standardised DTD for administrative information allows for the uniform representation of the existing raw data, and provides a solid base for further processing and retrieval of information. The tokens from the user's query are extracted from the conformation model, augmented by external lexicons and dictionaries (UMLS, WordNet), filtered by the user's personal characteristics and matched against the semantic internal domain ontology. The domain ontology constitutes the conjunctive link among the actual data and the user's query. Fig. 1 captures the major modules of the system and depicts the data flow among them. The major modules identified are:

- The Profiling module: it is responsible for the generation and update of the user profiles.
- The Conformation module: it is responsible for preprocessing the user query (given in natural language) to facilitate the extraction of the tokens, and thus the internal mapping.
- The Customisation – Presentation module: it is responsible for the customised delivery of the final outcome of the user's query.
- The Extractor Module: it extracts the information from the external information sources and imports it into the CITATION Content Base.
- The Information Representation, Indexing and retrieval Module; it:
  - Does the initial matching of the imported data to the Ontology Model
  - Maintains and augments the Ontology model, delivering mechanisms for editing, allowing the Domain Expert to upkeep the model.
  - Facilitates the semi-automated creation of relationships among the newly inserted and pre-existing entities.
  - Checks and verifies the integrity of the internal domain model.
- The internal Domain Ontology base: it represents and stores the Domain Specific Knowledge in a Uniform manner, despite of the domain peculiarities.

2.1 Representative Objects

As mentioned above the main objective of the CITATION project is the design and implementation of a user-oriented knowledge discovery system, based on a semantic formalism. CITATION investigates the creation, collection and distribution of resource descriptions, to provide transparent means of searching for, and using resources. To improve searching, filtering and processing of information on the CITATION information repository, a common effort is made in the direction of "metadata". Metadata are defined literally as "data about data", but the term is normally understood to mean structured data about resources that can be used to help supporting a wide range of operations [11].

In CITATION we have defined a metadata structure based on implementations of the Extensible Markup Language (XML). In particular, XML aims at providing a common syntax to emerging metadata formats. Based on these lines we have proposed the CDRS (Citation Document Representation Schema) that is used to describe the indispensable information that CITATION handles and manipulates. Metadata in CITATION help administer and manage the resources, (e.g. keep information about their location and acquisition). It is also used to help managing user access and to help the versioning of the resources [6-8].

2.2 Extractor Module

The administrative information that the CITATION system provides either legacy textual information or links to Web-based information, has to be indexed through the internal Ontology Model. The action of linking the initial pieces of information to the Domain Ontologies and the Retrieval of the information in response to the User Query, is undertaken by the Extractor module.

Although the CITATION platform is highly open and capable of manipulating novel information formats, our preliminary search revealed that the administrative information is currently vastly textual, thus special care is being paid in the text manipulation methodologies. Research in NLP (Natural Language Processing) for concept extraction is an undergoing task for many years [9]. Although a reasonable level of performance and success has been achieved, the actual use and dissemination of data extracted from free text is still very limited. Standard NLP- techniques such as lexical scanning morphological analysis and parsing are important tasks in the process of automated analysis of natural language texts. However NLP techniques can only process a text syntactically. To capture, represent and understand the knowledge contained in the text it is necessary to have a semantic framework.

2.3 Ontology Model & Domain Ontology

Although CITATION selects "health" as its application domain, it manages to provide uniform ontology specification capable of representing various kinds of administrative terms and information. This facilitates the efficient communication and exchange of information among administrations in a variety of subjects of concern, since the introduced model is highly open and customisable.

Ontology provides conceptualisation (i.e. meta-information) that describes the semantics of the initial data [12]. CITATION comes to facilitate this exact common understanding, by introducing a "Universal" model that allow for adequate representation of the semantics of the existing informative administrative content [13-19]. For the core of the Domain Model the Protégé paradigm was adopted [17].

Ontology Model provides all the necessary functionality for the capture and storage of the domain knowledge items (entities) and the relationships among those in various fields of practice (i.e. the CITATION pilot cases). The most important step on the definition of the Ontology model is the determination of the elements constituting the model; during the study of the CITATION Ontology Model, we have identified three major characteristics to be modelled, namely:

- The abstract Entities (classes/concepts)
- The Supported Relationships
- The Metadata (class containing the meta-information for the instance-entities)

2.3.1 Entities & Relationships

Modelling entities is a straightforward task, including definition of a metadata distinct set plus additional information on its behavior in the ontology model.

However, modelling relationships have been proven to be rather twisted: On the one hand, simple and comprehensive mathematical rules had to be established in order to set an adequate set of integrity constraints, while on the other hand conceptual taxonomisation of the relationships is facilitated to allow for standardisation in the manipulation of the overall model.

2.3.2 Semantic Characterisation

Analysing the user requirements, revealed initially three large semantic categories, namely: Hierarchical, Generalisation & Sequential. Each of the imported relations is denoted accordingly, to allow for automated manipulation later in the information retrieval process. Handling of more categories is vastly supported, since the architecture of the knowledge base [14] is highly adaptive.

2.3.3 Meta-information

Based on legacy external lexicons to each of the entities a prime definition (in English) is assigned. A whole set of weighted meta-data terms (i.e. semantically equivalent definitions of the entity) are assigned in each term in a weighted manner. The terms of the Universal Translation Dictionary (EuroWordNet [17]) corresponding to the prime definition are assigned to the CITATION translation table allowing for multilingual semantic matching.

3 Results

Using XML based implementation for representing the various information objects we are able to control a well-formated object, based on the special requirements of the project. XML allows for the explicit declaration of element types and representation of document structure in the Document Type Definition (DTD) or XML Schema. The core elements of CDRS, *Resource*, *Unit*, *SubUnit*, are used to split the initial objects to sub-objects based on the concept of its content. The correlation between two objects is achieved by defining an identical number for these objects, but the correlation between CDRS object and the ontology model is accomplished by using special elements such as *Class*, *Concept* and *Metadata*. By using CDRS as a representative object any resource can be described and managed in a more efficient and flexible way. CITATION Ontology Model is highly adaptable, capable of representing domain ontologies over a wide spectrum of administrative information. The rules and constraints identified manage to capture and categorise a significant portion of the administrative knowledge, creating a solid ambient for semantic-based information retrieval. Enhanced capabilities in conceptual information retrieval are the outcome, and the indications so far are stimulating. The recall accuracy and the precision of the results rely heavily on the completeness of the domain ontology

4 Conclusions

CITATION meets the users' (citizens, business, public authorities) needs for flexible and ubiquitous access of administrative information. It provides a platform with multifunctional dialogue interfaces and multi-lingual features, being an intelligent information tool on administration issues.

CITATION offers effective and transparent access to governmental services through information indexing and retrieval using a semantic oriented, ontology based approach. It introduces models that allow for adequate representation of the semantics of the existing healthcare administrative information content.

Acknowledgments

The CITATION project is partially funded by the EU (IST-29379).

References

1. Green paper on public sector information: a key resource for Europe.
2. Eysenbach G:Consumer health informatics. BMJ 2000;320:1713-6.
3. Musen M. Dimension of knowledge sharing and reuse:Computers and Biomedical Research 25 (1992): 435-467
4. Gruninger M, Fox M.S:Methodology for the Design and Evaluation of Ontologies. Proceedings of the Workshop on Bsic Ontological issues in Knowledge Sharing, IJCAI-95, Montreal
5. Erdmann M, Studer R:How to structure and access XML documents with Ontologies. Data & Knowledge Engineering 36(2001) 317-335
6. ISO 639 - Codes for the representation of names of languages. http://www.oasis-open.org/cover/iso639a.html
7. ISO 3166 - Codes for the representation of names of countries. http://www.oasis-open.org/cover/country3166.html
8. Getty Thesaurus of Geographic Names. http://shiva.pub.getty.edu/tgn_browser
9. Bateman JA:Ontology Construction and Natural Language. Proc Formal Ontology &KR, 1993.
10. Eurodicautom http://europa.eu.int
11. Metadata Architecture http://www.w3.org/DesignIssues/Metadata.html
12. Bateman JA:Ontology Construction and Natural Language. Proc Formal Ontology KR, 1993.
13. Gruber T:A Translation Approach to Portable Ontology Specifications. Knowledge Acquisition 1993; 5(2):199-220.
14. Guarino N, Giaretta P:Ontologies and Knowledge Bases: Towards a Terminological Clarification, in Towards Very Large Knowledge Bases. N.J.I. Mars (ed); IOS Press, Amsterdam, 1995: 25-32.
15. Karp P, Chaudhri V:Thomere J. XOL: An XML-Based Ontology Exchange Language.1999.
16. Erdmann M, Studer R:How to Structure and Access XML Documents With Ontologies.Data and Knowledge Engineering, Special Issue on Intelligent Information Integration.
17. Cognitive science laboratory | Princeton University | 221 nassau st. |Princeton, nj08542, WordNet http://www.cogsci.princeton.edu/~wn/online/ 2002 Jan. 27
18. Vossen P, Department of Computational Linguistics University of Amsterdam, Euro WordNet http://www.hum.uva.nl/~ewn/ 1999.
19. Hahn U, Romacker M, Schultz S: How knowledge drives understanding- matching medical ontologies with the needs for medical language processing. Artificial Intelligence in Medicine 15 (1999) 25-51

Clip Card: Smart Card Based Traffic Tickets

Michel Frenkiel, Paul Grison, and Philippe Laluyaux

Clip Card, 1501-1503 Route des Dolines,
F-06560 Sophia-Antipolis

Abstract. The introduction of information society technologies (IST) in administrative processes poses a range of problems of technical, legal and human nature, that need to be properly addressed for the evolution to be successful. Some of these problems (and a few solutions) are reviewed in the scope of replacing paper-based traffic tickets by smart cards.

1 Project Objective

Traffic violations constitute the most common offence in modern societies. 150 million tickets are distributed every year in Europe. In the average, only 60% are eventually collected, because the administrations are unable to process unpaid tickets within the legal timeframe. This figure varies greatly between countries, even between regions, spreading the idea that citizen are not equal when facing the law.
The cost of collecting the money of a ticket also varies. The figures lay between 6 and 12 in the surveyed regions.

Clip Card results from in-depth discussions with public authorities: city officials, justice departments and law enforcement agencies, in France and Italy.

With Clip Card, the paper based traffic ticket is replaced by a smart card recorded on the spot of the offence by the traffic warden and clipped to the windshield wiper. **The innovation of Clip Card lies in the fact that the smart card is the traffic ticket itself, the payment means and the payment receipt.**

The offender may pay the fine at payment kiosks with his credit card or, as today by checks, in cash or by credit card, in tax offices, tobacco shops, post offices, etc. Back-office processing (matching paid tickets, issuing late penalties) is automated on a large scale, using up-to-date telecom technology (SMS, GPRS, secure communications) thus enabling considerable savings of public money.

Clip Card helps improve security and fight crime: dangerous vehicles are reported for being removed, stolen vehicles and vehicles with many unpaid tickets are detected and reported for further action.

To process traffic tickets, the traffic warden is equipped with pre-personalised smart cards and a mobile terminal (a hardened, communicating PDA).
All parties find advantages in Clip Card:

R. Traunmüller and K. Lenk (Eds.): EGOV 2002, LNCS 2456, pp. 313–318, 2002.

- The traffic warden and the police officer who is released from administrative tasks and may concentrate on security missions.
- The offender who can pay his fine simply with his credit card at public kiosks. These kiosks could be combined with existing parking meters.
- The governments and the municipalities which save around 50% of the processing costs: in Europe, it means saving some 900 million per year.

2 Project Status

- Clip Card has received the PROGSI label (French ministries of Finance, Justice, Industry) and go-ahead for trial. Clip Card is financially supported by the European Commission. It has also received the "Label Innovation" (Telecom Valley) and a support from ANVAR.
- System architecture is complete, back-office and terminals development are in progress with industrial partners
- Real life trials are scheduled to start during the Fall 2002 in Cannes, Ventimiglia and Torino.

3 Functional Overview

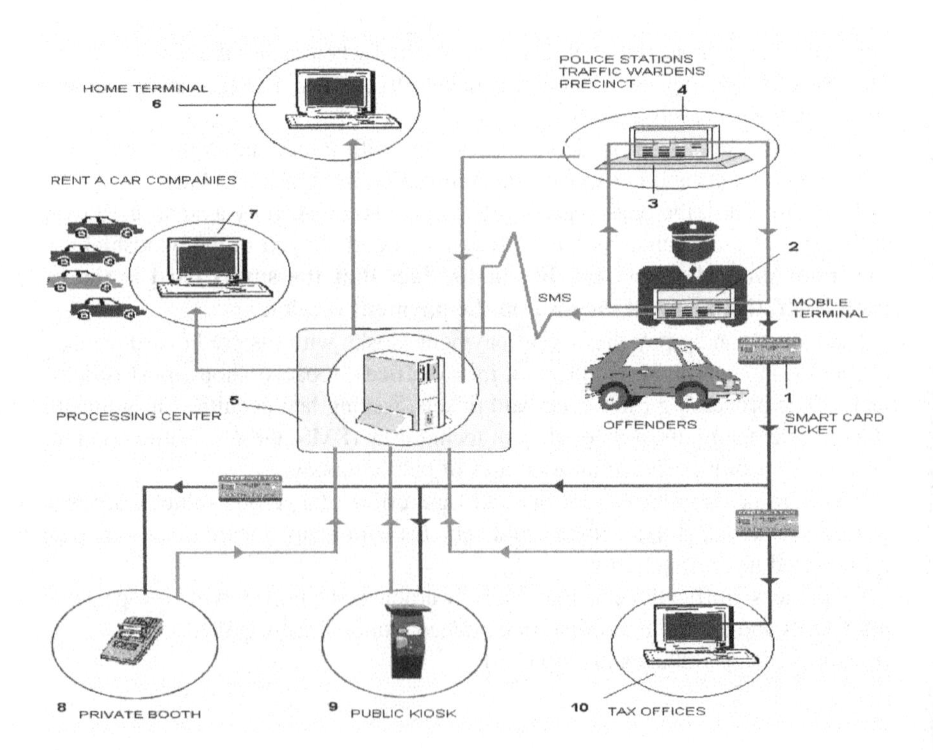

4 Interfaces with Administrations, etc.

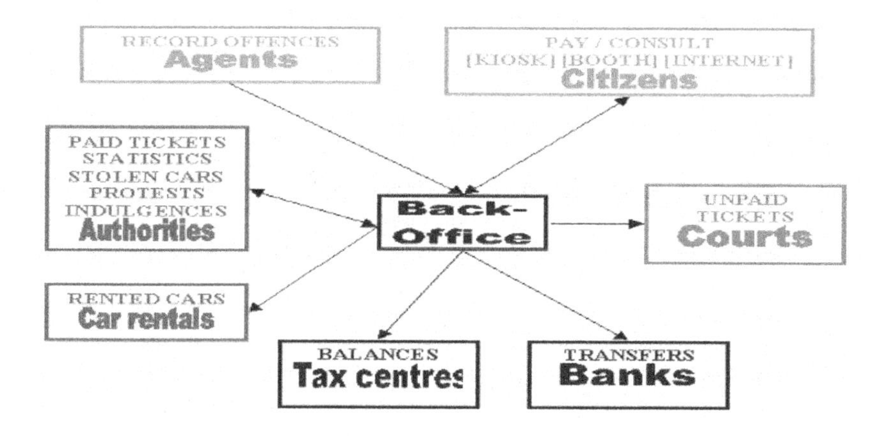

5 Key Questions Addressed

5.1 Technical

Availability, Reliability

- High availability (H24) required for the network
- Terminals on the network have availability and maintenance procedure comparable to payment terminals at points of sale
- Users and personnel are quite similar to those found at points of sale (customers-offenders are general public, sales clerks are administration personnel, tobacco shop personnel, etc.)
- Reliability must be very high for the mobile terminals and memory cards, as they are used in diverse, often rough conditions:
  - Indoor/Outdoor
  - In all weather (rain, sunshine, frost)
  - Day/night
  - Wardens are under stress, and constantly face a risk of aggression

Security Aspects

- payments: the same level of security as for self-service points of sales apply
- offence transmission requires a good level of security, but the network is entirely under control of Clip Card, and it presents a limited interest to hackers. As a full traceability of the transactions is required, mechanisms for detecting intrusion attempts are also present and freeze the most vital access to the system when an attempt is detected. There is a built-in high redundancy (as offences are both recorded in the mobile terminal and sent over the network), thus facilitating cross-checking and checkpoint/restart.

Cross-Border Payments

This may be the largest technical challenge, as it may require also an approval from the banking industry. In "normal" credit card operations, the "shop" serves several "customers", while all payments end up in the same bank account (the account of the shop itself): (arrows indicate the flow of money)

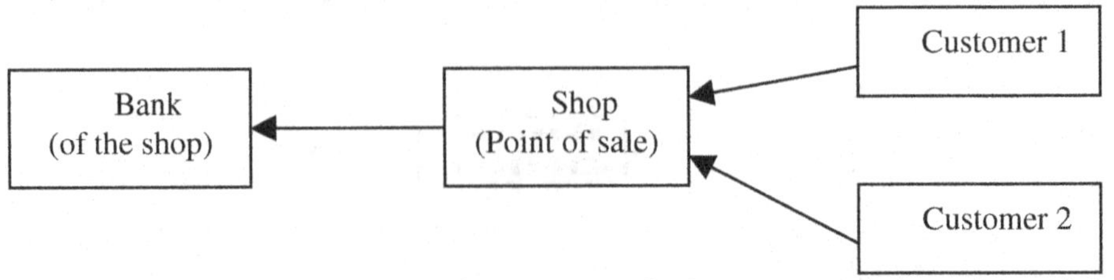

In Clip card, the payment transaction must credit the treasury of the emitter of the traffic ticket. For example, a driver travelling in France and Italy may decide to pay his traffic tickets collected in France and in Torino at once in a French tobacco shop:

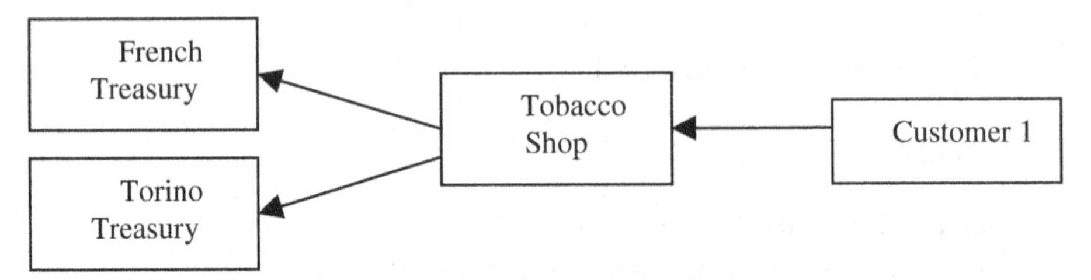

The point of sale terminals do not have the necessary capacity to hold all the programs needed to address the many treasuries (or banks). Regulations do not allow Clip Card or a third party to cash in the money of transactions, and to pay back the appropriate treasury. A solution has yet to be found.

5.2 Legal

Procedures of the Administration

The processing of traffic tickets may involve a number of different administrations. For example, in France:

- law enforcement people (*Police Nationale, Police Municipale, Gendarmerie*), who may all issue tickets, report either to the Ministry of Interior, the Ministry of Defence, or the City Mayor.
- Payments are handled by the Treasury (*Trésor Public*) and the Ministry of Finance
- Late penalties are decided by the Ministry of Justice (and collected by the Treasury)
- Installation of public equipment is decided by the municipalities, and must be approved by the *Préfecture* (Min. of Interior)

Before considering the deployment of Clip Card, and the necessary changes in the administrative procedures, the French administration requires a proof of the concept via a significant trial, to be conducted in a large city. This means that a first version of Clip card must be developed, and used in a non-disruptive way regarding the Administrations: this requires no direct connection with any data base of the Administration.

Administration Data Bases

To store and process traffic tickets, a computer program was developed by the Administration, and instantiated by the Min. of Interior once for the *Police Nationale* (WinAF) and once for the *Police Municipale* (WinPM). These two instances are disjoint.

The data bases are used to match payments, identify late penalties, process indulgences. Interface with the Justice Department regarding late penalties is done via paper and diskettes.

Clip Card Trial

To trial the Clip Card concept without changing the existing procedures, we will produce diskettes that will be given to the *Préfecture* to feed in traffic violations into one of the systems. And we will also produce diskettes to feed in payments. Tickets and payments will be matched by the systems in place (WinAF, WinPM) under control of the administration. Late penalties and indulgences will not even be known by the Clip Card system.

However, as the most costly operation today consists of recording the handwritten traffic tickets onto one of the administration data bases, the Clip Card trial remains valid to demonstrate its advantages.

Similarly, our smart card "must contain the exact information found on the present forms, and the payment terminal must print the present forms exactly". This puts strong requirements on Clip Card!

In the future, rather than develop IST solutions that match existing administrative procedures, it would be more efficient for administrations to adapt their procedures to the new possibilities offered by IST. This is how private enterprises adopted IST.

The situation in Europe regarding legal constraints varies greatly between Member States. It took one year of discussions in Paris, at the various ministries, to obtain an authorisation for launching a Clip Card pilot operation. It took just a few days of discussions with the staff of the mayors of Torino and Ventimiglia, to obtain the same in those Italian cities, without the need to involve any central administration.

We are discovering now the requirements of the other Member States.

5.3 Human

Two aspects must be considered:

- Usability, by all people involved: offenders, wardens, administration personnel
- Quality of work, and future prospects, for the wardens.

Usability

- For the offenders, it is not more difficult to pay a traffic ticket than to use a pay telephone, a cash machine or a Proton terminal. These facilities are now well accepted.
- For the administration personnel, Clip Card is totally transparent.
- For the wardens, usability is a key aspect, which has not always received adequate answers in the past. For example, Malaysia adopted in 2000 the smart card to hold citizen's ID. Police forces received a hardened PDA to read those cards. The operation failed because the ergonomic characteristics were not adapted to the situation:
  - Too small, hard to read characters
  - Too small, too many buttons
  - Too complex, too many menus

Similarly, the police in Madrid was equipped with PDAs to record traffic violations, and print the tickets. But the printers are too often out of order for the system to be well accepted.

Learning from these experiences, Clip Card offers a very simple user interface, which can be activated either via voice or via buttons. Wardens will have plenty of time to practice, before they use it on the streets, so that their confidence is high.

Wardens Quality of Work and Future Prospects

Clip Card was presented first to the syndicates of police forces. It was well accepted, because it allows a better, more gratifying usage of the police forces. Instead of doing paperwork in their office, they will do their job of maintaining law and order with the people.

Clip Card was then presented to some of the wardens who have been selected to test it. Their reaction was also positive, because they feel that the quality of their equipment is a recognition of their value. Also, replacing the pen and paper by a PDA is a gratifying opportunity to enhance working conditions, and to enhance the image of their work. But they demand that this is accompanied by adequate education, and that the equipment is fully operational and tested at the start of the pilot.

For the future, mobile digital equipments are ubiquitous, with new applications popping up every day. Police wardens will not carry an array of devices. But one single, robust and flexible device should be available to accommodate smart ID cards, digital camera, fingerprints capture, mobile phone (for voice and data), GIS, and thus interfacing administration data bases in total security. Secure communications, but also security of the warden, with a Help button and geo-localisation will probably soon be offered. As Clip card is the first such application, there is a high stake (and considerable future opportunities) on its mobile terminal, as it may well become the de facto standard to process those future applications.

VISUAL ADMIN – Opening Administration Information Systems to Citizens

Benoît Drion[1] and Norbert Benamou[2]

[1] AIRIAL Conseil, 3 rue Bellini, F-92806 Puteaux Cedex, France
benoit.drion@airial.com
[2] BFC, 137 avenue du General Leclerc, F-92340 Bourg la Reine, France
norbert.benamou@business-flow.com

Abstract. The *VISUAL ADMIN* project aims at creating an eGovernment solution that makes easier for citizens and businesses to interact with administration. Considering administration activities from the perspectives of both "customers" (citizens and businesses) and administrations, *VISUAL ADMIN* intends to provide citizens and businesses with an global online view on information relevant to them. It also intends to organise the flow of information between administration services for handling customers' cases, and to act as a portal for getting access to relevant online information services. The current paper focuses on the requirements identified in the early months of the Project and presents the rationale of the platform under development.

1 Need for Simplifying Administration Processes

It is often difficult for administrations to avoid slowness and heaviness in dealing with citizens and businesses. These customers frequently encounter difficulties because a single need is often handled through a series of procedures sometimes distributed among several public bodies; for instance, in France, installing new offices in an industrial area might require interacting with up to five administrations.

Moreover the organisation of the public sector is also complex, due to the devolution of powers at several geographical levels (national, regional, intermediate and local levels). Therefore an administration service might also have to collaborate with other public bodies to process the customer's case, and specialised entities can have a role for some procedures (registration of patents, ...).

Such a complex organisation sometimes causes a huge delay for citizens and businesses, as well as for civil servants; and generally speaking implies many risks and discomforts for the customer:

- Infringing some regulations by forgetting asking authorisation to one administration service;
- Long delays for completing his case;
- Lack of transparency that makes difficult for him to understand what is happening and at what stage his case is.

On the other hand, inside and between administrations themselves, the processing of a customer's case is not always managed in an optimal way from start to finish, for front

R. Traunmüller and K. Lenk (Eds.): EGOV 2002, LNCS 2456, pp. 319–325, 2002.

office services are sometimes structurally disconnected from back office information systems, databases or other services.

In this context, European governments have agreed on an initiative to simplify administrative procedures by going on line and by offering "eGovernment" services.

2 *Visual Admin* As a eGovernment Online Solution

The *Visual Admin* Project[1] has been launched in June 2001 by a European consortium gathering local governments and private companies with the objective of conceiving and experimenting an online eGovernment solution which could be set up by a local government as a Portal for efficiently supporting citizens and businesses in targeted policy areas.

Three local governments are involved in the project and will implement the *Visual Admin* solution in 2002:

- In France, the *Communauté d'Agglomération d'Agen* (CAA) is a grouping of 6 municipalities in the Agen conurbation. CAA will use *Visual Admin* to better serve both citizens, with respect to its city policy; and businesses, in the frame of its economic development policy. Several types of actors will be implied in *Visual Admin* operations: territorial authorities, state services, non-profit organisations and consular bodies serving the targeted customers, particularly citizens in precariousness end enterprises.
- In Italy, the *Comunità Montana Valle Maira* (CMVM) is a grouping of 14 municipalities in a rural area of the western Alps. CMVM will use Visual Admin to provide the "citizen" (in the broad sense of the term: resident and emigrated) and organisations representative of the civil society (associations, private organisations, trade unions) with information about natural research management and tourism in mountains areas, in the frame of the policies of water deployment and promotion of tourism.
- In Poland, the *City of Lodz*, the second biggest city in the country, will use Visual Admin to manage the co-operation between public bodies of different geographical levels, social partners and non-governmental organisations, in settling and communicating around EU integration policy. Moreover it will support the partnership between state authorities, social organisations and private business units concerning EU structural funds (principles, objectives and applying procedures).

The common vision shared by the 3 pilots for the *Visual Admin* application is the one of a Portal set up by a territorial government to support one or several policies for which the territorial government has legal power. The Portal enables Citizens and Businesses concerned by that policy to:

- Get access to relevant information and eProcedures either through the Web or Wireless information devices;
- Be supported in having their case processed by the territorial government and through it by the whole set of relevant administrations.

---

[1] The *Visual Admin* Project is co-funded by the European Commission through the IST – "Information Society Technology" research programme

3 Identified Requirements

To guide the design and development of the *Visual Admin* solution, the Project relied on a study of the needs of three pilot territorial governments which was conducted during summer 2001 through interviews, surveys and focus groups. Actors that were contacted are:

- Territorial government staff: political managers and administrative-technical managers responsible for the policies addressed in the pilot, information system manager, …;
- Managers of 3rd party public administrations with which the pilot should collaborate to implement *Visual Admin*;
- Citizens and enterprises being the administration's "customers".

Taking into account the series of needs expressed by the pilots and the regulatory environment for eProcedures in France, Italy and Poland, the current section presents a set of "abstracted" user requirements in relevant functional areas:

- Portal Organisation
- Case Processing
- Specialised eProcedures
- Case Monitoring and Supervision
- Security & Rights Management
- Accessibility
- Organisational Issues

3.1 Portal Organization

The Portal should be organized in areas that both contain ad hoc information managed by a large group of actors (the Territorial Government and 3rd parties such as other administrations); it could also embed information stored in 3rd party Web sites. The list of contact persons in all involved actors should be clearly identified in each Portal area and Citizens should be able reach them by *e-mail* through the Portal, with request for acknowledgement of receipt.

A first level of Portal areas concerns policies such as Economic Development, Social Welfare, etc..

Each Policy area should provide relevant "*Information*" such as a policy overview and the list of citizen needs that could be serviced within this policy. It should also support community development through an "*Open forum*" to discuss experiences and to comment on the existing initiatives, regulation changes, or other relevant contents.

A second level of Portal areas concerns specialized processes in the frame of a given policy, e.g. Enterprise Set Up Project, etc..

Each Process area should provide relevant "*Information*" such as a process overview (especially important for complex processes such as "Calls for Projects"), a Directory of actors, of their competencies and of contact persons, a Catalogue of all major document templates and forms to use (major formats such as PDF, Word, Excel, RTF or XML should be used), or a Guide about support measures and public aids that could concern citizens. It could also provide "*Self-Assessment*" facilities that could assist a citizen in positioning its need and identifying who to contact and which support measure to ask for. Finally, it should link enable Citizens to initiate a case through either eProcedures or traditional processes.

More generally *"Event-related advertisement"* such as announcement of a new Call for Projects, could be presented on the Portal Home page, in the related policy page and in the specific process sub-area.

3.2 Case Processing

The Portal should enable managing pilot specific workflows for processing Citizens cases which would gather Data forms, Files such as documents (Word, Excel, etc..), Maps (MapInfo format), Instantiation of document templates provided by the Portal, and Link with Web pages or components stored either on the Portal or on 3rd Party Web sites.

Depending on the specific workflow, the case initiator could be a Citizen or a civil servant. Actors involved in a workflow should be able to submit additional case information through either online data forms or formatted emails with attachments (filled in data forms or documents) and authenticated with digital signature. Workflows could also automate an eProcedure or the sending of e-mails with attachments such as data forms or documents.

Finally, a workflow could imply the creation of a private forum for supporting the communication between actors involved in the case processing.

3.3 "Official Tender" eProcedure

A generic type of eProcedures has been identified to support *"Official Tender"* as an alternative or complement to a paper-based request. The basic steps are:

- To package a digital proposal gathering an electronic request based on a "Request Template", the list of appendices (support information), and the set of appendices available in a digital form.
- To electronically send this proposal in a secure way (using digital signature)
- To receive an electronic acknowledgement of receipt with a registration number that must be used in any later communication, especially when sending "paper" appendices.
- To possibly receive an additional electronic acknowledgement of receipt when supplementary "paper" appendices are received by the administration
- To receive a "Eligibility Certificate" once the targeted administration has checked that all needed documents for a complete request have been delivered, mentioning also which organisations and which persons are in charge of processing the request.

Moreover, for official requests submitted in paper but prepared with electronic templates found on the Portal, the citizen may mention that (s)he authorizes the request processing to be monitored through the Portal.

3.4 Case Monitoring and Supervision

To monitor citizens' cases, a global "Case Status" indicator should be defined that reflects the workflow progress and is updated at each step of the workflow either automatically for steps performed through an information system or through follow-up forms or e-mails for steps performed externally.

A Citizen monitoring his/her own case should see both the "Case Status" and who is involved in its processing. Moreover, the citizen and all actors involved in the case could ask to be automatically notified via e-mail or fax of any change.

At a managerial level, Policy operators may use historical information about cases and workflows to assess needs for changes in work organization. Therefore, logs of cases handled through the Portal will be kept and will be used to create a data mart on which to base decision support facilities.

3.5 Security & Rights Management

Identification and authentication are critical to initiate or to get access to specific citizens' cases and associated forums. Adapted access rights should be granted by default to the case initiator and to the concerned citizen; they could also be granted to other users through a workflow template, when such a template exists, or by the initiator of the case.

Moreover official transactions with legal value must be secured with a digital signature carried out with a "secure signature creation device" and can be verified with "Qualified electronic certificates". For such purposes, the solution implemented must comply to domestic regulatory frameworks.

3.6 Accessibility

Accessibility first concerns the ease of adjusting the Portal to the Citizen. To achieve that, citizens using the Portal could "*Personalize*" their access by registering their interests and by selecting their preferred channel for later communication – email or fax, but perhaps also interactive chat or phone –. Voice recognition techniques could also support disabled people in getting access to the Portal.

Accessibility also concerns the ability to use the whole set of services from a single access point. Therefore the secure environment required for submitting an official request through an eProcedure should be easy to set up on a personal computer connected to the Internet, without the intervention of a specialist in information technologies. Moreover, to avoid the "digital divide" risk, this secure environment should also be made available to citizens in some public areas.

Finally accessibility concerns the possibility to involve users wherever they are. Therefore, to rapidly get in touch with a Citizen, civil servants should be able to capture on the Portal a text which would be automatically transformed into an SMS message. Furthermore, for civil servants intervening outside their office, mobile access will be supported through GPRS multimedia equipment.

3.7 Organizational Issues

An eGovernment project such as setting up a *VISUAL ADMIN* Portal goes beyond technology implementation and information management. It also implies setting up a set of support measures for both civil servants and citizens:

- Training and support about Internet, the Web and specific *VISUAL ADMIN* mechanisms should be offered to civil servants involved in the workflow processing, even outside the pilot itself.
- A Help Desk could be set up to help citizens in preparing requests. Minimal functionality should encompass Interaction management (email, interactive chat, etc...) and Case management (to follow up requests from candidates).

4 Rationale for the *VISUAL ADMIN* Platform

Taking into account the requirements presented above, the *VISUAL ADMIN* Portal is based on the combination of functionalities that significantly improve performance and quality of service:

- The Portal offers a single access point to the diversity of information systems available within the local government and in other related public bodies. It implies:
  - Information services for providing guidance to citizens and businesses on the targeted policy area;
  - Integration of existing databases and legacy applications of the administration and an online interaction between applications;
  - Interaction with existing eProcedures in public bodies.
- The portal includes a g-CRM (government's customer relationship management) module for handling the interaction of a customer with the administration workflow and for improving the quality and friendliness of the service. It implies:
  - Focusing on the customer's case rather than on constraints and needs internal to the administration.
  - Monitoring how each case is processed through the administration work flows in order to inform the customer on his case status.
- Mobile Access services are offered in order to provide citizens with a ubiquitous access to the administrative services through a multimedia mobile phone supporting GPRS or EDGE communication. The *VISUAL ADMIN* solution also offers an intuitive interface based on voice recognition for controlling access to administrative services and citizen's data.

5 Benefits Associated to *VISUAL ADMIN*

The *VISUAL ADMIN* Portal solution will thus enable a local government to:

- ***Better serve citizens:*** Relations between citizens and government services are often perceived as unbalanced with the citizen being more an anonymous item/case than a customer. By really involving citizens in their case processing in government services, the *VISUAL ADMIN* approach will improve people satisfaction with both the experience and the service quality.

Leveraging on local networks or communities of civil servants will also enable citizens to get a more comprehensive advice from the civil servants. Moreover, supporting civil servants that operate in nomadic work contexts will also bring closer together the citizens and the public service.

- ***Settle better working conditions for civil servants:*** By supporting the flow of information and work within and amongst government services, *VISUAL ADMIN* enables civil servants to decrease significantly time spent on low added value activities such as archiving or retrieving citizens' data, writing requests for additional information for other administration services, etc…
- ***Reduce government expenditures:*** By analysing bottlenecks in work organisation and supporting the redesign of processes thanks to workflow tools, and by reducing the time (and the cost) of data capture by civil servants.

e-Government Observatory

Freddie Dawkins

Senior Communications Manager, eGovernment Observatory,
GOPA Cartermill International, 6th Etage, 45, Rue de Trèves
B-1000 Brussels, Belgium

1 Description of Work in The e-Government Observatory

1.1 eEurope Context

The European Commission launched the eEurope initiative on 8th December 1999 with the adoption of the Communication 'eEurope – An Information Society for all'. The initiative aims at accelerating the uptake of digital technologies across Europe and ensuring that all Europeans have the necessary skills to use them.

The eEurope Action Plan sets ambitious goals concerning **"Government online: electronic access to public services"**.

The Lisbon European Council conclusions call for:

- efforts by public administrations at all levels to exploit new technologies to make information as accessible as possible.
- Member States to provide generalised electronic access to main basic public services by 2003.

The observatory takes as starting point one of the conclusions of the Sandhamn conference (Sweden on 13 and 14 June 2001):

"For eGovernment to be implemented successfully at the European level, policy should be developed that specifically addresses its European dimension. This is in addition to what is being achieved at the national level. Initially, such policy should be formulated to determine what European public administrations should do to make it easy for citizens and enterprises to transact business at the European level".

The Ministerial declaration produced as a conclusion of the eGovernment Conference in Brussels mandates IDA (Interchange of Data between Administrations) to set-up an *e*Government Observatory.

"Ministers agreed to encourage National Administrations and EU Institutions to establish a common view on which pan-European eServices are most essential on a European level, and to establishment of an eGovernment platform, building on the European Forum on eGovernment and the eGovernment Observatory"

The European Forum (www.eu-forum.org) on eGovernment and the eGovernment Observatory (www.europa.eu.int/ISPO/ida) work in a co-ordinated fashion and feed-back to each other.

R. Traunmüller and K. Lenk (Eds.): EGOV 2002, LNCS 2456, pp. 326–329, 2002.

1.2 IDA Context

The work described hereafter falls under the scope of Article 10 "Spread of Best Practice" as well as Article 9 "Interoperability with national and regional initiatives", of the IDA Interoperability decision [1]

IDA is well equipped to address this issue and has been consistently supporting work of relevance in the eGovernment domain.

1.3 Scope of Work

The subject matter of the Observatory is eGovernment domains, where IDA can add value focusing on the EU dimension. It helps to leverage national and regional initiatives of EU relevance, monitor other worldwide developments/trends and also promotes and learns from best practices.

Central to the Observatory is the gathering of information on state-of-the-art and trends concerning Information Society Technologies and their eGovernment applications being of relevance to IDA.

In this strategy, the IDA website plays a central role but the information is delivered using the most appropriate multimedia products and ensuring appropriate coordination. To date, the Observatory has used the website, electronic press releases, e-mail alerts, physical presentations in several EU Member States and wide-ranging electronic surveys, via Chgambers of Commerce, European Information Centres EICs) and others, to raise awareness of its work.

The information dissemination activities are coordinated with the promotion activities carried out in the context of the Information Society and of the eEurope/eGovernment initiatives.

An effective information network with the Member States Administrations is maintained, mainly through the National Government delegates to IDA, the Telematics for Administrations Committee.

The work addresses issues related to applications supporting the Internal Market, eDemocracy, public/private partnerships, and "pan-European eGovernment Services for European citizens and enterprises" at large.

The Observatory addresses both existing eGovernment initiatives and emerging trends in terms of applications, R&D technologies and commercial solutions which could have an impact to enable the effective deployment of the expectations raised by eGovernment.

2 Activities

The final objective of the Observatory is to proactively support IDA in the early identification and better understanding of emerging eGovernment areas where IDA shall play a role, complemented by Best Practice activities. Its outputs are so designed to contribute to the emergence of pan-European strategies.

---

[1] Decision no. 1720/1999/EC of the European Parliament and the Council of 12 July 1999 adopting a series of actions and measures in order to ensure interoperability of and access to trans-European networks for the electronic interchange of data between administrations (IDA) – OJ L203, 3-8-1999, p. 9.

The work involves the identification of relevant information, creating synergies with complementary initiatives in other DG's or Member States, packaging the information in the appropriate multimedia format(s), raising awareness and disseminating the information (web, information network with the Member States Administrations, meetings, documents, etc.). It also involves seeking feedback and spinning-off appropriate activities.

The information and its delivery has to combine clarity and wealth of content. It has to be appealing for people throughout Europe working on eGovernment or using its services. It supports the decision-making in Member States and European Institutions.

In order to achieve the objective above, the Observatory covers three major areas:

- Area 1: eGovernment Surveillance
- Area 2: In-depth information
- Area 3: Start-up

Area 1: eGovernment Surveillance

Central Repository and Continued Surveillance:
Being a virtual central repository where links with the most relevant eGovernment activities are kept. A "soft coordination" role is played. Surveillance of ongoing activities, trends, relevant technologies as well as developments in the private sector and information of relevance for the eGovernment domain. The information is disseminated through the IDA Web site[2] and a bi-monthly electronic newsletter is distributed along with regular meetings to exchange information and get feedback from Member States Administrations and involving as well key personalities in the domain.

An eGoverment "Tableau de bord" has been constructed on the website, and it monitors progress in national eGovernment plans as well as at the level of the Community.

National and Regional Initiatives and R&D Results:
Identify eGovernment national & regional initiatives in Member states as well as results from R&D projects of relevance for IDA. The strategy is to identify and give visibility to a few selected **"gemstones"** and then make proposals on appropriate actions in the different areas of activity of IDA (Best Practices, Generic Services, Common Tools, Security, content interoperability, etc.).

Area 2: In-depth Information

In-depth focused analysis is undertaken on a six monthly basis.
The first six monthly activities had as its focus:

On-line Services to Enterprises in Europe and the role of Public Private Partnerships.

---

[2] The Observatory provides the information but the actual delivery through the Website is subject of a separate project.

The objective is to identify the needs of enterprises concerning on-line services focusing on trans-border services. This involves analysing the electronic services already delivered as well as new services to be delivered. In order to understand those needs, both the offer side (administrations and others) and the demand side (enterprises) have to be analysed. Best Practices elsewhere world-wide will also be relevant.

The outcome will include possible activities to undertake at the Community level and proposed priorities.

The role Public Private Partnerships are already playing and the future role they can play in a Trans-European context need to be analysed. Other types of co-operation between the public and private sector are also relevant.

In the case of eProcurement, there is already some ongoing work on trans-border issues and Best Practices. The work will be a major input for the 2002 IDA Conference where all this work will be presented (September 19-20 in Brussels. It will also be an input for building IDA's policy for the future (a communication of the Commission is expected during 2002).

Each of the six monthly activities will ensure appropriate dissemination and exchange of information with national experts and will finish with a dissemination event.

The **Final Report** of the Observatory (after December 2002) will present a description of the work performed, the transborder catalogue and the recommendations issued by the project. The final report will be suitable for general dissemination, in particular via the IDA website.

References

For more information about eEurope see:
1. http://europa.eu.int/comm/information_society/eeurope/
2. http://europa.eu.int/comm/information_society/eeurope/documentation/index_en.htm

For more information about action line 3b of eEurope concerning "Government online: electronic access to public services" see:
3. http://europa.eu.int/comm/information_society/eeurope/actionplan/actline3b_en.htm)

Requirements for Transparent Public Services Provision amongst Public Administrations

Konstantinos Tarabanis[1] and Vassilios Peristeras[2]

[1] University of Macedonia, Thessaloniki, Greece
kat@uom.gr
[2] United Nations Thessaloniki Centre, Thessaloniki, Greece
per@untcentre.org

Abstract. In this paper we analyze the requirements posed by the Infocitizen project that attempts to make feasible the realization of a pan-European view for public service provision. The requirements are analyzed based on the project aims of conducting electronic transactions in multi-agent settings- e.g. multi-country involvement- in a transparent as possible manner for the citizen. Transparent public services provision for the citizen is posed as the requirement that both the inputs needed for the delivery of a service as well as the outputs produced by the service are respectively given and received in a transparent as possible manner for the citizen. That is, the citizen will only need to provide the input that cannot be automatically accessed from its relevant source and also the consequences of the delivered service will be automatically propagated to its relevant destinations. In order to achieve such intelligent provision of public services the forms of knowledge that need to be employed are also discussed.

1 The Problem Addressed by the InfoCitizen Project

As most other efforts in the area of electronic government, the InfoCitizen project addresses the "Service Provision" area [1][2] from those shown in Figure 1. The typical scheme for the provision of public services by a Public Administration (PA) involves a customer (citizen, business or other PA) who requests a service from the responsible PA. The PA then performs its internal tasks that may include interacting again with the customer during the various stages of process execution and at the end for service delivery.

The extent to which this simplified model of PA service provision is performed electronically has led to the conceptualization of the four levels of e-government reported in the literature [3]. The InfoCitizen project addresses the fourth level of e-government that of full service provision. Moreover, the project attempts to conduct electronic transactions in multi-agent settings- e.g. multi-country involvement- in a transparent as possible manner for the citizen. In the following we will elaborate on the requirements posed by InfoCitizen along its two main themes:

- transparent public service provision to the citizen and
- multi-agent setting of public service provision

R. Traunmüller and K. Lenk (Eds.): EGOV 2002, LNCS 2456, pp. 330–337, 2002.

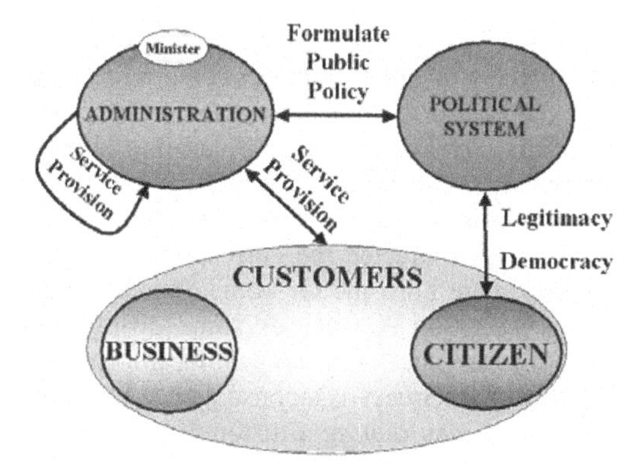

Fig. 1. InfoCitizen Scoping

1.1 Transparent Public Services Provision

Transparent public services provision for the citizen is addressed in InfoCitizen as the requirement that both the inputs needed for the delivery of the service as well as the outputs produced by the service are respectively given and received in a transparent as possible manner for the citizen.

Specifically, regarding the required input of a service, this should be automatically accessed when needed from its relevant source and the citizen will only need to provide anything that cannot be "pulled" in the above sense.

This "pull" strategy enables PA services with minimal input and precondition checks needed on behalf of the citizen-customer. That is, input that needs to be provided and preconditions that need to be checked in order to provide this service are, to the extent possible, automatically performed. However, we need to examine cases in which this "pull" strategy is difficult to apply and why.

In the other respect, the outputs of the service are defined in a broader sense than that of the delivered product or service. In this respect, as outputs of the service we also consider the consequences produced as a result of the service delivery itself. For example, when the PA service of a marriage is executed, the output of the service in a strict sense is the marriage certificate. However, outputs in a broader sense are all the by-products of the service. For example, the information that may need to be delivered to other agencies that document the change of marital status for the actors involved. The output may need to be delivered to the citizen, but its by-products (i.e. consequences) are required to be automatically "pushed" to the destinations (sinks) that are relevant.

The "push" strategy of the generalized output aside from alleviating the citizen from having to propagate the changes introduced by the public service that was just executed, also attempts to maintain consistency of the PA system as a whole across all countries considered. That is, any PA service that will be provided maintains consistency of all the individual PA systems, while any PA related question receives the same answer in all countries. For example, the marital status of a person receives the same answer anywhere posed.

This may mean that the consequences of the PA service provided are "propagated" to all relevant PAs in order to maintain a consistent state. As already described, under the term consequences we include all the after-effects that an administrative process could have on the PA systems of other countries. One part of this issue involves deciding when (and why) the local PA service has interest that is broader than the country where it is executed. In addition, one must determine how to decide the relevant PAs and with what knowledge will this decision be made. For example we could raise the question: who has to be informed that a person or couple has adopted a child in Greece? Is it the PA that has conducted the corresponding marriage? Is it the PA in which the child or parents were born in? Or both? Or is it some PAs in other countries? Also, perhaps the Adoption Act causes some different consequences in one country from another. For example, in Greece posting information regarding an adoption to the Ministry of Justice in Greece is a consequence not found elsewhere.

Furthermore, a PA service may require a different set of preconditions in order to be executed. This may also mean imposing pre-conditions to the PA service that are applicable in more than just the country of the PA that is responsible for its execution. For example: What if Greece does not have any age limit for adoption but Germany poses the age-limit of 21 for anyone interested in adopting a child? Or what will happen if a country gives the possibility of adopting a child to individuals, while another only to couples? In this case the overall PA system will be modeled in a "semi-consistent" state.

Taken together the InfoCitizen requirements of transparent PA service provision refer to transparent supply of input and delivery of output according to the knowledge management strategies of "pull" and "push" respectively.

As discussed an important issue in order to achieve transparent PA service provision involves the orchestration of the above "pull" and "push" strategies, that is, determining their applicability, the source of the "pull" and the destination of the "push" etc. These issues will be addressed by representing the aforementioned in a form of explicit knowledge. This knowledge has a control nature and must have a global scope. Such a role will be assumed by a "broker" or an "intermediate" that knows for example about the PA services in the countries considered. This "broker" will then route and facilitate all the information flows needed. It is important to emphasize at this point that this "broker" possesses mostly knowledge of a control nature and less that of data per se (e.g. personal data of the actors involved). It will contain control knowledge such as process descriptions and the relevant information that will enable the routing of the right piece of information to the right PA unit throughout the countries considered.

In order to build this "broker", schemes that model the control knowledge consisting of terms, pre-conditions, post-conditions, consequences, inputs, outputs etc. will be employed. Since such knowledge is expected to be related to a high-degree as PA systems have similarities amongst themselves, this knowledge will need to be modeled in a way that commonality in any of the above concepts is abstracted out. This "broker" is an important component of the InfoCitizen architecture.

This InfoCitizen requirement of transparent PA service provision may either define a new level of e-government above the four levels mentioned above or constitutes an advanced form of the fourth level. The fourth level of e-government, that is, conducting electronic transactions, can be achieved with a varying level of intelligence. For example, the simple exchange of the proper documents over a well-defined interface may lead to an electronic transaction. However, the InfoCitizen requirement of transparent PA service provision as analyzed above entails a high-degree of intelligence leading essentially to executable knowledge bases that support highly automated service provision.

1.2 Multi-agent Setting

The other InfoCitizen requirement, that of public service provision in a multi-agent setting also plays a central role for the project. This requirement has been motivated by the goal of the InfoCitizen project to facilitate a pan-European provision of PA services.

When looking at the internal tasks that are executed within PA in order to provide the service according to the previously discussed simple model for "Provide Service", it is often necessary for more PAs to participate other than just the PA responsible for the overall service. These associated PAs provide constituent services contributing to the main service. This gives rise to a process that has multi-PA agency participation and correspondingly multiple component processes. The form of participation of each PA agency and of its relevant component process is itself interesting to specify since it may vary to a large degree (e.g. simple provider of input, executer of part of the overall process, controller of part of the overall process etc.).

Multi-PA agency participation in the "Provide Service" process introduces more demanding coordination requirements for the responsible PA since the delivery depends in part on parties and processes external to the PA agency. Coordination requirements exist even in single PA agency case in order to orchestrate the steps of the process. However, in the multi-PA agency case the coordination requirements orchestrate the overall process both at the process step level as well as at the process-to-process level. The latter coordination requirements involve addressing issues such as "at what stage does another PA agency participate?", "which PA agency is responsible for this constituent service?", "what are its inputs and outputs?", "who controls this constituent process", etc. In order to address these coordination requirements knowledge again of a control nature is required.

In the previous discussion the implicit assumption is made that all such PAs (responsible and associated) are under the jurisdiction of the same country. InfoCitizen introduces the added complexity to a multi-agency "provide service" PA process in that the PA agencies involved are from the Public Administrations of more than one country. This added complexity affects the control knowledge discussed earlier in that:

- the control knowledge amongst Public Administration systems is different and may not even be compatible, and
- the terms used are different since on the one hand they are expressed in different languages while more importantly, differences amongst Public Administration systems lead to, among other things, slightly different definitions of linguistically identical terms.

These added requirements lead to the need of control knowledge at the meta-knowledge level in order to unify these distinct sets of control knowledge.

2 Knowledge-Based Intelligence for InfoCitizen e-Government

In order to achieve the intelligent provision of public services described previously in the analysis of requirements of the InfoCitizen project, certain forms of control knowledge and meta-knowledge need to be employed. Let us investigate the knowledge required as the level of intelligence of public service provision is gradually increased reaching the full level of InfoCitizen intelligence addressing transparent public service provision in a multi-agent setting.

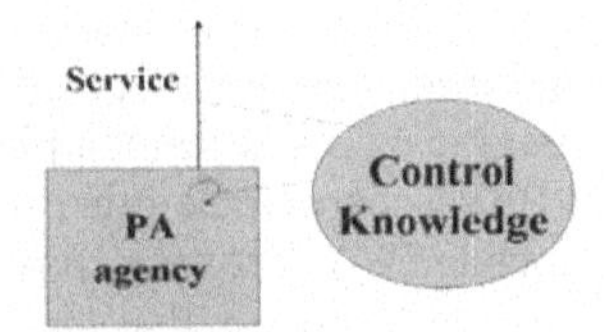

Fig. 2. Control Knowledge in a single PA agency setting

In the case of service provision by a single agency (Fig. 2), no other agencies are required in order to deliver the service. That is, any "push" or "pull" that needs to be executed lies within the domain of this agency. If the agency has achieved integration in its information systems environment, the "push" and "pull" actions are already part of its integrated operation. Otherwise, these "push" and "pull" actions need to be implemented, for example, as middleware applications. Such integration is more readily achievable since any organizational obstacles (e.g. who owns what information) are less pronounced. As a result, the knowledge that is required to implement the "push" and "pull" actions in this setting lies implicitly in the application that achieves integration and may not need to be explicitly represented.

In the case of service provision with the participation of multiple agencies all within the PA system of one country (Fig. 3), achieving coordination of the "push" and "pull" actions becomes a significant issue. Knowledge needs to be represented regarding the following:

1. preconditions for provision of a service,
2. consequences of a service together with the affected party, both within the agency but also in other agencies,
3. controls that apply in any of the steps involved in the provision of the service together with the agency responsible for applying these controls,
4. inputs of a service together with their source,
5. outputs of a service together with their destination.

In the case of service provision with the participation of agencies from several countries (Fig. 4), additional knowledge is needed in order to *unify* the distinct bodies of knowledge that address the previous case of multiple agency participation (see above list). Specifically, the additional knowledge that needs to be represented in this case pertains to the following:

1. knowledge that can reason with term correspondence
   a) at a language level (i.e. corresponding terms in different languages) and
   b) at a conceptual level

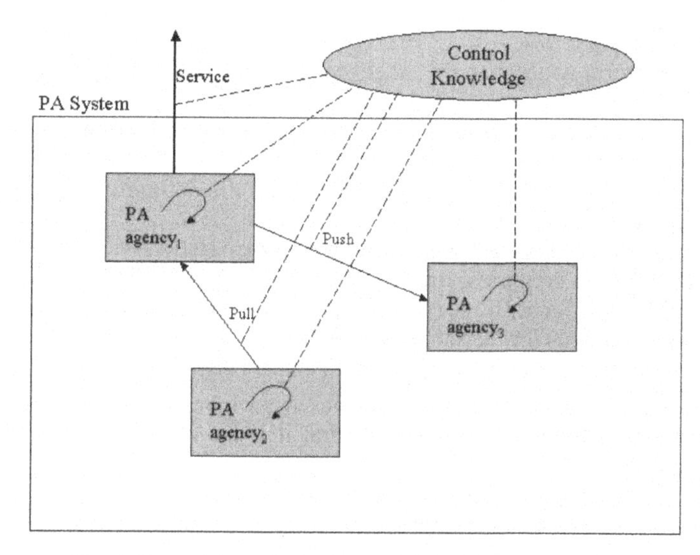

Fig. 3. Control Knowledge in a multiple PA agency setting

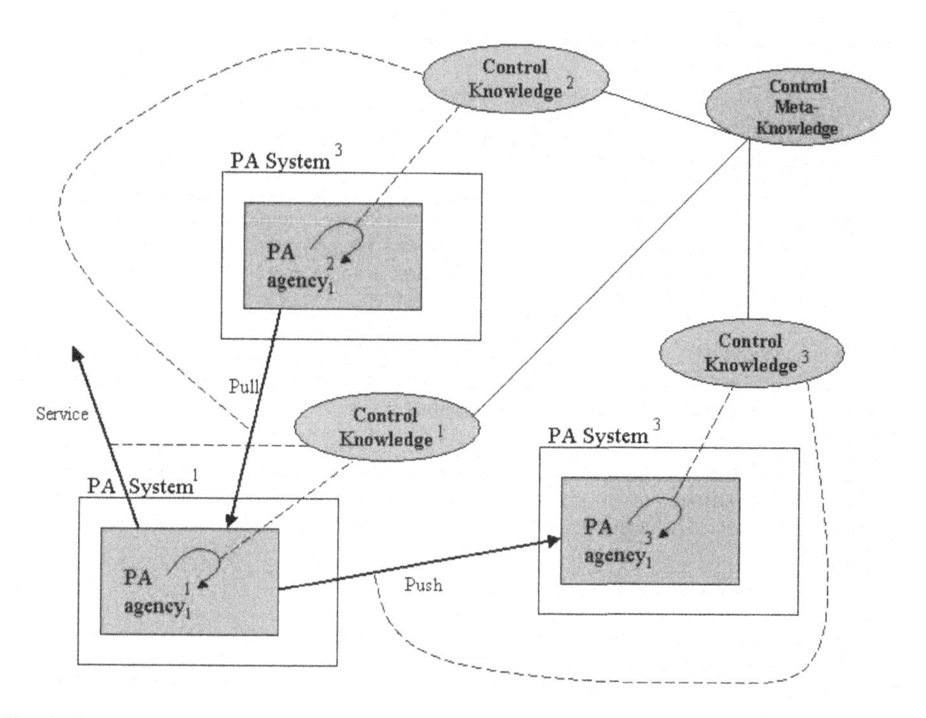

Fig. 4. Control Knowledge in a multiple PA agency setting corresponding to multiple PA systems

    i. identical terms,
    ii. one term wider than the other together with the definition of the associated surplus,
    iii. one term narrower than the other together with the definition of the associated deficit,
    iv. one term overlaps another together with the definition of the non-overlapping regions,
2. knowledge that can reason with cases where the same service is provided differently in different countries, that is, cases where any of the above corresponding pieces of knowledge differ – e.g. different consequences for the same service, different preconditions for the same service etc.

From the above it can be seen that the knowledge needed for this case, that is service provision with the participation of agencies from several countries, corresponds in essence to meta-knowledge, since in general it is knowledge that can reason about other knowledge.

The above described knowledge and meta-knowledge will be represented in appropriate structures. When examining the structure of the previously described types of knowledge, similarity can be detected to the structure of rules. This is attributed to the fact that a great deal of the control knowledge employed in PA is based on laws enacted by the government, which by nature are structured as rules.

The representation of laws has been studied in the area of legal information systems. Kralingen [4] proposes the use of a norm frame in order to model laws. The structure of the norm frame of Kralingen is shown below:

Table 1. Structure of the norm frame of Kralingen

1 Norm identifier	The norm identifier (used as a point of reference for the norm).
2 Norm type	The norm type (norm of conduct or norm of competence).
3 Promulgation	The promulgation (the source of the norm).
4 Scope	The scope (the range of application of the norm).
5 Conditions of application	The conditions of application (the circumstances under which a norm is applicable).
6 Subject	The norm subject (the person or persons to whom the norm is addressed).
7 Legal modality	The legal modality (ought, ought not, may, or can).
8 Act identifier	The act identifier (used as a reference to a separate act description).

Leff [5] proposes the use of an XML format to model rules. An excerpt of the structure is shown below:

```
<Rule Name= "RuleOne"……>
<LHS>
      <Check></Check>
</LHS>
<RHS>
      <Action></Action>
</RHS>
</Rule>
```

For the meta-knowledge that needs to be represented appropriate structures will also be sought.

3 Conclusion

In this paper we analyze the requirements posed by the Infocitizen project that attempts to make feasible the realization of a pan-European view for public service provision. The requirements are analyzed based on the project aims of conducting electronic transactions in multi-agent settings- e.g. multi-country involvement- in a transparent as possible manner for the citizen. In order to achieve such intelligent provision of public services the forms of knowledge that need to be employed are also discussed. In future work we will investigate initial implementation directions, such as specializing on existing XML and derivative XML (e.g. ebXML) schemas [7] for the Public Administration domain and following for this the UN/CEFACT Modeling Methodology [6] in order to ensure compatibility.

References

1. Peristeras V., K., Tarabanis, "Towards an Enterprise Architecture for Public Administration: A Top Down Approach", European Journal of Information Systems, vol. 9, pp. 252-260. Dec. 2000. Also in proceedings of the 8th European Conference on Information Systems, Vienna, vol. 2, pp.1160-1167
2. Tarabanis K., Tsekos Th., and Peristeras V., "Analyzing e-Government as a Paradigm Shift", to be presented in the "Annual Conference of the International Association of Schools and Institutes of Administration (IASIA)" 17-20 June 2002, Istanbul, Turkey.
3. European Commission, "List of indicators for benchmarking eEurope as agreed by the Internal Market Council" at http://europa.eu.int/information_society /eeurope/news_library/documents/text_en.htm.
4. van Kralingen, "*A Conceptual Frame-based Ontology for the Law*", Proceedings of the First International Workshop on Legal Ontologies, Melbourne, Australia, 1997
5. Lawrence L. Leff, *"Automated Reasoning with Legal XML Documents"*, Proceedings of International Conference AI and Law – ICAIL 2001, St. Louis, Missouri, USA, 2001.
6. UN/CEFACT's Modeling Methodology, Draft, TMWG/N090R10, UN/CEFACT, November 2001
7. ebXML Business Process Specification Schema Version 1.01, Business Process Project Team, 11 May 2001 at www.ebxml.org/specs/ebBPSS.pdf

CB-BUSINESS: Cross-Border Business Intermediation through Electronic Seamless Services

Maria Legal[1], Gregoris Mentzas[2], Dimitris Gouscos[3], and Panagiotis Georgiadis[3]

[1] Planet Ernst & Young SA
m.legal@planetey.com
[2] Dept. of Electrical and Computer Engineering, National Technical University of Athens
gmentzas@softlab.ntua.gr
[3] eGovernment Laboratory, Dept. of Informatics and Telecommunications, University of Athens
{d.gouscos,p.georgiadis}@e-gov.gr

Abstract. Business enterprises face significant obstacles in their quest to interact with public administrations and governments across Europe, such as bureaucracy, ambiguous procedures, functional disintegration, vague authority structures and information fragmentation. The recent trend towards the delivery of electronic services by governments ("*e-government*") and the development of integrated and customer-oriented mechanisms ("*one-stop government*") are efforts to overcome these problems. However, all related efforts focus on the national scene of each country and do not address the needs of businesses when they enter into cross-border processes. This paper presents the objectives, the overall approach and the architectural model of the CB-BUSINESS project, which aims to develop, test and validate an intermediation scheme that integrates the services offered by government, national and regional administration agencies as well as commerce and industry chambers of European Union and Enlargement countries in the context of cross-border processes.

1 Introduction

Business enterprises face significant obstacles in their quest to interact with public administrations and governments across Europe. The most common problems include bureaucracy, ambiguous procedures, functional disintegration, vague and/or overlapping authority structures and information fragmentation. The recent trend towards the delivery of electronic services by governments ("*e-government*") and the development of integrated and customer-oriented mechanisms ("*one-stop government*") are efforts to overcome these problems and radically revamp the services provided by European governments; see e.g. [1], [3], [5], [6] and [7].

At the European level, the *eEurope* initiative launched by the European Commission in December 1999 puts forward a concrete action plan for "Government On-Line (GOL)" with the aim to make public information more easily accessible and stimulate the development of new private sectors services based on the new data sources that become available; see [2].

However, related efforts both within and outside the European Union focus on the national scene of each country and do not address the needs of businesses when they

R. Traunmüller and K. Lenk (Eds.): EGOV 2002, LNCS 2456, pp. 338–343, 2002.

enter into cross-border processes. Certain initiatives such as EUREGIO (see [4]) have been set up in this direction, but their emphasis is rather on facilitating transactions between neighboring border regions and, in this context, supporting cross-border processes at the local level, than on explicitly providing support for "anywhere-to-anywhere" cross-border transactions. The critical need is to support, for instance, a company (possibly without core business competencies in exports) that occasionally enters into export procedures or has to pay foreign subcontractors, requiring information and/or needing to make transactions with the public administration and government organisations of another (any other, possibly non-neighbouring) country. In this area the problems of bureaucracy, ambiguity, vagueness and disintegration that business enterprises have to face when interacting with foreign governments, get more sharp than in the national setting. Such problems are insurmountable for SMEs with limited resources.

The CB-BUSINESS project presented in this paper addresses directly this situation as its primary objective is to develop, test and validate an intermediation scheme that integrates the services offered by public administrations as well as chambers of commerce and industry in the context of cross-border issues. The CB-BUSINESS project is a 24-month project co-funded by the European Commission under the 'Information Society Technologies' program and involves as partners Planet Ernst & Young SA, SchlumbergerSema, University of Athens, ComNetMedia, Greek Ministry of Economy and Finance, Instituto Tecnologico de Canarias, Bulgarian Chamber of Commerce and Industry, Paris Chamber of Commerce and Industry, Athens Chamber of Commerce and Industry and Chamber of Commerce and Industry of Romania and Bucharest Municipality.

2 CB-BUSINESS Objectives

The CB-BUSINESS project has three specific objectives.

Objective 1. Design a unified true "one-stop shop" service model for "Business-to-Government" interactions. CB-BUSINESS aims to develop a service model based on an intermediation scheme that:

- extends the "first-shop" (i.e. information counter) and "convenience store" (i.e. one location for different transactions) service models to develop a true one-stop government model_(i.e. one location that integrates many services necessary to satisfy concerns of specific client groups in specific events) that is transparent to the end-user company;
- focuses on cross-border processes necessary for administrative support of cross-country business searches, contacts and transactions, falling under the general theme of "cross-border entrepreneurship" and mainly initiated by EU businesses targeted at Enlargement Country markets; and
- is structured around end-user needs (rather than provider services – i.e. governmental processes) by grouping services around "business life episodes".

Objective 2. Develop a WWW-based intermediation hub that implements this service model and will act as a pivotal point of contact for EU and enlargement country enterprises. CB-BUSINESS aims to develop a system that will:

- Have an "intermediation hub" technical infrastructure able to accept user requests, identify the cross-border processes that have to be enacted, trigger and dynamically coordinate process workflows of individual service providers (administrations and chambers) and integrate the final results for delivery to end-users.
- Be accessible over the Internet – but also open to future extensions, both for multi-channel service delivery (e.g. through mobile phones, digital TV, call centres, etc) and to complementary infrastructures (e.g. banking and postage services)
- Provide a "seamless service feeling" for end-users, who can have access to cross-border services through single-stop, single-session, single-sign-on procedures at a low cost and up to a standard quality, as well as to decrease operational costs and increase quality of service for both government/administration and chamber service providers.

Objective 3.Prove the validity of the service model and pilot-test the WWW-based intermediation hub in various specific cases that strengthen European integration and facilitate cross-border processes between EU-enterprises and businesses in enlargement countries.

3 CB-BUSINESS Approach

CB-BUSINESS establishes an intermediation scheme between service providers and end-users, with the aim of transforming the many-to-many service provider/end-user communication mesh (in which each end-user should communicate with as many service providers as are involved in serving his/her need) into a many-to-one-to-many star-like communication topology.

In this architectural scheme, an end-user communicates with the intermediation hub as a single point of reference, and the latter handles all complexity of triggering and co-ordinating service provider workflows. Therefore, end-users enjoy a "seamless service feeling" whereas service providers avoid the burden of inter-organisational communication, since the hub undertakes all co-ordination procedures. This results in more structured tasks for service providers, who are enabled to better define their operational interfaces towards the intermediation hub and concentrate on the establishment and improvement of quality and performance dimensions for their process workflows.

Upon submission of a user request (see Figure 1) the CB-BUSINESS intermediation hub shall identify involved services, competent service providers and user input requirements and ask for the latter as appropriate. Upon provision of required input data from end-users, the intermediation hub undertakes forwarding of user input, triggering and co-ordination of process workflows of individual service providers, rendering internal details regarding service execution procedures or workflows transparent for end-users; however, upon user demand, the processing status of end-users' requests may be monitored and presented. Therefore, the CB-BUSINESS intermediation hub employs overall workflows of cross-border processes to trigger and co-ordinate individual service provider workflows, thus being able to deduce the progress status of user requests and report it as appropriate.

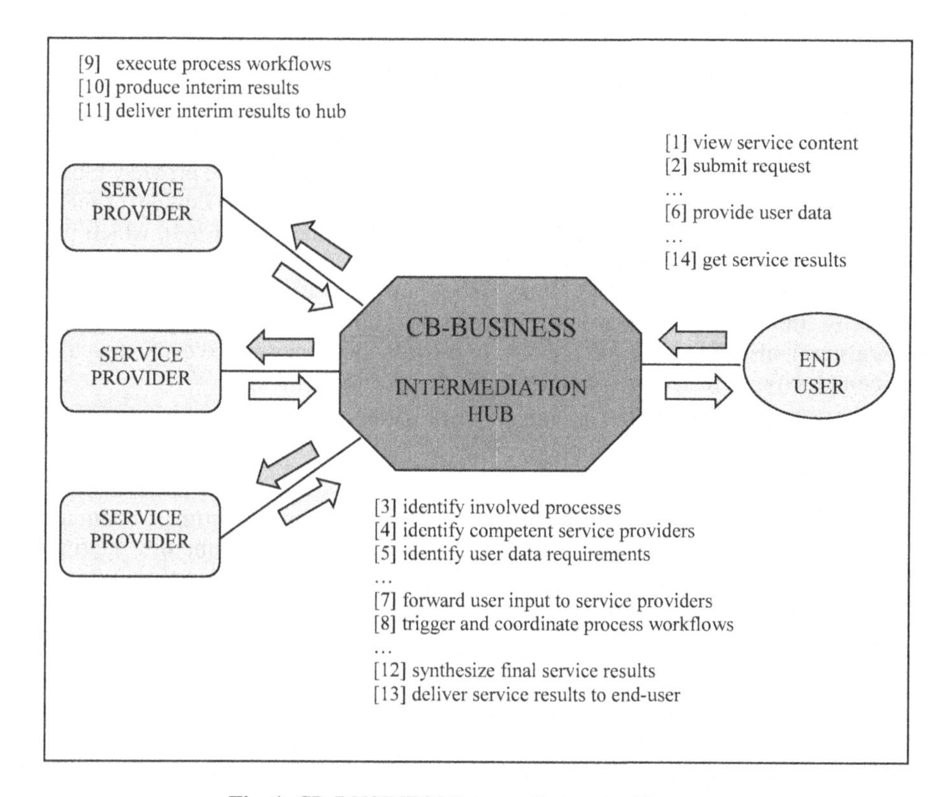

Fig. 1. CB-BUSINESS Intermediation Architecture

From a structural point of view, the CB-BUSINESS intermediation hub is deployed as an added layer on top of existing government and business service schemes (which indeed corresponds to the idea that the CB-BUSINESS service-mediary constitutes an added-value layer on top of existing service provision schemes). The intermediation hub employs standardized modules for interfacing with individual service schemes, and such standardized modules are also deployed in the latter as interfaces to the CB-BUSINESS hub. These modules (referred to as CB-BUSINESS interfaces) encompass all the technical infrastructures and specifications, functional conventions and operational arrangements which are necessary in order to establish information and control flows between the intermediation hub and individual service schemes. Two points should be made here:

- This structural architecture is based on the concept of standardized interfaces between the intermediation hub and individual service schemes, and is not dependent upon internal details of service scheme implementation. In other words, although the CB-BUSINESS intermediation hub itself and the CB-BUSINESS interfaces are ICT-based, CB-BUSINESS service providers may internally implement their service provision procedures with IT- and/or paper-based work schemes. This degree of independence allows even non-IT-enabled service providers to integrate their services with CB-BUSINESS, which is considered an important strength of the architecture.

- The star-like topology of the intermediation architecture described here, where there is a single central point of co-ordination, may also be generalized into a hierarchical-star topology without affecting the basic principles. In such a topology, individual service providers are grouped in clusters (according, for instance, with geographical, sectoral or mission level criteria), and each service provider cluster is coordinated by a corresponding low-level intermediation hub. Low-level intermediation hubs are then hierarchically coordinated by the central hub, which undertakes responsibility for global operation. In such a multiple-hub topology, end-users do not necessarily refer to the central hub but may access any hub of the hierarchy (the one most suited to their needs according to the service providers that it clusters); all hubs, however, are able to provide the same service content so that no complexity is created on the end-user side.

As far as the implementation architecture of the intermediation scheme is concerned, the following basic principles apply:

- the overall technical architecture of the intermediation hub is as much as possible compliant to current standards (with respect to information formats, communication protocols, etc) and open to emerging ones, while at the same time getting the most benefit out of technologies for e-service provision and delivery platforms
- the overall technical and functional architecture of the intermediation hub is modular with the aim of facilitating incorporation of alternative and complementary e-service platforms.

With these two principles in mind, CB-BUSINESS architecture employs the notion of a "**communication gateway**" for referring to the implementation of communication and co-ordination links between the CB-BUSINESS intermediation hub, end-users and service providers. A CB-BUSINESS communication gateway represents a specific communication channel (such as the Internet or WWW, fixed or mobile telephony etc) together with all technical standards (communication protocols, information content or information medium formats etc) necessary for exploitation of the channel, together with all associated operational dimensions (e.g. quality-of-service, security-of-service etc). It is clearly desirable to support as many different communication ways as possible, so that prospective user communities are broadened and accessibility constraints are relaxed. With this objective in mind, the CB-BUSINESS implementation architecture has been conceived as comprising of certain indispensable core modules, implementing the basic intermediation hub functionality, complemented by a number of optional pluggable peripheral modules, each implementing one specific communication gateway employed by end-users and/or service providers In this way, the core architectural design can ignore communication gateway details, which are handled by dedicated pluggable modules, and assume an **abstract communication model**; even more importantly, additional communication gateways may be incorporated into the system as new e-service provision and delivery platforms emerge, which is a fundamental prerequisite for keeping up with technological state-of-the-art and exploiting the benefits of technological advancements. Last but not least, this architectural modularity also allows for provision of non-IT-based communication gateways (such as voice telephony, facsimile or paper mail) which can be valuable for printed document transfer and non-IT-enabled end-users.

4 Current Project Status

As the CB-BUSINESS project started just a month ago (April 2002), the present paper is based on material from the project proposal. Currently, the CB-BUSINESS consortium is working on the analysis of service provider processes and end-user requirements, as well as on the specifications of the functional and technical architecture of the CB-BUSINESS system. It is expected that by September 2002 the service scenarios of the CB-BUSINESS intermediation hub will have been defined and analyzed in detail, and all functional and technical aspects of the system demonstrator will have been specified, so as to proceed with the technical design and solution development.

It should be stated that the next steps of the project implementation involve, also, the formulation of a set of architectural provisions on CB-BUSINESS service delivery platforms, process workflow ontology and operational interfaces. This is an important process in order to ensure the wide market acceptance and acceptability of the CB-BUSINESS system. The architectural provision shall be publicized in the form of Request for Comments, so as to acquire external feedback and promote awareness and consensus.

References

1. Atkinson, R., and J. Ulevich, Digital Government: The Next Step to Reengineering the Federal Government, Progressive Policy Institute Technology and New Economy Project, March 2000.
2. Commission of the European Communities, eEurope: An Information Society For All, Communication on a Commission Initiative for the Special European Council of Lisbon, March 23-24, 2000.
3. Dawes, S., P. 3Bloniarz, K. Kelly, and P. Fletcher , Some Assembly Required: Building a Digital Government for the 21st Century, Report of a Multidisciplinary Workshop, Center for Technology in Government, March 1999.
4. EUREGIO, *Das alltaegliche Europa*, Gronau, 2000.
5. Scheppach, R., and F. Shafroth, Governance in the New Economy, Report of the National Governors' Association, Washington D.C., 2000.
6. M. Sprecher, "Racing to e-government: Using the Internet for Citizen Service Delivery", Government Finance Review, October 2000, pp. 21-22.
7. West, D. Assessing E-government: The Internet, Democracy and Service Delivery by State and Federal Governments, Brown University, September 2000.

Bridging the Digital Divide with AVANTI Technology

Antoinette Moussalli[1] and Christopher Stokes[2]

[1] London Borough of Lewisham, Lewisham Town Hall, Catford, London
antoinette.moussalli@lewisham.gov.uk
[2] Fujitsu Consulting, Lovelace Road, Bracknell, Berkshire RG12 8SN
chris.stokes@uk.consulting.fujitsu.com

Abstract. Whilst e-Government provides many opportunities for local authorities to serve citizens more effectively, it also runs the risk of widening the digital divide and making non-IT users second-class citizens. AVANTI aims to address this problem by focusing on people who cannot or think they do not want to be involved in the Information Society. To enable and encourage these citizens to interact electronically, we are developing a user friendly on-screen assistant to serve their needs and help turn the vision of universal Internet access for all into a reality.

1 Why Do We Need AVANTI?

Participants in the Information Society receive numerous benefits – better interest rates on banking facilities, special deals on utilities and insurance, cheaper goods and services and enhanced customer service options. People who cannot use technology or who are reluctant to do so miss out on these opportunities, and hence receive a lower standard of service from commercial organisations.

The introduction of e-Government, providing electronic access to services provided by local or national government, also presents many opportunities: it permits council departments to be available twenty-four hours per day, seven days per week to respond to citizen queries, it enables councillors and politicians to canvass the opinions of their constituents in their own homes, and it allows citizens to participate far more fully in the running of their communities or their country.

However, in the same way as the technology-poor are missing out on commercial offers and opportunities, the introduction of e-Government runs the risk of leaving these citizens behind in the information revolution. Some people are unable to use technology due to disabilities – visual impairment, manual dexterity problems or learning difficulties. Some do not possess the required technology and may not see the benefit of purchasing it. Others have difficulties with language, perhaps because they do not speak or understand the local language. Many people simply find technology too complex and confusing.

The eEurope Action Plan is designed to speed up and extend the use of the Internet to all sectors of European society, and special focus has been given to addressing the digital divide, the gap between people who currently use and are comfortable with technology and those who cannot or do not use it. For example, European Member States are encouraged to adopt and promote the Web Accessibility Initiative to make

R. Traunmüller and K. Lenk (Eds.): EGOV 2002, LNCS 2456, pp. 344–349, 2002.

public sector web pages more accessible, particularly to visually impaired users. However, in establishing the AVANTI programme we decided that a more radical approach to accessibility was required rather than simply enhancing standard web pages.

AVANTI seeks to bridge the digital divide by creating an on-screen intelligent assistant which takes over the interface, so that rather than dealing with conventional pages of text, the interaction is a more natural conversation between the citizen and the computer. In this way we are seeking to simplify the interface for people uncomfortable with technology, and by introducing sophisticated natural language processing technologies and speech generation we aim to address the needs of many of the citizens who are currently disenfranchised. AVANTI is a research project, part funded under the European Commission's fifth framework programme. As such, we recognise that some of the components we are working on may not be perfected during the life of the project; however, we acknowledge that AVANTI is not the whole answer to this difficult problem, but rather a first and key step along the road towards an inclusive information society for all.

2 How We Are Approaching the Problem

Before engaging with technologists, to identify the products and solutions which could assist in meeting the needs of these users, we identified the key stakeholders and established a programme to understand their needs.

The most important stakeholders are the target users of the applications, the citizens who currently cannot or think they do not want to use technology. Each of the participating cities in the AVANTI consortium has established a local user group comprised of representatives from the local community, including, as far as possible, people either within the target groupings for the project or those who can understand and represent their interests. Where applicable, we are also involving support agencies for the target users, including council staff who work with these groups and local and national organisations. Initially we sought the opinions of these people on the problems they have with current technologies, and how we could best address these. As the project has progressed, we have continued to involve the local users by demonstrating prototypes of our applications and gathering their feedback. The local users will have a key role throughout the life of the project, steering the development work and identifying or selecting enhancements to ensure the final product meets their needs as far as is possible.

The second key group of stakeholders is comprised of local service managers, the council employees who have responsibility for the services we are aiming to deliver. For the project to be successful, it has had to be integrated not just into the systems of the council but also into the thought processes of the council staff. We consider it vital that this project is not seen as another project led by technologists, but one which has been designed in partnership between the council departments, the users and the technologists. Whilst we are working primarily with service managers whose departments will be impacted directly by the project, we have also sought to involve the other service managers from the council as part of the preparation for a wider-scale deployment of this technology once it is proved.

Finally, we have been keen to ensure that the solutions we identify are as generic as possible. We have therefore established AVANTI as a trans-European consortium to research and demonstrate this technology, with representation from the United Kingdom (Lewisham and Edinburgh), Sweden (Stockholm) and Latvia (Ventspils). This is enabling us to research the technology with different cultures and languages, and develop a range of applications interfacing with local city systems.

3 Involving the Technology Partners

Whilst the project is led by the users and steered by the city service managers, it has also been necessary for us to research, develop, integrate and use state-of-the-art technology solutions, and two of the world's leading technology companies have been involved since the inception of the project.

The lead technology partner selected was Fujitsu, one of the world's largest IT solution providers, with many years of experience in designing, developing and deploying strategic solutions for local government. Fujitsu is one of the pioneers in online advice and conversation based applications, having been researching the area in the private sector for several years.

In addition, Microsoft has also been involved in the project as an associate partner. Microsoft has been actively researching and developing several of the components we are seeking to deploy, including Microsoft Agent technology, enterprise application integration products and speech technologies.

From the requirements identified by the users, we identified the core components of the solution, together with a range of additional facilities which we are currently researching. The consortium continues to work closely with the research functions of both Fujitsu and Microsoft to keep current with the state of the art. The city partners are also engaging with local technology suppliers to develop and enhance the product, for example integrating alternative languages.

4 Components of the Solution

The overall aim of the project is to make the interface with the computer as natural as possible, so as to make it usable by people who traditionally would be uncomfortable with technology. We therefore decided to replace a standard Internet interface, with long blocks of text, hyperlinks, frames, tables and buttons, with a far simpler one based around an on-screen character or avatar. This character is able to speak to the citizen in his or her own language, understands their responses in natural language and acts accordingly. When the project was conceived, and from the feedback received from user panels, it was apparent that an ideal solution would be for the citizen to be able to speak to the computer, and this is a major research area for the project; however, the limitations of current technologies have made this difficult to deploy in the short-term, and a keyboard entry is currently being utilised for the initial work around the development of the first applications. By the end of the project we hope to be able to implement a speech based solution.

The primary technology being implemented for the user interface is based around Microsoft Agent, which enables us to deploy a character on the screen, control its animations and gestures and send text to it which can be displayed in a speech bubble and converted into text using a text-to-speech engine. The toolset we have developed also allows for alternative delivery mechanisms such as a simpler text interface delivered to a personal computer, interactive television or mobile device, but the primary focus of the project is delivery via an animated on-screen character.

Once the citizen has given a response – by typing or ultimately by voice input – the system then has to understand what they have said in order to drive the conversation or interact with council systems as required. This 'understanding' is achieved by a natural language processing component. Various technologies for this have been researched by Fujitsu, ranging from simply looking for a specific phrase through to complex intent matching algorithms. The solution adopted is context sensitive, searching for key words or phrases within the response appropriate for the context of the question. In addition, the natural language processing component is able to identify and extract numbers, money, dates, times and ages as part of a response, whether they are in figures or in words. The aim is to enable the citizen to enter their request or response in as natural a way as possible, and for the system to understand and act on it.

The conversation is controlled by a conversation management component, which identifies the next action to take in response to the input from the citizen. This could be as simple as giving a reply back to them through the on-screen character. It could also involve accessing local databases to store or retrieve information, or interfacing with local city systems and databases.

The integration with city systems is managed by an enterprise integration component. This is highly configurable to enable it to access a selection of databases directly. To enhance the integration with city systems, this component can also interface with Microsoft's BizTalk Server, which then has the ability to integrate with an even wider range of databases, applications and messaging systems.

5 Applications for the Technology

The four cities in the AVANTI consortium are implementing a number of demonstrators, illustrating the range of electronic services which could be opened up to technologically challenged users.

The London Borough of Lewisham is part of a consortium of five UK local authorities developing a solution to enable electronic access to council processes, from social services and housing through to education and benefits. From these, an initial set of processes concerned with housing benefits has been selected to implement using the AVANTI technologies. Many of the citizens applying for such benefits are in the target groups for the project – elderly, disabled or socially disadvantaged – and without AVANTI many would not be able to use the electronic access mechanisms being developed. The AVANTI project is making access available to these services using the on-screen avatar, which can be deployed either into citizens' homes or via public access kiosks and cybercentres.

In addition, Lewisham is developing an electronic consultation application for their citizens. Electronic consultation – conducting surveys and opinion polls electronically for citizens – is a key way to make local government more accountable and democratic, as local residents can give their opinions on key issues as often as required rather than waiting for the next council elections. However, without technologies like AVANTI, citizens who cannot or do not wish to use computers will increasingly become disenfranchised in the democratic process; their views will simply not be recorded or represented. Using the friendly accessible approach of AVANTI enables these citizens to participate fully in the democratic process. This could also provide a model for electronic voting accessible to all citizens.

The City of Edinburgh is deploying AVANTI in its library service, providing an accessible interface to their existing information technology infrastructure in this area. Citizens will be able to interact with an avatar character to find details about local library facilities, ask questions about the service and carry out tasks like reserving books and extending book loans. The library service is already installing public access computers in the libraries, and in addition to enabling easier access to library applications, there is the potential to offer additional AVANTI services through the same infrastructure.

The borough of Kista in Stockholm, Sweden, is providing access to local information using the AVANTI toolset. This will enable citizens or visitors to the area to find answers to a range of frequently asked questions, such as the location of local amenities. Later in the project, they are seeking to interface the avatar to a range of municipal, regional and national databases.

In Ventspils, Latvia, the toolkit is being proved in a different culture, and with considerable language challenges. Within the Ventspils local project, citizens will initially be able to ask questions and be led through processes concerning local taxation.

Across the demonstration sites, the toolkit will be tested in six different domains of knowledge, using at least four languages and a variety of cultures. By drawing together the experiences of users in the four cities, we will aim to identify the features of the AVANTI toolkit which best address the needs of the technologically challenged citizens, any parts which do not work for these users, and areas for enhancement as we seek to develop and exploit the project.

6 Future Enhancements

The initial focus of the work has been to produce the toolkit to enable the cities to develop conversation-based applications for their local citizens. During this development work, and from the feedback we have received from our citizen users, we have identified a number of enhancements which are currently being researched by the cities and the two industrial partners.

A key technology for making electronic services accessible is voice recognition, enabling citizens to speak to the computer rather than having to type. This is an area where there have been major advances over the past five years, and two possible approaches have been identified. The first uses trained voice recognition, where the citizen has to spend approximately fifteen minutes reading text to enable the computer to understand their voice pattern; once this is completed, however, the citizen will be able to speak naturally to the system with the voice recognition module converting this speech into text. The second option is speaker independent voice recognition, which requires no training and is therefore more suitable for wide-scale deployment.

However, this can only be used to recognise a limited vocabulary of key words, making the conversation less natural. We will investigate both of these options towards the end of the project.

Another key area for e-Government is security, ensuring that where access is provided to confidential information in council systems, this is restricted to authorised citizens. Whereas regular computer users are familiar with passwords, these are less suited to citizens who are not comfortable with technology. We are investigating password-free options for AVANTI, including smart cards and biometric technologies, and will aim to integrate these later in the project.

We are also exploring new access devices for citizens, including interactive television and mobile devices. Almost every home has a television, and with the introduction of interactive digital capabilities, increasingly these sets are being used for more than just passive viewing. Likewise, many citizens now have mobile telephones, which are now being used for more than just voice calls. In both cases, we are keen to try to develop and deploy a more accessible interface to council services using these devices, whilst recognising the limitations of these channels.

7 An Integrated Service Offering

There are numerous initiatives being progressed in e-Government, and for a successful implementation it is vital that they present an integrated offering to the citizens, rather than a disjointed approach. We must also be careful not to restrict users to interacting with councils in the manner we decide is best for them, but rather provide options to enable them to decide how they want to access the services they require.

AVANTI is not designed to be the complete solution to e-Government. There will be many citizens for whom it is not suited, and in addition there will be applications which cannot be delivered using an on-screen character due to complexity or sensitivity. We would therefore expect to deploy it as part of a wider solution, which may include more conventional Internet technologies, assisted service capabilities either in council offices or remotely using call centres, and video conferencing links.

8 Summary

We recognise that there are many citizens who currently cannot participate in the Information Society, and as e-Government becomes more pervasive they will increasingly be left behind and become disenfranchised. Through the AVANTI project, working closely with this group of citizens, we are seeking to understand their needs and develop solutions for them.

Over the course of a two year research project, we do not expect to solve all of the problems of including these citizens in the Information Society. However, we believe that the research we are conducting and the solution we are developing will be a vital step in informing the debate, and will provide models for to delivering services to this key group of citizens. Only by understanding and addressing their needs will local and national government be able to realise the vision of universal electronic access for all citizens, and deliver true e-Government.

An Integrated Platform for Tele-voting and Tele-consulting within and across European Cities: The EURO-CITI Project

Efhimios Tambouris

Archetypon S.A., 236 Sygrou Ave., Kallithea, 176 72 Athens, Greece
tambouris@archetypon.gr
http://www.archetypon.gr

Abstract. Tele-democracy is becoming increasingly important for local authorities in Europe. The EURO-CITI project aims to specify, develop and evaluate an integrated platform for two tele-democracy services, namely tele-voting for opinion poll petitions and tele-consulting. The technical developments are divided into those for operators at local authorities and those for citizens. The platform empowers operators at local authorities to initiate a call-for-vote on a local problem, to dynamically set-up secure networks of cities and initiate a call-for-vote on common problems, to monitor voting results and extract statistical information, etc. Regarding security and privacy, authentication/authorization solutions are proposed and a Public Key Infrastructure is specified. The trial sites for the EURO-CITI platform are three European cities, namely Athens, Barcelona and London Borough of Brent.

1 Introduction

Electronic government and tele-democracy are high in the agenda of the European Commission [1][2]. The benefits of both e-government and tele-democracy are now well understood by local authorities worldwide that launch relevant initiatives [3][4].

In the case of tele-voting for realizing opinion polls petitions, the application of technology provides some straightforward advantages (such as increased convenience and accessibility and reduced costs) but also a historic opportunity to re-establish some form of direct democracy. The concept of direct democracy suggests that all citizens decide via voting on their problems. This concept was abandoned as local communities were growing in size.

The aim of this paper is to present an integrated tele-democracy platform for tele-voting and tele-consulting services within and across cities. The technical infrastructure is deployed in Athens, Barcelona and London and will enable the respective Local Authorities (LAs) to conduct "intra-city" or "local" as well as "inter-city" or "network" tele-voting and tele-consultations. This platform has been developed within EURO-CITI [5][6], a research project partially funded by the European Commission under the IST programme [7].

This paper is organized as follows. In section 2, a general overview of the EURO-CITI architecture and respective tele-democracy services is given. In section 3, the

R. Traunmüller and K. Lenk (Eds.): EGOV 2002, LNCS 2456, pp. 350–357, 2002.

characteristics of the services are outlined while in section 4 technical details are presented. In section 5 the approach to security/privacy issues is outlined. Finally, in section 6 the conclusions and future work are given.

2 The EURO-CITI Platform: Architecture and Services

The main objective of EURO-CITI is to exploit the potential of on-line democracy by developing and demonstrating new transaction services, namely tele-voting for realizing opinion poll petitions and tele-consulting. The development of these services calls for a common underlying architecture to facilitate their implementation and fully exploit their potential. In this section, the EURO-CITI technical architecture is presented and the EURO-CITI services are outlined.

2.1 EURO-CITI Architecture

The EURO-CITI architecture consists of a number of platforms (one per city) that communicate over the Internet. This architecture is depicted in figure 1 in the case of three cities, namely Athens, Barcelona and London Borough of Brent.

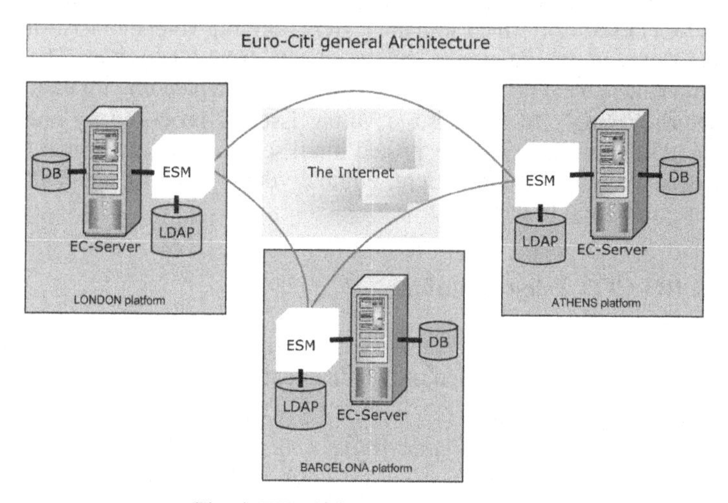

Fig. 1. EURO-CITI Architecture

The main components of the platform in each city are:

- The EURO-CITI (EC) Server where all applications for the operators and citizens reside.
- The EURO-CITI Security Manager (ESM) that is responsible for secure communications between platforms and authentication.
- The Lightweight Directory Access Protocol (LDAP) where all citizens and operators general information reside.
- The DataBase (DB) where all applications data reside.

Each local authority hosts one server. Those servers run the EURO-CITI services and are connected to sub-systems such as applications databases and LDAP repositories. EURO-CITI servers are able to communicate with each other thanks to the ESM component. Security requirements are fulfilled in order to provide citizens with trustworthy and secure services.

2.2 EURO-CITI Tele-voting

The EURO-CITI tele-voting application consists of two different tools:

- The *Tele-voting Administrative Tool* allows operators to initiate a call-for-vote, to invite other cities in a common call-for-vote, to determine the eligible voters, to initiate a call-for-vote in different languages etc.
- The *Tele-voting Service* allows citizens to vote, to request a call-for-vote, to extract statistics (if allowed by the operator), to switch between different languages etc.
- The EURO-CITI Tele-Voting service will be used for opinion poll petitions. In that context, three tele-voting scenarios have been identified by the participating local authorities as particularly important:
- "**Local** Voting". In this case, a voting issue is posted in one EURO-CITI server and eligible voters are citizens who are registered in that server.
- "Local Voting with European Scope". In this case, a voting issue is posted in one EURO-CITI server (termed initiator). Here, eligible voters consist of citizens who are registered in the initiator as well as citizens from other cities. These cities however must have been invited by the initiator and accepted that invitation.
- "Network Voting". In this case, a voting issue is proposed by one EURO-CITI server (termed initiator) and is posted in all servers (i.e. cities) that have accepted to participate in that voting. Here, eligible voters for each server are the citizens who are registered in that server.

2.3 EURO-CITI Tele-consulting

The Tele-Consulting module offers two types of services, Tele-Consultation and e-Forum. Each service is composed of two different tools:

- **Tele-consultation**
  - The *Tele-Consulting Administrative Tool* allows the operators to set up consultation campaigns.
  - The *Tele-Consultation Service* allows citizens to participate in consultation campaigns.
- **e-Forums**
  - The *e-Forums Administrative Tool* allows the operators to create new forums, to create new categories and to track the opinion given by the citizen in the different forums.
  - The *e-Forum Service* allows citizens to participate in the available forums by expressing their opinion or commenting on the opinion of other citizens.

In Tele-Consultation both "Local consultation" and "Local consultation with European Scope" scenarios are supported, where these scenarios have the same scope as in Tele-Voting. However, in e-Forums only "Local" scenarios are supported.

3 Characteristics

The main characteristics of the EURO-CITI integrated platform are:

- Intuitive, easy-to-use graphical interface for operators and citizens.
- Access from multiple devices for citizens.
- Authentication using multiple methods (login/password, smart cards, digital certificates).
- Security at the system level but also at the application level (in the case of tele-voting service).
- Ability to dynamically set up virtual private networks between cities in order to perform a common voting or consultation.
- Multilingual versions available for the operators to choose during installation.
- Multilingual content (e.g. postings) by operators are supported.
- Multilingual interface and content is available to citizens at any time.
- Archiving and auditing facilities are available to operators.
- Support of open standards e.g. Java, XML, WAP.

The specific characteristics of the tele-voting service are:

- Operators may create a new voting issue by inserting voting subject, options, duration, scope, category, keywords, URL for further information, multilingual information; by inviting other cities (in the case of network voting) and by determining eligible voters based on age, nationality and gender.
- The service supports multiple open voting issues at any time.
- Voting is secure and anonymous. No citizen is allowed to vote more than once for the same issue and no one is able to alter votes (democracy requirement). Also, citizens are able to verify their personal voting.
- Citizens are notified about forthcoming polls.
- Citizens are able to view the results of previous voting issues and the partial results of current voting issues (if the operator has enabled this option when creating the voting issue).
- Citizens are able to suggest a voting issue.

The specific characteristics of tele-consulting are similar with the relaxation of security constraints.

4 Development

The architecture used to develop the EURO-CITI platform is based on the J2EE standard. As an example, in figure 2 the software architecture for tele-voting is depicted. This architecture caters for a number of requirements (e.g. communication between services over ESM, communication of services with the Database and LDAP, access from multiple devices, support of multiple authentication methods etc.)

Each page of the resulting services is structured in three main parts (figure 3):

1. *Fixed part*: it includes the page head and the rest of components of the static design.
2. *The menu.*
3. *The page content.*

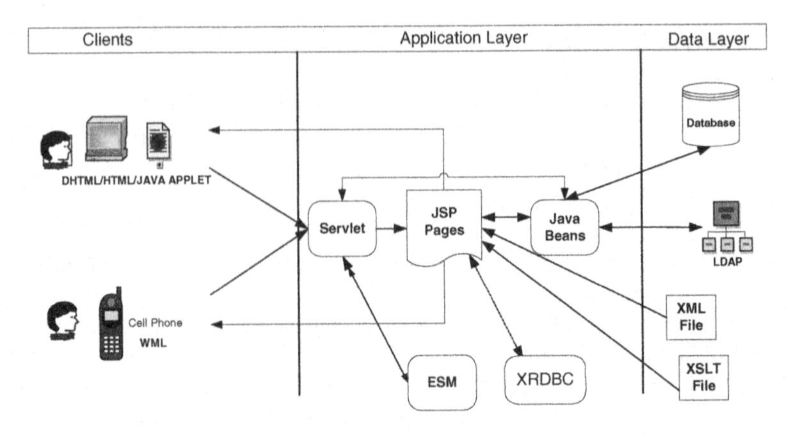

Fig. 2. Tele-voting Software Architecture

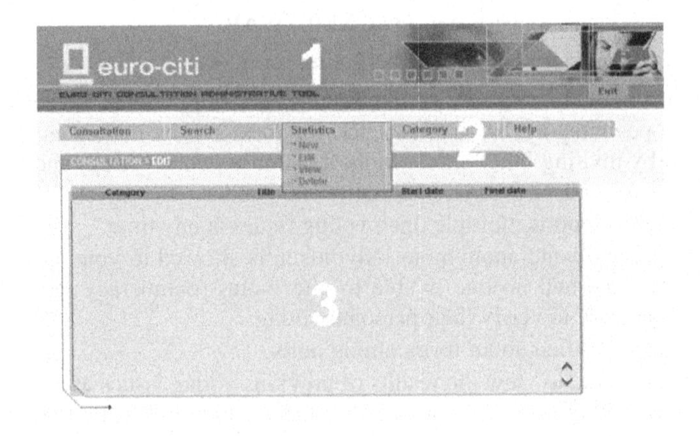

Fig. 3. Interface

5 EURO-CITI Security/Privacy Elements

Citizens access the EURO-CITI services through personal computers (Home PCs, kiosks) or WAP devices [8] using the Internet or wireless networks respectively. Links between the EURO-CITI nodes are protected by the following protocols (figure 4):

- From WAP devices to WAP gateway: **WTLS**
- From WAP gateway to EC servers: **SSL**
- From PC devices to EC servers: **SSL**
- From EC server to EC server: **IPSec**

Servers of different cities communicate with each other in the context of *network services*. A network service is launched by one city and is accessible from citizens of other cities. The participating cities can send their results to the city that has launched the service. In this context, citizens registered in a city and participating to a service of another city must be *remotely authenticated.*

For instance, if a citizen registered in London accesses a secure service proposed by Athens, the Athens server will have to ask to the London server if the citizen is authorized to access the service or not.

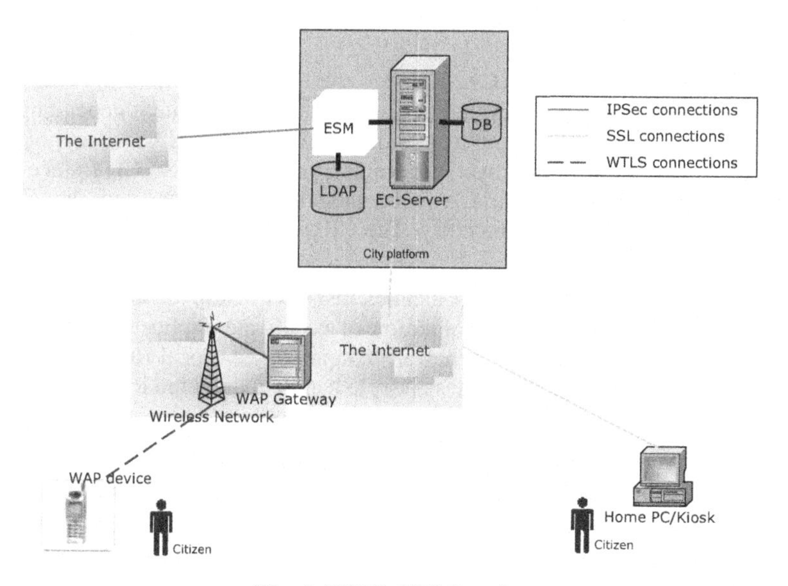

Fig. 4. EURO-CITI Security

With respect to authentication, two main methods are implemented: the login/password paradigm and the certificate-based authentication. These two solutions are combined with the use of Smart Card to realize four authentication solutions:

- **Simple login/password**: the citizen has to memorize his login/password pair. Citizens must use these credentials discretely in order to avoid their use by another person.
- **Login/password with smart card**: The smart card stores several login/password pairs. The citizen fills in the login/password window with a drag and drop application. The login/password pair can be provided to the citizen or stored in the smart card. Using the smart card is totally transparent for the EURO-CITI applications.
- **Certificates-based authentication**: certificates provide strong authentication with the use of complex cryptographic algorithms.
- **Certificate with smart card**: This is the strongest authentication method implemented. This authentication scheme is a two-levels authentication method. The use of the citizen's private key is protected by the card PIN code and it is never exported out from the smart card, thus enhancing a high security level.

In case a citizen owns a digital certificate, this certificate is stored either in a smart card or in the citizen's hard disk.

5.1 Public Key Infrastructure

The EURO-CITI architecture includes a Public Key Infrastructure (PKI) that manages digital certificates for citizens and web servers. A private PKI solution has been selected for managing citizen certificates while a public PKI solution handles EURO-CITI server certificates.

The EURO-CITI infrastructure includes all the key components of the following architectures:

The Certification Authority (CA) delivers, revokes, and renews the certificates. It implements the security policies that define the certificate content depending on both the certificate users and the future usages of certificates. The CA also archives certificates and private encryption keys (not implemented in EURO-CITI since data encryption is not required). The CA publishes the certificates and the *Certificate Revocation List* (CRL) in the directory. The CRL is the list of all the certificates that have been revoked.

The Registration Authority (RA) handles tasks *on behalf* of the CA. This mainly includes certificate applications, validation of certificate application, request of certificate suspension / revocation / renewal. In some cases, Local Registration Authorities can assist the RA in its task. These people handle locally the RA processes.

The Directory is a repository used to publish the EURO-CITI entities identities, like their name, first name, address, etc. The CA also uses the directory to publish certificates and CRLs. The EURO-CITI directory supports the LDAP protocol.

6 Conclusions and Future Work

The EURO-CITI platform equips local authorities with the necessary technical infrastructure in order to provide two important tele-democracy services: tele-voting for realizing opinion poll petitions and tele-consultations.

The trial sites for the evaluation of the EURO-CITI platform include three European cities, namely Athens, Barcelona and London Borough of Brent. For the evaluation, one hundred citizens of each city will be provided by smart cards while a significantly larger number will be provided by login/password credentials. The evaluation will include intra-city scenarios where, for example, citizens from one city will be able to vote on local issues. The evaluation will also include inter-city scenarios. In these scenarios, the operators at a city will propose a common call-for-vote and will invite other cities to join them. Upon acceptance, virtual private networks will be dynamically created and common votes will be possible for citizens across all participating cities.

Acknowledgments

The work presented in this paper was carried out as a part of the EURO-CITI project [6]. The EURO-CITI project (EURO-CITI IST-1999-21088) is partially funded by the European Commission under the IST programme [7]. The EURO-CITI consortium consists of the following partners: Archetypon S.A. (EL); University of Athens (EL); Schlumberger (F); T-Systems Nova (D); Indra Sistemas (E); Ajuntament de Barcelona (S); Municipality of Athens Development Agency (EL); London Borough of Brent (UK). The ideas expressed in this paper are those of the author and do not necessarily express the ideas of other partners.

References

1. European Commission: Public Sector Information: A Key Resource for Europe, Green paper on Public Sector Information in the Information Society, ftp.echo.lu/pub/info2000/publicsector/gppublicen.doc (1999).
2. eEurope2002: An Information Society For All, Action Plan of the European Commission, available at http://europa.eu.int/information_society/international/candidate_countries/doc/eEurope_june2001.pdf, [Accessed 14 May 2002].
3. Caldow J.: Cinderella Cities, Institute for Electronic Government, IBM Report (2002).
4. Telecities home page, 2002, http://www.telecities.org, [Accessed 14 May 2002].
5. Tambouris E., Gorilas S., Spanos E., Ioanidis A. and Gomar G.I.L.: European Cities Platform for Online Transaction Services: The EURO-CITI project, Proceeding of the 14th Bled Electronic Commerce Conference, vol. 1 (2001) 198-214.
6. EURO-CITI project, http://www.euro-citi.org, [Accessed 14 May 2002].
7. IST Home page, http://www.cordis.lu/ist, [Accessed 14 May 2002].
8. Tambouris E. and Gorilas S.: Investigation of tele-voting over WAP, SoftCOM2000, vol. II (2000) 643-653.

EURO-CITI Security Manager: Supporting Transaction Services in the e-Government Domain

A. Ioannidis, M. Spanoudakis, G. Priggouris,
C. Eliopoulou, S. Hadjiefthymiades*, and L. Merakos

Communication Networks Laboratory
University of Athens, Dept. of Informatics and Telecommunications
Athens, Greece

Abstract. Transaction services that enable the on-line acquisition of information, the submission of forms and tele-voting, are currently perceived as the future of E-Government. The deployment of these services requires platform independent access and communications security as a basis. This paper presents the methodology, network infrastructure and software kernel, which are used to achieve these objectives in the context of the EURO-CITI project. Well-known and established technologies such as SSL/TLS and IPsec are used. The internal design of the EURO-CITI Security Manager (ESM) kernel is discussed. This kernel is an advanced software platform, residing within EURO-CITI hosts. ESM supports the transaction services discussed in this paper but also takes provision for future services in the E-Government domain

1 Introduction

The main objective of the EURO-CITI project (realised in the context of the EU IST Programme) is to assess the potential of on-line democracy by developing and demonstrating fully-fledged pilots on transaction services such as tele-voting, electronic submission of forms and tele-consulting. The unified EURO-CITI architecture has the following technical characteristics:

1. Access from different end-points (home or public PC, kiosks and GSM/WAP handsets).
2. Support of different user access levels using network security and authentication-authorisation mechanisms.
3. Dynamic configuration and management of secure trans-european networks (IP-based virtual private networks, VPN) of EURO-CITI nodes.
4. Facilitation of provision of added value network transaction services.

A basic element in this platform is the EURO-CITI server, fulfilling the requirement of providing transaction services not only on the local level (e.g., a voting issue with local scope) but also on the European level (i.e., allowing citizens from other cities to participate in a voting scenario).The EURO-CITI server allows Local

* Contact author. Tel: +301 7275362, Fax: +301 7275601, E-mail: shadj@di.uoa.gr, Mailing address: University of Athens, Dept. of Informatics and Telecommunications, TYPA Bldg, Panepistimioupolis, Ilisia, 15784, Athens, Greece

R. Traunmüller and K. Lenk (Eds.): EGOV 2002, LNCS 2456, pp. 358–361, 2002.

Authorities to invite other Local Authorities and set-up a secure network, launch services on this network and, finally, drop the established associations.

EURO-CITI security services largely depend on network security mechanisms, authentication, authorisation and smart-card technology to meet the required level of privacy and security.

2 Basic Networking Requirements for the EURO-CITI Platform

Access to EURO-CITI services should be possible from home PCs or networked public PCs/kiosks via web browsers as well as through the industrially established Wireless Application Protocol (WAP). Additionally, the EURO-CITI platform should cater for network - level security and authentication using the login/password mechanism or smart cards.

Each EURO-CITI node is responsible for supporting and implementing services to local citizens but also to participate in a network[1] of EURO-CITI nodes which can be dynamically configured. Communication between servers should be performed through a well-defined interface providing a reliable and secure communications channel.

3 User Access Security Infrastructure

A basic requirement from the EURO-CITI platform is to provide support for citizens accessing the platform services using established and widely available tools. The selected medium for accessing the services was the web browser due to the high degree of penetration for almost all terminal platforms. Using a web browser to access services provides citizens with an interface they are most familiar and comfortable with. Other benefits include dynamic content and updates as well as platform independence.

Web browsers provide security by using the SSL and TLS protocols. SSL/TLS provides strong encryption and authentication at the application layer and is supported by most web browsers and also by some WAP-enabled mobile phones. SSL/TLS is considered a sufficiently strong security mechanism for the type of applications that are the targets of the EURO-CITI platform (sensitive but not critical).

Other options such as PPTP/L2TP and IPsec for citizen secure access have been considered. However, they are generally not available on as many platforms and using them requires special configuration.

Establishing a secure connection with a server requires trusting the server authenticity. This is accomplished through the use of server certificates, which are installed on the server (located within the EURO-CITI platform).

Citizen authentication is mainly supported through the use of username and password mechanisms. This is due to the lack of more secure yet abundant alternatives. However, for the few terminals equiped with smart card readers, the EURO-CITI server supports a special mode of the SSL/TLS protocol which provides simultaneous

[1] Here, the word network denotes a logical association between the involved parties

mutual authentication, instead of the more typical server-only authentication. This requires the presence of a user certificate to be installed on the smart card. The cards are issued by the Local Authorities to citizens wishing to use this authentication method and are prepared accordingly, using a Certification Authority (CA) present within the Local Authorities infrastructure. This CA is part of the EURO-CITI platform and is exclusively used for user authentication while accessing EURO-CITI services.

There is an additional issue concerning secure access through GSM terminals. WAP messages in GSM are secured using WTLS, which is translated to SSL/TLS at the WAP gateway before reaching the EURO-CITI server. During the translation, the data is momentarily decrypted and subsequently reencrypted. Fowever, it is considered that the security risks are acceptable for the applications supported by the platform.

4 Secure Communications between EURO-CITI Servers

For securing communications between EURO-CITI servers, the ESM (EURO-CITI Security Manager) component is used. ESM is responsible for creating and maintaining connections between Local Authorities wishing to cooperate. Connections are protected through the use of the IPsec protocol. IPsec provides a cross platform security mechanism, best suited for use by servers with sufficient processing resources. The security level offered by IPsec is generally agreed to be of the best available quality, with the potential of security every type of exchanged data whether it is application or connection control related.

ESM establishes IPsec associations between servers dynamically, for the duration of service availability (e.g. during periods where citizens are requested to vote for a particular issue). This is made possible through the use of an omnipresent channel (the P-Channel), responsible for controlling the establishment and termination of services as well as transferring additional control information. For the duration of a service between cooperating Local Authorities, a new separate channel is established, also protected with IPsec using a separate set of security parameters.

5 Services between Cooperating EURO-CITI Servers

The EURO-CITI platform supports two types of coperating services. The European Scope services allow users belonging to Local Authority A to use the service of Local Authority B. This requires that the Local Authorities have agreed to cooperate on a specific service and that Local Authority A wishes citizens registered to it, to participate to the service (e.g. a consultation issue) provided by Local Authority B. for this type of services, Local Authority A does not have any control over service execution and simply responds to requests from Local Authority B to authenticate its own citizens (Local Authorities are not capable of directly authenticating citizens registered elsewhere, such data remains private in each EURO-CITI server).

Network services are a second type of offered service. A networked service is essentially shared between multiple EURO-CITI servers and simultaneously provided to all citizens of participating Local Authorities. Information gathered during the service lifetime is also shared between all EURO-CITI servers. For example, several Local Authorities can start the same voting issue simultaneously. Votes which are collected are shared between the Local Authorities, so that at the end of the voting session everyone has the collected votes of all Authorities. This process is done ensuring that no citizen information is leaked and no citizen's vote can be revealed.

6 Conclusions

The use of widely available security technology such as SSL/TLS and IPsec allows for the creation of a flexible EURO-CITI base platform capable of reaching a wide audience and on top of which E-Government services can be deployed, improving upon the administrative procedures of Local Authorities and promoting direct democracy. Services can be offered supporting the cooperation between Local Authorities, while ensuring the security of transactions as well as citizen data.

The EURO-CITI platform has been developed in Java, ensuring platform and operating system independence. Where platform specific components must be accessed, as is the case with IPsec, abstraction layers and implementations have been created to facilitate integration with popular systems such as Windows.

References

1. T. Dierks, and C. Allen: "The TLS Protocol", IETF RFC 2246, January 1999.
2. S. Kent and R. Atkinson, "Security Architecture for the Internet Protocol", IETF RFC 2401, November 1998.
3. L. Barriga, R. Blom, C. Gehrmann, and M. Naslund, "Communications security in an all-IP world", Ericsson Review 2000, No 2, pp 96-107.
4. O. Kallstrým, "Business solutions for mobile e-commerce", Ericsson Review 2000, No 2, pp 80-92.
5. "WAP Architecture Specification, WAP Forum, April 30,1998.
6. Johan Hjelm, "Designing Wireless Information Services", Wiley, 2000.
7. IP Security for Microsoft Windows 2000 Server, http://www.microsoft.com/windows2000/docs/IPSecurity.doc, February 1999.

SmartGov*: A Knowledge-Based Platform for Transactional Electronic Services

P. Georgiadis[1], G. Lepouras[1], C. Vassilakis[1], G. Boukis[2], E. Tambouris[2], S. Gorilas[2], E. Davenport[3], A. Macintosh[3], J. Fraser[3], and D. Lochhead[4]

[1] e-Gov Lab, Dept. of Informatics and Telecommunications, University of Athens
{p.georgiadis,gl,costas}@e-gov.gr
[2] Archetypon S.A.
{gboukis,tambouris,sgorilas}@archetypon.gr
[3] International Teledemocracy Centre, University of Napier
{A.Macintosh,E.Davenport,J.Fraser}@napier.ac.uk
[4] City of Edinburgh Council
dave.lochhead@edinburgh.gov.uk

Abstract. Public transaction services (such as e-forms) although perceived the future of e-government have not yet realised their full potential. E-forms have a significant role in e-government, as they are the basis for implementing most of the twenty public services that all member states have to provide to their citizens and businesses. The aim of the SmartGov project is to specify, develop, deploy and evaluate a knowledge-based platform to assist public sector employees to generate online transaction services by simplifying their development, maintenance and integration with already installed IT systems. This platform will be evaluated in two European countries (in one Ministry and one Local Authority). This paper outlines key issues in the development of the SmartGov system platform.

1 Introduction

According to the European Commission [1] "*transaction services, such as electronic forms, are perceived as the future of electronic government*". Although a large number of initiatives have been undertaken at a local, regional or even national level, it is evident that these initiatives have not provided the expected results and in most cases public administration authorities have so far failed to exploit the benefits of using online transaction services, such as e-forms, in their processes. As stated in the eEurope initiative [2] "*eGovernment could transform old public sector organisation and provide faster, more responsive services. ... However this potential is not being realised.*"

The SmartGov project suggests that an advanced knowledge-based platform for transaction services and particularly e-forms will allow realising the potential of these online services. The development of this platform however requires experience and

* Project partially funded by the European Community under the "Information Society Technologies" Programme (1998-2002) (Project Number IST-2001-35399).

R. Traunmüller and K. Lenk (Eds.): EGOV 2002, LNCS 2456, pp. 362–369, 2002.

expertise at different levels such as technical expertise in diverse areas (e.g. knowledge management, Internet, XML, networks, user-interfaces etc.), expertise in the operation of public authorities at all levels that aim to provide online services, expertise in process models and process improvement but also social aspects such as the fears of public sector employees when facing new technologies. As a result of the problem's complexity, the SmartGov project believes that a European synergy of public authorities, universities and industry is required in order to specify and develop a platform that will allow the potential of e-forms to be unleashed. By conducting that research at a European level not only the best players will be involved but also the results will be better evaluated and also disseminated and exploited.

The rest of the paper is structured as follows: The second section outlines the objectives of the SmartGov project with special emphasis on the issue of trust in electronic services, the third section depicts the technical issues concerning the development of the SmartGov platform and applications, the next section provides a summary of the two pilot applications of the project and the last section concludes with the future plans.

2 SmartGov Objectives

The aim of the SmartGov project is to specify, implement, deploy and evaluate a holistic approach for online transaction services specific to the public sector. It will achieve this by developing a **knowledge-based platform** to assist public sector employees to generate **online transaction services** by simplifying their development, maintenance and integration with installed IT systems. It will capitalise on emerging standards (such as XForms by W3C) to create an **open architecture** that ensures **interoperability** between installed IT systems and to **develop new applications** to exploit that architecture. It will derive a knowledge management framework to facilitate both the deployment and acceptance of the online transaction services. Applications will be user-friendly requiring only basic IT skills -besides the necessary domain knowledge- to deploy and manage electronic services and will be tested in selected public administration application areas.

Based upon a thorough investigation of the state-of-the-art in online transaction services technologies, a survey of the current situation at Public Administration Authorities and an analysis of the user requirements for each of the user groups involved, the project will generate detailed specifications for the knowledge-based core repository and the SmartGov services and applications. The initial analysis has determined a number of potential technologies to be used for the knowledge repository, such as Data Bases, Data Mining, Data Warehousing, XML, XSL and XForms. On the whole, the core repository will contain the basic Transaction Service Elements (TSE), used to build electronic services along with domain specific information and knowledge for each TSE.

Furthermore, based upon the user requirements the project will develop services and applications to support the involved user groups in carrying out their tasks. As depicted in the next figure, the SmartGov platform will include services (e.g. the SmartGov agent and the Information Interchange Gateway) to enable the communication with existing or new 3rd party Information Technology Systems.

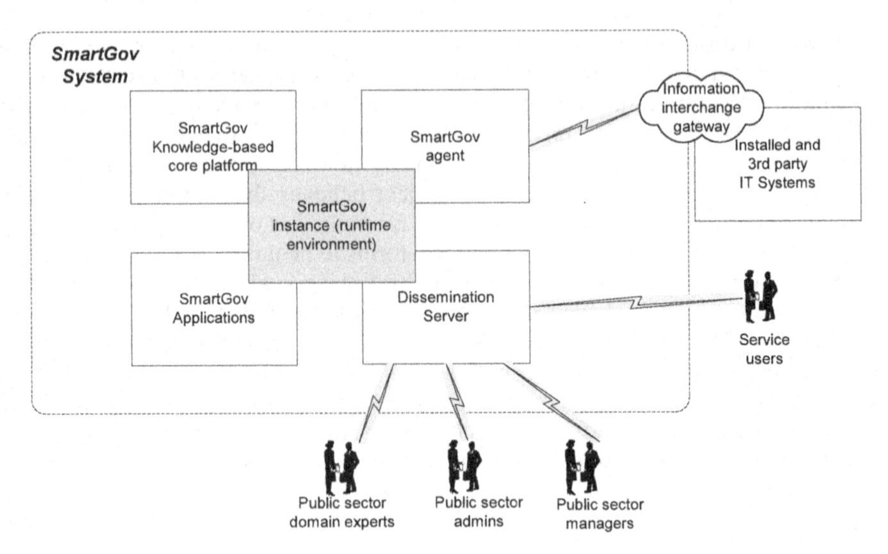

Fig. 1. Overview of *SmartGov* system

The platform will also include a dissemination service, to make available the SmartGov services to involved user groups, either internal to the public administration organisation (e.g. public sector domain experts, administrators and managers), or external user groups such as the end-users. Other envisaged SmartGov applications include administrative tools for capturing knowledge and for creating new transaction service elements.

User requirements analysis will also identify the end-user services that will be implemented and used to evaluate the SmartGov platform during the pilot application phase. In summary the SmartGov project will:

- Improve the working environment of public sector employees by equipping them with a knowledge-based platform, a set of relevant applications and a methodology allowing them to create and maintain e-services in an intuitive, user-friendly manner.
- Directly support staff involved in online transactions with citizens and businesses. The main features of the platform will include knowledge sharing and re-use, interoperability with installed IT systems, support of multiple access channels and full support of standards (such as XML and particularly XForms).
- Merge knowledge management principles with emerging standards on e-services (such as XForms) to enable administrators to capture and re-use their domain knowledge in the area of e-services.

A key issue in the development of electronic services is that of trust. Trust is an important resource in an e-services environment. While people dealing with commercial organisations are typically looking for financial integrity and confidentiality, when they deal with government agencies they expect not just integrity and confidentiality but also a level of transparency in the process that ensures trust in the service being provided. In SmartGov, representations of trust, trustingness and trustworthiness that take a more socially oriented approach required for public sector online transaction services, will be developed.

2.1 Trust in Electronic Services

To ensure that electronic services are designed, maintained, delivered and received effectively, it is important that people have trust in the components with which they interact.

Designers of services need to trust the procedures and tools that they use, particularly when redesigning existing services. They also need to trust the designers of other contributing or complementary services.

Deliverers of services have similar need of trust.

Clients (citizens and businesses) need to trust the behind-the-scenes people and procedures: trust that they are bona fide, trust that they will function as they are supposed to and trust that they will not misuse any information given by the client.

SmartGov, will endeavour to ensure that, in building models of electronic service delivery, models of trust will also be incorporated. Recent developments in social psychology suggest the value of studying situational trust, i.e. situations in which trust cues are provided by the situation or context as much as by the individual. This seems particularly relevant for services in which the various players may never come into direct contact with each other. This applies to many public authority services.

A comprehensive and informative analysis of *Trust formation in new organizational relationships* is offered by McKnight et al [3]. This report covers definitions of trust, the formation process, and the role of emotion in trust. Dibben [4] has decomposed business processes into a number of typical situations, and suggested what types of trust may apply in each of these.

There may also be particular relevance to SmartGov in studying "swift trust". The term "swift trust" was first used by Meyerson, Weick and Kramer [5] "to account for the emergence of trust relations in situations where the individuals have a limited history of working together". Relating this work to Dibben's examples of trust, learnt trust clearly does not exist in such scenarios and swift trust can arise as a result of situational trust.

3 Technical Aspects

Two main technology areas have been identified and will be addressed by the project: the knowledge-based core system and the applications and services.

3.1 The Knowledge-Based Core System

The project approach introduces and incorporates the key notion of the transaction service element (TSE), which is perceived as the main building block of transaction services. A TSE is the equivalent of a form field (such as the input space for a citizens id number or surname) but also contains metadata and domain knowledge that is attached by the form developer. Metadata may encompass the object's type, value range, multilingual labels, online help, while domain knowledge includes information about the relation of the object to other elements, legislation information, etc.

The knowledge-based platform provides a storage schema that is capable of storing and handling the services and the associated e-forms as well as the corresponding knowledge. The schema will be expandable and allow for the adoption of new services. This schema will be populated with Transaction Service Elements, forming thus the Transaction Service Elements Knowledge database (TSEKDB), which includes the essential elements for developing transaction forms along with all relevant information and knowledge. The domain knowledge embedded in installed systems will be used for the development of the TSEKDB.

Public sector employees interact with the TSEKDB through a user-friendly front-end (administrative) tool, which enables both the retrieval of already existing knowledge, as well as maintenance activities such as the addition of new knowledge, in an intuitive and user-friendly manner.

3.2 The Services and Applications

A Transaction Service (TS), within the SmartGov platform, is the equivalent of a form that contains a number of TSEs and some domain knowledge pertaining to the service as a whole. Under this scheme, development of a transaction service, consists of the following steps:

1. Selection of the appropriate TSEs to be included within the service
2. Decision of the layout that will be used to present the service to its users. This layout may be selected from within a standard template library (which may then be customised) or alternatively, any custom layout may be built from scratch.
3. Attachment of rules that govern the service, such as prerequisites for its usage, validation rules, triggering of other services etc.
4. Definition of MIS data and statistics to be captured for further processing.
5. When a transaction service has been developed, it may be deployed through service instantiation.

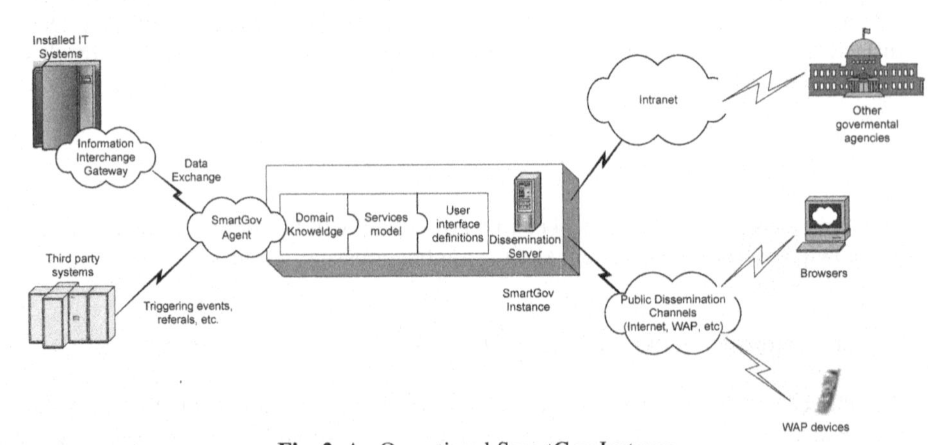

Fig. 2. An Operational SmartGov Instance

This procedure generates automatically a SmartGov instance, comprising of all web pages, forms, information repositories and programs needed to operate the service within the Web environment, wireless channels or any other supported service deployment infrastructure. The generated elements are installed on the Dissemination Server, which handles the presentation layer i.e. all interfaces with the applications users. The overall operation of an instantiated service is illustrated in Figure 2.

A service that has been deployed to the public may need to interact with an installed IT system in order to exchange data with it. All such communication is handled through the communication services, which include the SmartGov Agent and the Information Exchange Gateway. The Information Exchange Gateway is attached to the installed IT system and publishes an export schema, which contains all the data items that need to be accessed by services running within the SmartGov framework. The SmartGov Agent imports elements published within the Information Exchange Gateway's export schema within the SmartGov environment. Effectively, the Information Exchange Gateway encapsulates all peculiarities and idiosyncrasies of the installed IT systems, offering a uniform interface through which the SmartGov platform may communicate with virtually any IT system.

Besides providing the necessary link with the organisation's installed IT system, the SmartGov agent arranges for communication with third party systems the service should exchange data with, in order to access facilities that may complement or affect the running service. For instance, the SmartGov agent might provide linkage to document repositories where detailed instructions on form filling may be found, or support subscriptions to legislation databases, which emit alerts when legislation pertaining to the service operation is modified.

Service maintenance is also a major issue in operating transaction service environments that need to exchange data with installed IT systems. When a service undergoes modifications, for example due to legislation revisions, the electronic service published through the SMARTGOV instance must be 'in sync' with the organisation's private IT system, in order to carry out a full processing cycle for the service. In many cases, however, updating the private IT system may be quite cumbersome and time-consuming, while the 'front-end' part of the service, such as declaration submission, must resume operation rapidly. The SmartGov framework caters for these situations, by providing submission spooling mechanisms. These mechanisms allow for operating an electronic service and storing the submission data in a local information repository, until the organisation's back-end IT system is synchronised with the SmartGov instance. When the back-end IT system has been appropriately modified, the SmartGov instance may 'push' all collected submissions to the back-end, triggering thus the completion of the submission's processing cycle.

4 SmartGov Pilot Application

Pilot application will take place in two participating Public Administration Authorities: the General Secretariat for Information Systems, of the Ministry of Finance in Greece and the City of Edinburgh Council in Scotland.

4.1 General Secretariat of Information Systems

The General Secretariat of Information Systems (GSIS) is strategically oriented towards e-alignment of the services that it offers to citizens and businesses in taxation, customs and other application domains, exploiting the web as a major service delivery platform and interoperability technologies for integration with back-end IT infrastructures.

In this respect, exploitation of the SmartGov platform will present some substantial benefits for GSIS:

- GSIS employees at different levels will enjoy a user-friendly environment for implementing and maintaining e-services for the public (or even inter-organisational ones), as well as for transparently and seamlessly integrating these services with existing workflows and back-end IT infrastructures.
- Valuable domain knowledge associated with the e-services deployed through SmartGov will be preserved in a re-usable form that can be maintained as an organisational memory artefact.
- Various quality dimensions related to effectiveness, resource efficiency and reliability of the e-services deployed through SmartGov will be improved, whereas on the other hand the establishment and enactment of a performance management scheme is facilitated.

These benefits can substantially contribute to GSIS objectives for (a) achieving high quality of services towards citizens and businesses as a top-level strategic goal, as well as for (b) promoting Service Level Management (SLM) as a major operational policy.

4.2 City of Edinburgh Council

The City of Edinburgh Council (CEC) foresees a number of possible, different levels for the implementation of the SmartGov pilot application.

At the **national level**, with the establishment of the new Scottish Parliament and the Scottish Executive's commitment to modernising government through partnership working, there is an environment in Scotland that is conducive to testing new ways of sharing transaction knowledge and integrating service provision.

At the **city level**, the Smart City initiative in Edinburgh aims to provide a city portal that will be a single gateway to all relevant services and information.

At the **local community level**, there are many policy agendas and associated funding initiatives. The complexity of the mix is confusing and often leads to fragmented and disjointed efforts to take advantage of the opportunities presented.

The different conditions at city level and local level suggest that there is some value in running pilot applications at both levels, to compare the effectiveness of SmartGov in different situations. At the city level, policy is clear and SmartGov research can dovetail with the existing Smart City initiative. At the local level, there are particular challenges that SmartGov may find harder to address.

At the city level, the City of Edinburgh Council is in the process of establishing its Corporate Customer Service Model (CCSM), as a vital component of the Smart City. CCSM has many streams of activity, focused on people, process, technology and infrastructure. It presents an opportunity for SmartGov to make an impact on the Council's approach to designing and maintaining services efficiently and effectively. Staff at the International Teledemocracy Centre is working closely with Council staff to model existing processes, identify potential improvements and create a framework in which SmartGov principles can be applied. At the time of writing, several potential pilot applications have been suggested, such as citizens applying for housing benefits or businesses applying for licences to run bars.

At the local level, the West Edinburgh Community Planning Partnership area provides a challenging environment to test the SmartGov developments in conjunction with the local community. The area has:

- a strong infrastructure of community groups and local organisation with a history and experience of partnership working
- a local partnership organisation with multi-agency and cross sector representation
- an adopted Digital Inclusion Strategy
- a Community Learning Plan
- a number of access to employment initiatives

Key policy objectives of the Partnership are to increase access to learning, improve access to employment and enable social inclusion. These all require high levels of inter-agency trust and collaboration to be effectively delivered. At the time of writing, the Partnership is considering an appropriate application to do with access to learning and access to employment.

5 Current State and Future Work

SmartGov commenced on February 2002. So far, work has focussed on conducting a thorough investigation of the state of the art for e-services in the public sector, capturing user requirements and creating a high level set of system specifications. In the next phase work will continue to refine system specifications for each of the system components and to consequently implement them.

References

1. European Commission, 'Public Sector Information: A Key Resource for Europe', Green paper on Public Sector Information in the Information Society, ftp.echo.lu/pub/info2000/publicsector/ gppublicen.doc
2. eEurope 2002 An information Society for All Action Plan, 19-20 June 2000, http://europa.eu.int/information_society/eeurope/action_plan/actionplantext/ index_en.htm
3. McKnight, D.H., Cummings, L. and Chervany, N.L. (1995) Trust formation in new organizational relationships. Available at http://misrc.umn.edu/wpaper/WrkingPapers
4. Dibben, M.R. (2000). Exploring Interpersonal Trust in the Entrepreneurial Venture. London: Macmillan.
5. Meyerson, D., Weick, K. E., and Kramer, R. M. (1996). Swift trust and temporary groups. In: R. M. Kramer and T. R. Tyler, eds. Trust in organizations: Frontiers of theory and research. Thousand Oaks, CA: Sage Publications, 166-195.

Best Practice in e-Government

Josef Makolm

Austrian Ministry of Finance, Austrian Computer Society
josef.makolm@bmf.gv.at

Abstract. Evolution of e-Government is drifting. On the one hand, e-Government systems shift from information via communication and transaction systems to integrated systems. On the other hand, a partial trend from G2C and G2B systems towards G2G systems is observable. Different needs are to be met to build best practice solutions in the fields of G2C, G2B and G2G.

1 The Evolution of e-Government

1.1 From Information to Integration

As every system, e-Government is subject to the laws of evolution. Along the coordinate of time, a process of development becomes evident. Usually e-Government applications start as information applications, perhaps with components for communication. But this is just the starting point for further evolution because the availability of sufficient information creates the demand of interaction between citizens and authority. It is no longer sufficient to present application forms. Instead, it is necessary to enable the citizens to fill in these forms, to upload them to the authority and to receive an answer in electronic form, too. The e-Government system has to evolve to a transaction system. Starting from this point, the necessity to attach and upload documents comes up. These documents may be scanned documents coming from the paper world. Or they may be documents created by other electronic processes, e.g. a plan of a house or the actual balance-sheet of a company, which has to be transferred to the tax authority. This creates the demand for integration between the computer systems of applicants on the one hand, and of the authorities on the other hand, e.g. via XML structures. A "good practice" e-Government solution should meet these necessities.

A typical example for the evolution from an information system to a transaction system is the Austrian citizen information system "help.gv.at". It started as an information and communication system and brought life-situation related information to the citizens, supplemented by the ability to get answers to questions asked. As soon as the information in the system was more and more completed, the necessity to enable the citizens to submit their applications via the system came up. Now the system offers the ability to download several forms, to fill in these forms and to upload them to the authority. For example is it possible now to notify a dog or the migration of a dog to the authority, which is necessary for the matters of dog registration and dog taxation.

An example for the further evolution from transaction to integration is the possibility to upload a company balance-sheet to the Austrian company register. The

R. Traunmüller and K. Lenk (Eds.): EGOV 2002, LNCS 2456, pp. 370–374, 2002.

balance-sheet is created by the bookkeeping software and stored in an XML form. This XML data is checked by an applet, which can be downloaded from the authority's web site. If no error is detected by the applet, the XML form is uploaded to the authority.

1.2 From G2C and G2B to G2G

Citizens often have to accompany several documents when submitting their applications to the authorities. In many cases, these documents have been produced by other authorities and data relating to these documents is still available at the other authorities. Why should it be the duty of the citizens to run to several authorities for collecting their documents to complete their applications? Wouldn't it be a better way to let data run from one authority to the other? To make this offer to the citizens, it is the job of the authority to add necessary documents respectively data to the citizens' applications. And this collecting phase could be done by an electronic agent. This agent – located in a software driven workflow – could complete the citizens' applications before any official starts dealing with them. For example is it often necessary to add a birth or citizenship certificate to an application. These are typical cases to let data run and not citizens. The Austrian Citizen Card – available in 2003 – will help to realise this principle.

In any way, a broad social discussion – considering also the right of privacy – is a prerequisite for the described step of evolution. As a result of this discussion, political decisions are to be made.

2 Needs of G2C Systems

2.1 The Opinion Poll of the Austrian Computer Society

In the beginning of 2002, the conference "eGov Day" was organised in Vienna by the Austrian Computer Society[1]. As part of this conference, an opinion poll was done with the participants to find out the citizens' needs concerning G2C systems (a detailed analysis is planned but not yet published). To briefly sum up the main results of this survey, the following functions have been assessed and categorised as either "essential", "required", "useful" or "nice to have" (the order reflects the results of the opinion poll):

2.1.1 Essential Function

- Acknowledgement of successful receipt for the applicant (e.g. by e-mail).

2.1.2 Required Functions

- Indications pointing out whether a field has to be filled in mandatory or just optional;
- Choice between several paying modes for the user (e.g. credit card, internet banking, cash etc.);

[1] http://egov.ocg.at/egovday02.html

- Choice between several modes of delivery and the possibility to pick up the notice at the authority office;
- Pull-down menus, when certain content is necessary for form fields (e.g. a list of available documents in cases of document ordering);
- Possibility for the user to display the status of his application;
- Detailed information for the user concerning privacy, data protection and data security;
- Possibility of electronic submission of (scanned) plans or documents;
- After filling in the form and before submitting it, the possibility of a preview is necessary;
- All download forms have to be enabled as direct upload for electronic submission as well;
- After a form is accepted, it must be displayed with the possibility of printing;
- Possibility for the user to display his files kept by the authority;
- Online help for filling in the forms;
- The user can retrieve information concerning data transfer (encryption – technical security);
- At any time of an interaction (e.g. when filling in form), the user should be provided with adequate information on how many steps are finished and how many are still to be done (e g. page 1 of 5);
- Intelligent forms or help (form fields to be filled in change in context to typed input);
- Downloadable notice for further electronic processing by the user.

2.2.3 Useful Functions

- Personalization (e. g. after login the user's data and all his current proceedings are displayable);
- Possibility to log in as a test user to try out the system without any effect;
- Possibility of anonymous login (e.g. for submitting complaints and for displaying the status of ones anonymous complaints);
- Forms are designed in a way that these are not bigger than screen size.

2.2.4 "Nice to Have" Function

- Web information is displayable on the mobile phone.

3 Needs of G2B Systems

G2B systems are built for professional users. Satisfactory solutions have to meet the specific needs of this group of users:

- Transaction oriented system with login and logout function;
- Decentralised user and role administration;
- E-Application service;
- E-Delivery service.

Usually, professional users deal with more than one case. Therefore, they need transaction oriented systems with specific login and logout functions. Several roles have to be supported within one office. E.g. just one administrative key user is defined by the tax authority within the Austrian Tax Online System for a tax consultant's office. It is the duty of the tax consultant to assign the roles and authorisations within the tax online system to his employees: A secretary just may be authorised to submit e-applications to, or to receive e-notices from the tax authority while the tax consultant himself needs the authorisation to change his clients' booking within the accounts kept by the tax authority.

4 Needs of G2G Systems

G2G systems – like G2B systems – are built for professional users, too. For these inter-governmental systems, special needs are to be obeyed:

- Single login;
- Role based authorisation;
- Trusted inter-governmental authorisation;
- Workflow driven autonomous agents.

G2G respectively inter-governmental systems are used by government officials as part of their job. To avoid various logins and logouts when working with several systems, they need a single login function combined with trusted inter-governmental authorisation. Therefore, roles and authorisations for a person should be defined within the system the person logs in first. Other Systems just trust the authorisation handed over from this system. E.g. within the Austrian Ministry of Finance and its Tax Offices a Finance Portal offers the use of several systems: federal intranet, companies register database, real estates database, central registrations register, legal information system, intranet of the Ministry of Finance with the Tax Online System, the Online Tax Law Documentation System and several other internal systems.

At least, workflow driven autonomous software agents should be available for the routine checking and completing the incoming applications.

5 Technical Infrastructure – Outlook

Systems always evolve in layers, technical systems just as good as biological systems. After the process of evolution has formed a new layer, this layer works as starting base for further trials – and of course also errors – in a new evolutionary process. E-Government systems have their base in the existing internet infrastructure. Internet has exceeded the critical mass and is used as a matter of course. Secure data transfer and authentication techniques are available.

Partially governmental portals – connected by the Corporate Network Austria – are already active, Trusted Inter-Portal connectivity is starting. The Austrian Citizen Card with digital signature will be established in the beginning of 2003. A workflow system for the Austrian ministries is partially installed and will be completed in 2003. E-Payment methods, which meet the needs of the e-Government process, are discussed in a work group between banks and government. E-Application and e-Delivery service will be established in 2003.

As a conclusion can be said: There is a capable base and there are good promises for further evolution into effective e-Government services.

References

Maria A. Wimmer (ed.): Impulse für e-Government: Internationale Entwicklungen, Organisation, Recht, Technik, Best Practices; Tagungsband zum ersten eGov Day des Forums eGov.at, Band 158, Oesterreichische Computer Gesellschaft (books@ocg.at), Wien (2002)

e-Government Applied to Judicial Notices and Inter-registrar Communications in the European Union: The AEQUITAS Project

Carmen Diez[1] and Javier Prenafeta[2]

[1] Tools Banking Solutions, S.L. Paseo Independencia 32, 1. 50.004 Zaragoza. Spain.
diezc@tb-solutions.com
[2] Asociación para la Promoción de las Tecnologías de la Información y el Comercio Electrónico (APTICE). María de Luna 11. 50.015 Zaragoza. Spain.
jprenafeta@aptice.org

Abstract. The new technological advances should be accessible to the citizen, achieving this purpose through the development of informatic tools that speed up services' rendering by the Administrations, taking into consideration the security aspects in these communications. The European AEQUITAS Project aims to develop an Informatic Tool that shall, via TCP/IP networks, allow secure communications and transmissions of electronic documents between juridical operators, using electronic signature and certification. The herein paper, describes in detail this Project, that can be encapsulated within the so-called *Networked Government*

1 Introduction

e-Government can be defined as a new technology information and communications system application used to enhance both relations between Public Entities and the citizen, as well as those internal procedures within these public entities.

The most significant advantages of this application are reduction in costs, better information accessibility, internal management improvement, data storage and processing, and generally a quicker service. These aforementioned advantages justify the fact that governments worldwide, even in those not so industrially developed countries, foster e-government applications development. Moreover, other benefits can be highlighted, such as security data processing tools that guarantee higher levels of confidentiality and integrity as compared to the traditional data management systems.

The development and implementation of these mechanisms, do not only require an upgrade in the existing computing equipment, infrastructure and communications facilities, but also provide the basis for new regulations that facilitate and promote the use of these new technologies in today's administrative procedures with full legal guarantee.

R. Traunmüller and K. Lenk (Eds.): EGOV 2002, LNCS 2456, pp. 375–382, 2002.

Within the European Union, and mainly through the *e-Europe* initiative: *an Information Society for all*[1], the development of e-Government is fostered. The application of these new technologies in the public sector, as outlined during the European Council Summit held in Lisbon last 23rd and 24th March 2000, means that nowadays we are able not only to build the necessary mechanisms to render on-line services to the citizen, but also to realize those changes needed to use these tools in internal workflow processes.

Thus, within this frame, we can classify pursuant to the European Commission Information Society Project Office's point of view[2], a triple application of the information and communications technologies:

Open Government can be described as the development of government, ministry or other public organisms web sites whose aim is to make as much information as possible accessible to the general public.

Customer Orientated Government, aimed at offering, via the Internet, a series of services to the citizen through interactive platforms, adding authentication, integrity and confidentiality technical mechanisms needed in the different administrative procedures and processes. Generally speaking, we are referring to those mechanisms that can be applied to real estate registrars, tax payments, health and social security systems, ballot polls via Internet,... The objective is to offer 24X7 administrative services.

Networked Government. This application can be classified as the internal core part of the e-Government system. Being it's aim to integrate the new technologies in the Administrative management processes, to build intranets and to establish secure mechanisms with full guarantee when exchanging data and files amongst the different organizations and public entities at all levels, local, regional, central, or international.

The herein paper pretends to outline the TRUST FRAME FOR ELECTRONIC DOCUMENTS EXCHANGE BETWEEN EUROPEAN JUDICIAL OPERATORS (AEQUITAS) Project, set within the European Commission V Trust Frame Programme, located within the framework of the Key Action II, line II.4.2, called Large-scale trust and confidence and which envisages to extend, integrate, validate and prove technologies and architectures related to confidence in the context of large scale advanced settings for managerial or daily life. The validation should include, as a general rule, the evaluation of the legal consequences of the proposed solutions. More precisely, the objective of the AEQUITAS Project is to speed up procedures and increase the effectiveness of the quality service that the juridical operators, specifically Registrars and Attorneys, render to the society, through the use of secure and confidential Internet communications.

So as to reach this objective, the AEQUITAS Project pretends to develop a secure, confidential, scalable and interoperative system with all technical and juridical guarantees, as well as to transmit documents electronically between different European judicial operators. The following point gives more detail on the Project Participants, amongst which different Juridical Operators can be found.

---

[1] http://europa.eu.int/information_society/eeurope/index_en.htm

[2] *Public Strategies for the Information Society in the Member States of the European Union*, ESIS Report, September 2000. Available in http://www.irc-irene.org/documents/do-psismseu.html

2 Participants

2.1 Juridical Operators

- *Greffe du Tribunal de Commerce de Paris (France):* The Greffiers fulfill two functions, on the one hand they are secretaries of Commerce Courts and, on the other, they act as Trade registrars (www.greffe-tc-paris.fr)
- *Consejo General de los Ilustres Colegios de Procuradores de los Tribunales de España (Spain).* The Procuradores are professionals whose principal function is to represent litigants before the Courts and, in particular, to transfer documents (notifications, writs, ...) with which parties and Courts communicate (www.cgpe.es)
- *Ilustre Colegio de Registradores de la Propiedad y Mercantiles de España (Spain).* The Registradores are responsible for registering properties (real estate transfer rights), mercantile operations and personal property (www.corpme.es)
- *Câmara dos Solicitadores (Portugal).* The Solicitadores are professionals who council individuals regarding juridical enquiries. Given that, in Portugal, Notaries are public officials whose functions are limited to legalizing Deeds, the Solicitadores have the mission to assist in the drafting of documents which need to attain the level of Public Deed (www.camara-solicitadores.pt)
- *Land Cadastre and Registry of Lithuania.* It is a governmental profit-seeking agency engaged in the following main activities: administration of Real Property Register and Cadastre and he Register of Legal Entities, the registration of real property objects and rights in them, the appraisal of real property, and cadastral surveying (http://www.kada.lt/imone.html).

The AEQUITAS Project, considering those aforementioned classifications, can be encapsulated under the *Networked Government* definition, to the extent that some of the participants comply with public functions (Greffiers and Registradores), whilst the activities of the others (Procuradores and Solicitadores) are specially linked to tribunals, or to Notaries, entities that undertake public functions.

These juridical agents that participate in the project are an example a many other juridical operators in the European Union. The electronic transmission to be found between these juridical agents are realized through the AEQUITAS Informatic Tool.

2.2 Other Participants

- *Tools Banking Solutions, S.L. (Spain).* A state-of-the-art Company that develops new information technologies. It is responsible for the coordination and technical management of the Project, as well as of the drafting and implementation of the software (AEQUITAS Informatic Tool) (www.tb-solutions.com)
- *Asociación para la Promoción de las Tecnologías de la Información y el Comercio Electrónico (APTICE, Spain).* This organism gives support in the coordination process and is responsible for the diffusion of the Project and its results (www.aptice.org)
- *Universidad de Zaragoza (Spain).* This institution is responsible for the validation of the results obtained in the project from a legal perspective (www.unizar.es)

- *Vilnius Law Faculty (Lithuania).* This institution is in charge of investigating the current electronic signature and communications regulations in Lithuania. Moreover, it should give support during the trail phase of the AEQUITAS Informatic Tool, as well as in the diffusion of this tool (http://www.lta.lt/english.html).

3 Objective

The main aim of the AEQUITAS project is to achieve an improvement in the communications between European juridical operators, rendering more speed, security and confidentiality in these communications than those currently in use, and thus rendering a more effective and efficient service to the citizens.

So as to achieve this objective, a system has been designed to allow communications and telematic transmissions of documents between different participant juridical agents, in a secure, quick and confidentiality environment. This system is divided into the following elements:

- An **Association,** that groups, initially, the participant juridical operators in the Project. The members of this Association can be extended to the rest of participants in the Project, even though these are not juridical agents, as well as to third entities, not participants in the Project, namely, Judges, Secretaries and other juridical agents. Statutes have been drafted, pending approval as of today, that regulate not only composition of the Association, but also all those aspects related to it: nature, function, governing bodies, object, etc...This Association acts as a third trust party in the telematic transmissions that arise between the members. These transmissions shall be performed via the AEQUITAS Informatic Tool. For this purpose, it is necessary that each Association member entity have a LDAP directory with the digital certificates of its members pursuant to the structure defined by the Association. The Association shall be based on these directories to accredit the identity and signatures of its members.
- A **Web** site, with the following URL http://www.euro-aequitas.net, has been developed to disseminate the Project, and thus make it known and accessible to other users.

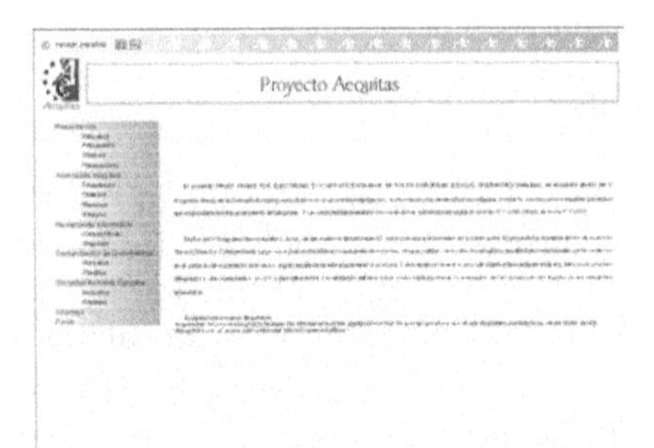

Fig. 1. Example of one of the AEQUITAS web pages

- A **Document Catalogue**, that shall allow the Informatic Tool to control the fulfilment of formal and signatory source requirements needed in order that these documents be accepted by the addressee. This document catalogue shall facilitate jurists from different countries to interpret these documents, described further on in more detail.
- An **Informatic Tool**, that allows for secure document transmissions, whether included or not in the aforementioned catalogue, and using for this purpose the so-called asymmetric key cryptography and electronic certification. A full description of this Tool is to be found further on.

4 Document Catalogue

The documents to be transmitted between the users can be either judicial or extrajudicial.

One of the tasks to realise within the AEQUITAS Project shall be the study of the existing communications flows between the different juridical operators, participants in the Project.

Nowadays, these information flows still prove to be scarce, but are bound to increase notably in the coming years as an effect of the more common enactment of substantive and procedural norms conceived to produce uniform effects throughout the European Union. Accordingly, those communications between the different European Registrars pursuant to the recently approved European Company Statute which is composed by a Regulation and a Directive that will come into force simultaneously on October 6th, 2004, have to be set; as well as those that take place pursuant to the European Council Regulation CE No 1348/2000, dated May 29th, regarding both civil and mercantile notices and transfers of judicial and extrajudicial documents in the Member States, enforced May 31st, 2001.

On the other hand, to be highlighted are the communications that exist between Proctors and Mercantile Registrars in the Spanish territory. These take place when the Proctors act as intermediaries and conduct Judicial Orders[3] issued by the Courts, handing these in at the Registrars, that once dealt with, return these to the Courts.

Once the communications flows have been studied, the documents sent in these communications shall be analysed. Work is performed using real documents, analysing its contents and so determine the data to be found in these, their signatories, as well as their nature and structure. In addition, formal and material requirements of these documents shall be determined, in compliance with the applicable legislation, and to be fulfilled so that these are accepted by the addressee and have juridical value.

Following the aforementioned document analysis, templates of these documents shall be developed in the XML language and integrated into the Informatic Tool. These templates shall show in the different fields that data that mandatory should be filled in the document. The users of the Tool should fill in and complete the templates that choose these data. In this way, there shall be a list of document templates from which to choose from and to be filled in by Tool users during the trial phase.

[3] Judicial Orders are used to order the issuing of certificates and testimonials and the practice of any action, whose execution corresponds to registrars of property, mercantile, naval, sales by instalments of moveable chattels, notaries, brokers or Court and Tribunal agents.

It is important to note that the objective is not to embrace all those documents currently transmitted between the Project's participant juridical operators, but to analyse a sufficient number of these so as to include them in the Tool and test them during the trial phase.

5 AEQUITAS Informatic Tool

5.1 Object

Tools Banking Solutions, S.L., within the AEQUITAS Project, shall develop a software application or informatic tool that shall allow, via TCP/IP networks, secure communications and transmission of electronic documents between the AEQUITAS Association members, using electronic signature and certification.

5.2 Users

The users of the Tool shall be the AEQUITAS Association members that shall use it to communicate with each other. Initially, therefore, the users shall be the juridical operators participants in the Project being that, for the moment, these are members of the Association.

Each of these juridical operators (Greffe du Commerce from Paris, Consejo General de los Ilustres Colegios de Procuradores de los Tribunales from España, Ilustre Colegio de Registradores de la Propiedad y Mercantiles from Spain, Câmara dos Solicitadores from Portugal and Land Cadastre and Registry from Lithuania) should be able to make available to their respective members (Greffiers, Procuradores, Registradores, Solicitadores, etc...) X-509 electronic certificates registered at their corresponding LDAP directories, which have to be adapted to the structure defined by the Association. These certificates shall be used in the Informatic Tool. There are two options: that the aforementioned users be constituted in a Public Key Infrastructure (hereinafter, PKI) and issue the electronic certificates to their members and publish them in the corresponding LDAP directories or else address an existing PKI so that it issues the electronic certificates and publish them in its LDAP directory. In both cases, these certificates should be accepted by the AEQUITAS Association.

5.3 Technical Structure

Architecture
The Tool has to main modules:

1. *Client Module*: that shall be installed in the Tool users' workstations (that, as we have already mentioned, shall be the juridical operators participating in the Project) to send documents between them. It a module that can be downloaded via Internet and compatible with Windows 98 or higher Operating Systems.

2. *Server Module*: there shall be a server at the Association and at each of the Association's member entities, in such a way that the internal communications in the same entity do not exit the corresponding entity's server. With regards to the communications between the members of the different participant entities in the Association, these shall use the Association server which acts as a third trust party, although it is possible to establish bilateral trust agreements between the organisms that are a part of the Association. This module is compatible with most frequently used Unix versions (Linux, Solaris, etc...) and with Windows NT or higher versions.

The servers used, that is, that of the Association and each of the others found in the different organizations, are STFIC servers, implemented by Tools Banking Solutions, S.L.

Certificates

The users shall be identified in the system via X-509 electronic certificates issued by each organization of which they are members of and accepted by the Association. Each organization shall administer the certificates issued by them, having to publish them in the LDAP directory that complies with the Association's defined structure, and revoke them when needed and publishing this act in their corresponding CRL.

The STFIC servers used by the system (that of the Association and those found at the Association member entities) shall have their corresponding server certificates.

5.4 General Workflow

1. The users shall certify via their electronic certificates so as to be able to access the Informatic Tool.
2. From the Informatic Tool, the document to be sent shall be selected. With this regards, there are two options:
   a) There is a possibility that the user draft the document using that software application that he/she wants (for example, Microsoft Word, specific management programme, etc...). In this case, the document has to incorporate or import the Tool. The user himself/herself shall determine the security level that he/she wants, that is, the message shall be signed and ciphered or only signed.
   b) It is also possible that from the Tool itself, one can choose, from a list of templates programmed in XML, a document drafted, complete and fill it in. In this case the security level is predetermined, setting as well as, the signatories needed. The Tool controls the compliance of all templates set requirements, preventing their transmission or warning the issuer and addressee of the irregularities detected.
3. Selecting the document to be sent, this document shall be signed electronically with the signatory's private key which shall be stored on a smartcard and duly protected with a password. This way, the identity and non-rejection of signatory, as well as document integrity, is guaranteed, being that once signed, any modification done to the document shall be detected. The documents can be signed by several signatories.
4. Furthermore, if required, the document shall be ciphered with a public key found in the addressee's certificate or in the organization's server of this addressee. This way, only the addressee shall be able to access the document contents guaranteeing thus confidentiality.

5. Once document is signed and, if required also, ciphered, it is sent to addressee and once delivered, a acknowledgement of receipt shall automatically be issued.
6. In any case, only those certificates accepted by the Association shall be accepted.
7. The servers validate the signatures of documents received by checking the certificate status through their corresponding CRL.
8. There is also the possibility that time stamping of signatures and communications acts be performed. Moreover, it is possible to make a copy or register of document contents sent be performed by a third trust party that could be the Association itself.

6 Conclusion

- The global accessibility of the new technologies to the citizens, can only be done through European projects, as the herein described, that in found within the so-called Networked Government.
- From a legislative point of view, it is necessary that there be a closer harmonization of the existing European State Member legislations.
- Moreover, from a technical perspective, it is totally necessary the setting of technical standards to achieve more compatibility between the different informatic systems used by the different European juridical operators.

The Concepts of an Active Life-Event Public Portal

Mirko Vintar and Anamarija Leben

University of Ljubljana, School of Public Administration, Gosarjeva 5, 1000 Ljubljana
{mirko.vintar,anamarija.leben}@vus.uni-lj.si

Abstract. Public Portals as common entry points to public services are becoming key elements of the future e-government infrastructure. In most countries, recent research in further development of public portals has been very intensive; however, approaches to the design of portal architecture and organization are still very diverse. In this paper, we will present current results of the research in progress aiming to develop prototype of an intelligent Life-Event Public Portal. We are focusing on the methodological aspects of the knowledge-based Life-Event Portals, which can provide much more efficient provision of e-services than conventional e-portals.

1 Introduction

In practically all European countries, numerous projects for development and implementation of e-government are in progress and Slovenia is no exception. The main objectives of these projects are to bring governments and citizens closer and to improve efficiency, effectiveness and transparency of its operation and increase the quality of its services. Through realization of these objectives, governments are starting to change their character from prevailingly power exercising institutions to pointedly more and more service providing and partnering institutions to the citizens and business community.

In Slovenia, a systematic approach to development of e-government started a few years ago. The basic information infrastructure for the implementation of electronic government is in place. The necessary legal framework was established by passing the Electronic Commerce and Electronic Signature Law in the year 2000. In the beginning of the year 2001, the Slovenian government accepted "The Strategy of e-Commerce in Public Administration of the Republic of Slovenia" for the development of e-government in Slovenia by 2004 [2] and the state's public web portal was introduced. This portal is designed to cover all three segments of e-government [8]:

- Government-to-Citizens (G2C): e-services for citizens provided by government,
- Government-to-Business (G2B): e-services for private sector provided by government,
- Government-to-Government (G2G): e-services for government provided by government.

Although a year has already passed since its introduction, at the time of writing this portal was still at the very initial state, meaning that it provides mainly information services and serves as a single entry point to the home pages of different administra-

R. Traunmüller and K. Lenk (Eds.): EGOV 2002, LNCS 2456, pp. 383–390, 2002.

tive bodies in Slovenia. It offers different information about the organization and functioning of public administration in Slovenia. In the first two segments (G2C and G2B), some e-services have already been introduced. The applications for the birth, marriage and death certificate are available for the public. For the businesses, some e-services concerning public procurement and public tenders are provided. For some other government services, downloadable application-forms are available, while some more sophisticated communication and transaction services still remain to be seen. Generally, communication with institutions and the relevant officials is available by e-mail.

2 Approaches to Provision of Services Using Web-Portals

A brief overview through the web-portals in different areas shows that we can define several levels of complexity of web-portals.

2.1 Simple 'Self-service' Portals

For these types of portals, it is characteristic that services are collected from different areas and administrative bodies and offered to the users via menus organised like shopping lists. Users are supposed to know exactly which services they need and which administrative bodies are responsible for their provision. Usually, most users don't have this information. In such situations, searching through endless lists of services and institutions may become a nightmare almost as unpleasant as the time-consuming chasing through offices in the traditional administration set up.

2.2 Life-Event Based Portals

Life-event based web portals are developed and organised according to the very realistic assumption that most users in a particular life situation do not know exactly which public services they need. For instance, the user only knows what he wants to achieve - to build a house, to start a business, to get married, etc. These situations are known as life-events. The system, i.e. the web portal, is supposed to have the necessary 'knowledge' to determine which services and administrative procedures are needed to be solved in order to assist the user in a particular life-event situation. To solve such a life-event, typically various administrative procedures at different administrative bodies usually have to be carried out.

Thereby, the system that guides the user trough the situation and helps him to identify the required services and their providers is needed. The web portal that includes such a system is called a *life-event portal* [5].

There are two types of life-event portals. The first is based on well-defined hierarchy of topics and life-events (*passive matrix of life-events*). The system allows user to select topics and subtopics and in this way guides him to particular life-event. When the life-event is selected, the information about required administrative procedures and the necessary assistance is offered. Examples of such portals are Austrian Internet

Service HELP [1] and Singapore e-Citizen [7]. These portals offer all information relevant to particular administrative procedure, e.g. competent administrative body, the documents to be presented, fees, and terms. In addition, adequate forms may be retrieved and filled. This information is manly offered in the form of a web page with links to other relevant web pages.

The second, user-friendlier type of life-event portal is based on so-called *active matrix of life-events*. The core system of such portals is a knowledge-based system. The knowledge-based system is a computer program based on inference mechanisms to solve a given problem employing the relevant knowledge [3,4]. The knowledge-based system in an active life-event portal (intelligent guide trough life-events) uses the pre-defined structure of particular life-event to form an active dialog with the user. In this way, the user is an active partner in the overall process of identifying and solving problems related to particular life situations.

3 The Architecture of an Active Life-Event Portal

Most of the 'supposed to be' active life-event portals available on the web today are still in the very initial state of development and are based on a very different architecture. In the next chapters, we will try to describe the basic building blocks of the prototype of the active life-event portal, which has been developed within our project [9]. An active life-event portal (portal based on active matrix of life-events) has the following main components [10] (Fig. 1):

- *Registry of procedures and forms* contains information required for implementation and execution of each type of administrative procedure (classification number, description, algorithm for execution, relevant administrative body, associated forms and documents – both input and output, the normative ground for its execution etc.)
- *Registry of normative regulations* contains all legal norms, which represent legal basis for the execution of administrative procedures.
- *Registry of life-events* contains data about topics and life-events requiring the governmental services. It also holds all data needed to define the decision aspect of life-events.
- *Electronic guide trough life-events* is a knowledge-based system that employs knowledge stored in above-mentioned registries. Together with the corresponding registries, it presents the communication interface of an active life-event portal.
- *Classification systems* are designed to classify administrative services procedures and forms for the entire public administration. The classification number of a procedure or form is its unique identifier.
- *Communication interface* is designed to enable an easy and user-friendly dialog between the user and the system through which all specific parameters of particular life-event of the user and his needs are defined.

The *communication interface* should meet three objectives of an active life-event portal. The first goal is to assist the user in selecting an adequate life-event. This can be achieved through the hierarchical structure of topics, which are supported by the portal. This structure helps user to identify the life-event that corresponds to his problem. The second goal is to identify the procedures needed to solve this life-event.

This could be achieved through the dialog with the user, based on the decision-making process, which is comprised in the structure of a life-event. This process results in the list of generic procedures. The third aspect is to identify an adequate variant of each generic procedure in this list. This is also a decision-making process, where the input parameters, needed to define the right version of the procedure, depend on the values obtained through the dialog with the user. For example, based on these parameters different supplements to the application form for the particular procedure are defined.

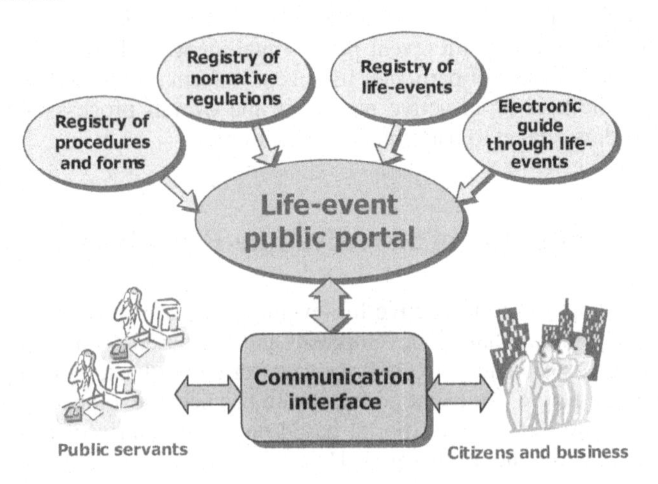

Fig. 1. The components of an active life-event portal

The decision tree through which communication interface of an active life-event portal should guide the users has, according to our architecture, three hierarchical levels (Fig. 2):

- Level of topics,
- Level of life-events,
- Level of administrative procedures.

Each hierarchical level of the decision tree is presented with specific model. For modelling of the decision making process related to the execution of a life-event, we decided to use selected concepts from eEPC (extended event-driven process chain) models [6].

3.1 Level of Topics

To establish hierarchical structure of topics, three types of topics are defined (Fig. 2):

- *Main topics* are topics at the highest level. When the communication interface is started, these topics are listed first. Main topics are always composed of at least two subtopics.
- *Subtopic* is composed of other topics (either subtopics or elementary topics). It can also include composed life-events.
- *Elementary topic* is composed only of composed life-events.

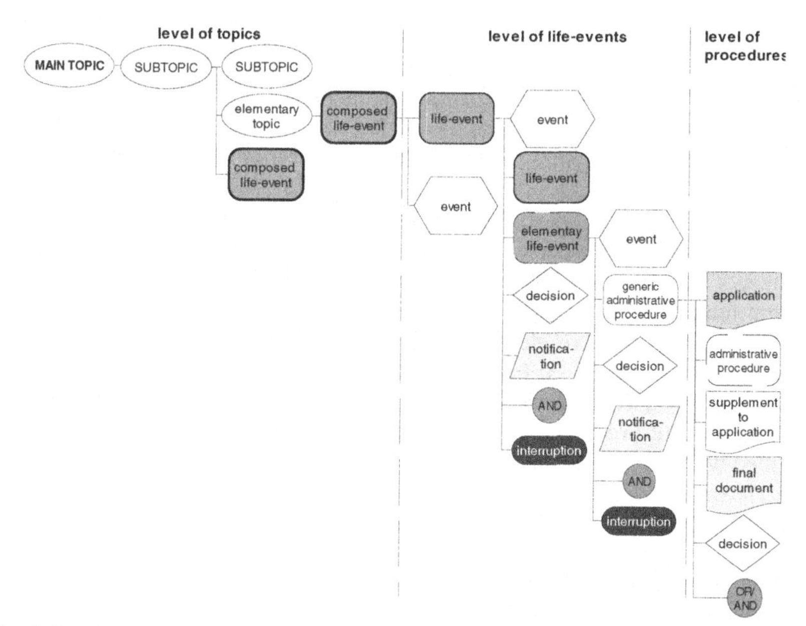

Fig. 2. Levels and main concepts representing decision tree of an active life-event portal

A model of a main topic is a tree. A root node of the tree is always a main topic, for which the model is designed. The leaves of the tree are always composed life-events, while other nodes are subtopics and elementary topics. The edges, connecting the nodes, are of the type 'is-composed-of'. They define the structure of the main topic.

3.2 Level of Life-Events

As this level includes a decision-making process, the models of life-events are more complex and more concepts are needed to define the structure of life-events (Fig. 2).

An *event* is a concept that helps to control the processing of life-event. Events occur in three different roles:

- As a start and end event: each life-event starts with start events (one or more) and ends with end events. The start event describes the impulse that initiates the life-event (e.g. 'I want to establish a business'); the end event describes the situation after the life-event is successfully finished (e.g. 'the business has been established').
- As a time event: time events identify the time at which a particular procedure or life-event is supposed to start. This time usually depends on some legal provision (e.g. 'in eight days after the applicant has been officially notified').
- As an event, describing in which state the life-event is at a particular point of time. With this type of events, the initiations of successive life-events or procedures are linked.

A *decision* is a basic concept in the life-event model as the decision-making process is defined by decisions and their *alternatives*. These two concepts define the active dialog with the user. A *notice* is another concept used in the dialog with user. It is usually connected with an *interruption*, which indicates, that processing of life-event is not successfully finished.

A *life-event*, an *elementary life-event* and a *generic procedure* define the actions, required to complete the overall process. With *AND-split* and *AND-join* points, the parallelism in the processing of life-event is modelled. The *control flow* indicates the direction in processing the life-event.

According to the structure of a life-event, three types of life-events are defined as follows:

- A *composed life-event* describes a sequence of life-events representing key steps in the processing of life-event. In the model of the composed life-event, only three concepts are used: an event, a life-event and control flow. It does not include a decision-making process.
- A *life-event* is the core element of the communication interface as it describes a decision-making process. In the model of the life-event, all above-described concepts may be used with the exception of composed life-event and generic procedure. Each life-event, included in the model of particular life-event, is further described with its own model.
- An *elementary life-event* is a special case of a life-event. In the model of an elementary life-event, all concepts may be used except other life-events. Instead of life-events, generic procedures are included. Consequently, the elementary life-events present the final step in the process of identifying the list of the generic procedures required to solve user's problem.

3.3 Level of Procedures

At the level of life-events, the generic procedures within life-events are identified. Variants of the same generic procedures are defined manly by the different documents that have to be presented with the application form to initiate a particular procedure. These documents (supplements to the application) can be understood as parameters of the generic procedure.

The concepts, used for modelling parameters of generic procedures, are shown in Fig. 2 and briefly described in the following.

An *administrative procedure* presents the generic procedure, for which the parameters are modelled. An *application* defines the document with which the procedure is initiated; the *final document* is the main output document (e.g. an official decision, a permit, etc.). An application and final document are always connected to an administrative procedure with the *data flow*.

Decisions and *alternatives* are used to define the structure of parameters: which *supplements to applications* are required depending on the chosen alternatives of the decisions. The structure of parameters is further defined with *AND/XOR-split* points. AND-split point is used to indicate that several supplements are needed simultaneously; XOR-split point indicates that one of stated supplements is required. The role of the *control flow* and *interruption* is the same as in the life-event model.

3.4 Data Model of Communication Interface

Key entities of the communication interface are shown in Fig. 3 and briefly described in the following.

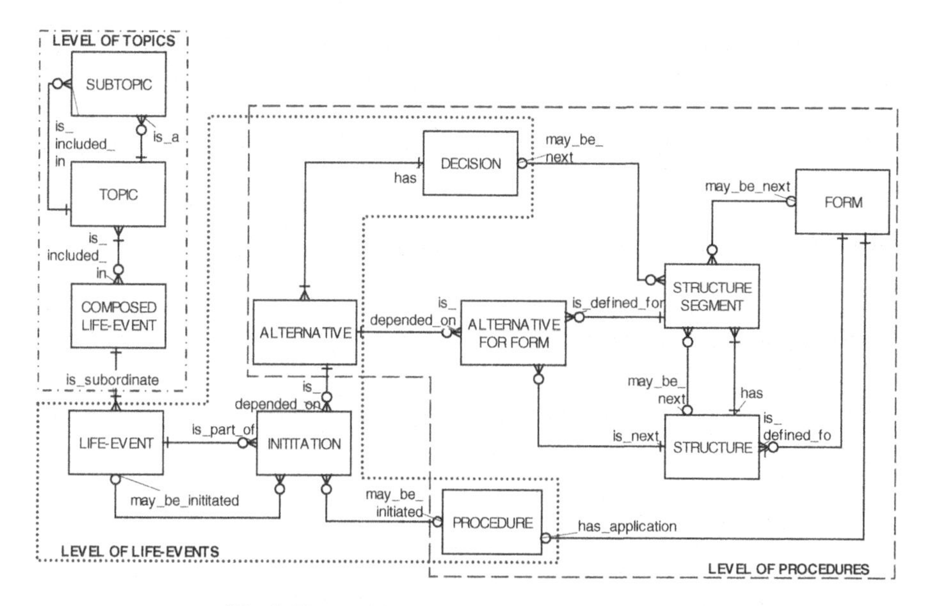

Fig. 3. Key entities of the communication interface

- Entities TOPIC and SUBTOPIC define the structure of topics, together with entity COMPOSED LIFE-EVENT they define the *level of topics* in the communication interface.
- Basic data of life-events are kept in entity LIFE-EVENT. Basic data about decisions and their alternatives and about events and notices are stored in entities DECISION and ALTERNATIVE. Data in entity INITIATION along with relationships define that part of the communication interface, which results in the list of generic procedures for particular life-event. In these relationships, the structure of life-event and basic life-event is described. The *level of life-events* in the communication interface is defined in this part of the data model.
- The basic data about administrative procedure and adequate forms are stored in entities PROCEDURE and FORM, which present the key entities in the registry of procedures and forms. Data in entities STRUCTURE, STRUCTURE SEGMENT and ALTERNATIVE FOR FORM along with the relationships define that part of the communication interface, which helps to identify the variant for generic procedure and to display relevant information for each procedure. The *level of procedures* in the communication interface is defined in this part of the data model.

4 Conclusions

On the basis of the presented concepts and mechanisms, we have developed a working prototype of an active life-event based web portal. The prototype was developed with the use of ASP (active server page) technology and ORACLE data-base management system. During further research and development, we will try to refine and generalize the modelling mechanisms and implement them for further development of the Slovenian public web portal. We believe that the concepts described in the paper can considerably contribute to the development of new solutions, and better ways of providing e-services to the citizens and companies. They are general enough to be easy implemented in developing other public web portals.

References

1. Austrian Internet Service HELP. http://www.help.gv.at. (February 2002)
2. Government Centre of the Republic of Slovenia for Informatics: The Strategy of e-Commerce in Public Administration of the Republic of Slovenia for the Period from 2001 until 2004. http://e-gov.gov.si/e-uprava/english/index.jsp. (February 2001)
3. Jackson P. Introduction to Expert Systems (third edition). Addison Wesley Longman Ltd., Harlow, (1999)
4. Klein M., Methlie L.B. Expert Systems: A Decision Support Approach. Addison-Wesley Publishers Ltd., Workingham, (1990)
5. von Lucke, J. Portale für die öffentliche Verwaltung: Governmental Portal, Departmental Portal in Life-Event Portal. In: Reinermann H., von Lucke J. (eds.): Portale in der öffentliche Verwaltung. Forschungsinstitute für öffentliche Verwaltung, Speyer (2000)
6. Scheer A.W.: Business Process Engineering, Springer-Verlag, Berlin-Heidelberg-New York-Tokyo, (1996)
7. Singapore e-Citizen. http://www.ecitizen.gov.sg. (February 2002)
8. Slovenian public web portal. http://e-gov.gov.si/e-uprava/english/index.jsp. February 2002.
9. Vintar M. et all. Report on the project: Development of public web portal (in Slovene). University of Ljubljana, School for Public Administration, Ljubljana, (December 2001)
10. Vintar M., Leben A. A Framework for Introducing e-Commerce Concepts in Public Administration (in Slovene). In: Proceedings of Days of Slovenian Administration, Portoro, (September 2000)

New Services through Integrated e-Government

Donovan Pfaff[1] and Bernd Simon[2]

[1] Goethe-University Frankfurt, Department of Electronic Commerce, Mertonstrasse 17, 60054 Frankfurt, Germany.
pfaff@wiwi.uni-frankfurt.de
http://www.ecommerce.wiwi.uni-frankfurt.de
[2] SAP Deutschland AG& Co. KG, Neurottstrasse 15a, 69190 Walldorf, Germany.
Bernd.Simon@sap.com
http://www.sap.com

1 Introduction

The New Public Management initiative in the 90's had a tremendous impact on the principles of public administration. Cost transparency and customer orientation have become strategic goals. The public sector is still in motion: eGovernment is a new trend that also progresses the idea of customer orientation. International studies document that eGovernment has become a well known phrase in many countries worldwide. There are, however, significant differences in their respective development. Many administrations use the internet technology just to provide information[1]. The opportunity to generate additional revenues through integrative IT-solutions is rarely used[2]. The challenge of performing additional tasks within a declining budget forces governments to develop new ideas in order to increase revenues or to reduce costs. This could be a future business for administrations. In the past, most administrations tried to realize e-government by establishing their own website. Current initiatives concentrate on transactional aspects trying to connect specialized systems with the web. Strategic concepts focusing on architecture and service portfolios are becoming more and more important[3]. In addition to the political and administrative part of eGovernment, there is also a commercial aspect of services. In particular those commercial services associated with payment processes require and demand integrated transactions.

The first section of this paper documents the technical requirements for an integrated eGovernment framework. Legal aspects like data security or digital signature will not be part of the discussion. Subsequent sections will describe the architecture of an integrated solution. The last section presents proposals on new services of public administrations.

---

[1] KPMG (2001), p. 14

[2] an Accenture Study on EGovernment in Germany in 2002 documents that only 31% of public sector employees are convinced that EGovernment is a opportunity to receive additional revenues.

[3] e.g. BundOnline 2005 in Germany, uk.online or help.gv in Austria.

R. Traunmüller and K. Lenk (Eds.): EGOV 2002, LNCS 2456, pp. 391–394, 2002.

2 Requirements for an Integrated Framework

A media break free communication between the individual IT-systems of one organization (like account system, reporting or document management) requires integration. In addition to this internal integration, special business scenarios having an effect on several institutions require an external or cross-authority integration as well. The best way to handle a business scenario "moving from city A to city B" is to realize a horizontal integration by connecting both systems directly. The web service of ordering a museum ticket for a special city in state government portal can serve as an example for a vertical integration.

Both kinds of process-oriented integration and the resulting collaboration scenarios are success factors for future business in the public sector. Business processes must be aligned to a common goal. It is not sufficient to provide an isolated web service. A common information base is needed in which web services are available when needed. All applications have to provide the functionality via interfaces to enable administrations to use the appropriate web services in any environment and to arrange innovative business processes using existing functions. In order to ensure interoperability between heterogeneous platforms, web services and interfaces must offer common, open technical internet standards enabling an cooperation between different technical platforms. These open technical standards are for example J2EE und Microsoft .NET. Usually these standards are HTTP (hyper text transfer Protocol), XML (Extensible Markup LANGUAGE), SOAP (Simple Object Access Protocol), WSDL (Web Services Description LANGUAGE) and UDDI (Universal Discovery, Description, and Integration).

A complete and perfect solution supports different standards to ensure a connection between applications and web services. An integrated eGovernment framework must be flexible in making changes and adding new process and components. A fast adjustment to rising volume of data requires a scalable architecture. The adding of new web services and processes, as well as compatible extensions of existing messages and patterns, must be possible without an interruption of business operations. This flexibility ensures the implementation of new services in a quick and efficient way.

Apart from the technical requirements, security plays an important role in an integrated e-government. Security means protecting and maintaining values like Integrity, Authenticity, Confidentiality and Availability. User and role administration, secure system management and digital signatures are keystones in maintaining these values.

3 Architecture

An ideal eGovernment landscape provides integrated business processes from the web portal to the backend systems. In order to implement such scenarios, it is necessary on the one hand to prevent media breaks and, on the other hand to provide basic functions such as security or workflow. It makes sense to provide these functions in a kind of middle office platform (e.g. mySAP CRM). The middle office should include functionality to connect and design specialized systems as well as tools to develop specific web services. Application Programming Interfaces (APIs) provide a data exchange between midoffice and specialized systems using XML. Irrespective of the relationship (Government-to-Consumer; Government-to-Business or Government-to-Government) it should also be possible to maintain central business partner master data for all kinds of customers as a prerequisite for a central customer account.

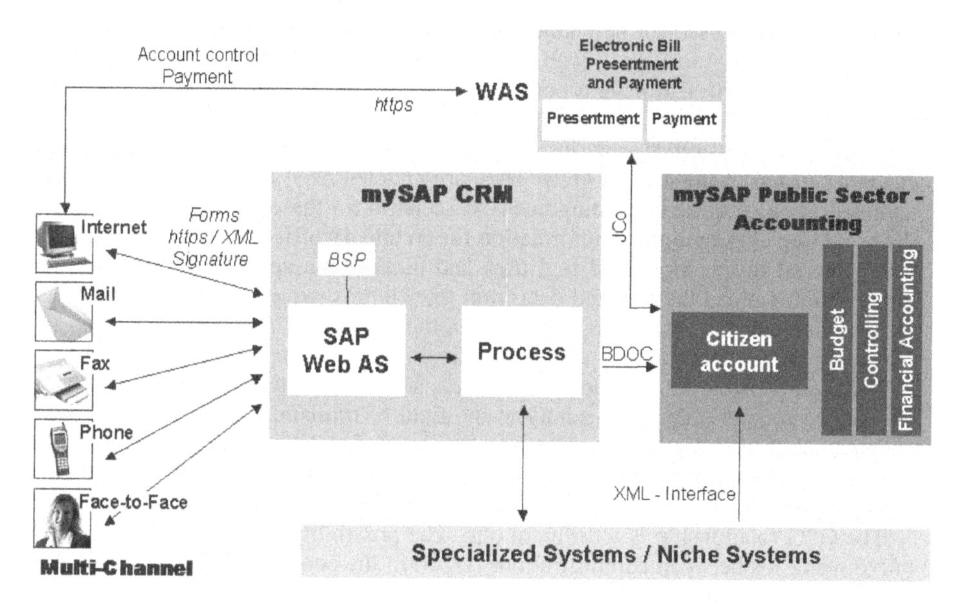

Fig. 1. Integrated eGovernment framework from SAP AG, Germany (www.sap.com)

When a customer uses the web service, the data coded by means of an electronic signature arrives via a secured Internet protocol (HTTPS) at the intermediate and goes from there over an interface into the correct specialized procedure.

When a fee occurs it must be passed on to the citizen's account. If the specialized procedure should not include invoice functionality, the invoice item in the customer account must be activated through the middle office by means of a billing engine. At that point an open item is created and presented to the customer through the web services (Electronic Bill Presentment and Payment) and made available for payment. The customer receives an overview of all open items and outstanding debits with the possibility of making the payment (for example invoice, debit or credit card). As current studies confirm the invoice is the most popular way of payment in the net. In order to also meet these demands within eGovernment, an integration of an Electronic Bill Presentment and Payment (EBPP) solution should be targeted. EBPP describes the electronic representation of the invoice, including a payment function.[4] Thus on the one hand a process without any media breaks is made possible for the administration and on the other hand the customers can pay using their preferred method. The open and balanced items are updated in the financial accounting of the administration. This scenario is supplemented by the documentation of the business transaction in an electronic records management system.

[4] Spann/Pfaff (2001), p. 509

4 New Services

The eGovernment framework presented above offers the advantage that central issues of e-government like security, electronic workflow, online payment are already resolved and can be used for new kinds of services. It is important that administrations develop creativity for new web services and extend their service portfolio. eGovernment could also have a commercial touch and help to refinance the framework.

So-called professionals (notaries, car dealers, architects etc.) represent an important customer group in the area of G2B. They are characterized by frequent contact with administration. A quite interesting business scenario for these customers could e.g. be the providing of geographical information for architect's offices. Instead of seeing the map material in the local land registries and making the necessary copies, the customer can download the required data from the administration portal using the special web service and pay the respective fees via their customer account. The official in charge no longer needs to deal with this business.

It is also possible to develop new business scenarios or to optimize existing services for customers (G2C), especially in the field of tourism. Big events e.g. a world championship offer various opportunities for local, federal or state government. Online services like ticket reservations, souvenirs, bar licenses or trading licenses could create additional value for customers and additional revenues for the administration.

The G2G relationship is a different one. The possibility of implementing collaborative scenarios between administrations (G2G) in the context of joint up government or one stop government is dependent on the level of networking and integration between administrations. For instance a central unit, such as the German "Bundeskasse" could offer their framework as a solution provider not only to the assigned state authorities but also to other public administration e-commerce-services. In order to exchange the data records in a secure way, a standard data exchange format (XML pattern) will have to be established.

Literature

1. **KPMG** (2001), "Verwaltung der Zukunft - Status quo und Perspektiven für eGovernment", Report.
2. **Spann, M. / Pfaff, D.** (2001), "Electronic Bill Presentment and Payment (EBPP)", Die Betriebswirtschaft (DBW), 61, 509-512

Risk Assessment & Success Factors for e-Government in a UK Establishment

A. Evangelidis, J. Akomode, A. Taleb-Bendiab, and M. Taylor

School of Computing & Mathematical Sciences, Liverpool John Moores University,
Byrom Street, Liverpool, L3 3AF, England, UK
{cmsaevan,j.o.akomode,a.talebbendiab,m.j.taylor}@livjm.ac.uk
http://www.cms.livjm.ac.uk

Abstract. In a quest to modernise their activities and underpin their public-private partnerships, many governments around the globe have initiated their local eGovernment programmes. In this regard, best-practice, emerging Information Communications Technology (ICT) and e-business potential are leveraged to provide 24*7 access to online public services, ranging from online tax forms, to online voting. Whilst much may have been achieved towards developing and supporting one-stop shop to a range of online government services, more research is required, for instance, to provide a seamless integration and interoperation of these services, their integration with legacy systems, and risk management strategy. Based on an ongoing research focused on risk modelling and analysis of eGovernment web services, this paper introduces a categorisation of the main generic risk factors. The paper only elaborates on the first two categories of the risk factors and develops a set of potential success factors for eGovernment.

1 Motivation for the Research

Over the recent years, many governments around the globe have initiated their eGovernment strategies to exploit: ICT, e-business models and best-practice. The main interest is to improve operations and to support their citizens through the use of ubiquitous web access technology to public sector and eGovernment digital content. Currently, the way the public sector is implementing eGovernment may be viewed from two sides: (i) the development of eGovernment Web portals, providing various services to its customers, for example UK Online [1]; (ii) the development of Web sites 'dedicated' to a particular service by a single governmental department, for example TAXISnet ([2], pp.5-8).

At the moment, most governments around the world, including the UK government wish to take a step further and they have made (or are making) plans with tight deadlines to fully integrate all or at least most of the governmental services. An example could be made of the government of Canada ([3], p.5) and the UK government ([4], p.6). This 'next generation' of eGovernment is likely to be the result of a partnership between the public and private sector. The ongoing research seeks to identify risk factors in all aspects of eGovernment and develop a framework for risk assessment. The situation may serve as a guide to assist people working towards the next genera-

R. Traunmüller and K. Lenk (Eds.): EGOV 2002, LNCS 2456, pp. 395–402, 2002.

tion of implementation and management of eGovernment projects to derive increased benefits.

Due to confidentiality the names and exact locations of the collaborating establishments are omitted but the quality of the material presented remains unaltered.

2 eGovernment Risk Factors

eGovernment projects are inherently complex, sharing similar risk factors with their e-business counterpart projects. Implementing eGovernment as a major development may not be easy since it may involve many factors of risk that could threaten the success of the project. Adequate risk assessment and management procedures may help in avoiding major pitfalls, though sometimes failures cannot always be predicted precisely.

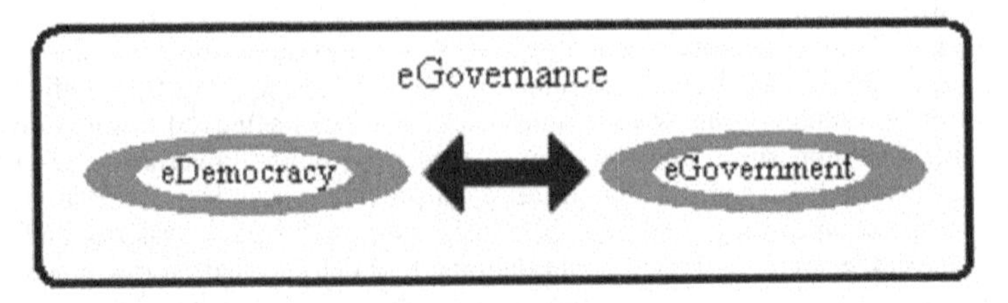

Fig. 1. eGovernance Perimeter

Before attempting to elicit the main generic eGovernment risk factors, it may be prudent to specify the environment in which eGovernment resides. Figure 1 shows that eGovernment finds itself within the perimeter (or boundary) of 'eGovernance' and interacts with the 'eDemocracy' concept. The latter may be defined broadly as the improvement of effectiveness and efficiency of democracy through the use of ICT [5].

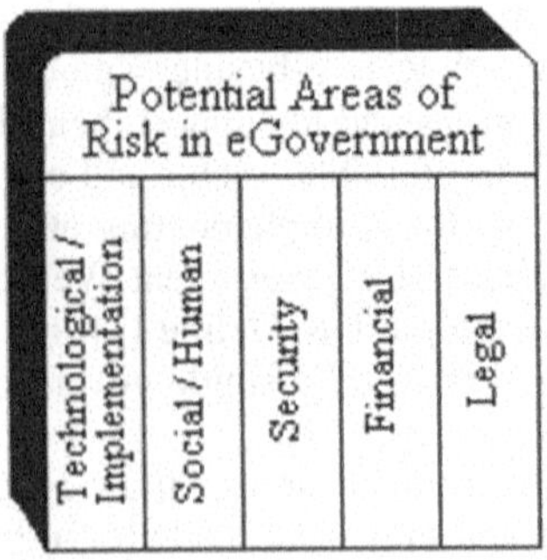

Potential Areas of Risk in eGovernment

Technological / Implementation | Social / Human | Security | Financial | Legal

Fig. 2. The Main Generic Risk Elements (or Factors) of eGovernment

The research indicates that the main generic eGovernment risk factors may be initially classified into the five categories [(i) technological/implementation; (ii) social/human; (iii) security; (iv) financial; (v) legal] shown in Figure 2.

A more detailed analysis of the first two categories of risk elements is presented below i.e. with regard to: (i) technological/implementation; (ii) social/human. Also, brief discussions are advanced on the latter three generic risk elements.

2.1 Technological and Implementation Risk Factors

Since eGovernment relies on the use of ICT, almost all eGovernment projects are ICT-centric. Thus, one of the main sources of risks in eGovernment-related projects lies on the technologies used and the way the technologies are implemented in order to serve the purpose a particular project is meant for. The most common technological and/or implementation risk factors identified are presented and discussed.

(i) *Risk of gaining quick political advantage:* eGovernment may be a rather new and trendy concept for most politicians, as such they often want to encourage the implementation of eGovernment projects quickly while they are still in office, in order to gain better political advantage because the alternative may be worse for them. The situation may lead to inadequate planning that may eventually result in the failure of the project. Experience from one UK establishment shows that such failure may be due to inadequate planning in linking technological capabilities with the requirements of the establishment, and it may be viewed as being propelled by an attempt to satisfy political request and instructions quickly.

(ii) *Risk of lack of adequate identification and classification of services:* The eGovernment concept is basically a way for the Governments (local or national) to offer their customers (citizens, businesses, other governments) better services. This research reveals that in this UK establishment there is a lack of: (a) adequate identification of service requirements; (b) adequate standard classification of the services the eGovernment system is meant to satisfy. This implies that the eGovernment (or IT) staff find it difficult to classify the services and therefore have problems in designing and developing the IT infrastructure. A clear example of this issue is the service of *Council tax* disregards, where – for instance – there is the case of students who are eligible for full exemption from paying the council tax and there is also the case of lone-parents who are eligible for a huge deduction from the council tax. In this case, it is obvious that without resolving the type of classification with city council officials it will be unclear to the IT staff on how to implement a workable eGovernment system to satisfy the need.

(iii) *Risk of lack of adequate methods for requirements identification*: Requirements identification, analysis and specification for a business or an organisation form an essential part that leads to the successful implementation of most ICT projects ([6]; [7]). There are various methods to achieve this goal of identification and analysis, which are based on the premise of Systems Analysis and Design (SAD) or Knowledge Elicitation (KE). In most ICT projects, this stage of requirements identification, analysis and specification are either ignored, shabbily carried out or inappropriate methods are employed. The situation often leads to project failures in ICT. In the case of the UK establishment investigated, it was found that no clear formal method of requirements identification and analysis or specification and classification was used before the next stage of technical implementation.

(iv) *Risk of lack of proper design and maintainability:* A design with risk elements employed in an ICT project can also lead to inappropriate implementation not minding the sophistication level of the hardware and software used. This situation often creates more risks, as the resulting system may not deliver the performance level expected from it. Eventually, maintainability of the system may become another added problem as the original 'product' fell short of its quality in performance. This represents one aspect of the main problem experienced by the public establishment(s) investigated in the UK with regard to eGovernment projects.

(v) *Risk of lack of resources or expertise by the public sector:* In most cases the public sector does not have all the resources or expertise to accomplish a full scheme of eGovernment and therefore collaboration with the private sector is of paramount significance. As such some parts of eGovernment projects may be outsourced to private establishments ([8], p.5). Experience from one public establishment (a city council) in the UK shows that such partnerships are not always healthy and can be susceptible to problems. In this case, there was a lack of mutual trust and understanding by both parties (public and private), as the private organisation did not honour all of its promises and the city council had to find a new business partner.

(vi) *Risk of integration with legacy systems:* A holistic approach towards eGovernment implies that everything relating to governmental services has to be integrated under one single eGovernment umbrella. The implication being that all governmental and non-governmental data, information, systems, services and other necessary items have to exist and interact in a common platform of communication. In the case of this UK establishment the IT staff found it difficult to achieve this goal when they attempt to integrate new IT systems with legacy systems. The main drawback was that in some cases, the whole system infrastructure (i.e. legacy system) in a locality (or a public establishment) may be so outdated that it may be quite difficult and frustrating or almost impossible to fit it into a modern technology platform of the desired eGovernment project.

(vii) *Lack of adequate business/project management skills:* The research divulges that lack of adequate leadership is a major eGovernment risk factor ([8], p.4; [9]), especially if such leadership is with regard to business or project management. Fundamentally, eGovernment is based on the use of information and communications technology as such technical experts of ICT are often viewed as the ones to lead and make the project a reality. On the other hand, eGovernment may be viewed as a grand business project that requires adequate business/project management skills. When an IT expert with little or no business management experience is placed in a leading managerial position of a huge eGovernment project, there may be an increased possibility of the risk of failure. Equally, someone with business and project management skills alone with no background in ICT may not be the appropriate person to head or lead an ICT (or an eGovernment) project. A knowledge of both sides (business/project management and ICT) by one person or group is likely to provide better leadership skills for successful eGovernment projects.

(viii) *Risk of large ICT projects:* Most eGovernment projects are huge information systems projects. The research reveals that the larger the project of IT, the more the likelihood of failure due to size and complexity. It may be advisable for eGovernment practitioners to opt for small projects ([8], p.2) or sub-divide a large project into manageable parts. This implies that in a large eGovernment project major activities of the project are to be identified clearly and appropriately grouped, in order to minimise the risk of failure. This situation enables a proper allocation of suitable staff to each of the major activities or group of activities as well as the use of inappropriate technology, which may lead to major implications or risks ([6]; [10]).

(ix) *Lack of formal risk management strategy:* Risk management involves a procedure for monitoring and controlling risks. It serves as an essential management tool (or strategy) for supporting project managers. This UK public establishment (city council) did not have a formal risk management strategy in place. The IT staff had to confront every newly discovered project threat on the basis of makeshift arrangements. The research discloses that the absence of adequate risk management strategies increases the chances of overall project failure.

(x) *Risk of lack of adequate security measures*: The issue of security in ICT projects is not only interesting, it is also vast and covers both: physical, technical and human aspects [11]. This section will only discuss briefly, the technical implications of security in ICT projects with regard to: confidentiality, integrity and availability ([12]; [13]). These three properties and their susceptibility form the core of technical security in ICT systems ([14]; [15]; [16]). Due to a combination of internal politics and expertise, security issues relating to the items discussed is an aspect of risk factors being experienced by the public establishment(s) investigated in the UK with regard to the issue of eGovernment project.

2.2 Social and Human Risk Factors

eGovernment projects are often huge and complex to implement. They usually affect the way people live and interact in a society. The belief is that there may be several social and human implications when implementing and using current available facilities of eGovernment or when developing the next generation of eGovernment. Major risk elements in this aspect are presented below.

(i) *Risk of lack of IT skills in the public sector:* The research findings indicate that social and human risk factors are capable of hindering progress in eGovernment projects, especially due to lacks of people with IT skills ([8], p.4; [9]). Also, experience from the UK establishment indicates that employees from other departments (not the IT department) do not understand the new systems. Therefore, they are either unwilling to cooperate with the IT staff or they simply cannot use the new systems. Similarly, people from the IT department of this city council complained of under-staffing (another major social risk factor), as they do not have enough human resource to carry out their tasks on time in the eGovernment project.

(ii) *Political risks*: Politicians are the people initially responsible for most projects of eGovernment implementation. The politicians decide whether or not a public establishment should proceed towards eGovernment, then they may provide the funding for the project. In most cases of eGovernment success or failure, the politicians may be in the position to receive the praise or blame ([8], p.2).

(iii) *Risk of lack of adequate collaboration arrangement*: Findings from the UK establishment have revealed a rather embarrassing issue. It may be reasonable to assume that within a national environment for accomplishing projects of eGovernment, the various local governments will and should collaborate and assist each other, in order to achieve a common goal of success. Unfortunately, this research reveals that this is not the case. Neighbouring local city councils do refuse to join forces due to political and cultural problems and beliefs. Certainly, this kind of behaviour from local city councils may damage any progress in eGovernment at a national level.

(iv) *Risk of digital divide and lack of adequate education*: Sadly for the officials or proponents of eGovernment, not all people have access to a computer or to the Internet [17]. Various reasons can be attributed to it, ranging from insufficient funds to lack of other resources but whatever the explanation may be, they form a modern social problem, for example, the so-called digital divide. As long as this digital divide exists and grows eGovernment projects may continue to be doomed to failure. Another factor of risk is that of the uneducated citizens. More specifically, many people may be able interested and willing to access eGovernment services, but they may not be sufficiently educated to be able to do so confidently unless they receive some level of training.

(v) *Risk of bureaucracy and fear*: Research findings from the UK establishment (city council) indicate that one major threat to eGovernment initiatives is the level of bureaucracy practised in the establishment. For instance, when staff from the IT department of the establishment approach employees from other departments for information to help them to develop the project further within the organisation, they often received unnecessary excuses as to why the required information cannot be made available. Consequently, it can be said that bureaucracy within an establishment - mostly due to fear of redundancy - can hinder eGovernment programmes.

(vi) *Risk of lack of citizens' understanding and privacy*: Another major social risk factor affecting eGovernment projects is the erroneous comprehension of the needs of customers (mainly the citizens) by the eGovernment developers ([8], p.5; [9]). There is no point in creating eGovernment programmes when the customers do not understand the delivered services or they are not delivered in the proper manner.

2.3 Financial Risk Factors

Financial risk factors in eGovernment are mainly related to lack of funding, especially lack of cross-agency project funding. Government funding models are not often set up to fund many of the eGovernment projects that are cross-agency or cross jurisdictional in nature [18]. The research in the UK establishments reveals further financial risks, but are out of the scope of this paper.

2.4 Legal Risk Factors

The legal aspects of eGovernment may be viewed as being still fluid. Such risks may include the issue that some transactions may not be processed electronically ([19], p.17), the enhancement of current laws ([20], p.23), the reluctance in reforming legislation ([21], p.213). Unfortunately, this discussion exceeds the scope of this article and may be presented in the future.

3 Potential Success Factors for eGovernment

The first two [(i) technological/implementation; (ii) social/human] generic risk factors may provide a clear explanation to the potential risks involved in the implementation of eGovernment projects and their use. Based on the discussion advanced, the more specific elements of risk can be turned around to create potential success factors, as shown in Table 1, for the improvement of eGovernment implementation and operation. The table can be used as a checklist (or guide) to support the successful implementation and operation of eGovernment projects.

Table 1. Checklist of Success Factors for eGovernment

Potential Success Factors for eGovernment	
Technological /Implementation	**Social / Human**
✓ Have clear targets for eGovernment projects	✓ Educate and train public sector employees
✓ Have standardised and classified services	✓ Invest in human resources for better ICT
✓ Improve relationship between public and private sector	✓ Assess any political cost/implication
✓ Address problem of integration	✓ Avoid any cultural and collaboration problems
✓ IT project management & Business management experts should lead an eGovernment project	✓ Address the issue of digital divide
✓ Sub-divide a large project	✓ Reduce bureaucracy and eliminate fear
✓ Employ standard method of requirements identification and systems analysis	✓ Understand the needs of the customer (citizen, private sector, other governments)
✓ Have a strategy for risk management	✓ Educate and train customers
✓ Consider all aspects of security for the system	✓ Employ standard methods of social intervention
✓ Educate and train staff	✓ Identify required services of the establishment
✓ Develop a plan for maintainability	

4 Conclusion

It should be recognised that eGovernment is not entirely a technology phenomenon. It is about re-inventing and re-organising the way service providers (public/private) and citizens (or customers) interact in the society. Consequently, an attempt has been made to give an insight into an ongoing risk assessment work carried out in some public establishments in the UK, regarding eGovernment implementation and management. From a holistic point of view a classification of the generic risk factors associated with eGovernment has been presented with a more elaborate discussion on the first two factors in Figure 2. The issue of security in eGovernment encompasses both physical, social and technical facets and can be enormous and quite exciting to discuss. A detailed presentation of the research findings on the components of security risks in eGovernment has been deliberately avoided in this paper, but an aspect of the associated technological security risk elements are discussed succinctly in section 2.1(x). Based on the results of the investigation carried out, Table 1 is presented as a checklist of potential success factors for eGovernment. It is hoped that the details in the table will be useful to practitioners, researchers and others interested in the domain of eGovernment implementation and management.

References

1. UK Online, http://www.ukonline.gov.uk. Crown copyright (2002)
2. Gouscos, D., Mentzas, G., Georgiadis, P., Planning and Implementing e-Government Service Delivery. Presented at the Workshop on e-Government in the context of the 8th Panhellenic Conference on Informatics, Nicosia, Cyprus, 8–10 November 2001
3. Government Of Canada, Government On-Line And Canadians. Canada (2002)
4. UK Prime Minister, Modernising Government. UK Cabinet Office (1999)
5. Watson, R., T., Mundy, B., A Strategic Perspective of Electronic Democracy. Communications of the ACM, vol. 44, no. 1, (2001) 27-30
6. Akomode, J., Moynihan, E., Employing Information Technology Systems to Minimise Risks in an Organisation. The 3rd International Conference of Business Information Systems BIS'99, 14-16, April 1999, University of Economics in Poznan, Department of Computer Science, Poland, (ed) Abramowicz, W., ISSN 1429-1851
7. Avison, D., Shah, H., The information Systems Development Life Cycle: a first course in information systems, McGraw-Hill (1997)
8. OECD, The Hidden Threat to E-Government. PUMA Policy Brief No.8 (2001) 5
9. West Sussex County Council, Implementing Electronic Government Statement – Risks. http://www.westsussex.gov.uk/e-government (2001)
10. Caldwell, F., Keller, B., Managing the Risk of US Voting Lawsuits. (ref. No. FT-12-8903), 11 January 2001
11. Akomode, J., Potential Risks in E-Business and Possible Measures for an Enterprise. The 10th Annual BIT Conference, 1st/2nd November 2000, (E-Futures), (ed) Hackney, R., (CD): ISBN 0 905304 33 0, No. 07
12. Mercuri, R., Voting Automation (Early and Often?). Communications of the ACM, vol. 43, no. 11 (2000) 176
13. Phillips, D., Von Spakovski, H., A., Gauging the Risks of Internet Elections. Communications of the ACM, vol.44, no.1 (2001) 73-85
14. Pfleeger, C. P., Security in Computing. Prentice-Hall (1997)
15. Briney, A, Got Security. http://www.infosecuritymag.com/articles/1999/julycover.shtml (1999)
16. Di Maio, P., Security Plans. http://www.mi2g.com/press/030400.htm (2000)
17. Stahl, B., C., Democracy, Responsibility, & Information Technology. Proceedings of the European Conference on e-Government, Trinity College Dublin (2001) 429-439
18. Keller, B., Baum, C., Identifying and Addressing Inhibitors to E-Government. (ref. No. COM-13-1123), Gartner Group, 8 March 2001
19. US Department of Labor, E-Government Strategic Plan. Office of the Chief Information Officer (2001)
20. Hagen, M., Kubicek, H., One-Stop-Government in Europe: Results from 11 National Surveys. Hagen, M., Kubicek, H., (eds.), Bremen, University of Bremen, ISBN: 3-88722-468-x (2000)
21. Klee-Kurse, G., One-Stop-Government in Finland. In One-Stop-Government in Europe: Results from 11 National Surveys, Hagen, M., Kubicek, H., (eds.), Bremen, University of Bremen, ISBN: 3-88722-468-x (2000)

Quo Vadis e-Government? – A Trap between Unsuitable Technologies and Deployment Strategies

Tamara Hoegler[1] and Thilo Schuster[2]

[1] Research Center for Information Technologies at the University of Karlsruhe (FZI), Haid-und-Neu-Strasse 10-14, 76131 Karlsruhe, Germany
hoegler@fzi.de
[2] Cit GmbH, Kirchheimer Strasse 205, 73265 Dettingen/Teck, Germany
thilo.schuster@cit.de

Abstract. In Germany, eGovernment stagnates more than it progresses towards the electronic era of administrations. Lacking deployment strategies and acceptance problems concerning Smart Cards hinder its progress, resulting in a great gap between the targets and the actual state of eGovernment. This article describes a method for the stepwise introduction of electronic signatures placed on Smart Cards in public administrations and gives an approach for the deployment of eGovernment.

1 Introduction

According to the Resolution of the German Federal Cabinet[1], all transactions taking place between citizens and the Federal Administration should be available on-line up to the year 2005. Therefore, approximately 3.700 laws have to be adapted for the Internet[2], and public administrations have to hurry up with reducing the discrepancy between their targets and the actual state of eGovernment that seems to be more a standstill than a progress towards the electronic era of administrations.

Several problems accompany the deployment of eGovernment: On the one hand, all transactions between citizens and administrations need a legally valid signature that depends on rigid legal basic conditions. On the other hand, most of the administrative authorities do not have any strategy for the deployment of eGovernment solutions yet.

2 Digital Signature and Smart Cards – The Right Technological Pre-requisites for e-Government?

According to the German law, an electronic signature is valid legally if it meets the criteria described in the German signature decree[3]. Therefore, the electronic certificate is placed on a Smart Card and the user has to obtain the Smart Card from a trust cen-

[1] Date: 14th November 2001.
[2] Pricewaterhouse Coopers, PwC Deutsche Revision: Die Zukunft heißt E-Government, August 2000, p.18.
[3] The so called *Signaturverordnung*.

R. Traunmüller and K. Lenk (Eds.): EGOV 2002, LNCS 2456, pp. 403–406, 2002.

ter in a highly secure, but lengthy ordering and identification process. Because of disadvantages like the complicated and fussy application of Smart Cards, the need for special equipment[4] (mostly not included in today's hardware and software configurations) and an installation that can result in all kinds of technical compatibility issues, Smart Cards are bound up with acceptance problems within the population as well as administration[5].

A stepwise approach for the deployment of the digital signature is a solution for this dilemma, because experience in deploying such a system can be gained easily, employees of the administrative authorities can become accustomed to it slowly and the risks involved in using Smart Cards can be minimized.

Step 1: The Importance of Forms

The first step towards a fully integrated eGovernment system is to provide a downloadable and printable version of forms, that are established in the administration and that guarantee a collection of data within the law. After the download, the user fills out a form (e.g. a PDF document), signs it and sends it to the administration by mail. After the arrival of the mail, an administration workflow begins.

A different representation of the data (as it could be the case when using a printed form, e.g. a PDF document) could result in the complete refusal of the request. Therefore, the employees have to be instructed to accept these self-produced or self-printed forms as well as the standard paper forms.

Step 2: Using an Intelligent Screen Dialog

The next step towards eGovernment is an electronic form that can be equipped with fields (that can be filled out on a client) and where the data can be validated by using an expression language. This process gives rise to the need of a well-known and stable client environment that cannot always be guaranteed in the Internet application area. An interesting approach is an intelligent screen dialog, which can be flexible due to an advanced data validation mechanism. It is based on the user's input and offers online help facilities. In order to be independent from browser specific features, server based dialog technologies (like e.g. ASP or JSP) are suitable.

The user signs the (PDF) form electronically and sends it via email to the person in charge. The signature is verified and the files are stored safely for archive purposes. Afterwards, the (PDF) form can be transferred manually to the back-end processing system. Therefore, this process allows providing a large number of eGovernment applications without waiting for a close integration with the back-end systems.

Step 3: The Electronic Inbox

The next step is to establish an electronic path from the public Internet dialog to the administrative employee. The user can sign the form electronically and send it to the person in charge. The data can be transferred either using e-mail or a browser-enabled

---

[4] Including a Smart Card reader, the corresponding device driver and certain application software that is necessary to be able to use an electronic signature.

[5] The focus of our deployment strategy lies on overcoming the acceptance problems of Smart Cards *in administrations*.

database application. This step allows providing a large number of eGovernment applications without waiting for a close integration with the back-end systems.

Therefore, the *electronic inbox* is a final step towards a fully integrated solution – an electronic workflow without a breach of the medias involved. Further steps are the creation of a signed response to the Internet user or extending the electronic workflow in order to include non-administrative organizations.

3 Strategies Concerning the Procedural Model for the Deployment of eGovernment

Aside from the digital signature, the deployment strategy for eGovernment solutions is a serious problem. A survey, carried out between May and November 2001 at the FZI[6], proves this fact. The following paragraphs describe the survey and the approach for an optimized eGovernment deployment strategy.

3.1 Basics and Methodology of the Survey

The survey's target was to get an overview of the current state of eGovernment in Germany. It was carried out by email in the form of a questionnaire and included 424 cities and municipalities[7]. A return ratio of 16.2% was achieved, cities with 10.000 to 50.000 inhabitants sent almost 80% of the returns.

3.2 Results

The survey points out, that apart from financial aspects a practice-oriented deployment strategy becomes a substantial factor of success for eGovernment solutions. A deployment strategy can reduce or even avoid financial risks and personnel bottlenecks, which were mentioned as main causes for the occurring problems.

Fact is, that most of the municipalities (79,7%) have neither a general eGovernment strategy nor concrete strategies concerning the deployment of online services (63,8%) or Municipal Information Systems (46,4%). Only 10% fall back on standardized methodologies of external providers like computer centers.

Containing the latest findings concerning critical and success factors for the deployment of eGovernment[8], this survey represents the basis of the procedural model, which is adapted to the needs of German administrations.

3.3 A Procedural Model for the Deployment of eGovernment

For a successful deployment of eGovernment a holistic strategy is needed, covering all aspects mentioned as essential during the survey. None of the analyzed models

---

[6] Reiter, Markus: Entwicklung eines Vorgehensmodells zur Einführung von E-Government in der Öffentlichen Verwaltung. Master Thesis at the Institute of Applied Informatics and Formal Description Methods, 2001.

[7] All located in the German states Baden-Wuerttemberg and Bavaria

[8] e.g. planning of applications, identified obstacles and priorities

(the Requirements Pyramid of Masser[9], the Classification Scheme of the BSI[10] and the Procedural Model of WIBERA[11]) covers all these aspects.

Covering all these aspects, our procedural model represents a holistic approach for the deployment of eGovernment by giving e.g. concrete contributions to the settling of targets during eGovernment projects as well as detailed guidelines for the definition of the strategy or the selection of suitable online services. Furthermore, the model includes support for the realization of workshops, the training of employees, the intern and extern marketing, for the security and for the financing of the project.

4 Summary and Conclusion

In this article, clear approaches regarding the deployment of eGovernment were given and a procedural model was presented, which is adapted to the needs of the German administrations. Practical requirements, established in 424 German cities and municipalities, form the basis of the procedural model.

The next few years and new technologies like the mobile signature will show if public administrations will overcome its lethargy or if the state of the art of eGovernment will stagnate at the status of its infancy.

Since it is independent of the Smart Cards in its first stages, our systematic approach to a fully integrated eGovernment system reduces project risks like mis-investments and acceptance problems because the employees are lead slowly to a new working style. The acceptance of such systems in public is reasonably higher since – in their initial stages – they can be used without having electronic certificates or a Smart Card equipment. This approach also reduces the time-to-market, which can be an important factor in satisfying the public expectations.

[9] Masser, K.: Kommunen im Internet. Neuwied: Luchterhand Verlag, 2000. Page 79.

[10] Bundesamt für Sicherheit und Informationstechnik, translated: Federal Agency for Security and Information Technology

[11] WIBERA is an agency for economic advice and auditing, situated in Germany.

A New Approach to the Phenomenon of e-Government: Analysis of the Public Discourse on e-Government in Switzerland*

Anne Yammine

fög – Forschungsbereich Öffentlichkeit und Gesellschaft, Universität Zürich, Wiesenstrasse 9, 8008 Zürich, Switzerland
Anne.Yammine@access.unizh.ch

Abstract. EGovernment is commonly approached with a technical emphasis. In contrast to this perspective, we will take eGovernment into consideration as a phenomenon of communication. Our perspective is based on the assumption that the introduction of eGovernment has to be concomitant with a process of discussion in order to have lasting effects on a society. In contrast to the internet, eGovernment is approached more pragmatically through the public discourse. This line of argumentation will be highlighted by analysing the Swiss media discourse on eGovernment.

1 A New Perspective

EGovernment is commonly defined as "the use of information technology, in particular the internet, to deliver public services in a much more convenient, customer-oriented, cost-effective, and altogether different and better way[1]". In opposition to numerous technical and applied studies which are based on this definition, we would like to adopt a new perspective through which eGovernment is considered as a phenomenon of communication.

2 Assumptions and Basic Hypothesis

We assume that the introduction of a new technology is always concomitant with a process of reflection and discussion involving society as a whole and therefore postulate that eGovernment cannot be a factor of change until it causes a communicational process generating an overall positive attitude towards its implementation.

---

* This research is part of a doctoral thesis and of a research project conducted for the Swiss Federal Government, in which we will analyse the public discourses on eGovernment in Switzerland and on an international level.

[1] Holmes, D.: e.gov. e-business. Strategies for Government. Nicholas Brealley Publishing, London (2001) 2

R. Traunmüller and K. Lenk (Eds.): EGOV 2002, LNCS 2456, pp. 407–410, 2002.

The introduction of the internet generated great expectations which were especially linked to the so-called dot-com economy flourishing within the internet hype. However, this hype collapsed in the fall of 2000, when the NASDAQ crashed and the dot-com sector slithered from a state of euphoria to one of despair[2]. Characterised by the context of the dot-com crash, the style of the eGovernment-discourse is more technique-orientated and overall precautious in its judgments and expectations.

Through combining these two assumptions, we can formulate a basic hypothesis which puts forward that, in comparison to the internet, eGovernment does not go through an initial phase of hype and euphoria and has not managed, at this point of the process, to create an overall positive reflective process within the society it affects.

3 Methodology and Research Question

Our analysis of the media discourse on eGovernment is based on the so-called methodology of issue-monitoring which allows us to capture a discussion systematically within a defined arena, to generate respective issues and to analyse these in regard to their relevance and their dynamics of communication[3]. The discussion on eGovernment represents a specific discourse, concentrated within one big issue which is part of the internet-discourse. In regard to the structures of attention of the media system, we are interested to know, under which circumstances, in which forms and with which contents the issue of eGovernment is generating attention in the Swiss media arena[4].

In order to grasp the resonance of the eGovernment issue, we have chosen a sample ranging from the most representative Swiss to the foreign newspapers[5].

---

[2] From that point onwards, the nature of the internet discourse changed into a more problem-focused and technically-orientated discussion, see: Imhof, K., Kamber, E.: Das Internet als Phänomen der massenmedialen Kommunikation. Vortrag anlässlich der Jahrestagung der Deutschen Gesellschaft für Publizistik- und Kommunikationswissenschaft (DGPuK) und der Österreichischen Gesellschaft für Publizistikwissenschaft (ÖGK). Wien (1st June 2000).

[3] see: Imhof, K., Eisenegger, M.: Issue Monitoring: Die Basis des Issues Managements. Zur Methode der Früherkennung organisationsrelevanter Umweltentwicklungen. In: Röttger, U. (ed.): Issues Management. Theoretische Konzepte und praktische Umsetzung. Eine Bestandesaufnahme. Westdeutscher Verlag, Opladen Wiesbaden (2001) 257-278.

[4] Our plan is to apply our analysis as well to the international media arena. For this publication however, we have chosen to focus on the Swiss media arena for reasons of publication terms.

[5] *Switzerland*: Neue Zürcher Zeitung, Tages-Anzeiger, Der Bund, Basler Zeitung, Neue Luzerner Zeitung, Berner Zeitung, Le Temps, Blick, Sonntagsblick, Sonntagszeitung, Facts, Cash, HandelsZeitung, Weltwoche, Mitteland-Zeitung, l'Hebdo, dimanche.ch; *France*: Le Monde, Le Figaro, Les Echos, L'illustré, La Tribune; *Austria*: Das Wirtschaftsblatt; *Germany*: Die Tageszeitung, Frankfurter Allgemeine Zeitung, Der Spiegel; *United Kingdom*: Financial Times, The Daily/ Sunday Telegraph, The Guardian, The Economist, The Times and Sunday Times; *USA*: The New York Times, The Los Angeles Times, The Wallstreet Journal, The Washington Post, Newsweek; *Canada*: The Standard, National Post, The Toronto Star, The Vancouver Sun, Calgary Herald, The Gazette.

4 Results of Analysis[6]

In 1999, the discourse on eGovernment in Switzerland generated significantly less resonance than it did in other national media arenas[7], such as in the USA, Canada, the United-Kingdom or Austria where the discussion on the internet spread more rapidly to the administrational and political system. Since the summer of 2000, Switzerland caught up with the international discourse dynamic however with less articles per newspaper. This tendency is overall maintained until the first quarter of the year 2002.

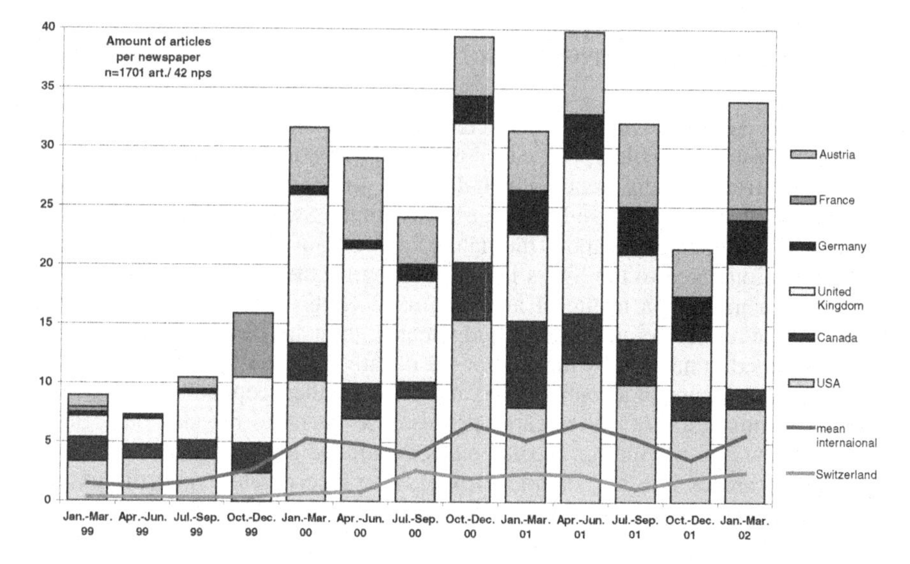

Fig. 1. Analysis of eGovernment-discourse dynamics 1999-2002. The figure shows the amount of articles per newspaper we found on eGovernment in Switzerland and within a few selected foreign media arenas. The mean value international represents all the foreign arenas summarised into one international arena which can be compared to the Swiss media arena.

For our categorical content analysis of the Swiss discourse, we have chosen to analyse all the articles focussing on eGovernment for the period of 2000 till the end of March 2002 which equates to a total of 276 relevant articles[8].

[6] In the presentation of our research results, we will focus on the Swiss discourse on eGovernment which can be situated into an international context through illustrating the relevance of articles on eGovernment issues in an international comparison.

[7] Previously to 1999, the internet discourse in Switzerland is related to the administrational system, however eGovernment is never mentioned as such.

[8] Our content analysis focuses on four categories which are applied to all the articles: per article we distinguish a maximum of six positions for each of which we determine the following categories: 1) *temporal perspective*: present-, past- or future-oriented; 2) *judgement/ evaluation*: positive, negative, neutral, ambivalent; 3) *focused problematic aspects* such as the social consequences of eGovernment and its technical and legal challenges; 4) *concrete applications* such as eVoting, eTaxes, eCensus and the "guichet virtuel".

From the point of view of the *temporal perspective*, the Swiss discourse on eGovernment focuses on the present process in 80% of all cases. In 50% of all cases, eGovernment is positively *judged*, but never euphorically. The overall discussion is lead in a controversial manner: most of the positive *judgements* are counterbalanced by negative or ambivalent *judgements* on the same aspect. Two third of the discourse is *focused* on the *problematic aspects* of the implementation of eGovernment, on its technical and legal challenges and on its effects on the administrational and political process. Social consequences are only evoked in 5% of all cases.

5 Conclusion and Prospect for the Continuation of the Research[9]

In regard to our first assumption concerning the implementation process of a new technology within a society, we can conclude provisionally that eGovernment has caused a controversial discussion which does not yet generate enough general positive expectations in order to have a lasting influence on the Swiss society as a whole.

Our second assumption about the nature of the eGovernment-discourse has been confirmed. Compared to the Swiss internet-discourse, the eGovernment-discussion is rooted much more in a technical and pragmatic context from its early beginnings. Though there are numerous positive judgements, they are never really euphoric and can be equalled in number by the sum of the negative and ambivalent positions which indicates the presence of a controversial and overall rather sceptical discourse.

In the continuation of our research process, we intend to broaden our categorical content analysis to the foreign media arenas mentioned in figure one and to integrate the political arenas into our sample as well as to incorporate expert discussions on eGovernment taking place within specific off- and online publications and circles.

References[10]

1. Donges, P., Jarren, O.: Politische Öffentlichkeit durch Netzkommunikation. In: Kamps, K. (ed.): Elektronische Demokratie? Perspektiven politischer Partizipation. Westdeutscher Verlag, Opladen Wiesbaden (1999) 85-108
2. Gisler, M., Spahni, D. (eds.): eGovernment. Eine Standortbestimmung (2nd edition). Verlag Paul Haupt, Bern Stuttgart Wien (2001)
3. Imhof, K.: Digitale Agora? Das Internet und die Demokratie. In: Standortbestimmung Internet. Bern (27. November 1997) 121-130
4. Jarren, O.: Internet – neue Chancen für die politische Kommunikation?. In: Politik und Zeitgeschichte, Vol. 40. Beilage zur Wochenzeitung "Das Parlament" (25. Sept. 1998) 13-21
5. Siedschlag, A., Bilgeri, A., Lamatsch, D. (eds.): Kursbuch Internet und Politik. Vol. 1: Elektronische Demokratie und virtuelles Regieren. Leske + Budrich, Opladen (2001)

---

[9] At this point of our research, we are not able to validate or reject our basic hypothesis definitively, because it would imply a comparison of the internet and the eGovernment discourse which we are unable to provide in this context of publication. However, we can give some indicators which support our two initial assumptions.

[10] This list of references only contains titles which are not already mentioned in the footnotes.

Self-regulation in e-Government: A Step More

Fernando Galindo

University of Zaragoza, 50009 Zaragoza
cfa@posta.unizar.es

Abstract. The paper presents an initial guide for the construction of codes of practice for e- government

1 Introduction

If it is necessary to create public-private Associations and frame regulations/codes of practice for electronic commerce[1], how much more so in the field of electronic government where services and functions are of vital importance and different to applications in electronic commerce[2]. Those functions may be briefly summarised in the following terms.

The field of electronic government[3] comprises a triple mission:

- To ensure open government and transparency in the activities of government agencies by designing web sites and portals to provide information and involve the general public in the doings of the administration.
- To provide on-line services enabling citizens to use the Internet to pay taxes, access registries, make applications or undertake procedures, elect their representatives, express their opinions, participate in administrative decision-making processes, and so on. All such online services and activities must of course guarantee the authenticity, integrity and confidentiality of the communications channels established by the State in a manner appropriate to official procedures in a democracy.
- To ensure the interconnection of government agencies. This means sharing workflows and infrastructure between agencies, regardless of the country where each is located, using modern information and communications technologies.

---

[1] See: GALINDO, F., "Public Key Certification Providers and E-government Assurance Agencies. An Appraisal of Trust on the Internet", in TJOA, A.M., WAGNER. R.R., (eds.), *12th International Workshop on Database and Expert Systems Applications*, Los Alamitos, IEEE, 2001, pp. 348-349

[2] On the particulatities of the applications in e government and public administrations see: LENK, K., TRAUNMÜLLER, R., WIMMER, M., "The Significance of Law and Knowledge for Electronic Government", in GRONLUND, A., *Electronic Government: Design, Applications & Management*, Hershey, Idea Group Publishing, 2002, pp. 61-77

[3] *Public Strategies for the Information Society in the Member States of the European Union*, ESIS Report, September 2000. Available at http://www.irc-irene.org/documents/do-psismseu.html. For a general discussion on electronic government, see: GRONLUND, A., *Electronic Government: Design, Applications & Management*, Hershey, Idea Group Publishing, 2002

R. Traunmüller and K. Lenk (Eds.): EGOV 2002, LNCS 2456, pp. 411–418, 2002.

On this view of the possible content of electronic government, there is little more that need be said in support of immediate regulation, even though actual implementation of e-government systems is still in its infancy. Having said this, however, we should note that the infrastructure required cannot be confined to the general proposal for e-commerce, but must be customised to fit the special conditions existing in the sphere of electronic government due to the services and functions involved.

In light of the above, in this paper we shall concentrate on the following issues:

- the presentation of a code of practice and an organization for electronic commerce as reference (2).
- an initiative launched to seek solutions for the practical problems arising from the spread of electronic government and establish appropriate regulatory frameworks (3).
- an initial proposal for the possible use of service charters to serve as a basis for the adaptation of electronic government applications. The contents and functions of these instruments are similar in the field of the public administration to the codes of practice used in electronic commerce (4).
- a short conclusion (5), finally.

2 The Code of Practice

2.1 Introduction

The internet's early history of develop in academic, military and industrial research carried out in the United States during the 1960s had a marked impact on the initial characteristics and common culture of the network. ICANN itself is a direct product of this tradition[4]. Now that internet use has spread to citizens all over the world, however, it has become necessary to progress further with the construction of a culture that takes the use of the network by a diverse community, rather than narrow academic, business or military interests into account. The whole of society now has a stake in the operation and governance of the internet. Naturally, domain names, one of the main fields of internet regulation, will remain an area of regulatory concern, but it has now become necessary to safeguard other rights such as data protection, ensure implementation of reasonable commercial practices, oversee internet content and establish dispute resolution mechanisms, to give but a few examples.

This is the context of the organisational and regulatory framework outlined in this section, which is representative of the will and opinion of citizens and social organisations. This organisation, which was established in April 2000 under the name APTICE (Asociación para la Promoción de las Tecnologías de la Información y el Comercio Electrónico – *Spanish Association for the Promotion of Information Technologies and E-Commerce*), has drawn up its own code of practice and implemented it through the creation of an independent institution, AGACE (Agencia para la Garantía del Comercio Electrónico – *Agency for the Guaranteeing of E-Commerce*).

---

[4] See on this history and regulation: GALINDO, F., "Autorregulación y Códigos de práctica en Internet", in CAYON GALIARDO, A., (ed.) *Internet y Derecho*, Zaragoza, Monografías de la Revista Aragonesa de Administración Pública, 2001, pp. 30-40

2.2 The APTICE Association: Membership

The Association for the Promotion of Information Technologies and E-Commerce, (APTICE: (www.aptice.org) was founded in Zaragoza, Spain, in April 2000.). The association currently has 83 members comprising private individuals, enterprise (telecommunications companies, banks, communications media, etc.) and public institutions, and is the fruit of a year-long period of debate and preparation by its founders (individuals, businesses and the Aragonese Development Institute, an independent government agency). APTICE reflects the conclusions reached from the joint R&D activities undertaken at the University of Zaragoza by companies and research teams, mainly associated with the Philosophy of Law Department.

2.3 The APTICE Code

APTICE has framed its code of practice/conduct for e-commerce in consultation with all of its members. The code was thus drafted in a spirit of consensus. The quality seal and guarantee infrastructure created in parallel with the code are charged with its implementation using the organisational machinery specifically designed for that purpose, which is embodied in the AGACE Agency.

The Code of Practice[5] is intended to provide a self-regulatory tool for the use of companies and public institutions in their relations with users, whether be citizens, other businesses or government agencies carrying out electronic transactions with subscribers. The code has been prepared on the basis of prevailing legislation in Spain and the European Union, taking into consideration the practices required by other similar codes worldwide, expert opinion on the issues involved and the experience of companies operating in the e-commerce industry.

To achieve its purpose, the APTICE Code of Practice contains seven general principles covering the key elements for building trust between the parties entering into on-line transactions over the internet and defining service quality and improvements needed in the activities and internal procedures of businesses and public institutions. These principles are as follows:

Principle #1: Identification of the Organisation.- In accordance with this principle, any organisation subscribing the APTICE code of practice must provide sufficient activities regarding its nature and activities in its web pages. The organisation must also comply with the domain name requirements established by the internet's central domain registries and with registration requirements established by legislation governing intellectual and industrial property. The future need for the use of advanced electronic signature systems and server authentication certificates is also provided for, as well as monitoring of legislation applicable to the establishment and its commercial activities.

Principle #2: Guarantees concerning claims and performance.- This principle requires that key commercial information (e.g. prices, delivery conditions, product descriptions, warranties, and many others) be displayed in the web site, together with instructions and procedures for carrying out on-line transactions, customer service details, and information concerning logistics, usability of web pages and contractual and extra-contractual liability.

---

[5] The full text is available in http://www.agace.org/en/index.html,

Principle #3: Security and technology infrastructure.- This establishes mandatory security policies for subscribing organisations.

Principle #4: Data protection.- This principle requires organisations subscribing the code of practice to comply fully with the Spanish Data Protection Act.

Principle #5: Content quality.- This centres on issues such as the organisation of illegal and offensive content, the protection of children and advertising practices.

Principle #6: Rules for out-of-court dispute resolution: The APTICE code of practice categorically requires subscribers to adhere to out-of-court dispute resolution systems. In principle, these would be bodies such as Consumer Arbitration Tribunals (business to citizen relations) and Chambers of Commerce (business to business relations). The intention is to ensure that any possible dispute that might arise between a company or public institution and customers (be they private individuals or other legal entities) is resolved as quickly and smoothly as possible, without the need for action in the courts, which are not sufficiently adapted to the exigencies of the new technology industries. APTICE will act as a mediator in disputes, which it will seek to resolve amicably or by referring cases to the most appropriate arbitration tribunal in the circumstances.

Principle #7: Requirements for the implementation of the APTICE Code of Practice.- This principle establishes the requirements that must be met by a company or institution intending to implement the code. These requirements refer, in particular, to the preparation of procedures manuals and records in accordance with the instructions provided by the institution or agency responsible for the performance of Code compliance audits. This principle will therefore allow auditors accredited by APTICE (to date only AGACE is an accredited auditor) to carry out appropriate procedures to examine the activities of an organisation and assess its compliance with the rules enshrined in the code of practice and, accordingly, its readiness to receive the award of the quality seal. Principle #7 also includes the sanctions mechanism established for cases of non-compliance by any subscribing organisation with its obligations upon accepting the code of practice as a guide for its own conduct.

The enumeration of these principles brings us to the need for a mechanism to implement the code of practice. This mechanism is now in place and comprises the quality seal and the organisational structure for its implementation. This infrastructure goes under the name of AGACE (*Agency for the Guaranteeing of E-Commerce*).

2.4 Guarantee Agencies

For some time now, it has been clear that the practicalities of the internet require the existence of specialised services, known as quality seals[6], to underpin the reliability of on-line transactions. Seals thus cover a different range of issues from guarantees referring to the identity of the parties entering into electronic transactions and the security of the messages exchanged between them. In view of the similarity of the goods

[6] It is common for standards agencies to award seals to companies manufacturing products or providing services subject to accepted quality standards. See: MOLES I PLAZA, R.J., *Derecho y calidad. El régimen jurídico de la normalización técnica*, Barcelona, Ariel, 2001, p. 29

affected, it seems appropriate in terms of the European legal tradition for such services to be able to guarantee both public and private operations.

Such services are already being implemented through initiatives such as TRUSTE[7], which concentrates on compliance with data protection regulations, BBB on line[8], which is concerned with on-line trading practices, and Web Trader[9], which designs codes of practice for on-line businesses.

The AGACE[10]e-commerce guarantee initiative bears a certain similarity to the above initiatives, but differs in that it is concerned with the various aspects of reliability and trust taken as a whole. Accordingly, the AGACE seal will be awarded only to those e-commerce activities that demonstrate compliance with the requirements of the APTICE code of practice, as described above.

AGACE has only recently commenced its activity, having carried out pilot consultancy work in the fields of e-commerce and e-government.

3 Organisational Initiatives

This point will present the main features of two different initiatives, all of which seek to safeguard the principles and values of public service in the design and implementation of electronic government applications. These are FESTE and AEQUITAS.

The first of these initiatives, FESTE (www.feste.org) was launched in November 1997 by the Spanish General Council of Notaries, the General Council of Legal Practitioners and the University of Zaragoza. The acronym FESTE is drawn from the name *Fundación para el Estudio de la Seguridad de las Telecomunicaciones* (Foundation for Secure Telecommunications Research). In accordance with its institutional object, FESTE provides security services for electronic communications under the provisions of prevailing legislation governing signatures and in light of the practices established in democratic legal systems for the activities of notaries public and commissioners of oaths[11].

The second initiative is AEQUITAS. This project is the result of a proposal made by the Spanish and French registrars, and the Spanish and Portuguese solicitors. The initiative has been awarded the status of a European Union project under the Information Society Technologies Programme (IST). The objective is to prepare the ground for the interconnection of the basic electronic government infrastructure (i.e. public key infrastructure and electronic communications certification services) that is already operational in participating organisations. The AEQUITAS initiative is also backed by APTICE.

Let us briefly describe the current status of AEQUITAS initiative.

The objective of the AEQUITAS project is to build the basic infrastructure for the mutual recognition of public key certificates in electronic government applications designed for the interconnection of government agencies.

---

[7] http://www.truste.com/

[8] http://www.bbbonline.com

[9] http://whichwebtrader.which.net/webtrader/

[10] http://www.agace.org/en/index.html

[11] See http://www.feste.org/.

The project's full name is *Trust Frame for Electronic Documents Exchange Between European Judicial Operators,* AEQUITAS project[12]. It is partially funded by a European Commission grant awarded within the V Framework Programme in the IST area, which is concerned with the information society[13]. The project is included in line II.4.2 (key action II), headed *Large-scale trust and confidence,* which is intended to extend, integrate, validate and pilot trust technologies and architectures in the context of advanced large-scale scenarios in business and everyday life.

From a legal point of view, the objective of the project is to enhance the quality of the services provided to society by judicial operators, and in particular by registrars, solicitors and their equivalents, by incorporating secure, confidential on-line communications into their activities.

AEQUITAS' specific objective is to ensure that e-mail messages exchanged between Spanish registrars and French *greffiers,* between Spanish and Portuguese solicitors, and between all such judicial operators are secure. The same infrastructure will subsequently be available for electronic communications with the citizens of the three countries involved. This security is sought not only in terms of the encryption programs and techniques employed in the exchange of messages, but also in terms of public trust and confidence in the institutions designated by the State to underpin the operation of the technology. This means laying the foundations for trust in law rather than relying on the *ad hoc* systems generated by the Internet itself following the ICANN philosophy practised by firms such as VeriSign, which we have already discussed in section 3.

In this context, one of AEQUITAS' key objectives is to create an Association formed by relevant stakeholders (e.g. professional organisations, companies, other associations, private individuals and legal entities). The future members of this Association hold that its mission should be to establish rules, certification practices and codes of practice in order to guarantee that the use of the Internet for electronic government purposes is in line with the practices and customs proper to a democratic society, in which the involvement of the citizen in public affairs and consensus-building in relation to issues with a profound impact on society are the ultimate gauge against which to measure the activity of government.

The Association will be created in the certainty that organisations of this kind are particularly well-suited for this and other tasks necessary for the proper functioning of the on-line world, such as involvement in the administration and running of high-level and territorial domains, the creation of voluntary accreditation systems for certification service providers, official recognition for firms specialising in the development and implementation of electronic government applications, clear technical standards and so on. In short, the Association's mission is to undertake tasks that are currently the province of multinational businesses.

---

[12] The use of the name AEQUITAS in connection with secure electronic communications in judicial matters goes back to 1997, when the first AEQUITAS project was promoted by the European Union as part of the INFOSEC programme. The contents of the 1997 project were, however, unrelated to the present undertaking. In fact the subject of the first AEQUITAS project was *The Admission as Evidence in Trials of Penal Character of Electronic Products Signed Digitally*. The content of the project is available in http://www.cordis.lu/infosec/src/study11.htm,.

[13] See http://www.tb-solutions.com/es/radff348.shtml. The project leader is the Spanish firm Tools Banking Solutions (www.tb-solutions.com).

4 Codes

The progressive regulation of electronic government by the institutions referred to and as a result of the initiatives described in the preceding section cannot, of course, take the place of general legislation in this area. As in the case of the APTICE Code, these are private rules or codes of conduct implemented by the same institutions as draw them up. They are intended to bind members and inform individuals or legal entities seeking the services of such institutions of the consequences of adhering.

The terminology used by these institutions to refer to the sets of rules they have created differs widely. FESTE, for example, uses the American-sounding expression "declaration of practices and certification policies" to refer to the basic regulations governing the certification service it has set up. This solution is similar to the terminology employed by other certification services. The AEQUITAS project, on the other hand, will resolve the issue of finding an appropriate title for its regulations in the future, in view of the action undertaken by the planned Association following its incorporation, which is currently in progress. The main concern at present is to draft the Association's statutes.

With a view to the direction that the initiatives described are likely to take in relation to the spread of electronic government applications, it may be of interest to explain here that an alternative to private regulations already exists to specify content in the field. This involves framing service charters, an instrument recognised in Spanish administrative law through Royal Decree 1259 of 16 July 1999 governing service charters and quality standards in government agencies.[14].

In accordance with article 3 of Royal Decree 1259/1999, "Service charters are defined as written documents comprising the instrument by which the agencies of the Spanish national government, Regional institutions and Social Security Management Entities and Services inform the public of the services entrusted to them and the quality standards established for the performance thereof, as well as the rights of citizens and users in relation to such services."

Pursuant to article 4 of Royal Decree 1259/1999, service charters must set out the following content:

1. General and legal content
   a) Identification and object of the agency or organisation providing the service
   b) Services provided
   c) Specific rights of citizens and users in relation to the services
   d) Collaboration or involvement of citizens and users in the improvement of services
   e) Up-to-date description of regulations governing each of the services and facilities provided.
   f) Availability and access to a complaints and suggestions book and, in particular, complaints procedures, response periods and effects of complaints
2. Quality standards
   a) Minimum quality standards, including at least the following information:
      1. Maximum periods for handling procedures and/or providing services
      2. Communication and reporting mechanisms for both general and personal information
      3. Public offices and opening hours

[14] For references to service charters see http://www.igsap.map.es/docs/cia/cartas/cartas.htm.

b) Information required to gain access to the service and under the best possible conditions.
c) Quality assurance, environmental protection and health and safety systems in place, where applicable
d) Quality assessment indicators

3. Other content
a) Postal addresses, telephone numbers and e-mail addresses of all offices where each service is provided, including a clear indication of routes and, where applicable, the nearest public transport
b) Postal address, telephone number and e-mail address of the unit responsible for framing the service charter
c) Other matters of interest concerning the services provided

Clearly, the general requirements for service charters contain basically the same items as those included in the APTICE Code, but with certain adaptations to fit them for their intended use as instruments by which government agencies undertake to perform public services in accordance with regulatory requirements. Specifically, charters must identify the government agency concerned, and refer to the quality of service content, the dispute resolution mechanism (complaints and claims procedures) and the procedure followed to frame and implement the charter in question (as provided in the remaining part of Royal Decree 1259/1999).

5 Conclusion

It would appear, then, that the contents and undertakings applied to the field of the public administration and electronic government are parallel and analogous to the APTICE Code of Practice described above.

The parallel solution to the APTICE e-commerce Code in the field of electronic government would therefore be to draw up a general or framework service charter setting out the basic requirements for any government agency or service entering into on-line transactions or relationships with citizens, other government agencies, companies and institutions.

This would not only be possible and a step forward in the self-regulation of electronic government, but would be in harmony with the democratic legal system as a whole.

UK Online: Forcing Citizen Involvement into a Technically-Oriented Framework?

Philip Leith and John Morison

School of Law, Queen's University
Belfast BT7 INN, N. Ireland, UK
{p.leith,j.morison}@qub.ac.uk

1 Introduction

UK Online is a centralised initiative which attempts to structure the nature of Government-citizen interaction, part of which is to expand notions of "citizen involvement" using technological approaches. The UK Online initiative lies within a general process of "modernisation" that is driven by the UK Government's White Paper *Modernising Government*[1]. We suggest that this project – along with other UK e-Government projects – which advertise a avowedly neutral strategy of developing ICT in government actually involves an attempt on the part of Government to structure and control a *new space* that is opened up.

The supposed "neutral" form of structured communication being planned by government is adversely affected by two forces: first, the desire of government to control the communication, but secondly, the technical challenges resulting from programming an interface to allow this communication - in particular the problems caused by "business model" communications when used by government.

Our critique of the e-Government project is not that it is not a useful – and perhaps efficient – way to move forward, but that it is inspired both by a mythology of the business model and the citizen as customer, rather than the citizen as citizen. We develop a conception of 'governmentality' to account for this re-ordering of relationships and urge a re-thinking of how UK on-line is developed.

2 Governmentality

The general movement from government to governance that is observable throughout the developed world stresses how today the state is involved more in "steering" than "rowing". The governmentality approach that is associated with the later work of Michel Foucault and other critics provides an important way of theorising about this general process.[2] This perspective helps to explain how government today is not so

---

[1] HMSO Cmd 4310 1999.

[2] See, for example, M. Foucault, *Power: The Essential Works 3* Harmondsworth: Penguin 2000) and "Governmentality" in *The Foucault Effect: Studies in Governmentality*, G. Burchell, C. Gordon and P. Miller (Hemel Hempstead, Harvester Wheatsheaf 1991) p.87-104 and *Technologies of the Self: A Seminar with Michel Foucault* edited by L. Martin, H. Gutman and P. Hutton (London, Tavistock, 1988). See also N. Rose *Powers of Freedom: Reframing Political Thought* (Cambridge, Cambridge University Press, 1999).

R. Traunmüller and K. Lenk (Eds.): EGOV 2002, LNCS 2456, pp. 419–423, 2002.

much as a set of institutions but as a domain of strategies and techniques through which different forces and groups attempt to render their particular programmes operable. The practice of government is seen as involving engagement with the many networks and alliances that make up a chain or network which translates power from one locale to another.

From within this theoretical framework, we study the UK government's "modernisation" programme and its emphasis on "information age government" where, in the words of the White Paper:

Government must modernise the business of government itself, achieving joined up working between different parts of government and providing new, efficient and convenient ways for citizens and businesses to communicate with government and receive services. [3]

The UK Online initiative seeks to put this into practice. We set this initiative into an continuum which represents the evolutionary process that changes the nature of communication between the state and the citizen and the very structure of the state bureaucracy itself. As Figure 1 suggests, the UK programme's ambitions are set at the highest level – stage 6 of the evolutionary move towards citizen/state interaction:

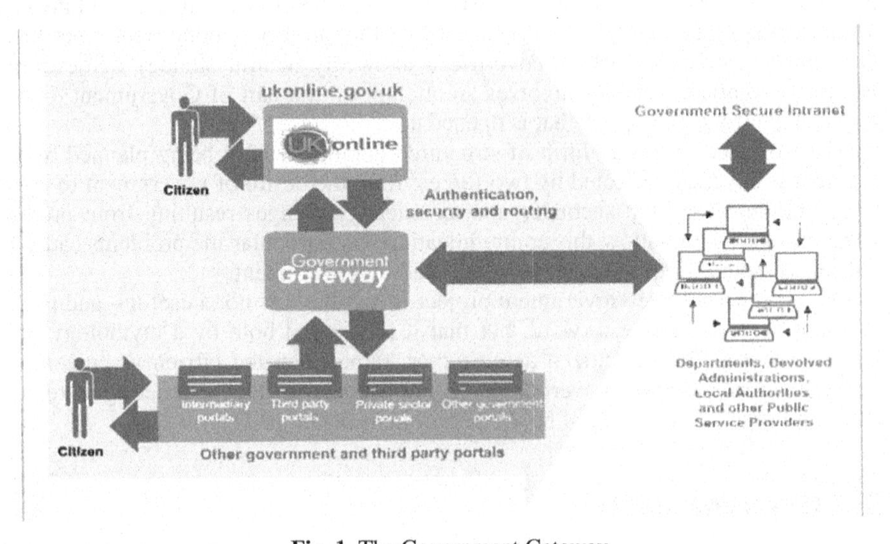

Fig. 1. The Government Gateway

We argue that the assumptions within this model must be challenged and its feasibility and desirability questioned. In particular, it is important to query the simple assumption that putting elements of government business on-line will necessarily enhance democracy by improving openness and participation. This, we assert, is subject to challenge.

Within UK Online the provision of services is actually quite limited: the promised revolution in government-citizen interaction is more equivalent to the transition from traditional shop to self-service supermarket rather than anything more fundamental. In relation to the promised transformation in consultation and democratic engagement

[3] Chapter 5, para. 5

which is intended to "*democratise the very institutions of democracy themselves*", the reality can be seen as amounting to little more than providing an electronic suggestion box where government structures the interaction and dominates the process. The governmentality framework allows us to see the UK Online revolution as an attempt to re-engineer Government-Citizen relations generally rather than as a merely neutral technical development as it is presented by its proponents in Government.

Indeed we take issue even with the idea of a neutral technology that is simply being harnessed by government in a general process of improvement. We argue instead that the whole idea of technicality (that is, simple use of business methodologies and techniques) is being used within a wider political process.

3 Technicality

The reader of any of the White Papers dealing with technology change within Government will find the same call to arms – that business methods demonstrate user-friendliness and efficiency and that the models of IT practice within business should be transferred over to government. The largest project to be so announced was that of the computerisation of the UK welfare department, the DSS, in the 1980s. We have argued [4] that the project was problematic for a number of reasons:

- lack of technical understanding of the 'software crisis' (by which we mean problems found in large scale implementations)
- usage of a business model which was inappropriate for the kinds of processing being undertaken
- an attempt to force processing into a computer-suitable algorithm.

and later analysis demonstrated that our analysis was correct[5]. The DSS computerisation project was over-budget, under-performed and the technical framework imposed by programming and other considerations has led to the system being difficult to use and requiring various 'work arounds'. Also, as with current proposals, the claimed advantage that staff could be moved to more interesting work has not proven to be true; indeed, staff are usually deskilled by these implementations, serving the failures of the system more than the public.

The current e-Government project is also being pushed by IT models from business. Thus, the NAO report on e-Government to date[6] picks up on the optimism of business:

"The Oracle Corporation saving of £71 million through ... web-enabled ... functions ... and British Gas productivity improvements ..."[7]

This is despite the clear evidence that Government usage of technology is problematical. Reading the computing trade press demonstrates that projects are difficult

---

[4] "Law in a Changing Technological Society", in Law, Society and Change, Livingstone S. and Morison J. (eds), Gower, Aldershot, 1990. R.Geary & P.Leith.

[5] "From Operational Strategy to Serving the Customer: Technology and ethics in welfare law.", in International Review of Law Computers and Technology, 2001. R. Geary & P.Leith.

[6] "Better Public Services through e-government", Report by the Comptroller and Auditor General, HC 704-1 April 2002.

for government to implement. Rarely are IT projects successful – more usually they are late, over-budget and under perform. And they impose problems on future developments and transformations of government simply due to the expense of re-programming and re-engineering such expenses systems.

We are interested in why government is so poor at programming this face between the citizen and the state. Obviously, the software crisis is not unknown in the commercial world, but it seems to be particularly problematic in government usage of technology. We suggest that many of these problems arise from the difficult relationship of government to citizen and the difficulty of programming systems which properly effect this relationship: that is, systems which are flexible and not overtly formalist [7]. The glory of the business model is that complex transactions can be reduced to a relatively simple one of buy/sell. The IT systems developed have one usually one goal – to reduce costs in an economic transaction. When we move to government, though, the transaction is much more complicated – in ways which are not yet clear, but which do not support the view that the citizen is a customer. The rhetoric of e-government, though, remains that the citizen is a customer.

As one small example, we can look at the limitations on government use of information which are not present in the commercial world. In a commercial business relationship, that business would seek to reduce costs of storage and processing of information. A common database would be set up where information is verified, stored and processed – all with one aim in mind. In government, the amounts of information, the diversity of that information, the lack of funds to ensure it is up-to-date and verified[8], and arguments about 'big brother' and the collation of these diverse information sources, mean that the single pseudo-commercial relationship which the government want to implement via UK-online will always be technically problematical and difficult to overcome.

4 Connecting Governmentality and Technicality

If Government was simply a commercial relationship with the citizen, where taxes are collected, information provided, forms downloaded, then the strategy which is being outlined in e-government in the UK would be appropriate. However, a government has wider concerns. At a time when the health of orthodox, representative democracy is at best uncertain and voter apathy and general disenchantment with the answers offered by "big government" challenge us to widen and deepen democratic engagement it is important that we think more seriously about how to use this new space between citizen and government to enhance democratic engagement. While the approach of UK Online *may* bring some efficiency gains as it distributes services it misses an opportunity to use new technologies to facilitate more direct forms of democracy and reinvigorate traditional representative democracy. While technology cannot return us to the direct democracy of classical Athens, it may be that it can

---

[7] Formalism is the achilles heel of computing. See Formalism in AI and Computer Science, Ellis Horwood, 1990. Philip Leith.

[8] A complaint being made by the Information Commissioner with respect to many government data banks – including the police national computer system.

begin to supply an answer to Anthony Gidden's question of 'How can we democratize democracy[9]?'

5 Conclusion

We conclude by arguing the case for a much more limited but more realistic version of UK online which rather than seeking to structure and control the whole nature of government-citizen interaction, instead facilitates the more democratic possibilities that e-government can allow. We are endorsing a normative ideal of democracy as political communication and suggesting a model of democratic process where citizens must engage with one another and with government in a public space where they make proposals and criticise one another in an effort to persuade all parties of the best solutions to collective problems. This is active democracy where citizens are involved as much as is practical in making these fundamental decisions about how to live together as well as with simply transacting for services. Here it is important that the processes should be fair, open and, most of all, inclusive. This involves facilitating models of government and citizenship beyond those of service supplier and consumer to engineer both in technical terms and in a democratic sense a better approach to online government.

[9] *The Third Way: The Renewal of Social Democracy* London: Polity 1998 at p.72. This question keys into a variety of concerns about the nature of modern democracy and its ability to deliver the legitimacy that is required to underwrite government in a changing world. To take just one critic from a wide range proffering versions of this complaint, Barber argues that liberal democracy is a '"thin" theory of democracy, one whose democratic values are prudential and thus provisional, optional, and conditional——means to exclusively individualistic an private ends. From this precarious foundation, no firm theory of citizenship, participation, public goods, or civic virtue can be expected to arise'. *Strong Democracy: Participatory Politics for a New Age* (1984), p.4.

Data Security: A Fundamental Right in the e-Society?

Ahti Saarenpää

Professor of Private Law, Institute for Law and Informatics,
Faculty of Law University of Lapland, Box 122 , 96101 Rovaniemi Finland

Abstract. The birth of the modern network society and the strengthening of the idea of the constitutional state in Europe have occurred largely at the same time. This juxtaposition, although more accident than design, obligates us to examine from the legal perspective the tension between the array of opportunities (e.g., convergence) and new risks which the network society brings and the legal effectiveness of the constitutional state. The prevailing attitude towards data security offers an illuminating example of the new encounter between technology and law. A look at legislation and legal practice in this area reveals a variety of approaches. I present these and go on to argue for a position whereby data security can and should be assessed in terms of fundamental rights. We have a right to data security in the information infrastructure. At the same time, it must be pointed out that, if we are to avoid the potential risks involved, data security must quite literally be security, whereas legal regulation strives for certainty. When we attempt to forestall risks, the degrees of security and certainty needed at any given time should, in the final analysis, be assessed from the standpoint of fundamental rights.

1 A Changing Society and a Changing State

Society is changing profoundly. We have largely completed the transition to mass use of IT and data networks, and most of what at one time was the province of experts is now routine. The use of IT and networks is part of our daily lives. We can justifiably speak of "the network society" or, as it is already being dubbed, "the e-society." It is a society that has emerged rapidly, in the space of less than ten years. Few had the foresight to anticipate the change before it was upon us. (1)

A second significant change we are witnessing is *the rebirth of the constitutional* state and a concomitant decline of the administrative state. We now speak of the principle of *the constitutional state* as one of the guiding European legal principles. And, true to this commitment, we are now taking more seriously human rights, fundamental rights and the guarantees that these rights will be realized. At least this should be the case.

The changes in society and the state seem to have gone hand in hand for some time. After all, the approval of the *European Personal Data Directive* came at a time when both the development of IT and improved protection of fundamental human rights were prominent issues. While a superficial look at these two developments might suggest that the legislator had realized the significance of fundamental rights in the network society at the time, a broader historical perspective reveals that we are dealing with accident rather than design. The lengthy drafting process that preceded

R. Traunmüller and K. Lenk (Eds.): EGOV 2002, LNCS 2456, pp. 424–429, 2002.

the enactment of the Directive was completed at a time when the use of open networks was already becoming extensive, so much so that even the experts were taken by surprise. We can hardly speak of these trends in IT and fundamental rights as an instance of joint or even coordinated development: chance has merely juxtaposed two different developments. Yet, these are changes that urge Law, in particular Legal Informatics, to assess afresh the relationship between law and technology.

Today we simply cannot any longer be content only to state that we live in an *e-society* and a *constitutional state*. We must probe deeper and identify the problems that impinge on - or might impinge on - the relationship between the network society and the constitutional state. We find ourselves faced with a complex cluster of legal issues: we must not only ponder the need for and potential of legislation that meets the requirements of the new network society but also analyze the extent to which the constitutional state needs to be renewed. For example, do we have legislative techniques that are sophisticated enough for the network society and does the network society require that we rethink what we mean by *the constitutional state*? (2)

In another context, I have gone so far as to assert that one of the distinctive legal features of our time is the scarcity of justice. For example, convergence, the commodification of knowledge, network communities, and globalization combine to make it harder than ever to achieve the effectiveness of rights through conventional legislative and procedural means. Promoting interests crucial to networks tends to restrict freedoms. Effectiveness is sought in the form of increased control and special procedural measures, which restrict even fundamental rights. The tension is plainly visible if only we want to see it. Moreover, many fundamental issues have been and continue to be overlooked, and we run an increased risk that the scarcity of justice will worsen. (3)

The assessment of *risks* has long traditions in different branches of Law, and we will do well to bear this in mind when considering the legal problems of the new information infrastructure. We also need legal risk analysis. We carried out such an analysis in the data security report written in 1997 at the *Institute for Law and Informatics* at the University of Lapland. The analysis identified 20 different risks, ranging from risks affecting currency and payment transactions to a "*drafting risk*", which occurs when the legislator fails to realize the legal problems of the network society in time, or fails to notice changes taking place in the overall situation when implementing individual legislative measures. The report also dealt with the notion of an information war front and center. (4) Many found the issue bewildering at the time; fewer would find it as mystifying today.

2 Establishing a Position on Data Security

Of the many pivotal issues that emerge in this area, I have opted to focus in my presentation on *legal data security* in the network society characterized by a new *information infrastructure*. From the legal point of view, both data security and infrastructure are old as well as new issues. Indeed, the various provisions specifying the form and confidentiality of documents have promoted data security for centuries. Similarly, significant infrastructures have either been provided for in law or been under the direct control of public authority. The wealth of telecommunications directives in the European Union provides compelling evidence of this tradition.

It is appropriate at this point to provide a brief definition of *legal data security*. It can be described as the totality of laws, regulations, guidelines and practices whose purpose is to guide and assess the implementation of data security and to evaluate the various threats to it. Data security is very much - very much indeed - a legal issue.

This general description is of course only a basis for a closer examination of the issue. In looking at data security from the legal point of view, I see at least six perspectives that can be distinguished both in theory and in practice: these are data security as a tool, a technological development, a market, a system, an element of justice, and a fundamental right.

The tool perspective is the most traditional of the six positions: data security is seen as merely an aid in realizing more important objectives and, accordingly, is not given much weight when it comes to legislation. What we see here is essentially the distinction drawn by *Lord Snow* between humanistic and technical cultures: that which is less well known is less interesting. The consequence in the present case has been that the issue of data security has been addressed through unsystematic regulation of a general nature and, accordingly, has ended up being governed by sector-specific practices. The development curve of many a data security firm reflects this trend.

The technological perspective is a more modern one. After getting a start, the legislator trusts in favorable technological development. We even speak of the *technological imperative*. It is easy to find examples of this type of development in the network society, as the legislator hastens to come up with solutions dealing with network activities. This seems to be the case in many countries, as different states vie to be the frontrunners in the e-society. The Finnish electronic identity card, with no market for a lack of services and established standards, serves as an isolated but striking example.

Yes, the market. The market approach is to show confidence in the market and their ability to ensure adequate data security. Issues to be regulated are left to the market, which is expected to follow development so dependably that the legislator has no need for more detailed specifications. An illustrative example of this view in Finland is the Act on Privacy and Data Security in Telecommunications, which allows users to equip their connections with any of the security systems available. Although the provision also means that so-called strong encoding is permitted, it says even more about the view on the data security market I have mentioned here.

The system perspective represents another modest step forward legally. Adherents of the position see data security as an essential element of data systems. Data security must be taken into consideration in the planning of such systems. From the point of view of legislative technique, what we have here is a technical norm of sorts. A good example of this in Finland is Section 18 of the Act on Openness in Government Activities, which *deals with good information management practice*. The section obligates all units in the public sector to plan their information systems to take into account not only content but also data security. The complexly worded provision has as yet been applied very little in practice. On many occasions, I have referred to it as the most important provision in the Finnish public sector, where it would function as a central guiding provision in information-related work.

The legal perspective looks upon data security as a right of an individual or community - a right that must be rendered effective. While Section 18 of the Act on Openness in Government Activities clearly includes some of these features, the perspective is even more apparent in the data security provision of the Act on the Protection of Personal Data. Personal data must be duly protected in order for our *right to privacy to be safeguarded.* This principle, also seen in the Personal Data Directive, is significant in practice because it steers us towards applying the principle of proportionality in the name of the protection of privacy more so than the more conventional efficiency of economic production.

The fundamental rights perspective takes us a step further towards the new ways of thinking characteristic of the new constitutional state. It is not sufficient that we assess data security strictly as a legal issue. We must ask whether data security is one of the fundamental rights. To the best of my knowledge, the answer to this question – at least in Finland – would be affirmative. Three main arguments can be presented:

First, we must remain mindful of the link between the protection of personal data and data security. The two go hand in hand. The protection of personal data as a fundamental right requires sophisticated data security, thus making data security more than just a conventional right.

Second, it is also easy to see that the fundamental right to having one's affairs processed without undue delay by authorities and the courts as well as the fundamental right to *good government* require appropriate data security when working with information and making use of data systems in different ways. Here, too, data security is more than just a conventional right.

Third, we must take into account that we are changing over to electronic government and in fact have done so already to some extent. With this change and the development of electronic commerce, we increasingly exercise our fundamental rights - for example, and, in particular, privacy - in network environments. This observation, if no other, indicates that in the new information infrastructure data security has become both a substantial legal principle and a metaright, i.e., a fundamental right which is an ideological and actual precondition for the realization of statutory fundamental rights. Our right to self-determination can only be realized if we create the requisite data security information infrastructure.

What I have said here views the issue from the perspective of the individual. This of course is the primary approach we should take to data protection. But we cannot forget communities, private and public communities. There, too, the hierarchy of rights and metarights is relevant. The functionality of the free economy in the network society is largely dependent on the level of data protection and data security (information and commodity security). There is simply no way around this fact.

3 Security and Certainty

At this point, it will be appropriate to pause and consider a special issue - the relationship between security and certainty. On the one hand, we speak of data or information security. The key concept here is security, which is fairly transparent. On the other hand, we talk about legal certainty, which is a somewhat less accessible concept.

Yet, legal certainty is one of the pivotal elements of the constitutional state. It is often associated with the concept *rule of law.* The very expression *legal certainty* embodies a strong purpose - being able to anticipate the outcome of activities and legal decisions in legal life. Legal certainty refers to this predictability, which we pursue through an advanced legal culture.

Here I will bring in the concept of risk to illustrate the relationship between data security and legal certainty. Data security and legal certainty are the means we use to forestall the principal risks to information and information processing in the network society. The relationship among the three can be depicted in the form of a triangle:

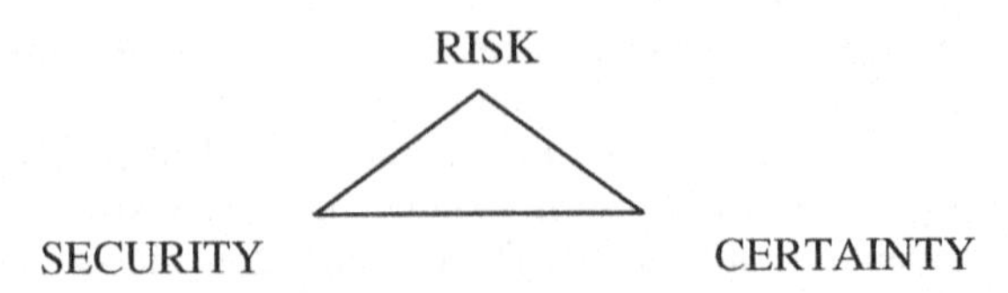

Fig. 1. The relationship among security, certainty and risk

The number of risks we face has not, to my knowledge, decreased. Quite the contrary. However, this is not the place to go into an extensive and detailed assessment of the different risks. Instead, I will focus on two risks that we covered in our report but which were not given the same scope. These are the risks affecting fundamental rights and the risk of excessive control.

The potential risk to *fundamental rights*, that is, the jeopardizing of our fundamental rights in making use of the information infrastructure, is a factor whose significance is growing as we exercise these rights to an increasing extent on networks, and open networks in particular. In *e-government* in the e-society, we are guided us towards exercising our fundamental rights on networks. We are increasingly network dependent. This is a situation which makes it essential that we assess the significance of the risk to our fundamental rights; this is not something we have spent much time doing. (5)

The risk of *excessive control* is crucially connected with the risk to fundamental rights. For example, the protection of personal data in Finland is nothing less than a fundamental right. The issue has to do not only with unforeseen risks that come to light when data security breaks down but also, and above all, with the increase in conscious control. Advances in IT and the resultant convergence enable control both on and of the network. Observations to this effect have been made in government in Finland. Efforts are made to increase control of the individual, accompanied by appeals to the need for heightened efficiency in bureaucracy and protection of the individual. The legal problems such developments entail are not always noticed or at least are not acknowledged. On balance, we have every reason to speak of a risk of excessive control.

Both the risk to our fundamental rights and the risk of excessive control illustrate that a tension - a growing tension - obtains between the developing network society and its *e-government,* on the one hand, and the constitutional state, on the other. If we are to rise to properly defend the rights of the individual, we must be prepared to reassess the constitutional state and its legal culture. (6)

4 Conclusion

For decades now, personal data protection has had the thankless task of being a buffer against many failings of knowledge, understanding and education that have accompanied the juridification of our increasingly technological society. We can still encounter those for whom data protection as a means for regulating the protection of personal data is either not a particularly important matter or is a downright curse word used to denounce what is seen as a hindrance to the smooth functioning of bureaucracy and the satisfaction of curiosity.

Data security became a pivotal tool in implementing the protection of personal data quite early on. Data protection and data security began to go hand in hand. For example, in Finland the 1987 Personal Data File Act had a provision on individual data security, albeit only one of the fifty or so provisions in the Act. The idea was to protect people through procedures that maintain and promote data security in the processing of data which someone has the right to process. The development of data protection legislation has in fact expressly linked data protection and data security. Appropriate data protection requires appropriate data security. Yet, this fact was not readily noticed as long as the new data security remained focused on technology.

Today, data protection and data security continue to go hand in hand, but their relationship has changed or at least is changing. Data protection still requires data security. But as a form of security data security has progressed to being a meta-level fundamental right. Realizing the significance of this metaright is becoming one of the cornerstones of the development and existence of the constitutional state in Europe. We have the right to expect that our fundamental rights will be realized in a network environment with the requisite data security. Similarly, in assessing the development of the constitutional state, we have every right to make the quality of the digital infrastructure the standard for the new legal culture. It is one of the basic pillars of certainty. And *certainty* is what we pursue in the constitutional state when undertaking to protect the individual.

References

1. Duff, Information Society Studies, 2000
2. Aarnio - Uusitupa (ed): Oikeusvaltio, 2002
3. Saarenpää: Law, Technology and Data Technology p 41 pp. Judicial Academy of Northern Finland Publications 4/1999
4. Saarenpää - Pöysti : (ed) Tietoturvallisuus ja laki, 1997
5. Saarenpää: Personal Data Protection and the Constitutional State p. 37 pp. In Judicial Academy of Northern Finland, Publications 3/2001.
6. Ferrajoli: Fundamental Rights p 1 pp in International Journal for the Semiotics of Law vol 14 (2001)

Legal Design and e-Government: Visualisations of Cost & Efficiency Accounting in the *wif!* e-Learning Environment of the Canton of Zurich (Switzerland)

Colette Brunschwig

University of Zurich, Department of Law, Freiestrasse 36,
8032 Zürich, Switzerland
colette.brunschwig@rwi.unizh.ch

Abstract. This paper applies Legal Design, a new field of inquiry, to discuss the form and contents of an E-Learning environment recently implemented by the Canton of Zurich (Switzerland) to enhance the training and development of public administration staff. It is argued that there is a need to visualise this environment more effectively. Working from basic notions of Legal Design and E-Government, the paper uses a set of clearly defined text visualisation rules and a multi-stage procedure adopted from visual communication to visualise one key module of the learning environment with a view to achieving a greater degree of iconicity and thus to meet established didactic and mnemotechnic criteria more successfully.

1 Legal Design and e-Government

Legal Design is still very much a new field. It deals with conceiving, creating and assessing visualisations of contents and materials that are either purely legal, or financial and economic ones, for example, that have a legal basis. Legal Design is applied in various contexts (research, teaching and practice) and shares a frame of reference with other disciplines, such as jurisprudence, economics, visual design and history of art.[1] In this paper, e-Government refers to those measures and actions public administration takes to employ modern information and communication technology, respectively the e-learning environments based on such new technologies, for the purposes of staff training and development.[2]

---

[1] On Legal Design, see Colette BRUNSCHWIG: Visualisierung von Rechtsnormen. Legal Design, Diss. Zürich 2001, Zürcher Studien zur Rechtsgeschichte, Vol. 45, ed. Marie Theres Fögen [et al.], Zürich 2001, p. 1ff.

[2] On the notion of e-learning, see, for example, Andrea BACK/Oliver BENDEL/Daniel STOLLER-SCHAI: E-Learning im Unternehmen. Grundlagen, Strategien, Methoden, Technologien, Zürich 2001, p. 28ff., and Michael KERRES: Multimediale und telemediale Lernumgebungen. Konzeption und Entwicklung, 2., completely rev. ed., München [et al.] 2001, p. 14.

R. Traunmüller and K. Lenk (Eds.): EGOV 2002, LNCS 2456, pp. 430–437, 2002.

2 Starting Point

Since March 2002, the administration of the Canton of Zurich has been providing its staff with an online course on its reform of public administration (the reform bears the name *wif!*, short for „wirkungsorientierte Verwaltungsführung", i.e. ‚effect-oriented public administration'; cf. http://e-learning.wif.zh.ch; visited on 21 March 2002).[3] The learning environment consists of two modules: the first explains the principles of effect-oriented administration, while the second considers the Canton's steering instruments in action. This paper focuses on one of the submodules of the second module: Submodule 2e „Cost & Efficiency Accounting" (Kosten- und Leistungsrechnung or KLR in German; referred to hereafter as CEA). As regards the steering instruments, the principal learning objectives are that learners know how the various instruments are deployed and that they understand the instruments' mode of operation in the various phases of the controlling process. In terms of teaching methodology, the steering instruments are explained by way of reference to the Yearbook of Statistics of the Canton of Zurich, which serves as a model.

How has the *wif!* e-learning interface been designed? First, the *wif!*-logo appears at the top of the screen on the left. Beneath it, there is a vertical image bar, featuring a separate image for each submodule. The title bar „Wirkungsorientierte Verwaltungsführung" (i.e. ‚effect-oriented public administration') and the crest of the Canton of Zurich have been positioned to the right of the vertical image bar. Beneath the title, there is a navigation bar that learners can use to help them decide whether they wish to access the learning modules, any background information or the *wif!*-glossary, or whether they intend to order printed materials, mail an online competition entry, request help or call up links to online services of the Canton of Zurich. The central space on the site's screen features the syllabus and learning materials in multimedia format. Finally, there are additional navigation functions at the foot of the screen, enabling learners to move around in the learning environment.

The *wif!* e-learning environment contains quite a large number of different visualisations, such as charts, tables, diagrammes, images and animations. By contrast, Submodule 2e-1, which deals with what CEA is and for what purpose it is required by a local government office, and Submodule 2e-3, which considers the differences between CEA and general (financial) accounting, only feature a single image each in the vertical image bar to the left of the learning environment at the center of the screen: the fingers of a human hand holding the stem of a ring with a soap bubble sitting on it. Even at a second glance, this image has nothing to do with CEA. Therefore, in terms of didactics, the image is unnecessary: „Employing illustrations, animations and video sequences must not be an end in itself and should not distract [the

[3] In order to save space, I shall not present any screenshots of the *wif!* e-learning environment. I would like to refer the reader to the above mentioned Internet address. I would like to thank lic. rer. publ. HSG Sandra Vetsch, co-project leader of the *wif!* staff, for providing me with the information I needed to write this paper.

user] from the essential – the contents – (which is however often the case). Employing visuals must always help achieve pedagogic goals."[4]

3 Objective

The purpose of this paper is to visualise CEA.

4 Learner Features and Need Analysis

According to the introductory page of *wif!* e-learning, the learning environment and its materials are addressed to the staff of the public administration of the Canton of Zurich. Given the disparate nature of the members of this target group, an enormous effort would be required to identify what kind of learners these are and what their actual learning needs are.[5] It is beyond the scope of this paper to collect empirical data on these needs, particularly in view of visualising CEA. This means that the following visualisations have been conceived without taking the needs of all the various learners into either specific or detailed account.

5 Visualising Cost & Efficiency Accounting

5.1 Visualising Text As a Multi-stage Procedure

Visual communication has developed a multi-stage procedure to visualise text. The first stage involves identifying the topic of the text that is going to be visualised (receptive understanding). Then, the topic of the source text has to be processed (productive-creative procedure) through a heuristic phase, a conceptual phase and a production phase. I have discussed this procedure elsewhere in the context of visualising legal norms[6] and shall apply it here to conceive visualisations of CEA, respectively of what CEA is, not least since this has a legal basis in the public law of the Canton of Zurich.

5.2 Conceiving the Visualisation

Identifying the Topic: What Cost & Efficiency Accounting Is

In this phase, I shall set out to understand the text I intend to visualise by referring to the relevant literature. Thus, this phase involves consulting specialist texts on the topic treated in learning module 2e-1 (what is Cost & Efficiency Accounting).

---

[4] Egon DICK: Multimediale Lernprogramme und telematische Lernarrangements. Einführung in die didaktische Gestaltung, Nürnberg 2000, p. 91.

[5] On determining the addressees of an e-learning environment, see KERRES, see footnote 2, p. 52 and 135ff., and Ludwig J. ISSING: Instruktions-Design für Multimedia, in: Information und Lernen mit Multimedia und Internet. Lehrbuch für Studium und Praxis, ed. Ludwig J. Issing and Paul Klimsa, 3., completely rev. ed., Weinheim 2002, p. 159f.

[6] See BRUNSCHWIG, see footnote 1, p. 80ff. and 217ff.

Processing the Topic: What Cost & Efficiency Accounting Is

Heuristic Phase. The purpose of the heuristic phase is to collect materials well-suited to working out what CEA is. Due to the at times fairly abstract *wif!*-learning text, this involves spelling out its contents and meaning. On the one hand, this involves finding text material specifically on CEA; on the other hand, it means looking out for images that are somehow connected with CEA and that could be used as models for visualising what CEA is.

Textual concretisations of CEA. Although various cantonal decrees mention CEA (cf. Budget Act § 18 I, Fiscal Administration Ordinance § 16 and Global Budget Ordinance § 19), no further details are given. The legal regulations merely specify the purpose and preconditions of CEA, who is responsible for it and how CEA should be conducted.

„Cost and Efficiency Accounting consists of three main elements or principal areas: cost type accounting, cost centre accounting and cost unit accounting.“[7] Cost type accounting concerns the costs incurred by particular services.[8] Cost centre accounting explains where costs are incurred.[9] In cost unit accounting, costs are attributed to the various services, respectively cost bearers, such as services, products etc.[10] In connection with CEA, NADIG uses visual metaphors, such as „**cash-river**“,[11] „**value-drain**“ and „**value-inflow**“.[12]

Visual interfaces of CEA. Basically, the materials in learning module 2 are illustrated with activities of the Office of Statistics of the Canton of Zurich, in particular of the Yearbook of Statistics. This is also the case where CEA is explained. The Cost Accounting Sheet (hereafter CAS) in learning module 2e-2 visualises what CEA is in the form of a table by drawing on the Yearbook of Statistics.[13] The CAS breaks down the cost types and their amounts (cost type accounting) that arise in the context of the Yearbook of Statistics (book and CD-ROM format). Further, the CAS attributes these costs to the cost bearer, i.e. the Yearbook of Statistics of the Canton of Zurich (cost unit accounting). It also gives details of which costs have been incurred by which cost centre within the Office of Statistics of the Canton of Zurich. Finally, it details revenues from Yearbook sales, which in turn indicates the extent of cost recovery.

NADIG and SCHELLENBERG present charts or graphic overviews to visualise CEA.[14] To the best of my knowledge, there are no illustrations in the relevant literature that visualise Cost & Efficiency Accounting.

---

[7] Aldo C. SCHELLENBERG: Rechnungswesen. Grundlagen, Zusammenhänge, Interpretationen, 3., rev. and exp. ed., Zürich 2000, p. 267.

[8] Cf. Linard NADIG: Kostenrechnung als Führungsinstrument. Grundlagen, Zürich 2000, p. 26, and SCHELLENBERG, see footnote 7, p. 267.

[9] Cf. NADIG, see footnote 8, p. 26f., and SCHELLENBERG, see footnote 7, p. 267f.

[10] Cf. NADIG, see footnote 8, p. 27, and SCHELLENBERG, see footnote 7, p. 267f.

[11] NADIG, see footnote 8, p. 13.

[12] NADIG, see footnote 8, p. 18.

[13] On the features of tables, see Steffen-Peter BALLSTAEDT: Wissensvermittlung. Die Gestaltung von Lernmaterial, Weinheim 1997, p. 137ff.

[14] Cf. NADIG, see footnote 8, p. 28, Ill. 1/12, and SCHELLENBERG, see footnote 7, p. 268, Ill. 65; on charts, see BALLSTAEDT, see footnote 13, p. 107ff.

Conceptual phase. This phase involves selecting representative concrete illustrations of the text material that is going to be visualised from the materials (i.e. texts and visuals) collected in the heuristic phase. The materials that have been found are then ordered and attention is paid to identifying those texts and images that are typical of CEA.

From the textual material collected, I have chosen the visual metaphors „cash-river", „value-drain" and „value-inflow". As regards the visual sources of what CEA is, it is worth noting that the tabular CAS, contained in Submodule 2e-2, presents cost types and allocates them to the Yearbook of Statistics in book and CD-ROM format respectively. Besides, it is important that the CAS specifies which costs encumber which cost centre within the Office of Statistics. Moreover, it is quite decisive that the table and charts keep cost type accounting, cost centre accounting and cost unit accounting separate. Given the modest amount of text and visual material chosen, it is unneccessary to order it any further.

Production phase. During this phase, details have to be worked out and rules for visualising text have to be observed. As regards working out details, the concrete text material that has been found and selected is visualised, i.e. taken into account in designing the visualisation. Besides, the visual material that has been found and selected is used inasfar as it fits the subject matter of what CEA is. Finally, I shall also lay open the rules for visualising text that I have observed in designing the visualisation of CEA.

Before carrying out these steps, I should like to describe the visualisation of what CEA is. Essentially, these are sequences of moving images (animation). Within the scope of this paper, I shall limit myself to one animation.

Animation of the external costs of the Yearbook of Statistics in book form.

- Scene 1: Shows a printing press running at the printer's office of the Cantonal Stationery and Resources Office (KDMZ). The press is printing the Year*book* of Statistics. The letters „KDMZ" can be seen, indicating that the printing press is located at the KDMZ.
- Scene 2: Picture of the building where the Office of Statistics is located. The house is marked with the sign „Statistisches Amt des Kantons Zürich" (Office of Statistics of the Canton of Zurich). A river filled with banknotes and coins is flowing out of the building entrance to the right towards the open entrance of a building where the KDMZ (Stationery and Resources Office) is located. This building is clearly marked with the sign „KDMZ" and the sum of „CHF 80'880" (Swiss Francs) can be read in the „cash river".
- Scene 3: Shows a member of staff of Mendelin AG (private limited company) dispatching a sales promotion letter to Yearbook customers at a post office. Apart from a fictitious address, the text on the front of the letter reads „Yearbook Customer".
- Scene 4: Shows another cash river flowing out of the entrance of the Office of Statistics to the right straight towards the open entrance of Mendelin AG. The building is marked with a sign that reads „Mendelin AG" and the sum of „CHF 800" (Swiss Francs) is written in the cash river.

- Scene 5: Shows a post office worker of „Swiss Post“ handing a list of potential customers to a member of staff of the Office of Statistics. The exchange takes place in a room at the Office of Statistics, marked as such.
- Scene 6: Shows a cash river flowing out of the entrance of the Office of Statistics to the right towards the entrance of a „Swiss Post“ building. The sum of „CHF 1565“ (Swiss Francs) is written in the cash river.

These details show that the metaphor of a river of cash – or „cash flow“ as it is often called – has been integrated into designing the animations. Further, the animation will also consider concrete contents of the tabular CAS. More specifically, this means that some of these contents will be transferred directly into the animation, i.e. without designing any further images, whereas others require visual representation. Thus, the animation refers only indirectly to Submodule 2e-1. What remains to be done is to establish which text visualisation rules apply to the visualisation conceived so far. Given the limited scope of this paper, I shall only consider Scenes 1 - 4 of the animation by way of example.

Text Visualisation Rule with reference to Scenes 1 and 2: The rule of visual semantic repetition says that the designer repeats the contents and meaning of the text that is going to be visualised (source text) in the visualisation (target image).[15] In the CAS, it says, „KDMZ, Druckkosten Jahrbuch [...] CHF 80’880“, i.e. ‚Stationery and Resources Office, Yearbook Printing Costs [...] 80’880 CHF (Swiss Francs)’. What this implies is that the printing office at the Stationery and Resources Office of the Canton of Zurich (KDMZ) prints the Yearbook and that printing costs amount to CHF 80'880. Scenes 1 and 2 provide a visual repetition of this fact. It follows that the text visualisation rule of visual semantic repetition has been observed. This rule also applies to the cash river flowing from the entrance of the Office of Statistics to the open entrance of the Stationery and Resources Office, since the visual metaphor of the cash river, which is significant for CEA, is repeated visually.

Text visualisation rules with reference to Scenes 3 and 4: If the rule of visual association is applied, this means that the source text is visualised by related image contents in that this association is accounted for either by experience, knowledge or meaning.[16] Even if the target image has different contents than the semantics of the source text, it is nonetheless related in terms of experience, knowledge or meaning to the semantics of the source text. According to the rule of experience-based visual association, visual representation occurs so that the components of the target image are related in time and/or space with those resulting either directly or indirectly from the textual source.[17] On the CAS, we read „Mendelin AG, Werbemailing [...] CHF 1’600“, i.e. ‚Mendelin AG, Dispatch of Promotion Materials [...] CHF 1'600’. What this means is that Mendelin AG takes charge of dispatching promotion materials to Yearbook customers and that mailing costs for the book and CD-ROM version amount to CHF 1'600. Scene 3 (showing a member of staff of Mendelin AG dis-

---

[15] Cf. Werner GAEDE: Vom Wort zum Bild. Kreativ-Methoden der Visualisierung, 2., improved ed., München 1992, p. 92 and 101.

[16] Cf. GAEDE, see footnote 15, p. 34, 91 and 103.

[17] Cf. GAEDE, see footnote 15, p. 94f., 100 and 103.

patching sales promotion letters to Yearbook customers at a post office) is related to the contents of CAS in spatial terms in that CAS implies that the dispatch of promotion materials takes place at a post office and nowhere else. Experience shows that a member of the mailing company is responsible for doing this. It follows that the text visualisation rule of experience-based visual association has been observed. Further, the rule of visual repetition applies again with regard to the cash river flowing from the entrance of the Office of Statistics to the entrance of the building where Mendelin AG has ist premises (see explanation above). As the visualisation refers to the book version of the Yearbook of Statistics, I have reduced the costs to CHF 800, whereas I have considered the other half of the costs in the corresponding visualisation of the CD-ROM version (not covered in this paper); thus the two visualisations represent a sum total CHF 1'600.

6 Findings

The animation discussed above and that has been conceived with a view to establishing what CEA is, is an illustration. Its degree of iconicity is substantially greater than that of the tabular used in the *wif!* e-learning environment.[18] As such, the animation is much more concrete and graphic than the other, tabular mode of representation, and thus meets one of the criteria, among others, for assessing the didactic[19] and mnemotechnic[20] value of presentations and learning materials far more persuasively. It needs to be added that the weakness of the visualisation presented here is that it has been conceived separately from real learners' needs and that producing it requires a great deal of expense and labour.[21] In producing the animation, importance has to be attached that is does not run too fast and that it can be stopped, winded forward, rewinded and repeated. Like this, there is no danger that learners could not grasp the sequence of moving images.[22] Should producing the animation lead to the expenditure of too much effort, „the presentation of a sequence of single images, respectively illustrations"[23] would be possible, too.

---

[18] On the abstractness of diagrams compared to realistic images, cf. Wolfgang SCHNOTZ: Wissenserwerb mit Texten, Bildern und Diagrammen, in: Information und Lernen mit Multimedia und Internet. Lehrbuch für Studium und Praxis, ed. Ludwig J. Issing and Paul Klimsa, 3., completely rev. ed. Weinheim 2002, p. 66.

[19] On the didactic principle of vividness, cf. Beate BRUNS/Petra GAJEWSKI: Multimediales Lernen im Netz. Leitfaden für Entscheider und Planer, 3., completely rev. ed. Berlin [et al.] 2002, p. 22.

[20] Cf. Bernd WEIDENMANN: Multicodierung und Multimodalität im Lernprozess, in: Information und Lernen mit Multimedia und Internet. Lehrbuch für Studium und Praxis, ed. Ludwig J. Issing und Paul Klimsa, 3., completely rev. ed. Weinheim 2002, p. 52 and 61.

[21] Cf. ISSING, see footnote 5, p. 164.

[22] Cf. KERRES, see footnote 2, p. 229, and Bernd WEIDENMANN: Abbilder in Multimediaanwendungen, in: Information und Lernen mit Multimedia und Internet. Lehrbuch für Studium und Praxis, ed. Ludwig J. Issing und Paul Klimsa, 3., completely rev. ed. Weinheim 2002, p. 95.

[23] KERRES, see footnote 2, p. 181.

References

1. Andrea BACK/Oliver BENDEL/Daniel STOLLER-SCHAI: E-Learning im Unternehmen. Grundlagen, Strategien, Methoden, Technologien, Zürich 2001
2. Steffen-Peter BALLSTAEDT: Wissensvermittlung. Die Gestaltung von Lernmaterial, Weinheim 1997
3. Colette BRUNSCHWIG: Visualisierung von Rechtsnormen. Legal Design, Diss. Zürich 2001, Zürcher Studien zur Rechtsgeschichte, Vol. 45, ed. Marie Theres Fögen [et al.], Zürich 2001
4. Beate BRUNS/Petra GAJEWSKI: Multimediales Lernen im Netz. Leitfaden für Entscheider und Planer, 3., completely rev. ed. Berlin [et al.] 2002
5. Egon DICK: Multimediale Lernprogramme und telematische Lernarrangements. Einführung in die didaktische Gestaltung, Nürnberg 2000
6. Werner GAEDE: Vom Wort zum Bild. Kreativ-Methoden der Visualisierung, 2., improved ed., München 1992
7. Ludwig J. ISSING: Instruktions-Design für Multimedia, in: Information und Lernen mit Multimedia und Internet. Lehrbuch für Studium und Praxis, ed. Ludwig J. Issing und Paul Klimsa, 3., completely rev. ed. Weinheim 2002
8. Michael KERRES: Multimediale und telemediale Lernumgebungen. Konzeption und Entwicklung, 2., completely rev. ed., München [u.a.] 2001
9. Linard NADIG: Kostenrechnung als Führungsinstrument. Grundlagen, Zürich 2000
10. Aldo C. SCHELLENBERG: Rechnungswesen. Grundlagen, Zusammenhänge, Interpretationen, 3., rev. and exp. ed., Zürich 2000
11. Wolfgang SCHNOTZ: Wissenserwerb mit Texten, Bildern und Diagrammen, in: Information und Lernen mit Multimedia und Internet. Lehrbuch für Studium und Praxis, ed. Ludwig J. Issing und Paul Klimsa, 3., completely rev. ed. Weinheim 2002
12. Bernd WEIDENMANN: Multicodierung und Multimodalität im Lernprozess, in: Information und Lernen mit Multimedia und Internet. Lehrbuch für Studium und Praxis, ed. Ludwig J. Issing und Paul Klimsa, 3., completely rev. ed. Weinheim 2002
13. ID.: Abbilder in Multimediaanwendungen, in: Information und Lernen mit Multimedia und Internet. Lehrbuch für Studium und Praxis, ed. Ludwig J. Issing und Paul Klimsa, 3., completely rev. ed. Weinheim 2002

The First Steps of e-Governance in Lithuania: From Theory to Practice

Arūnas Augustinaitis and Rimantas Petrauskas

Law university of Lithuania
Ateities 20, LT-2057 Vilnius, Lithuania
{Araugust,rpetraus}@ltu.lt

Abstract. The article provides theoretical analysis of e-governance steps in Lithuania, based on comparative analysis of conceptual documents, strategies and plans, also draws conclusions on shortcommings thereof, and practical analysis of the issue. Existing two theoretical concepts of e-government in Lithuania are examined against selected theoretical and methodological criteria, while practical evaluation of e-governance steps in Lithuania is measured by experimental research of internet communication quality between citizens and government in Lithuania. In particular the research is targeted at websites of different public institutions and their feedback to citizens.

1 Introduction

"The new situation requires active work to meet new challenges in the field of high technologies, the internet and new communication services and to prepare a successful merge between the old and new economies" (Erki Liikanen). Seeking to improve the quality of public administration services provided to the citizens and increase their communication via the internet, a comprehensive research on different public institutions websites and their feedback to citizens in different countries has been carried out in the EU since 1999 [1]. Unfortunately, scientific research on similar topic in Lithuania has just started and has not preceded the drafting and enactment of the e-government foundation documents in Lithuania, the Lithuanian concept of e-government and the strategy of developing an information society in Lithuania. This article attempts a theoretical comparison of these two conceptual documents, explains different paradigms of e-government, which have resulted in the dualism of concepts of e-government, examines them against selected theoretical and methodological criteria. Practical measurement of e-government concepts is carried through the research of the quality of the Internet connection between the public institutions and the community.

2 Theoretical Perception of e-Government in Lithuania

The topic of e-government is one of the most discussed questions both at national and at European Union (EU) levels recently [2]. It is connected both with the paradigm-

R. Traunmüller and K. Lenk (Eds.): EGOV 2002, LNCS 2456, pp. 438–445, 2002.

atic change of the governance institutions in the global context and with the impact of networked environment on the shift of living. The conception of e-government is not a simple or just a technological problem as it may seem from the first sight. This conception embodies several key structural complexes of analysis: ***methodological, political, sociological and strategic***.

From the methodological point of view the most important thing is to gauge the general perception of e-government and the role it gets in the sphere of national governance and public administration. The role of e-government may be defined as:

- ***Complementary factor***, which in one way or another organically complements the existing formal bureaucratic structure of governance.
- ***Subsidiary factor***, which plays only subsidiary attendant role.
- ***Synoptical vision*** or a "big brother" - type political abstraction, the extreme case of which is understood as an absolute control of governance by technical means.
- ***Substitute orientation***, when e-government increasingly replaces conventional forms of public administration and gives them new functional potential. It leads to absolute conceptions related to the technological rule of on-line democracy.
- ***Management factor*** as development of technologies and methods of public administration.
- ***Social factor*** as public service, which may be defined as a formula "on-line welfare for everyone".

From the political point of view e-government is some kind of democracy mechanism, because it is related with public information processes, dissemination of civic information and development of the public sphere. In a democratic society the shapes of e-government acquire value content through the mechanisms of communication. ***In fact democracy coincides with the principles of organization and management of communication processes***. From the social point of view the development of e-government is related to the tendencies of knowledge society. This relation reflects the impact of e-government on the so-called ***access to lifestyle***. It is the entrenchment and further development of networked life changes, where the key factors are the impact of Internet, problems of digital divide, governance attitudes towards global influences, political will and state information programmers.

Some kind of generalization may be made following the research in this field in Lithuania, but there are only the initial results, which require further systematic and deeper research projects [3].

It is not sufficient to make researches of direct social aspects of e-government. Civic information model is not conceptualized yet, nor is the model of the development of public sphere and there is a clear ***deficit of social ideas, self-government and community conceptions*** penetrating into the complicated process of entrenchment of legal environment and democratic values. Researches show a vast need for social ideas and social changes, but conservative stereotypes of public administration do not allow apply elements of e-government in the daily practice effectively and fast enough [4].

Another question is the practical realization of presumptions of e-government research. Is the idea of e-government acceptable to governance and citizens? In the first case one should bear in mind that in public administration there ***prevail conservative governance methods***, based on systematic conception and the dominance of formal positions. At present, there are ***two alternative e-government concep-***

tions prepared, which notwithstanding some from the first sight common formulation have different conceptual bases (Table 1).

Table 1. Comparison between e-government conceptions of Lithuania

Feature	*Centralized model*	*E-government for citizens*
Global context	-	+
Environment of changes	-	+
Integration into knowledge society	-	+
Education priorities	-	+
Accessability to information	+	+
Quality and development of citizens' interactions	-	+
Cohesion of citizens and governance institutions	+	+
Level of democratization	-	+
Quality of decisions	+	+
Optimization of public administration	+	+
Transparency of public administration	+	+
Individualization of public services	-	+
Governance depersonalization	-	+
Unification of public services	+	-
Departmentalism	+	-
Public spirit	+	+
Eurointegration	-	+
Integrated systems of state registers	+	+
Integration of state communication and data transmission networks	+	+
Common system for person identification	+	+
Social data system	+	+
Changes in legal environment	+	+
Data protection	+	+
Free general access to public information	+	+
Management of information processes	+	-
Development of e-business	+	-
Computer literacy	+	+
Compatibility of e-government projects	+	-
Standardization of e-government	+	-
Open formats	+	+

The first is traditional systematic approach which emphasizes high technological centralization and standardization. It continues the soviet regime principles which where implemented in such programmes as State automated control system or State automated scientific technical information system. E-government is understood as specialized activity or even a business sphere, which is governed by departmental principle. Centralized orientation is reflected in such projects as Single Governance Internet Portal, State Administrative Information System, infrastructure for electronic signature, integration of state registers and other similar projects, which present e-government as a cohesive normative system, hierarchically and functionally covering all the state ruling and public administration. The systematic approach is more or less tried out and better understood, and it leaves less possibilities for the unexpected.

The second approach towards e-government is based on tendencies of knowledge society and knowledge economy. It emphasizes diversity, decentralization, pluralism, personalization and oneness. Cultural and social priorities are more important than bureaucratic and normative centralization. This model is oriented towards future and is not void of uncertainty and certain elements of risk. Both approaches underline the need for society networking, but each treats this phenomenon in a totally different way.

There is the third approach towards e-government – a ***pragmatic one***. The main feature of it is drift and some kind of deconceptualization. It deals only with partial tasks, which are offered by life. While deep in argument over the strategic dominance at the conceptual level, natural processes of formation of e-government in different Lithuanian governance and public administration spheres are going on, such as specialized public systems "Customs Rates in the Republic of Lithuania", "Vehicle Queues at Lithuanian Boarders", statistic data about the social and economic state of country, tourist information etc.

3 Quality of Communication by Internet between Citizens and Government

3.1 Experimental Methodology

The skills of the most government institutions are not sufficient for the applications of the new principles of communication and this to a certain extent hinders the development of the knowledge society [5]. The lack of application skills of separate institutions and individuals as well as the shortage of technical possibilities in the sphere of application of modern information technologies becomes obvious when assessing their application of the Internet to receive and disseminate the information. In order to see the extent of the Internet application in the offices of central and local public authority of Lithuania, starting January 2001 the Law University of Lithuania carried out an experimental research of the internet intelligence of some of the Lithuanian government institutions. The results of the research demonstrate the quality of communication by internet between citizens and governance. This research was carried out in two stages:

Research stage 1 (Jan of 2001). The research was carried out in two directions: by analyzing the websites of the highest rank central government institutions and examining how effectively the ministries' officials use the e-mail for communication with citizens.

Research stage 2 (Feb – March of 2002). The research was expanded to all ministries and municipalities. Citizens' e-mails with the some questions were sent to e-mail gates and personally to administration officers of these institution.

3.1.1 Research Stage One

The first part of the research was conducted according to the research methodology developed by Phil Noble (Sweden) and Bernhard Lehmann (Germany) to examine the Internet intelligence of the EU states' politicians and officials [1]. This methodology is based on the specific analysis of the politicians' websites and the feedback from the community. The subject of their research was the four categories of the government institutions: offices of the presidents or prime ministers, offices of the parliament or the government, ministries of economy, social security or education. Depending on the category of the institution a questionnaire of 9-13 questions was developed for each of the four categories of institutions and according to the answers given to the submitted questions the institution of every category was assessed in the scale of 25 points. The research carried out by the ***Law University of Lithuania*** analysed the internet intelligence of the Seimas (The Parliament) of the Republic of Lithuania, the Office of the President and the Ministry of Economics, as well as the Ministry of Social Security and Labour.

The generalizing assessment of the Seimas was 19 points in the 25-point scale. If compared with the EU states [1], the internet intelligence of the Seimas is quite high. The Internet intelligence of the Office of the President assessment was good enough and equaled 17 points for the scale of 25 points. The results of the Ministry of Economy and the Ministry of Social Security and Labour was comparatively low and equaled 11 points.

The subject of the second part of the research of Stage 1 was the officers of different ministries of the Republic of Lithuania. In the process of this research, e-mails were sent to ten random officials of every ministry with a simple inquiry and then the results were analyzed. In order not to arouse the officials' suspicion about the research the question was simple. Moreover, the answer to this kind of message gives information about the way the ministry officials communicate with the ordinary people, the feedback style. Taking into consideration the fact that the ministers' and vice-ministers' e-correspondence is administrated by appointed people, they were not included into the list of addresses. Computer science specialists were also excluded.

Ten of thirteen of the Lithuanian ministries were examined. They were the ministries that indicated their officials' e-mail addresses. The answers were of different type: comprehensive (indicating the exact time and place of the reception, contact telephone numbers, fax numbers etc.), short, laconic or even of question type.

Information of officials' replies from the different ministries is illustrated in percent in Fig. 1. The findings show that only 11 percent of the respondents use e-communication. 82 percent of the ministries' officials did not respond to the inquiry.

3.1.2 Research Stage Two

The research was developed in the second phase (February-March, 2002). The websites of all ministries and local authorities were analyzed and the research was carried out in order to see how the ministry and local authority officials use the e-mail for the communication with citizens. It was decided to develop the research including all of the 13 Lithuanian ministries and 60 local authorities. The quality of the communication by the Internet between citizens and governance was examined by sending e-mails via e-mail gate or to administration officers of the institutions.

3.2 Analysis of the Websites

After reviewing the information on the Swedish, German, Finnish and Spanish local authorities' websites, 24 criteria, divided into 4 groups, were chosen. These 4 groups criteria help to assess the quality of the information displayed on the websites for the society:

I. Basic information about the local authority (7 criteria).
II. Important information for the citizens (10 criteria).
III. Entertainment information (4 criteria).
IV. Globalisation of the website and technical advantages (3 criteria).

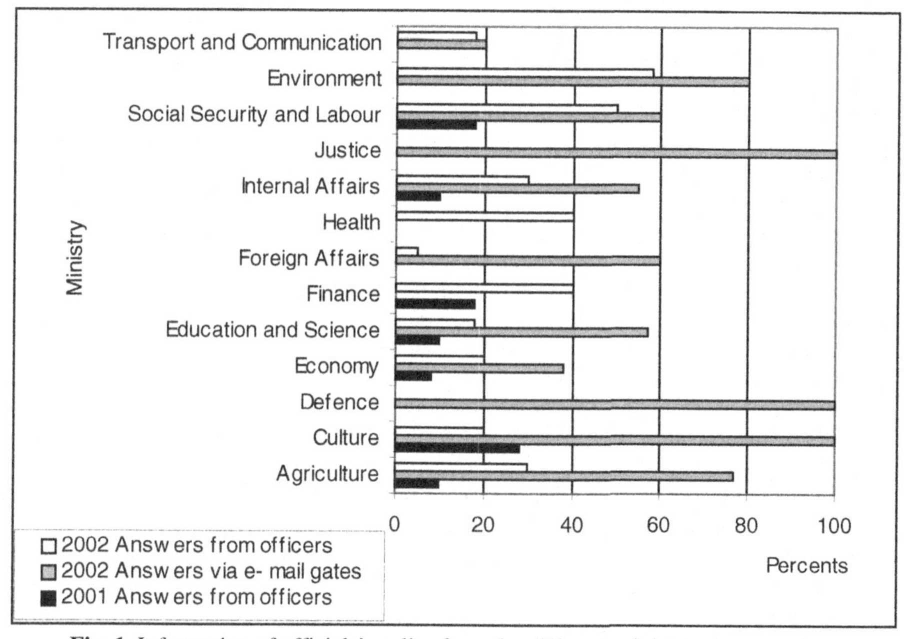

Fig. 1. Information of officials' replies from the different ministries in percentage

The ministries' websites of Belgium, Spain, Austria and Sweden were also reviewed. It should be mentioned that the ministries of the indicated countries do not have a uniform structure displaying the information on the Internet, due to the fact that the ministries are in charge of different spheres of activities. Therefore the following 14 basic criteria for the assessment of the information on ministries' websites may be distinguished: the structure of the Ministry; E-mail gate of the Ministry; personal e-mail address of the Minister; E-mail addresses of the Ministry staff; regulations of the Ministry; documents and drafts of legal acts; international relations; press releases; news; reception schedule; comments; FAQ, version in a foreign language; information on the website revision.

After reviewing the websites of the local authorities it was noted that 25 out of 60 Lithuanian local authorities still do not have their own websites designed. About 39% of the local authorities websites meet less than half of the indicated website information assessment criteria, 30% of the websites meet half of the criteria. Approximately 23% of the local authorities websites meet more than half of the criteria. Only 9 % of the analysed websites almost fully meet the criteria.

All the 13 ministries have their own websites that meet half or more of the basic criteria.

3.3 Research on the Internet Communication

In carrying out this research, typical citizens' questions were sent, thus imitating the inquiries of the citizens. 5 questions were designed for each ministry and local authority and were sent to the e-mail gate of these institutions. Also 1 universal and simple question was designed for administration officers of the ministries and local authorities and was sent directly to their personal e-mail boxes. While composing these questions, it was decided to take into consideration the list of frequently asked questions in the website of each institution.

613 e-mails with questions for the investigation of the feedback between citizens and governance were sent. 370 questions to e-mail gates and 243 questions were sent to administration officers of the ministries and local authorities. 47% answers via e-mail gates and 28% answers from personal e-mail boxes of the administration officers were received. 68 answers were received to 243 questions addressed to the personal e-mails of the administrative staff working in the Government, ministries and local authorities of the Republic of Lithuania (28 % of answers).

Through the e-mail gateway of the Government of the Republic of Lithuania 40 % of answers were received, whereas only 22 % of answers were received from the administrative staff of the Government of the Republic of Lithuania. 60 % of the answers came through the e-mail gateway of the ministries of the Republic of Lithuania (figure 1). Answers to all questions were received through the e-mail gateways of the ministries of Justice, Culture and National Defence. 33 % of the answers were received from the administrative staff of the Lithuanian ministries. The greatest number of answers (60 %) was received from the administrative staff of the Ministry of the Environment.

Through the e-mail gateways of the local authorities of the Republic of Lithuania 45 % of answers were received. Through the e-mail gateways, answers to all questions came from 6 authorities. 26 % of answers were received from the administrative staff of the local authorities.

The quality of the answers from e-mail gates was rather high. Only 10 % of the answers were not answered in essence, i.e. they referred to another person or institution. The rest of the answers were rather exhaustive and informative. The worse situation was with the answers from administration officers. About 69 % of the questions were not answered in essence i.e. they asked to indicate the position of the inquirer, to specify the details of the question or other. The rest of the answers were not very elaborate or informative.

4 Conclusions

1. Lithuanian government makes a significant attempt to concentrate on the problems of e.government. Processes of looking for paradigmatic alternatives for two different e.government conceptions, which can be characterized on the basis of different theoretical criterions, is in progress. The first Lithuanian e-government conception is based on a centralized model of state administration. The second one is oriented towards citizens and social multiplicity (diversity).

2. The findings of the preliminary research show, that the highest Lithuanian government institutions, those of the Seimas of the Republic of Lithuania and the Office of the President, website Internet intelligence is close to the average level of the EU states.
3. The research carried out in 2002 demonstrated that the quality of the Lithuanian ministries' websites as well as the community feedback has greatly improved. Only 35 Lithuanian local authorities out of the total number of 60 have their own websites, but only one third of those satisfies the website information assessment criteria.
4. Many more replies, and of higher quality, were received by communicating via the e-mail gate of the ministries and local authorities than by sending e-mail directly to the institution officials.

References

1. Noble P., Lehmann, B.: Interactive Internet Study of EU Governments. Amsterdam - Maastricht Summer University http://www.amsu.edu/jac/ default3.htm. (2000).
2. Holmes Douglas. E-Gov. E-Business strategies for government. London: Nickolas Brealey Publishing, 330 p. (2001).
3. OSF - Lithuania. Public Policy Projects. http://www.politika.osf.lt/index.en.htm (2002).
4. Augustinaitis A. Informacijos visuomenės savivaldos tendencijos. Informacijos mokslai, 2000, (14), p.18-45.
5. Petrauskas R. Informacinių technologijų taikymas viešajame administravime. Vilnius: LTU, 65 p. (2001).

The Role of Citizen Cards in e-Government

Thomas Menzel and Peter Reichstädter

Chief Information Office Austria, Ministry of Public Services
{thomas.menzel,peter.reichstaedter}@cio.gv.at

Abstract. Citizen Cards serve as a central item in e-Government for the identification of the acting persons. Establishing the model of a Public-Private-Partnership, the Citizen Card will not be uniform, produced and issued by a public authority in Austria, but consist of diverse chip cards, issued by public or private organisations. These cards can be used for e-Commerce as well as in legally binding electronic communication with the administration. In this paper we discuss the basic requirements all these cards must meet, and present a typical e-Government session (a citizen applies to any authority using a signed XML form and web transport, and in return receives the official and legally binding decision of the public authority.)

1 Purpose of the Card

In the course of extension of functionality in e-Government-applications, direct communication between citizens and administration via Internet will be established. Because of the openness of this medium, special elements for secure identification and transaction, which are mandatory requirements for legally binding communication with public authorities, must be used in official filings with public authorities. Existing security breaches can be avoided by using electronic signatures for identification and authentication.

As also in conventional administrative proceedings, applicants must prove their identity once by showing, for instance, photo identification when filing their applications. Electronic filing requires proof of identity in an equally functional way. According to the European Signature Directive [SigDir01] and the Austrian Act on Electronic Signatures [SigAct99], electronic signatures, and in some sensitive cases secure electronic signatures[1], are required for official communication. Chip cards represent the state of the art technology for secure storage of signature creation data. If an electronic signature is used in e-Government, then the card to store the signature should be one based on "Concept Citizen Card". This concept is part of the Austrian e-Government-Strategy and covers items beyond the card itself, including specifications about issuance, applications and connected tools, like secure viewer, hash algorithm, Security Layer as interface, etc. The purpose of this concept is to develop an electronic identification card for the citizen's use on the data-highway by means of electronic signatures in a secure public-key-infrastructure.

[1] The Austrian Signature Act uses the term „secure electronic signature", which demands the same quality of an electronic signature as the term of the signature directive "advanced electronic signature based on a qualified certificate".

R. Traunmüller and K. Lenk (Eds.): EGOV 2002, LNCS 2456, pp. 446–455, 2002.

2 Requirements for All Citizen Cards

The concept for a Citizen Card, prepared by A-SIT (in charge of examining technical standards for electronic signature products) and CIO Unit (responsible for e-Government strategy), does not expressly designate a certain card or type of card. Only minimum requirements are defined. All kind of cards which fulfil such minimum requirements can be used as a type of Citizen Card in accordance with the concept. The minimum requirements are based on legal guidelines provided in the Signature Act (as last amended) with respect to data security and the integration of the ZMR-number for accurate identification. Whereby Signature Act and Order determine here on one hand procedures and components, which are regarded as suitably for secure electronic signatures, and on the other hand confirmation bodies (A-SIT) evaluate the commercial products in this sector taking care of the fulfilment of all legal safety requirements. Citizen Cards must meet exactly the same criteria as all other cards, which are used in the framework of secure electronic signatures.

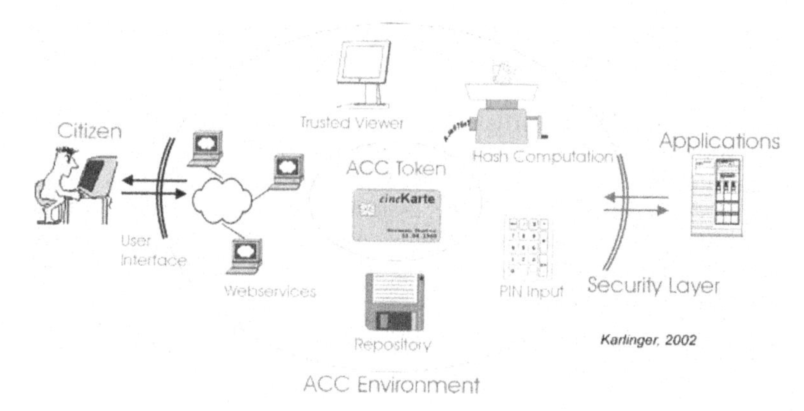

Fig. 1. Components for a Secure Signature Environment [KarG02]

Further the concept considers items that, while not obligatory are recommended since they substantially facilitate the use of the card. For example, info-boxes are recommended on the card to store user data, which is very often needed in connection with an e-Government session. The specification of a concept and the lack of a specific defined card should lead to a rapid and far spreading use of Citizen Cards, since a significant increase of chip cards with signature function, e.g. bank cards, is to be expected in the next years. All these cards should also be usable in e-Government and therefore be consistent with the Concept Citizen Card.

2.1 Secure Electronic Signature

The employment of secure electronic signatures in communication with the administration is the key element for the guarantee of security in e-Government. The Austrian Signature Act defines a secure electronic signature as an electronic signature that

a) is allocated solely to the signatory;
b) allows the signatory to be identified;
c) is created using devices under the signatory's sole control;
d) is linked with the data to which it refers to in a way which allows any subsequent change to the data to be identified; and
e) is based on a qualified certificate and is created using technical components and procedures which comply with the security requirements of the present federal law and the orders issued on the basis thereof.

A substantial security characteristic, especially important in meeting the requirement in (c), above, entails the suitable selection of the storage space for the signature creation data. The exclusive access of only the authorized signatory is best guaranteed by current technology by storing the signature creation data on a chip card with a crypto processor. In this manner, the encoding of the hash values for the signature is executed on the card, so the signature creation data never leaves the card; this is guaranteed by the architecture and operating system of the card processor. The development of other storage media, as for instance USB token or SIM cards for mobile phones, is likewise considered by the Concept Citizen Card. If these other media components feature sufficient security, they should also be applicable vis-à-vis the Concept Citizen Card.

Each card that achieves sufficiently secure electronic signatures is suitable for the Concept Citizen Card. The Austrian signature law equally applies to e-Commerce and e-Government, so that cards that are used for e-Commerce solutions and are consistent with the principles of the Concept Citizen Card in this context can likewise be used in e-Government as Citizen Cards. Although the legal basis is technology-neutrally formulated, at present only digital signatures fulfil all requirements. They are based on the following technical elements:

- Public key cryptography, also called asymmetric cryptography, represents the mathematical basis.
- Hash procedures ensure integrity of the data and permit an efficient signature creation process.
- Certificates bind the technical items like cryptographic codes to the identity of the signatory.
- Signature algorithms ensure the overall technical security of the electronic signature.

It is important to note that the legal framework permits both the application of RSA and Elliptic Curve Cryptography (ECC) as mathematical algorithm for signature creation. Therefore the exclusive use of ECC is not mandatory according to the Concept Citizen Card and RSA may also come into operation. But the longer period during which elliptic curves will be secure against brute force attacks, the relief of the chip hardware, and the shorter length of the value of a signature at the printout favor the use of ECC over RSA, in order to rollout future-oriented cards. [Certi98], [ZhIm98]

The private key, used to create a secure electronic signature, must be used exclusively for the signing process. This is necessary to ensure that only conscious declarations of will are signed. A clear separation between the process of signing and other key-based applications is essential. These card-based applications for authentication and content decryption also need access to a private key. Therefore, at least one further private key is stored on the card, which is used for these other applications in the Concept Citizen Card. Thus a separation of the signature process from all other processes, which are likewise based on asymmetrical cryptography, is ensured.

2.2 Personal Identity Link (PID)

Another feature is necessary to support a unique relationship between card and card holder. Each person living in Austria is listed in a Central Population Register and a unique, high-quality and life-constant identification number called a ZMR-number is assigned to every individual. This number is also used in the Concept Citizen Card for exact identification. During the initial registration process both public keys of the cardholder are sent to the Central Population Register. There these keys are bound to the card holder's ZMR-Number and the complete dataset is signed by the authority – this is the PID – and sent back to the registration authority, who stores it on the card for the cardholder's disposal. Therewith an exact identification of all persons living in Austria is guaranteed. The following diagram illustrates the structure of the PID:

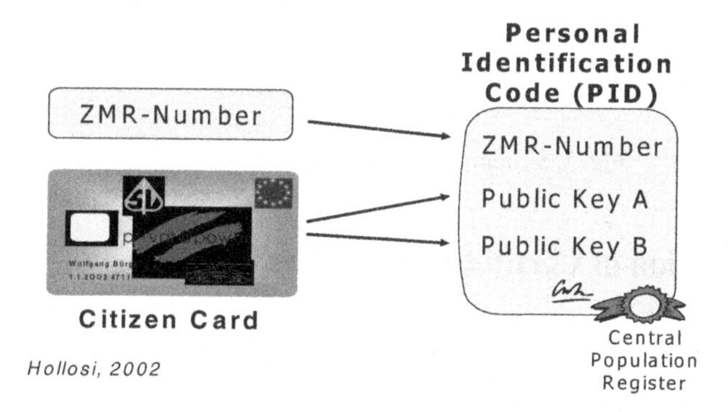

Fig. 2. Pattern of the Personal Identity Link

Due to data-protection demands, the PID in this form is not stored in the individual administrative files of a citizen. It is only used for identification purposes at the portal, but never stored in the back office databases, because storing this identification-number in databases would allow easy matching of unrelated databases (e.g. health care with income tax). Starting with the PID, a different context-dependent identification number (cd-id) is generated for each administrative procedure and only the hash-value of the cd-id is stored in the database. This procedure prevents the transmutation of cd-ids back to ZMR-numbers and PIDs in general and the conversion of different cd-ids of the same citizen. [HoLe02] The legal basis for the storage of the PID within the sphere of the card owner is outlined by the 2002 amendment to the administrative reform law 2001. The process is outlined in Figure 3.

2.3 Optional Info-Boxes

Elements needed during an e-Government session, apart from the keys and PID of the cardholder, consist mainly of rights and pointers to information. Supplements, which are needed in many different administrative procedures, as for instance for legal authorizations, birth certificates or similar documents, are at the citizen's disposal in electronic form. All these documents are also secured by an electronic signature of the issuing authority. They can be stored in info-boxes on the same card, on another card, on the PC of the card holder, or on other personal document safes anywhere on the Internet. For practical reasons the Citizen Card should supply the most essential documents directly, so that they can be used by the card holder anywhere without the need to connect to an external resource. Which information a card holder stores on his card is his decision.

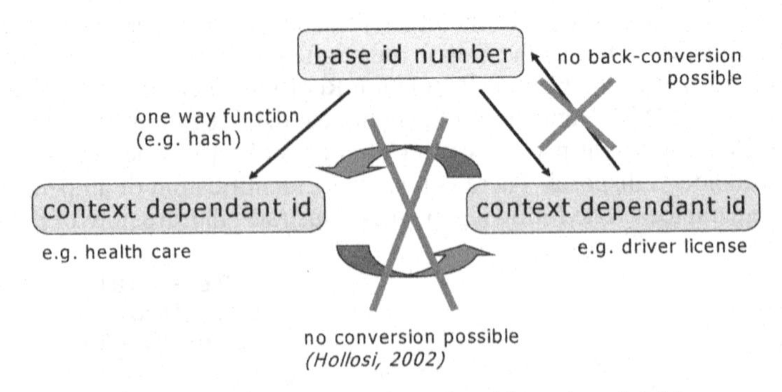

Fig. 3. System of context dependant id-numbers (cd-id)

3 Provision of Certification Services

The qualified certificate represents a substantial item of the secure electronic signature and so also of the Citizen Card, as it links the identity of the signatory to his key pairs. No public authority in Austria offers certification services currently or plans to do so in the near future. These services are provided entirely by private companies. In the context of a Public-Private Partnership, basic certification services (the generation and publication of a certificate in the directories and the execution of a revocation) are to be provided by the private certification service providers active in the market [RTR02], who offer qualified certificates. The registration process can be done by the public authority based on the principle of One-Stop-Shopping in e-Government. Figure 4 describes this process.

4 Possible Uses for Citizen Cards

The Concept Citizen Card, which must meet the minimum criteria mentioned above, should produce a scenario in which, within a short time, many different card types can also be used as a Citizen Card for communication with the administration. The trend in the context of the chip cards points towards the increasing integration of the electronic signatures on chip cards. So it is to be expected that most cards will also be signaturable, beginning in the year 2005, and therefore usable as Citizen Cards. Further the use of the secure electronic signatures is not limited on the Citizen Card to

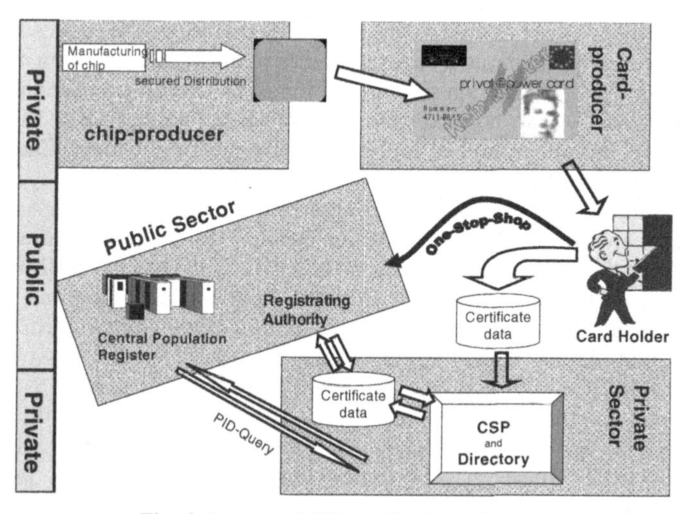

Fig. 4. Process of Citizen Card enrollment

communication with the administration. The signatures created with Citizen Cards can also be used for private legal transactions of the card holder, as a side effect increasing security in E-Commerce, E-Banking and other private sector electronic communication. An open list of possible Citizen Cards shows the versatility of the Concept Citizen Card:

- Electronic Identity Card
- Student-Service-Cards
- Social Insurance Cards (eCard)
- Cards for Chamber Members
- Service (ID) Cards for Officials in Public Administration
- Bank Cards
- Citizen Cards of other States

Fig. 5. Samples of Citizen Cards

5 Technology Independence

The versatility of the Concept Citizen Card also determines the definition of interfaces, which yield varying card types and specifications. The role of the Security Layers is seen as one of a standardized interface between various Citizen Cards and applications, which guarantees that applications can be developed without knowing or using the current technology. Therefore it is the responsibility of the Security Providers (e.g. certification service provider...) to fulfil the needs and requests of the applications, as well as to integrate a card or another safe signature creation device into the e-Government process. The interface definition of the Security Layers also allows the Concept Citizen Card to be adaptable to future developments, e.g. PDAs, mobiles, and to support the integration of foreign technologies (e.g. Citizen Cards from other EU Member States, etc.).

The information exchange between application and Security Capsule containing the Citizen Card can be implemented, for example by TCP/IP-bindings; coding the items between them takes place by means of XML [BPSM00], [TBMM01], [BiMa01] a fundamental technology based on an international specification and recommendation [ISO-10646], [Unic96] with the advantage of system- and platform independence as well as multilingual capability. XML also has the characteristics of being signaturable [EaRS02], established as a document format within the e-Government process, and used for the administration and integration of supplements, authorities or generally for the definition of procedure/process identifiers and personal identity links [Holl02], [Karl02]. Base services forming the building blocks for portals, market places, etc, are therefore implemented with future-oriented concepts and products such as Java, XML, TCP/IP or HTTP as well as open systems.

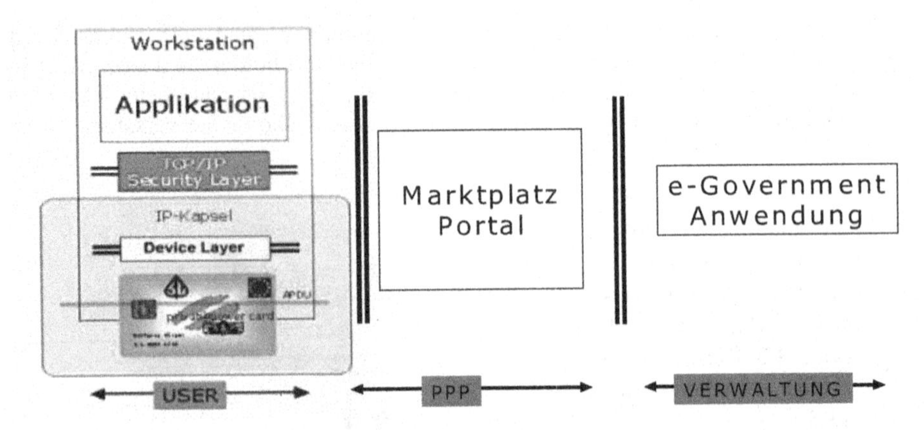

Fig. 6. Interfaces within e-Government process

6 Scenario of a Typical e-Government Session

There is no model of a 'typical' e-Government session; the session always depends on the e-Government process itself, whereby parts and phases have to be considered within a typical e-Government session. However, there are at least a few essential steps within the model that will always be found in e-Government.

The steps in the figure below are present in most e-Government procedures:

Authentification
OnLine dialog
HTML to XML conversion
Inclusion of enclosures
Creation of signed Data
Sending of signed data block

Gateway/Portal and
Back-Office process

Informing the applicant with
SMS, Mail, of delivery-process
Delivery of notifications/ records

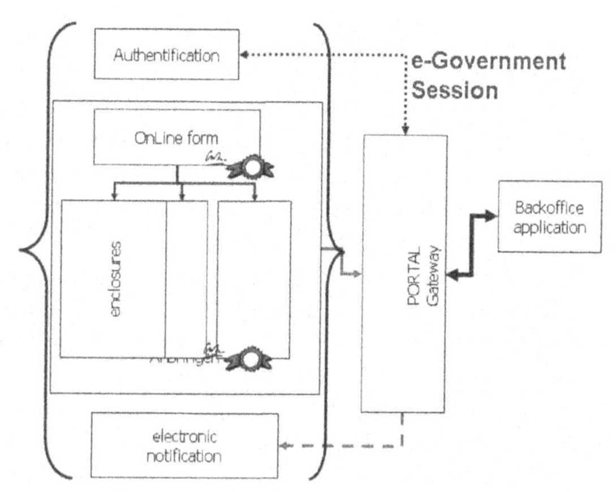

Fig. 7. Essential structure of typical e-Government session

After a (positive) authentification while using the citizen card and the functionality of the Security Capsule for determination of the authorizations, a TLS-secured session (including connection encoding) is established between the citizen and the Gateway/Portal. During the OnLine dialog, while completing HTML-forms with the WebServer application, there can be different types of help-systems activated, depending on problems and failures recognized when completing the e-forms. Some of the fields can either be completed with the information stored on the citizen card (to retrieve the stored data, a PIN may be required) or, when using personalized portals, with information from the personal data sheet. The e-form data is transmitted to the Web server, creating a specific XML-file, which file is then returned to the Web-Browser. This step is necessary due to the fact that the Security Capsule accepts only Input-data in a specific predefined XML format, before data can be signed through the capsule.

In some cases it may be necessary to electronically incorporate payment-files, mandates and other notifications/records (tax notification, criminal record) into an e-Government session. These enclosures can either be included as XML-structures or, until these enclosures are standardized and defined as XML structures, as .tif files. Finally, the XML e-form data and all necessary enclosures are put into an XML-Container, signed by using the electronic signature of the citizen stored on the citizen card together with the Security Capsule. Then they are sent (using the previously established TLS-session) to a virtual inbox of the portal, and forwarded to the relevant administrative office (i.e. interface to Backoffice applications) for further processing. The citizen receives a status message showing inbox-delivery.

Possibilities must exist for the citizen to place status queries regarding the executed e-Government procedures in order to further transparency in administrative proceedings. Further, there must be a possibility of communication with public authorities at key points with the goal of continuing incomplete administrative proceedings.

Depending on the preferences set during delivery service registration, the applicant will receive via some delivery service - after being informed through SMS, E-Mail, VoiceMail, fax – a signed notification in the form of an XML-Data Container (XML Data itself and XSL Style sheet). The XML structures underlying the notification from the public authorities must be standardized in respect of the XML-Tags that are employed (individual: family name, :given name, :occupation,...) [ReHo02],[HR-XML] in order to enable further automatic processing.

If the delivery of the notification does not occur within the TLS-session in which the e-form data is deposited, it will be necessary to consider a Delivery Service between administration units and citizens responsible for notification, logging and the delivery process itself. Such service will have to include the authentication of the administration signature within the notification XML record as well as the handling of receipts for delivery and decoding of the notification data. If the delivery process cannot take place electronically due to a technical or organizational reason, then it must take place by conventional means.

The individual steps depend on the e-Government procedure itself. In the sense of comprehensive administrative efforts however some steps toward standardized procedure definitions still need to be taken, including supplying documentation for the structures and groups of items as well as defining e-forms and electronic 'enclosures'. The free availability of these defined structures and documentation on (public) information servers (e.g. http://reference.e-government.gv.at) is absolutely necessary.

Together with the working group of the Austrian provinces in the context of the e-Austria initiative of the Austrian federal government, some progress has already been made. However, it will be important to look beyond national boundaries, and to define and standardize structures in the international context to prepare for and implement comprehensive administrative e-Government procedures throughout e-Europe.

7 Conclusion

Currently diverse new technical components, protocols and models are developed together with proven methods of administrative computer science to an efficient and constantly standardized IT system for modern administration. In addition to matching these new technologies with organizational concepts and legal bases (the signature law is a solid base for e-Government processes; other laws must be adapted for the sake of efficiency), a uniform concept must be considered for the systems used in the entire public administration in order to avoid interruption in communication between citizens and agencies or between the agencies and administrative units themselves.

The Citizen Card and the technologies of e-Government mentioned in this paper should be used to establish an efficient application of information technologies between the two ends of the interfaces - the citizen and the administration. By providing new methods of communication and information processing to support the work of the administration, computer science is able to do its part to contribute to the overall plan of administrative reform.

References

[BiMa01]: P. Biron, A. Malhotra: XML Schema Part 2: Datatypes, W3C Recommendation, 2001.

[BPSM00]: T. Bray, J. Paoli, C. M. Sperberg-McQueen, E. Maler: Extensible Markup Language (XML) 1.0 (Second Edition), W3C Recommendation, 2000.

[Certi98]: Certicom, The Elliptic Curve Cryptosystem for Smart Cards, http://www.certicom.com/resources/download/ECC_SC.pdf, 1998.

[EaRS02]]: D. Eastlake, J. Reagle und D. Solo: XML-Signature Syntax and Processing. W3C Recommendation, 2002. - http://www.w3.org/TR/2002/REC-xmldsig-core-20020212 .

[HoLe02]: Hollosi, A, Leitold, H, An Open Interface Enabling Secure e-Government - The Approach Followed with the Austrian Citizen Card, In: Proceedings of IFIP TC6 and TC11 Working Conference on Communications and Multimedia Security, CMS'2002, Portoroz, Slovenia, 26.-27. September 2002, in print.

[Holl02]: A. Hollosi, XML-Spezifikation der Personenbindung, März 2002.

[HR-XML]: HR-XML Consortium - http://www.hr-xml.org/.

[ISO-10646]: ISO/IEC 10646-1:2000. International Standard -- Information technology -- Universal Multiple-Octet Coded Character Set (UCS) -- Part 1: Architecture and Basic Multilingual Plane. UTF-8 is described in Annex R.

[Karl02]: G. Karlinger, Protokoll zur Abfrage der Personenbindung beim ZMR durch einen ZDA, März 2002.

[KarG02]: G. Karlinger, Anforderungen Bürgerkarten-Umgebung, http://www.buergerkarte.at/konzept/spezifikation/aktuell/Anforderungen%20Bürgerkarten-Umgebung.20020412.pdf, 2002.

[ReHo02]: P. Reichstädter, A. Hollosi, XML-Spezifikation der PersonData Struktur, Mai 2002.

[RTR02]: An official list of all Austrian Certification Service Providers is available at: http://www.signatur.rtr.at/en/directory/index.html.

[SigAct99]: (Österreichisches) Bundesgesetz über elektronische Signaturen (Signaturgesetz - SigG), öBGBl. I Nr. 190/1999, idF: BGBl. I Nr. 152/2001, download at: http://mailbox.univie.ac.at/thomas.menzel (follow the link "docs" in the left frame)

[SigDir01]: Directive 1999/93/EC of the European Parliament and of the Council of 13 December 1999 on a Community framework for electronic signatures, Official Journal L 013, 19/01/2000, p. 12 – 20.

[TBMM01]: H. Thompson, D. Beech, M. Maloney, N. Mendelsohn: XML Schema Part 1: Structures, W3C Recommendation, 2001.

[Unic96]: The Unicode Consortium. The Unicode Standard, Version 2.0. Reading, Mass.: Addison-Wesley Developers Press, 1996.

[ZhIm98]: Zeng, Y, Imai, H, Efficient Signcryption Schemes On Elliptic Curves. In: Papp, G/Posch, R, Global IT Security – Proceedings of the XV. IFIP World Computer Congress, OCG-Schriftenreihe, 1998, p. 75-84.

Indicators for Privacy Violation of Internet Sites

Sayeed Klewitz-Hommelsen

Fachhochschule Bonn-Rhein-Sieg, Grantham-Allee 20, D-53757 Sankt Augustin, Germany
Sayeed.klewitz-hommelsen@fh-bonn-rhein-sieg.de

Abstract. The purpose of the SAD (system for automated privacy check) project was, to scan public websites to explore, if there are any procedures or information found, which may potentially violate the privacy of the users. The criteria were to define, which information may indicate privacy relevance. To avoid misuse of the system, their where actions to be implemented. The system should support responsible persons with hints to potentially critical information.

1 Introduction

When the European Union started to harmonize the European privacy law, all member countries had to harmonize their privacy laws[1]. Germany was rather late in reforming its privacy law in 2001[2]. The following, however, will focus not on the specific regulations of privacy law, but rather on some aspects of the principles of privacy law.

Public administrations are searching for new ways to interact with their clients, especially with their citizens. This implies that the internet and particularly the world wide web are being used as a platform for new communication structures. Also the information issues of governments and administrations have changed. The trend is towards more, direct and complete information to the public. New processes are being established, new ways of participation and new partnerships are being developed. The number of web pages produced by public administrations and governments has become nearly uncountable.

Nevertheless, presenting content in the web and communicating with external partners always has an aspect of privacy protection for the involved persons. Every piece of personal data presented and every form on a web page gives or gathers information and often personal data. The idea of the "sad"-project[3] was to identify potentially

---

[1] Directive 95/46/EC (Directive 95/46) of the European Parliament and of the Council of 24 October 1995 on the protection of individuals with regard to the processing of personal data and on the free movement of such data (http://europa.eu.int/eur-lex/en/lif/dat/1995/en_395L0046.html), and the regulation (EC) No 45/2001 (regulation 45/2001) of the European parliament and of the council of 18 December 2000 on the protection of individuals with regard to the processing of personal data by the Community institutions and bodies and on the free movement of such data http://europa.eu.int/comm/internal_market/en/dataprot/news/reg45-2001en.pdf

[2] Bundesdatenschutzgesetz von 2001 (http://www.rewi.hu-berlin.de/Datenschutz/DSB/SH/material/recht/bdsg2001/bdsg2001.htm)

[3] SaD = **S**ystem zur **a**utomatisierten **D**atenschutzprüfung (system for automated privacy check), http://sad.inf.fh-bonn-rhein-sieg.de

R. Traunmüller and K. Lenk (Eds.): EGOV 2002, LNCS 2456, pp. 456–459, 2002.

critical data or personal data and to give the author or the responsible person guidance in specifically checking his website for such relevant content. The idea included two separated questions to answer:

1. Would it be possible to identify such potential critical pages?
2. Would it be possible to find these pages using an automated procedure?

The objective was to test a special site, generate a report for the responsible person and explain for each item the reason why this item was marked.

2 What Should Be Identified?

European directive 95/46 on the protection of individuals with regard to the processing of personal data and on the free movement of such data and European Regulation 45/2001 for the processing of personal data by Community institutions and bodies and on the free movement of such data expect the controller of the processing of personal data to take responsibility of the following:

- The fair and lawful processing of personal data. Where not specified, this implies that explicit and legitimate purposes exist, that the subject has to consent to the processing of his data. (Art. 4 Regulation 45/2001, Art. 7 paragraph a) Regulation 45/2001)
- Informing the data subject about the purposes of the data collection (with exception in Art. 12 Paragraph 2 Regulation 45/2001)
- Providing appropriate security for the processing of personal data (Art. 17 Directive 95/46)
- The controller has to provide the data subject with information about his identity (Art. 10 and 11 Directive 95/46).

All these aspects offer at least the partial possibility of implementing automated procedures.

3 What Can Be Identified – Potential Indicators?

First, one must remember that all content-related exploration is language dependent. That means that there will be parts that must be designed to deal with different languages. For example, one can check every page on which we find a form, whether fieldnames indicate their potential use. These potential fieldnames are looked up in a table containing "suspicious" keywords[4]. This is a typical case of language dependency. As you can easily see, it is not really difficult to handle the language specific parts. So the method is language independent and can be easily reimplemented for each language you wish.

The idea, then, was to look for pages that use forms, because in that case information from the user is typically requested. Furthermore, we expect specials hints "near" the form about the privacy strategy and the intended use of the information. Will the gathered information be forwarded to any third party? When will the collected data be

[4] i.e. name, address, street, ZIP-Code etc.

erased? The method of scanning for the presence of such information leads via links to the area of forms and to keyword scans.

Will the transferred information be secure? Often forms request passwords without any security measures directly over the internet. So we tried to check if any kind of encryption was used during the information transfer. Secure sockets layer and transport layer security, shttp (secure hypertext transfer protocol) protocols are checked.

If the web server tries to establish cookies on the client computer, we check the timestamp of the cookie to determine if the duration makes sense with respect of the service used. As one can see, we reach here a barrier of context understanding, which a simple scan-machine can't overcome. At least the person who requested the report will be able to answer the question and so with this feedback method, a first approach to solving the problem can be found.

In addition to several information obligations, one must check for a correct address field somewhere on the web page. So the site is tested by keyword scan for typical links to the provider information[5] and similar expressions.

The possibility for the user to come into contact with the website provider is an important feature. So one action to prove this possibility was to scan the whole site for applied email addresses, rank them and send a test email to the most-used address. The scanner waits up to one week, to get an answer to his test mail. If no answer is received, it is marked in the report as a severe problem.

All of these aspects provide indicators for at least a threat on privacy protection. It does not mean, however, that the concrete page bears any danger for personal data. But in any case the offer of personal data (i.e. addresses of employees) or the input of personal data always carries the opportunity for violation of privacy within itself.

4 A Word About the Technical Details

The scanner was built on a standard Linux server. First the customer has to access a web server[6], where he enters a root web address and his email address (where the report should be sent). Then the scanning process is started. In the demonstration model we used a mix of standard tools. The entered website is completely downloaded by wget[7], a utility which retrieves files from the web. The limitation of this tool is that it doesn't interpret java script. So all links or menus generated by a java script are skipped.

After this part the result is analyzed by some shell scripts and in some special cases additional perl scripts came into use. So, for example, the entire email analysis is done by a perl script. The server was of moderate power (PC architecture, PIII 500 MHz, 128 MB, 60 GB) and nevertheless performed acceptably.

To reduce the download traffic, the system tracks whether a specific site was scanned the last time and starts a new scan only after a pause of several weeks. Should the same site be requested again, the system uses the earlier download.

---

[5] Art. 5 of the directive 2000/31/EC of the European Parliament and of the Council of 8 June 2000 on certain legal aspects of information society services, in particular electronic commerce, in the Internal Market ('Directive on electronic commerce') http://europa.eu.int/eur-lex/en/lif/dat/2000/en_300L0031.html

[6] http://sad.inf.fh-rhein-sieg.de

[7] wget: http://www.gnu.org/software/wget/wget.html

5 Actions against Potential Misuse

One of the misuse possibilities we saw was that someone could scan the website of someone else. So we restricted the start of a check to people whose return email address had the same domain as the domain to be checked. This way, first the owner of the most-used email address gets the information on who had started a scan on his website. Furthermore, we plan to implement a kind of stop list for special well-known domains. So the famous free mail providers have thousands of users, with the correct domains in their email-addresses. Neverthcless, we don't expect most people to be entitled to request a scan for this domain. In such cases the scan starts when the operator sets the job to run level.

6 Final Remarks – Perspectives

During the project many additional aspects appeared and couldn't be integrated into the project. The most important was the idea of a kind of privacy/suspicion index. Such an index, if reasonable to define, could make web sites comparable on an abstract layer and motivate web site providers to optimize their sites under privacy aspects.

The project also showed that the service my become expensive, if many users use it. When the amount of text data we download from a site accumulates to a respectable volume, possibilities of financing such a service have to be evaluated.

A lot of technical details still haven't been solved. Java script and Java for example, can't be ignored nowadays. As special problem appeared with the typical use of several domains in conjunction with one domain. At this point, a change from www.domain.com to special.domain.com would not be followed. The implementation of more intelligence than just using a simple lookup table with keywords still seems to be challenging.

Verifiable Democracy a Protocol to Secure an Electronic Legislature

Yvo Desmedt[1] and Brian King[2]

[1] Florida State University,
desmedt@cs.fsu.edu
[2] Purdue School of Engineering and Technology, IUPUI campus,
briking@iupui.edu

Abstract. The manner in which a legislature votes is similar to a threshold signature scheme, and the power to sign legislation is similar to possessing shares to sign. The threshold k denotes the quorum number, the minimum number of legislators required to be present in order for legislature to be passed. Here we discuss techniques to ensure a secure electronic legislature.

1 Introduction

In democratic organizations, at a given time the number of legislators varies, while maintaining the relationship that a majority of the legislators can pass legislation. The integral part of democracy is the mechanism, which allows transfer of power from a set of n legislators to a subset of these n legislators. In a physical legislature, this mechanism is trivial to achieve. However, in an "electronic legislature", problems will arise. This paper will discuss how to achieve a verifiable democratic government using secret sharing techniques. We first proposed a verifiable democracy protocol in [1]. Subsequently, in [3], an alternate proposal was offered. However their proposal requires an administrator, which is not realistic. We will detail various problems which must be overcome, introduce the requirements for a verifiable democracy, discuss attempts at solving the verifiable democracy problem, and outline a protocol. Due to space limitations we omit technical details, such details will be included in an extended version of this paper.

The interest in developing an electronic government varies. One is that remote voting is desirable. Another is current events, in particular, terrorism attacks. As is often speculated in the media, in the September 11th terrorist attack, potential targets had included White House and/or the Capitol Building. Immediately following this attack, a second terrorism attack occurred, the mailing of anthrax spores to U.S. legislators. This attack successfully stopped the U.S. House of Representatives from meeting, and restricted the contact of the U.S. Senate. Fortunately, the House was able to meet within a few days and the contamination was limited to the Hart building, an office complex for senators. If the contamination had actually reached the congressional building, then the stoppage caused by the attack would have been much

R. Traunmüller and K. Lenk (Eds.): EGOV 2002, LNCS 2456, pp. 460–463, 2002.

greater. Many democratic organizations from legislatures to board of directors are susceptible to terrorist attacks. A solution to this problem of terrorism is to develop a distributed electronic legislature.

In an electronic government, the legislature's ability to pass or to not pass legislation should be thought of as the legislature digitally signing (with some secret key) the legislation or not signing the legislation. The power held by each legislator to vote on legislation will need to be a digital key, or rather should be thought of as a *share* of the legislature key (the one that will generate this legislature signature).

When considering an electronic government, we ask "will such a government be as representative as the physical government in place?" The danger of using a distributed electronic government is that the mechanisms for reigning-in legislative abuse is not necessarily in-place due to lack of the physical proximity of participants. The concern for the possibility of cheating among participants in an electronic government is warranted.

2 Background: Tools and Terminology

Suppose Alice wishes to send to Bob a *signature* of message M Alice applies a hash function $h()$ to M, so that $m=h(M)$. Alice sends to Bob M and $S=Signature(m)$, whereupon Bob can verify the signature. If the signature is verified then Bob accepts the message. Some examples of signature schemes that can be used in this protocol include the *RSA signature scheme* [4] and the *El Gamal signature scheme* [2].

In a *k out of n threshold sharing scheme* [5] the secret key $\mathbf{K}$ is shared out to n participants, so that any subset B of k participants can combine their shares and construct $\mathbf{K}$ while any subset of cardinality $\leq k-1$ gain no information about the $\mathbf{K}$. In a *k out of n threshold signature scheme*, the signing key $\mathbf{K}$ is shared out to n participants so that any k participants can sign a message M. We denote S_i as participant P_i's *partial signature*. *Verifiable signature sharing* is a cryptographic sharing technique that allows a holder of document to distribute shares of the signature of the document to proxies (participants), so that the proxies can later reconstruct and sign the document (if they wish).

In an electronic voting scheme, if a voter leaves data/information such that this data allows others to verify that the voter's vote has been counted, we say that the voter has left a *receipt*. A voting scheme is said to be *receipt-free* provided that no information is left by the voter, which allows others to verify the voter's vote.

We represent the legislature by $\mathbf{A} = \{P_1, \ldots, P_n\}$. We use $\mathbf{A}_t$ to represent the legislators present at time t, thus $\mathbf{A}_t \subseteq \mathbf{A}$. Here n is the size of the original legislature and n_t is the number of legislators present at time t. A *session* is a continuous period of time for which the legislators present $\mathbf{A}_t$ can vote on legislation and that the set of participants present remain fixed. The threshold k_t represents the threshold required to pass legislation at time t, for example in a legislature for which majority rules $k_t = \lfloor |\mathbf{A}_t|/2 \rfloor + 1$.Every time the legislature $\mathbf{A}_t$ changes, some type of redistribution of shares will need to take place.

The following describes requirements for a verifiable democracy. Due to space limitation we have omitted the arguments as to why each is a requirement.

- The transfer of signature power needs to be temporary.
- The transfer of signature power needs to be done blindly.
- The participants from At, when given an opportunity to act on legislation must know that the outcome ("sign" or "not sign") is a result of their decision and not a result of bad faith on the part of the participants who had transferred them the power to sign.
- No set of participants should gain any information about a motion made during an illegal session.
- In a representative government it is the right of a citizen to know how their representative voted. In such a case we are referring to a voting scheme that requires a receipt.

3 Verifiable Democracy Protocol – A Democratic Threshold Scheme

We start with a k out of n threshold scheme. At time t, n_t will represent the number of participants present, m_t will represent the message, and k_t will represent the dynamic threshold. A quorum exists provided $n_t \geq k$. Whenever $n_t \geq k$, we will naturally assume that $k_t \leq k$ and $k_t \leq n_t$. During the set-up, the legislature is empowered with a secret key so that any k out of n can compute the secret signing key. If $n_t \geq k$ we proceed with the protocol, if $n_t < k$ then there are not enough legislators to pass the legislation. At any time t, a message/law m_t, may be proposed. $\mathbf{A}_t$ represents the set of participants present at time t, $n_t = |\mathbf{A}_t|$.

Legislative key generation. A secret key **K** is distributed to the n participants so that a "blinded message/law" can be signed in a k out of n threshold manner. In addition to distributing shares of **K** this distributor generates ancillary information. This ancillary information will be broadcasted to all, i.e. public record. The nature of the ancillary information is dependent on the verifiable sharing scheme that is used.

Blinding message. The participant P^* who proposes message m_t, blinds m_t before they present it to the legislative body $\mathbf{A}_t$.

Transfer of Power -- Partial Signature Generation TPSG. As long as n_t exceeds (or equals) k the message will be considered for signing. If so k participants in $\mathbf{A}_t$ are chosen and they generate partial signatures for the blinded m_t.

Transfer of Power -- Partial Signature Distribution TPSD. Each of the k participants share out their partial signatures in a k_t out of n_t manner to $\mathbf{A}_t$ (we will refer to these k participants as *partial signature distributors*). Each participant in $\mathbf{A}_t$ has received k shares, whereupon they compress the k shares to one share. In addition to distributing partial signatures, the partial signature distributors will also distribute ancillary information that allows the legislative body $\mathbf{A}_t$ to verify the correctness of the partial signatures of the blinded m_t.

Transfer of Power -- Partial Signature Verification TPSV. The ancillary information provided in TPSD is first verified by each legislator in $\mathbf{A}_t$. Upon verification the ancillary information is used by each legislator to verify the correctness of their "share of the partial signature of the blinded m_t". The verification procedure is devised so that with overwhelming probability it can be determined that a recipient has received a valid share this is achieved via a "verification and complaint" protocol. If a verification fails then a complaint will be raised, at that time a cheater has been detected, what remains is a protocol to determine whether the cheater is the "partial share distributor" or the "complainer". The consequence is that the completion of this stage with no complaints implies that the signature power for the message has been transferred to $\mathbf{A}_t$ such that any k_t can sign the message.

Unblind the message. The message is revealed to the legislature. Who reveals the message? P^* could. Or if one utilizes a trusted chairperson as in [3], then the trusted chairperson could reveal m_t. In [1], the protocol utilized RSA signatures and so the legislators themselves could unblind the message without the legislators revealing their partial signature of m_t.

Decision -- vote on m_t. The legislators decide whether to vote for or against m_t.

Partial Signatures Sent PSS. If any legislator wishes to vote for, the now unblinded, m_t they send their share of the partial signature of the blinded m_t.

Verification of the signature -- determining the passage of m_t. If k_t or more participants have sent their partial signatures then the message may be passed. If so, the combiner selects any k_t of the sent partial signatures and verifies the correctness of these partial signatures using the ancillary information provided within this protocol. For each one of these invalid partial signatures the combiner selects one of the remaining partial signatures sent and verifies it. If the number of valid partial signatures is less than k_t then the message m_t is automatically not passed. We have adopted a receipt-required version of the verifiable democracy protocol. The partial signature sends (PSS) together with the partial signature verification (PSV) implies k_t "valid votes". Who can play the role of the combiner? Any person, collection of people, or even the legislators can verify.

Message passed. The message is passed if a signature of m_t can be computed and there were k_t "valid votes"' sent and verified. A vote for m_t is a valid partial signature.

References

1. Y. Desmedt and B. King. Verifiable democracy. *IFIP TC6/TC11 Joint Working Conference on Communications and Multimedia Security (CMS'99)*, Kluwer Academic Publishers, 1999, pages 53-70.
2. Taher El Gamal: A Public Key Cryptosystem and a Signature Scheme Based on Discrete Logarithms.*CRYPTO 1984,* pages 10-18.
3. H. Ghodosi and J.Pieprzyk. Democratic Systems. *ACISP 2001*,pages 392-402
4. R. Rivest, A. Shamir, and L. Adelman. A method for obtaining digital signatures and public key cryptosystems. *Commun. ACM*, 21, pages 120-126, 1978.
5. A. Shamir. How to share a secret. *Commun. ACM*, 22, pages 612-613, Nov., 1979.

Arguments for a Holistic and Open Approach to Secure e-Government

Sonja Hof

University of Linz, Institute of Applied Computer Science,
Division: Business, Administration and Society, University of Linz, Austria
Eracom Technologies Switzerland AG,
sonja.hof@eracom-tech.com

Abstract. Security is widely acknowledged as one of the most important aspect for a successful e-Government implementation ([1], [2], [3]). Every month, new vulnerabilities and attacks are discovered and published. New security standards, patches, "fashionable" abbreviations pop up almost daily. How should a user trust the security and integrity of an e-Government portal in such an environment? This contribution starts with an overview of the technical as well as social aspects of security for an e-Government portal. They are introduced one by one and it is shown how to tackle them by emphasizing their drawbacks as well as their differences to the better-known world of e-Business. The paper finishes with by pointing out the requirements for the next generation e-Government platforms by proposing a simple and uniform approach security in its whole picture.

1 Introduction

Starting the implementation of a "secure" e-Government portal is an ambitious project. There are huge sets of different security aspects that have to be considered, solved and implemented ([1], [2], [4]). Some of them are still a matter of research, for which no "off-the-shelf" solutions exist. With each aspect, the paper shows their inherent problems and points out existing solutions or the lack of them. It also shows the individual drawbacks of them, as well as the drawback of this approach as it appears, if one puts all the different aspects and solutions together. Using different solutions for different aspects, instead of a complete solution that covers the full picture, hinders citizens to "identify" themselves with the system.

This paper gives an overview of the different aspects that have to be inspected for a secure e-Government portal. This is achieved by looking at the problem from two different angles: technical and social. As a third separate entity, this paper looks at the anonymity required for some processes, e.g. e-Voting. Finally, the paper concludes with a proposal for a uniform approach for a holistic view on e-Gov security.

R. Traunmüller and K. Lenk (Eds.): EGOV 2002, LNCS 2456, pp. 464–467, 2002.

2 Technical Security Aspects

2.1 Authentication

Secure electronic authentication is of utmost importance for every e-Government project. The sureness to communicate with the appropriate partner, and not with an impersonator is an inherent feature of e-Government. On the technical side there are three different dimensions for authentication: how, who and when.

There exist many different techniques and approaches to securely authenticate a user electronically. They range from purely software solutions, e.g. passwords, to mixed solutions, e.g. transaction numbers, to hardware solutions, e.g. smartcards.

The second dimension to authentication defines who has to authenticate himself to the system. E-Government allows for communication between the government (one part of it) and a second party, e.g. citizen or a juristic person.

2.2 Privacy

Similar to the different authentication requirements, different tasks require different levels of privacy. Some tasks require no privacy, for others it is sufficient to ensure data privacy, and again other tasks require that even the act of using the e-Government portal has to be considered to be private.

Additionally, privacy has to be divided into short and long-term privacy. Short-term privacy has to ensure the privacy of the data while the task is active, e.g. transmission privacy. On the other side, long-term privacy deals with the time after the task has been completed. It deals with the data after it has been stored on the portal. Technical solutions in this area include SSL, file or disk encryption solutions.

2.3 Different Security Levels of Data

As mentioned above, it is important that data is only accessible to authorized users. Additionally to this restriction, there is a second dimension that restricts the access rights valid users actually have. This dimension defines the tasks that may be done with the data. Existing systems define multiple layers.

2.4 Data Integrity and Safety

Another important aspect of an e-Government portal is data integrity. The portal has to prevent tampering the data during transmission to and from the user, as well as to deny tampering the data once it is stored on the portal itself. The second aspect introduces a long-term security problem. Technical solutions in this area include digital signatures, unforgeable audits...

A further technical aspect regarding data is its safekeeping. Once transmitted and acknowledged, data should be stored in such a way that it is reconstructible that a data item was submitted, even if it lost for some reason. Technical solutions in this area include databases, transactions and digital signatures.

2.5 Quality of Service

For many tasks on an e-Government portal the quality of service (QoS) is of minor importance. Of course, responsiveness and availability are not to be neglected. However, compared to other requirement, they are of minor importance.

In contrast to that, there exist tasks, for which QoS is of utmost importance, e.g. e-Voting. In such cases it has to be guaranteed that a valid user is able to finish his task, e.g. to cast his vote. Technical solutions include firewalls or load balancer.

3 Social Security and Anonymity Aspects

E-Government portals have to cope with two different dimensions while coping with security. On the one hand there are technical security aspects as described in the previous chapter. On the other hand there are social aspects that influence security on quite a different level. Solutions to these issues are to be found on a social level.

3.1 Trust

Users tend to be careful in trusting the government to keep their data safe. Especially getting their trust for anonymous e-Voting is difficult. We think that the only way to gain their trust is to present them an open system, i.e. to allow other parties to give their opinion about the system, e.g. McKinsey or the Chaos Computer Club. In the end, all effort spent into an increased trust pays back.

3.2 Privacy and Simplicity

Privacy has, besides technical dimensions, also social aspects. Different users consider different data as private or public. The direct approach to always use private transmission is not feasible, as the necessary overhead is not always acceptable.

Simplicity seems to be an obvious aspect. Of course, an average user will never "understand" the security measurements and policies of an e-Government portal. However, he may know a "might-be" expert or try to inform himself with some literature. While the overall problems of course remain, one can decrease the "subjective" difficulty by trying to decrease the number of involved systems, e.g. a single authentication mechanism for all services.

3.3 Anonymity

E-Government portals have a peculiarity that distinguishes them from e-Business portals: Anonymity. For a set of tasks, the portal has to obey two contradicting rules: authentication and anonymity. Ready technical solutions in this area do not exist.

4 Conclusion

As shown in the previous sections, securing e-Government cannot be seen as a single process, but as a collection of hugely different tasks, systems, processes and requirements. In contrast to e-Business portals, e-Government has to go one step further regarding security. Additionally to technical security, it has to make this step also in the area of social security.

Seen from a technical side, an e-Government portal has, besides additional requirements (e.g. anonymity), to offer a higher level of security. The risks involved with a security leak are just too big. Most e-Business portals life with the fact of lost data and illegal transactions. E-Government portals just cannot accept this.

On the social side, the difference between e-Government and e-Business sites is even larger. For an e-Gov portal "trust" is of utmost importance, as a "user" has no way of circumventing the portal by not using it or by using an alternative one. Additionally, the user has also to cope with the fact, that even though he may not use the portal, the decisions made with its help, e.g. a vote, influence him directly.

Recapitulated, a secure portal does not only have to be secure, but also has to be "trusted" to be secure. To reach this goal, we propose to focus on global and open solutions to secure e-Government portals. As pointed out in the previous sections, we think that this goal can best be reached using an approach that has two distinct properties. First, it has to be open in the sense that third parties can evaluate its security. There should be as few restrictions as possible on what may be evaluated. Second, the security solution should consist of as few as possible building blocks, as each building block has to be "trusted", in order to trust the whole system.

References

1. J. von Lucke, H. Reinermann. Speyerer Definition von Electronic Government; http://foev.dhv-speyer.de/ruvii
2. Maria Wimmer and Bianca von Bredow. 2001. E-Government: Aspects of Security on Different Layers. In Proc. of the International Workshop "On the Way to Electronic Government", IEEE Computer Society Press, Los Alamitos, CA, pp. 350-358
3. Maria Wimmer and Bianca von Bredow. 2002. A Holistic Approach for Providing Security Solutions in e-Government. In Proc. of the HICSS 35. (ISBN 0-7695-1435-9)
4. Gerhard Weck. 2001. Zertifikate, Protokolle und Normen, Stolpersteine auf dem Weg zu einer PKI. In Patrick Horster (eds.). Elektronische Geschäftsprozesse, pp. 132-154 (ISBN 3-936052-00-X)

Supporting Administrative Knowledge Processes*

Witold Staniszkis

Rodan Systems S.A., Puławska 465, 02-844 Warszawa
witold.staniszkis@rodan.pl

Abstract. We present the general knowledge management model typical for central as well as local government agencies pre-requisite for successful implementation of knowledge management initiatives. We show that knowledge management systems comprising intelligent workflow management features are necessary to provide sufficient level of support for administrative knowledge processes. We conclude with a brief presentation of the ICONS project aiming at providing a KMS platform for e-government.

1 Introduction

The common fallacy of the IT side of the KM scene is focusing on the purely technological view of the field with the tendency to highlight features that are already available in advanced contend management systems. Such systems are commonly referred to as corporate portal platforms or, more to the point, as the knowledge portal platforms. From the KM perspective, as discussed in [McElroy1999], such claims may be justified only with respect to a narrow view of the field **focusing on distribution of existing knowledge throughout the organization**. The above views, called by some authors the "First Generation Knowledge Management (FGKM)" or "Supply-side KM", provide a natural link into the realm of currently used content management techniques, such as groupware, information indexing and retrieval systems, knowledge repositories, data warehousing, document management, and imaging systems. We shall briefly refer to existing content management technologies in the ensuing sections of the report to show that, within the above narrow view, the existing commercial technologies meet most of the user requirements.

With the growing maturity of the KM field the emerging opinions are that **IT support for accelerating the production of new knowledge** is a much more attractive proposition from the point of view of gaining the competitive advantage. Such focus, exemplified in stated feature requirements for so called "Second Generation Knowledge Management (SGKM)", is on enhancing the conditions in which innovation and creativity naturally occur. This does not mean that such FGKM required features as systems support for knowledge preservation and sharing are to be ignored. A host of new KM concepts, such as knowledge life cycle, knowledge processes, organizational learning and complex adaptive systems (CAS), provide the underlying conceptual base for the SGKM, thus challenging the architects of the new generation Knowledge Management Systems (KMS).

---

* This work has been supported by the project ICONS IST-2001-0324226

R. Traunmüller and K. Lenk (Eds.): EGOV 2002, LNCS 2456, pp. 468–471, 2002.

Government agencies constitute a rather specific environment from the point of view of the knowledge management requirements. We attempt to identify the knowledge creation and dissemination processes in e-government, calling them administrative knowledge processes, and then, we discuss the KMS features required to provide IT support for administrative knowledge management initiatives.

2 Knowledge Management in Public Administration

The schematic view of the knowledge management cycle typical for a public administration agency is shown in figure 1. The view closely follows the Popper's three world model [Popper1971], with the bottom level corresponding to the realm of physical and abstract objects existing in the environment, the middle layer corresponding to perceptions, skills and attitudes of employees (tacit knowledge), and the upper layer representing the knowledge resources (explicit knowledge) maintained and disseminated in an organization.

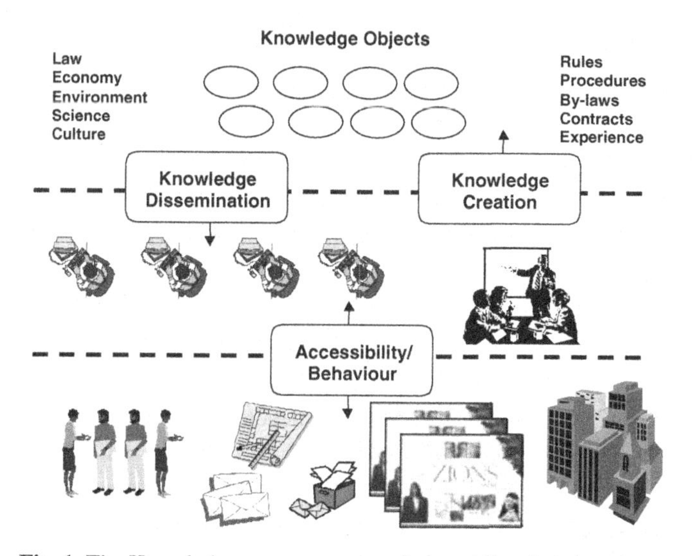

Fig. 1. The Knowledge management cycle in public administration

The accessibility and behavior are the principal characteristics of a public administration agency experienced by the environment (society, other organization), whereas the tacit knowledge [Nonaka1995] determines the actions of agency's employees. The knowledge management cycle is based on the one hand on externalization of tacit knowledge to create explicit knowledge artifacts to be accessible to others in the internalization process. Thus, although indirectly, the knowledge management cycle determines the quality of work in a public administration agency.

3 The Knowledge Management System Reference Architecture

A Knowledge Management System (KMS) is an IT platform supporting knowledge management processes taking place in an organization. A KMS reference architecture developed as the starting point of the IST ICONS project [ICONS2002] is presented in figure 2.

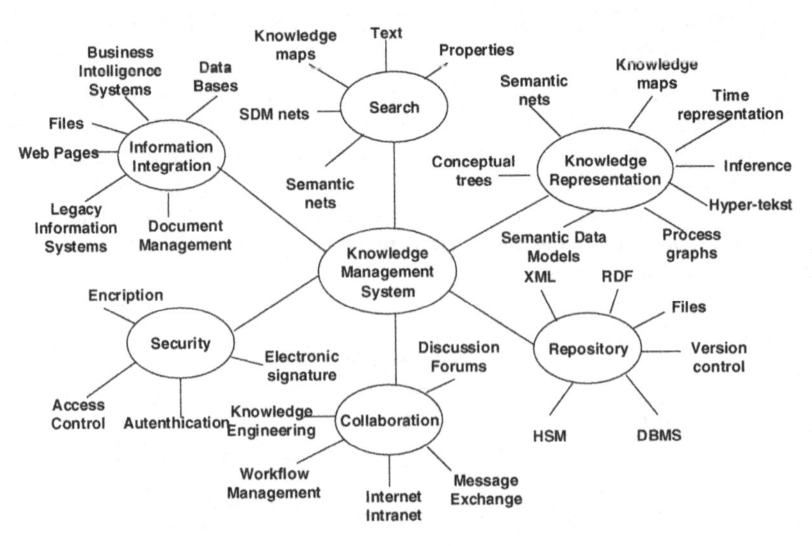

Fig. 2. The KMS reference architecture

The user functions clustered in the principal KMS features may play varying support roles within the knowledge management processes. Collectively, the sum of user requirements for a given principal feature, defined within the distinct knowledge management processes, represents the user requirement set for a given principal KMS feature. The reference architecture has been developed on the basis of a feature analysis of knowledge management systems presented in [KMForum2001, KMForum2001_D11, KMForum2001_D11a, KMForum2001_D12].

The KMS features, grouped into six principal feature sets, represent our current views pertaining to the KM technology requirements. Some of the features are already common in the advanced content management systems, referred to as the corporate portal platforms, some other are subject to the on-going KMS research efforts.

The Domain Ontology features pertain primarily to knowledge representation including the declarative knowledge representation features, such as taxonomies, conceptual trees, semantic nets, and semantic data models, as well as the procedural knowledge representation features exemplified by the process graphs. Time modeling and knowledge-based reasoning features pertain both to the declarative and the procedural knowledge representations. Hyper-text links are considered as a mechanism to create ad hoc relationships between content artifacts comprised in the repository.

Taxonomies provide means to categorize information objects stored in the content repository. Categorization classes may be arbitrary hierarchical structures grouping information objects selected by the class predicates. Class predicates are defined in the form of queries comprising information object property values or as full text queries comprising key word and/or phrases. Categorization classes are not necessarily disjoint.

Semantic networks provide means to represent binary 1:1 relationships, expressed usually as named arcs of a directed graph, where vertices are information objects belonging to any of the information object classes. Normally, the linked object classes are determined by the binary relationship semantics of the corresponding named arc. An example of a simple semantic net may be a binary relation Descendants defined as a subset of the Cartesian product of the set of Persons.

Business processes are usually represented by process graphs, typically by the Event-Condition Petri Nets or by directed graphs. Petri Net representation allows for expressing richer process semantics, in particular the pre-and post-conditions for process activities. The process specification must also be supplemented by the set of role definitions, one definition for each process activity, to enable the workflow management engine to properly assign tasks to KMS actors. The process graph representation should comprise a set of process metrics and, possibly, performance constraints and exception conditions.

References

ICONS2002, The ICONS Project Description, www.icons.rodan.pl, 2002.

KMForum2001, Weber, F., Kemp, J., Common Approaches and Standarisation in KM, EKMF Workshop on Standarisation, Brussels, June, 2001, www.knowledgeboard.com.

KMForum2001_D11, Kemp, J., Pudlatz, M., Perez, P., Ortega A.M., KM Technologies and Tools, European KM Forum, IST Project No 2000-26393, March, 2000, www.knowledgeboard.com.

KMForum2001_D11a, Kemp, J., Pudlatz, M., Perez, P., Ortega A.M., KM Terminology and Approaches, European KM Forum, IST Project No 2000-26393, March, 2000, ww.knowledgeboard.com.

KMForum2001_D12, Simpson, J., Aucland, M., Kemp, J., Pudlatz, M., Jenzowsky, S., Brederhorst, B., Toerek, E., Trends and visions in KM, European KM Forum, IST Project No 2000-26393, April, 2000, www.knowledgeboard.com.

KPMG1999, KPMG Consulting, Knowledge Management Research Report 2000, November, 1999, www.kpmg.co.uk.

McElroy1999, McElroy, M.W., Second-Generation KM, Knowledge Management, October 1999.

Nonaka1995, Nonaka, I., Takeuchi, H., The Knowledge Creating Company, Oxford University Press, 1995, New York, USA.

Popper1972, Popper, Karl R., Objective Knowledge, Oxford University Press, 1972, London, England.

IMPULSE: Interworkflow Model for e-Government

Aljosa Pasic, Sara Diez, and Jose Antonio Espinosa

SchlumbergerSema, Albarracin 25, 28037 Madrid
{aljosa.pasic,sara.diez,jose-antonio.espinosa}@sema.es

Abstract. New applications such as Virtual supply chains or E-government stress importance of application integration in heterogeneous environments. Workflow based approach is particularly interesting for public administrations, which traditionally tend to have hierarchic streamlined processes. We present a model for interworkflow complex service execution where control and co-ordination is performed by a supervisor workflow hub.

1 Workflow Tools in a Cross-Agency Process

Workflow Management Systems (WFMS) provide mechanisms to define businesses processes and to automate their execution in a way that isolates the business logic and the integration of different systems. In the past, WFMS have focused on homogeneous and centralised environment within the boundary of a single organisation. However, in newly created Internet paradigms such as B2B E-commerce or E-government, WFMS should support collaboration between different organisations. The challenge is therefore, to create new models for interaction of heterogeneous systems where internal workflow process details are abstract for other organisations and where enactment of different WF is performed in a co-ordinated way. Therefore, we will use interworkflow definition given in [1] to describe a tool that takes care of coordination of each workflow among different organisations.

On the other hand, a new concept closely related to enterprise application integration (EAI), has been built: BPI, Business Process Integration. BPI usually appears with another common acronym: BPM, Business Process Management, defined as: "The concept of shepherding work items through a multi-step process. The items are identified and tracked as they move through each step, with either specified people or applications processing the information. The process flow is determined by process logic and the applications (or processes) themselves play virtually no role in determining where the messages are sent." [2]. Our interworkflow model combines the best of a workflow tool with a powerful integration platform where CORBA and XML are the technological pillars under which the current model rests. The FORO Connector Architecture [3] defines a standard architecture for connecting the FORO Workflow Engine with external applications and legacy systems. These external applications will be invoked through common application drivers called connectors. The FORO Connector Architecture represents the implementation of the Interface 3 suggested by the Workflow Management Coalition [4] and [5].

The workflow interoperability standards define the mechanism that workflow product vendors are required to implement in order that a workflow engine makes

R. Traunmüller and K. Lenk (Eds.): EGOV 2002, LNCS 2456, pp. 472–479, 2002.

requests to another workflow engine to effect the selection, instantiation and enactment of known process definition by that other engine.

The requested workflow engine should also be able to receive back status information and the results of the enactment of the process definition. As far as possible this is to be done in a way that is "transparent to the user". Workflow enactment services provides the runtime environment in which process instantiation and activation occurs utilising one or more workflow management engines, responsible for interpreting and activating part, or all, of the process definition and interacting with the external resources necessary to process the various activities. Therefore, in our approach we distinguish the external resources such as humans or other software tools invoked to perform particular tasks from enactment services that are used to address other workflow engines.

Improving Public services (IMPULSE) is an IST programme project where two different workflow engines (Staffware and Foro-wf) are used in order to implement and validate this approach. The project will provide workflow enactment services and some of the external resources involved in sub-agencies will, in their turn, be implemented as other workflow enactment services (WES nesting).

2 e-Government: Transactions across Different Systems

The term E-Government, according to definition of World Bank [6] refers to the use by government agencies of information technologies (such as Wide Area Networks, the Internet, and mobile computing) that have the ability to transform relations with citizens, businesses, and other arms of government. Naturally, this interaction is implemented through web sites, but less obvious is the actual stage of e-government background implementation: the actual integration and transaction of procedures.

Therefore, some definitions of e-government are more restrictive, focusing on different stages, and making it the public sector equivalent of e-commerce.

In this paper we will limit our scope to what in [7] was described as Stage 2: Enabling inter-organizational and public access to information.

A number of governmental services is listed as promising candidates to be implemented as E-government services where workflow based integration forms the heart of back office system:

- claims processing and management;
- bid and proposal routing and tracking;
- handling of customer service and complaints;
- grant and scholarship award, approval, and processing; and
- human resource recruitment and hiring.

Any of above services contains processes that can be spread over various information systems, moreover they can be spread over various workflow systems. This type of services, which have been described in IMPULSE user scenarios [8] as "complex services", usually include several already available "simple" processes, which reside on the local (single agency managed) workflow (WF) or GroupWare (GW) systems. On the other hand, any service or sub-service of different governmental agencies that contains same variables (such as "change address" service), we defined as the "common services".

From the implementation point of view any service that requires the two or more WF and/or GW to be linked together, it is necessary to:

- Deal with the existing implementation problems where a physical boundary exists (Workflow and/or Groupware servers are not at the same site).
- Deal with the logical and organisational differences.

The end-to-end view describes the possible obstacles and differences that might exist between departments or organisations and offers an approach for dealing with them.

Furthermore, in our project each single agency WF/GW administration is responsible for publishing and maintaining its own service catalogue (with processes that have been solicited by interworkflow service model), depending on the policies and resources at its disposal. Global management and responsibilities of the IMPULSE interworkflow system, as well as responsibility for the design of the complex service, have also been left out of the scope of this project.

Finally, in figure 1, we present an excerpt from one of user cases in the project. The service "Apply for individual benefits" from Canary Island Government, spreads over three agencies and connects to Staffware and Foro-wf workflow management systems.

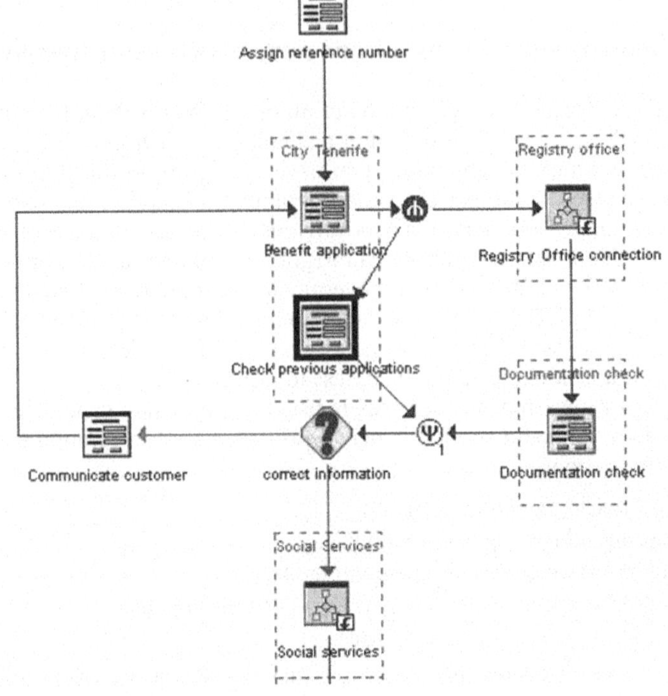

Fig. 1. A part of the complex service implemented with IMPULSE interworkflow tool

3 Interworkflow Modelling

Interworkflow process editor that we use is actually enhanced designer tool from one of the WFMS. The main difference is that sub-process, such as Registry office connection in the figure 1, can be inserted and defined as the external process that starts and ends with invocation of connector to an agency based workflow.

In [9] the modelling focuses on three main points: 1) start and end of work linkage; 2) structures for controlling the workflow and 3) exchanged data. Another model of coordination module between different WFMS is presented in [15] and has been named workflow mediation. However cooperative interworkflow models are more suitable for virtual enterprise and similar applications while public administrations are characterised by hierarchic structure of streamlined procedures. In the line with so called "one-stop-shop" idea, we propose supervising interworkflow hub, which will act as a control and co-ordination tier between single window for citizens or companies and separate agency managed workflow and groupware tools. Furthermore, control layer of this hub will also take care of connection, data translation and high level role mapping, while co-ordination layer is implemented by the existing workflow engine (with minimum modifications). This brings obvious investment savings.

IMPULSE model is presented in the figure 2.

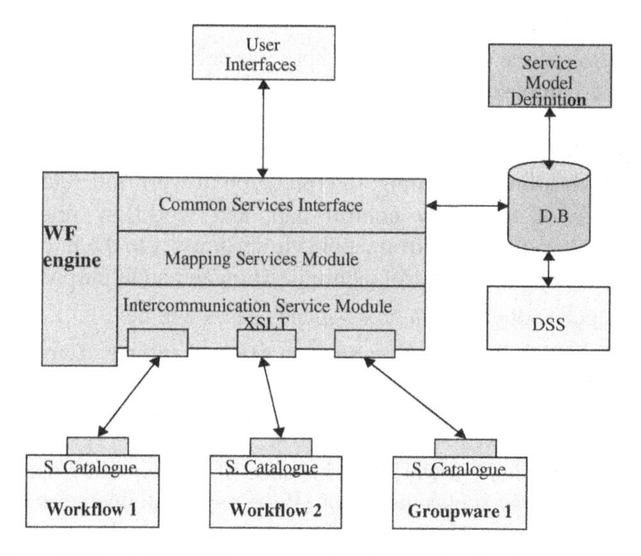

Fig. 2. IMPULSE model for interworkflow control and coordination in e-government applications

In our example from figure 1, we are making multiple data exchanges: data (citizen data request) is sent from the interworkflow to the WFMS1, from where it goes to the checking procedure that can take place in the same agency. In parallel, registry with previous applications is checked by IMPULSE interworkflow, which for this case acts as WFMS2. Except for execution of this task, IMPULSE is also responsible for running control agent that periodically checks control status and output data (if any) from the WFMS1. Finally, when both IMPULSE (WFMS1) and WFMS2 finished their sub-process, the first one will perform conditional step and will eventually pass the work to the WFMS3 (Social services).

Therefore, we are combining co-operative and hand-over types of linking.

Data exchange is based on XML-based interoperability specification [10] but it is able to interact with other XML markup vocabularies. The full overview of workflow interoperability and XML frameworks is given in [12]. In our interworkflow data model items may represent the properties of the process instance (workflow control or workflow relevant data), and/or any application related data associated with invoked applications during process enactment (application data) or they can represent process instance states. Also, various types of exceptions are defined, including temporary and fatal error types. However, the open problems remains, as also described in [13] where interworkflow approach was investigated in the healthcare domain, that each possible interleaving of steps that might occur at execution time, has to be predicted at process designing time, which leads to extreme complexity of design process.

The attractive characteristics of IMPULSE runtime server is that it will provide also runtime execution, based on existing WFMS engine. This reduces costs for companies that already have Staffware or Foro-Wf workflow engines. In IMPULSE architecture interworkflow environment is implemented as a layer around this existing engine that is responsible for:

- Interpretation of the complex service process definition
- Control of complex service process instances: creation, activation, suspension, termination…
- Navigation between process activities, which may involve sequential or parallel operations, deadline scheduling, interpretation of workflow relevant data
- Maintenance of workflow control data and workflow relevant data, passing workflow relevant data to /from applications or users to the interworkflow layer
- Supervision actions for control, administration and audit purposes

A typical sequence of events in cross-agency cases will be:

1. When the IMPULSE service requires an external service, it notifies the IMPULSE Intercommunication layer.
2. The IMPULSE Intercommunication layer *sends* the request to the Connector of the chosen WFMS or GW.
3. The Connector initiate a workflow instance (if the requested operation is a process in a workflow) or executes the action (if the requested operation is in GroupWare).
4. The domain server performs this part of the IMPULSE service on behalf of the IMPULSE server.
5. When the service part is completed, the Connector receives the results of the process. It forwards these to the Intercommunication layer.
6. The consumer Intercommunicatio*n* layer hands the results to the corresponding IMPULSE workflow queue or it translate it (if necessary) and sends it to the following domain Connector.

4 Connector Agents with External WFMS/GW

IMPULSE tool accesses external applications for tasks and services execution, via standardised software applications named connector agents. These agents, positioned between the WFMS and the communication layer, can be suitable for a specific application or for a class of common applications, and they run on different platforms. Each agency WFMS must also include a connector agent (that can be easily installed) to support communication with IMPULSE server and other domains. This requires co-ordination logic that copes with invocation across different platforms and network environments, together with a means of transferring workflow relevant data in a common format or transferring it to the case instances in the individual application environments. In addition, the connector also takes care of commands for requesting the cancelling of a linkage between instances.

Connector agent also negotiates the information needed by elements, specifying name and type of the information element (variable or document). Once known the information required by IMPULSE engine, it sends the information elements and invokes external enactment services.

The basic protocol to follow is quite simple:

get_required_informationDef
set_input_information
need_moreInfo
get_returned_informationDef
get_output_information
release

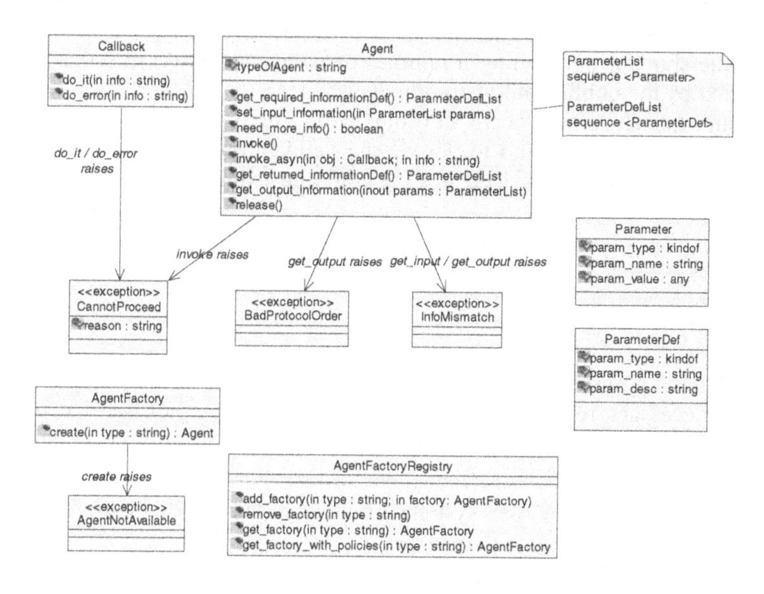

Fig. 3. IMPULSE connector agent architecture

IMPULSE connector agent architecture, presented in figure 3, is compliant to the WfMC specifications of workflow interfaces.

One of the common requirements for E-government applications is the citizens (in C2G services) or business (in B2G services) is the need for messaging of the case instance status. IMPULSE tool have several ways to interact with the case initiator: while the information is sent to public administration through web interface (Java server pages with special Tag library), the opposite way communication, that can be resolution data or interim status report, is either sent by an e-mail notification or it is placed in case initiators virtual tray. While resolution data sending is triggered by the case competition, interim status report is more dubious to design and implement because of the productivity trade-off. The proposed solution in connector agent is similar to a "courier" that periodically visits individual agencies involved in the case execution.

5 Further Work

In order to improve service design related communication between governmental agencies and their workflows which are handling data from multiple registries, a mechanism for representing semantic properties of governmental data is needed. Relationships between concepts and terms in the domain is crucial for designing complex services such those we mentioned in this paper. Some work on this subject, relying on a domain ontology that encapsulates the concepts in the small business initiation domain was done in [11].

As the public administration moves towards the competitive environment, adopting some of the business world paradigms, it is to expect a growing number of public-private partnership for performing a number of independent services that nowadays are completed by the public administration agencies.

This would introduce new needs into our model, such as the mechanism to compare different service offers and to derive and validate quality of service (QoS) parameters. The work done described in [15] could be exploited in order to meet this requirements.

6 Conclusion

From a functional point of view, the supervising interworkflow hub contain sufficient information necessary to establish, enact and control a complex service public administration exchange relationship. More specifically, the prototype we delivered in the IMPULSE project supports:

- connecting the relevant WFMS systems of different public administration agencies.
- designer interface for complex services
- Tag libraries for interface between "one-stop-shop" (single e-government window) and IMPULSE server
- the interactions needed to complete the complex services (mapping, translation...).
- the cross-agency status control.
- activities surrounding the service provision/consumption: e.g. remuneration, auditing.
- the termination mechanisms of the service enactment.

However, further research on ontology in public administrations and work on distributed knowledge representation is needed in order to design more effectively these complex cross-agency services.

References

1. K. Hiramatsu, Ken-ichi Okada, Yutaka Matsushita, H. Hayami: Interworkflow System: Coordination of Each Workflow among Multiple Organizations. CoopIS 1998: 354-363
2. ITQuadrant, Inc. ; "Glossary of terms"; November 2000.
3. Jose Antonio Espinosa, Antonio Sanz Pulido: IB-a workflow based integration approach, September 2001.
4. David Hollingsworth; "The Workflow Reference Model; Document Number TC00 – 1003"; Workflow Management Coalition; January 1995.
5. Workflow Management Coalition; "Workflow Management Application Programming Interface (Interface 2 & 3) Specification version 2.0"; Document Number WFMC-TC-1009; Workflow Management Coalition; July 1998.
6. http://www1.worldbank.org/publicsector/egov/definition.htm
7. Clay Wescot: E-government: enabling Asia-Pacific governments and citizens to do public business differently, july 2001
8. IMPULSE consortium: Improving public services requirements, April 2001
9. Interworkflow application model: WfMC-TC-2102 specification
10. Workflow standard interoperability Wf-XML binding: WfMC-TC-1023 specification
11. Nabil R. Adam, Francisco Artigas, Vijayalakshmi Atluri, Soon Ae Chun, Sue Colbert, Melania Degeratu, Adel Ebeid, Vasileios Hatzivassiloglou, Richard Holowczak, Odysseus Marcopolus, Pietro Mazzoleni, Wendy Rayner, Yelena Yesha: E-Government: Human-Centered Systems for Business Services
12. Michael zur Muehlen, Florian Klein: AFRICA: Workflow Interoperability based on XML-messages
13. Reichert M., Dadam P.: A framework for dynamic changes in WFMS,
14. Kamath Mohan and Krinith Ramamritharm: Failure handling and coordinated execution of concurrent workflows
15. Vassilis Cristophides, Richard Hull, Akhil Kumar and Jerome Simeon: Workflow mediation using VorteXML, March 2001
16. Justus Klingemann, Jurgen Wasch and Karl Aberer: Adaptive outsourcing in Cross-organisational workflows, ESPRIT project X-flow

Visualization of the Implications of a Component Based ICT Architecture for Service Provisioning

René Wagenaar and Marijn Janssen

School of Technology, Policy and Management, Delft University of Technology
Jaffalaan 5, NL-2600 GA, Delft, The Netherlands
Tel. +31 (15) 278 1140/8077, Fax. +31 15 278 3741
{Renew,MarijnJ}@tbm.tudelft.nl

Abstract. The planning and subsequent nationwide implementation of E-government service provisioning is faced with a number of challenges. Initiatives are confronted with a highly fragmented ICT-architecture that has been vertically organized around departments and with hardly any common horizontal functionality. It is anticipated that in the long run, an architecture based on generic, standardized components in the form of Web services will lead to a more flexible provision of government services over electronic channels. This paper reports on the use of a simulation environment for communicating the advantages of such a component based approach to ICT decision makers within local government.

1 Introduction

Public administrations should stay closer to citizens' every-day life, and act more proactively rather than reactively. Governmental organizations are challenged to provide more customer-oriented products and services [2,4]. Customers can be targeted through multiple channels, such as web-based, call centers and physical offices in the municipality hall. In order to exploit these channels in a coherent, efficient and effective way, the need to restructure administrative functions and processes is clearly felt to support coordination and cooperation between different departments. Legacy information systems within governmental organizations, however, often restrict the development towards new customer-oriented processes.

In general, the current situation is such that each governmental organization has developed its own information systems in rather isolation, and that for each product or service a separate information systems exists. No generic architecture is available that enables communication between front office and back-office applications, between back-office applications or with systems outside the own organization. Beneath being monolithic packages, enterprise information systems have been criticized that they often impose their own logic or business process view on an organization and lack flexibility and adaptability in today's dynamic environment [1]. Currently, pleas have been made for more open, flexible architectures constructed of relatively small

R. Traunmüller and K. Lenk (Eds.): EGOV 2002, LNCS 2456, pp. 480–483, 2002.

components, which can be configured to support a limited number of functions [3]. Governmental organizations are relatively slow in adapting such approaches as they lack sufficient insight into the pros and cons of such an approach. Gaining commitment is further complicated due to the large number of stakeholders that are involved such as politicians, process owners, information managers, ICT-departments, administrative departments etc. Some stakeholders might have a natural resistance against or not trust new initiatives; some might have too limited knowledge for decision-making or lack experience with ICT.

In this paper we will investigate the strength of simulation as a communication vehicle for the evaluation of a component based ICT architecture for E-Government service provisioning, as it allows to understand the essence of business systems, to identify opportunities for change, and to evaluate the effect of proposed changes on key performance indicators.

2 Project Description

Dutch municipalities are free to design their information architecture and to choose appropriate software vendors. Often there is no central management and departments can buy their own applications for each process. As a result, municipalities have a highly fragmented ICT-architecture, consisting of legacy systems for each product they offer. In short, there is an *interoperability* problem between applications within a municipality, but also between municipalities.

The VNG, the Dutch association of municipalities, has launched a number of initiatives to develop *communication standards*. The most important initiative in this respect is the creation of standards for the GBA, the Dutch authentic registration of data about all residents living in a particular city or village. Many applications, such as the passport or drivers license renewal request, and address changes need to use data of the GBA and consequently need to communicate with the GBA.

Apart from the interoperability problem, the Dutch software market for municipalities can be characterized by lock-in through a duopoly. Due to this dependency, municipalities have either to invest heavily for developing their own customized applications or pray that one of these companies will ship out software or customize an existing software package that will serve their needs. To counter this lock-in situation, the information managers of the Dutch cities with more than 100,000 inhabitants have joined forces in a cross-municipality Information Management Council (IMG) under supervision of the VNG. The goal of this council is to search for more open, flexible architectures that can support multiple municipalities. The IMG council initiated the "*AnalysePilot*" aimed at developing a reference architecture that should provide guidance for the development towards a component-based architecture. Such an architecture should not only bring online the 290 products currently provided in the municipalities' portfolio, but also support existing distribution channels. One of the goals of this project was to support management in their decision-making about the potential of a component-based ICT architecture.

Generally, a component based ICT-Architecture considers information systems as combinations and integrations of different software components. The manageability increases as large components can thus be constructed from smaller components. Each single component can be replaced by another component without affecting the others. Components can run from different computer platforms and interact with each other through standardized interfaces. Components can communicate with each other directly or through *middleware.* The advantages of using middleware over direct interaction are that fewer connections need to be established and maintained and that changes need only be made at one place in the overall ICT-architecture. Besides, middleware enables interoperability and portability between components; it prevents lock-in and allows users to select components based on criteria such as quality, costs, functionality etc. rather than to accept offerings from a few vendors.

Process analysis of the existing processes at a number of municipalities was used to determine which type of generic components could be suggested for the architecture. We focused on modeling the tasks for one particular process, the renewal of driver licenses.

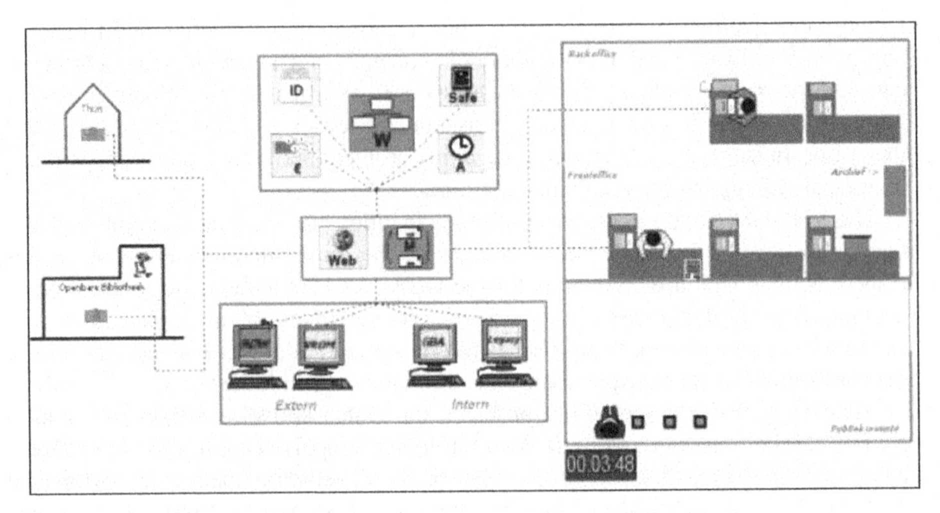

Fig. 1. Simulation model of Component based ICT architecture

3 Visualization Support

Simulation can be used as a communication instrument to stakeholders, since it allows to identify opportunities for change, and to evaluate the effect of proposed changes on key performance indicators without changing reality 5. The philosophy behind simulation is to develop a dynamic model of the problem situation, experiment with this model, and experiment with alternatives for the problem situation. In combination with visualization it can be used for creating understanding among non-experts for the impact of new process and ICT designs. As such, it was chosen by us in this project to show the Dutch municipalities which impact a generic component based ICT architecture could have on the execution of multi-channel service provisioning. Hitherto, two models were developed, an "as is" (current situation) and a "to be" situation (component-based and with multiple channels for service provisioning). Figure 1 shows a screenshot of the "to be" situation, with the new channels on the left side of the figure and the ICT-architecture based on components in the middle part of the figure. The functions of the components were simulated to ascertain that the model was independent of specific implementations. Municipalities should be able to replace any component with better ones when necessary, consequently only the functions of the components were simulated. Two new channels were added to the simulation, citizens using a computer at home and citizens using a terminal in the library, both connected to the Internet and requesting a driver's license. It shows that on a conceptual level a component-based approach is flexible and open enough to support a multi-channel approach. Future mobile channels can be supported using this approach as well.

4 Evaluation

Simulation models were presented to the information managers and management of the Dutch municipalities involved in this project. They agreed that a component-based approach created an open, flexible architecture because new components can be added to and generic components can be shared by multiple services and applications. They became also aware of the need for development of interface standards to counter potential drawbacks, such as time intensive wrapping of legacy systems, and issues concerning the maintenance of the ICT architecture.

Important *limitations* of the research are that we did not test the component based approach in practice, but limited the research to building simulation models to prove the concept. The purpose of this paper was to show that visualization based on simulation strongly supports in communicating new ICT concepts to the intended decision makers. It will be further used to assist the Dutch government in their efforts towards a more flexible, scalable and manageable ICT infrastructure.

References

1. F.J. Armour, S.H. Kaisler, and S.Y. Liu: A big-picture look at Enterprise Architectures. IEEE IT Professional, 1, 1, (1999) 35-42
2. R. van Boxtel: Actieprogamma Elektronische Overheid, The Hague, Ministerie van Binnenlandsezaken en Koninkrijksrelaties (1999) (in Dutch)
3. M. Fan, J. Stallaert, and A.B. Whinston: The adoption and design methodologies of component-based enterprise systems. European Journal of Information Systems, 9, 1, (2000) 25-35
4. K.J.L. Layne, Developing fully functional E-government: A four stage model. Government Information Quarterly, 18 (2001) 122-136
5. A.M. Law, and D.W. Kelton: Simulation Modeling and Analysis. New York: McGraw-Hill (1991)

Author Index

Lecture Notes in Computer Science

For information about Vols. 1–2371
please contact your bookseller or Springer-Verlag

Vol. 2372: A. Pettorossi (Ed.), Logic Based Program Synthesis and Transformation. Proceedings, 2001. VIII, 267 pages. 2002.

Vol. 2373: A. Apostolico, M. Takeda (Eds.), Combinatorial Pattern Matching. Proceedings, 2002. VIII, 289 pages. 2002.

Vol. 2374: B. Magnusson (Ed.), ECOOP 2002 – Object-Oriented Programming. XI, 637 pages. 2002.

Vol. 2375: J. Kivinen, R.H. Sloan (Eds.), Computational Learning Theory. Proceedings, 2002. XI, 397 pages. 2002. (Subseries LNAI).

Vol. 2377: A. Birk, S. Coradeschi, T. Satoshi (Eds.), RoboCup 2001: Robot Soccer World Cup V. XIX, 763 pages. 2002. (Subseries LNAI).

Vol. 2378: S. Tison (Ed.), Rewriting Techniques and Applications. Proceedings, 2002. XI, 387 pages. 2002.

Vol. 2379: G.J. Chastek (Ed.), Software Product Lines. Proceedings, 2002. X, 399 pages. 2002.

Vol. 2380: P. Widmayer, F. Triguero, R. Morales, M. Hennessy, S. Eidenbenz, R. Conejo (Eds.), Automata, Languages and Programming. Proceedings, 2002. XXI, 1069 pages. 2002.

Vol. 2381: U. Egly, C.G. Fermüller (Eds.), Automated Reasoning with Analytic Tableaux and Related Methods. Proceedings, 2002. X, 341 pages. 2002 .(Subseries LNAI).

Vol. 2382: A. Halevy, A. Gal (Eds.), Next Generation Information Technologies and Systems. Proceedings, 2002. VIII, 169 pages. 2002.

Vol. 2383: M.S. Lew, N. Sebe, J.P. Eakins (Eds.), Image and Video Retrieval. Proceedings, 2002. XII, 388 pages. 2002.

Vol. 2384: L. Batten, J. Seberry (Eds.), Information Security and Privacy. Proceedings, 2002. XII, 514 pages. 2002.

Vol. 2385: J. Calmet, B. Benhamou, O. Caprotti, L. Henocque, V. Sorge (Eds.), Artificial Intelligence, Automated Reasoning, and Symbolic Computation. Proceedings, 2002. XI, 343 pages. 2002. (Subseries LNAI).

Vol. 2386: E.A. Boiten, B. Möller (Eds.), Mathematics of Program Construction. Proceedings, 2002. X, 263 pages. 2002.

Vol. 2387: O.H. Ibarra, L. Zhang (Eds.), Computing and Combinatorics. Proceedings, 2002. XIII, 606 pages. 2002.

Vol. 2388: S.-W. Lee, A. Verri (Eds.), Pattern Recognition with Support Vector Machines. Proceedings, 2002. XI, 420 pages. 2002.

Vol. 2389: E. Ranchhod, N.J. Mamede (Eds.), Advances in Natural Language Processing. Proceedings, 2002. XII, 275 pages. 2002. (Subseries LNAI).

Vol. 2390: D. Blostein, Y.-B. Kwon (Eds.), Graphics Recognition. Proceedings, 2001. XI, 367 pages. 2002.

Vol. 2391: L.-H. Eriksson, P.A. Lindsay (Eds.), FME 2002: Formal Methods – Getting IT Right. Proceedings, 2002. XI, 625 pages. 2002.

Vol. 2392: A. Voronkov (Ed.), Automated Deduction – CADE-18. Proceedings, 2002. XII, 534 pages. 2002. (Subseries LNAI).

Vol. 2393: U. Priss, D. Corbett, G. Angelova (Eds.), Conceptual Structures: Integration and Interfaces. Proceedings, 2002. XI, 397 pages. 2002. (Subseries LNAI).

Vol. 2394: P. Perner (Ed.), Advances in Data Mining. VII, 109 pages. 2002. (Subseries LNAI).

Vol. 2395: G. Barthe, P. Dybjer, L. Pinto, J. Saraiva (Eds.), Applied Semantics. IX, 537 pages. 2002.

Vol. 2396: T. Caelli, A. Amin, R.P.W. Duin, M. Kamel, D. de Ridder (Eds.), Structural, Syntactic, and Statistical Pattern Recognition. Proceedings, 2002. XVI, 863 pages. 2002.

Vol. 2397: G. Grahne, G. Ghelli (Eds.), Database Programming Languages. Proceedings, 2001. X, 343 pages. 2002.

Vol. 2398: K. Miesenberger, J. Klaus, W. Zagler (Eds.), Computers Helping People with Special Needs. Proceedings, 2002. XXII, 794 pages. 2002.

Vol. 2399: H. Hermanns, R. Segala (Eds.), Process Algebra and Probabilistic Methods. Proceedings, 2002. X, 215 pages. 2002.

Vol. 2400: B. Monien, R. Feldmann (Eds.), Euro-Par 2002 – Parallel Processing. Proceedings, 2002. XXIX, 993 pages. 2002.

Vol. 2401: P.J. Stuckey (Ed.), Logic Programming. Proceedings, 2002. XI, 486 pages. 2002.

Vol. 2402: W. Chang (Ed.), Advanced Internet Services and Applications. Proceedings, 2002. XI, 307 pages. 2002.

Vol. 2403: Mark d'Inverno, M. Luck, M. Fisher, C. Preist (Eds.), Foundations and Applications of Multi-Agent Systems. Proceedings, 1996-2000. X, 261 pages. 2002. (Subseries LNAI).

Vol. 2404: E. Brinksma, K.G. Larsen (Eds.), Computer Aided Verification. Proceedings, 2002. XIII, 626 pages. 2002.

Vol. 2405: B. Eaglestone, S. North, A. Poulovassilis (Eds.), Advances in Databases. Proceedings, 2002. XII, 199 pages. 2002.

Vol. 2406: C. Peters, M. Braschler, J. Gonzalo, M. Kluck (Eds.), Evaluation of Cross-Language Information Retrieval Systems. Proceedings, 2001. X, 601 pages. 2002.

Vol. 2407: A.C. Kakas, F. Sadri (Eds.), Computational Logic: Logic Programming and Beyond. Part I. XII, 678 pages. 2002. (Subseries LNAI).

Vol. 2408: A.C. Kakas, F. Sadri (Eds.), Computational Logic: Logic Programming and Beyond. Part II. XII, 628 pages. 2002. (Subseries LNAI).

Vol. 2409: D.M. Mount, C. Stein (Eds.), Algorithm Engineering and Experiments. Proceedings, 2002. VIII, 207 pages. 2002.

Vol. 2410: V.A. Carreño, C.A. Muñoz, S. Tahar (Eds.), Theorem Proving in Higher Order Logics. Proceedings, 2002. X, 349 pages. 2002.

Vol. 2412: H. Yin, N. Allinson, R. Freeman, J. Keane, S. Hubbard (Eds.), Intelligent Data Engineering and Automated Learning – IDEAL 2002. Proceedings, 2002. XV, 597 pages. 2002.

Vol. 2413: K. Kuwabara, J. Lee (Eds.), Intelligent Agents and Multi-Agent Systems. Proceedings, 2002. X, 221 pages. 2002. (Subseries LNAI).

Vol. 2414: F. Mattern, M. Naghshineh (Eds.), Pervasive Computing. Proceedings, 2002. XI, 298 pages. 2002.

Vol. 2415: J.R. Dorronsoro (Ed.), Artificial Neural Networks – ICANN 2002. Proceedings, 2002. XXVIII, 1382 pages. 2002.

Vol. 2416: S. Craw, A. Preece (Eds.), Advances in Case-Based Reasoning. Proceedings, 2002. XII, 656 pages. 2002. (Subseries LNAI).

Vol. 2417: M. Ishizuka, A. Sattar (Eds.), PRICAI 2002: Trends in Artificial Intelligence. Proceedings, 2002. XX, 623 pages. 2002. (Subseries LNAI).

Vol. 2418: D. Wells, L. Williams (Eds.), Extreme Programming and Agile Methods – XP/Agile Universe 2002. Proceedings, 2002. XII, 292 pages. 2002.

Vol. 2419: X. Meng, J. Su, Y. Wang (Eds.), Advances in Web-Age Information Management. Proceedings, 2002. XV, 446 pages. 2002.

Vol. 2420: K. Diks, W. Rytter (Eds.), Mathematical Foundations of Computer Science 2002. Proceedings, 2002. XII, 652 pages. 2002.

Vol. 2421: L. Brim, P. Jančar, M. Křetínský, A. Kučera (Eds.), CONCUR 2002 – Concurrency Theory. Proceedings, 2002. XII, 611 pages. 2002.

Vol. 2422: H. Kirchner, Ch. Ringeissen (Eds.), Algebraic Methodology and Software Technology. Proceedings, 2002. XI, 503 pages. 2002.

Vol. 2423: D. Lopresti, J. Hu, R. Kashi (Eds.), Document Analysis Systems V. Proceedings, 2002. XIII, 570 pages. 2002.

Vol. 2425: Z. Bellahsène, D. Patel, C. Rolland (Eds.), Object-Oriented Information Systems. Proceedings, 2002. XIII, 550 pages. 2002.

Vol. 2426: J.-M. Bruel, Z. Bellahsène (Eds.), Advances in Object-Oriented Information Systems.Procedings, 2002. IX, 314 pages. 2002.

Vol. 2430: T. Elomaa, H. Mannila, H. Toivonen (Eds.), Machine Learning: ECML 2002. Proceedings, 2002. XIII, 532 pages. 2002. (Subseries LNAI).

Vol. 2431: T. Elomaa, H. Mannila, H. Toivonen (Eds.), Principles of Data Mining and Knowledge Discovery. Proceedings, 2002. XIV, 514 pages. 2002. (Subseries LNAI).

Vol. 2435: Y. Manolopoulos, P. Návrat (Eds.), Advances in Databases and Information Systems. Proceedings, 2002. XIII, 415 pages. 2002.

Vol. 2436: J. Fong, C.T. Cheung, H.V. Leong, Q. Li (Eds.), Advances in Web-Based Learning. Proceedings, 2002. XIII, 434 pages. 2002.

Vol. 2438: M. Glesner, P. Zipf, M. Renovell (Eds.), Field-Programmable Logic and Applications. Proceedings, 2002. XXII, 1187 pages. 2002.

Vol. 2439: J.J. Merelo Guervós, P. Adamidis, H.-G. Beyer, J.-L. Fernández-Villacañas, H.-P. Schwefel (Eds.), Parallel Problem Solving from Nature – PPSN VII. Proceedings, 2002. XXII, 947 pages. 2002.

Vol. 2440: J.M. Haake, J.A. Pino (Eds.), Groupware: Design, Implementation and Use. Proceedings, 2002. XII, 285 pages. 2002.

Vol. 2442: M. Yung (Ed.), Advances in Cryptology – CRYPTO 2002. Proceedings, 2002. XIV, 627 pages. 2002.

Vol. 2443: D. Scott (Ed.), Artificial Intelligence: Methodology, Systems, and Applications. Proceedings, 2002. X, 279 pages. 2002. (Subseries LNAI).

Vol. 2444: A. Buchmann, F. Casati, L. Fiege, M.-C. Hsu, M.-C. Shan (Eds.), Technologies for E-Services. Proceedings, 2002. X, 171 pages. 2002.

Vol. 2445: C. Anagnostopoulou, M. Ferrand, A. Smaill (Eds.), Music and Artificial Intelligence. Proceedings, 2002. VIII, 207 pages. 2002. (Subseries LNAI).

Vol. 2446: M. Klusch, S. Ossowski, O. Shehory (Eds.), Cooperative Information Agents VI. Proceedings, 2002. XI, 321 pages. 2002. (Subseries LNAI).

Vol. 2447: D.J. Hand, N.M. Adams, R.J. Bolton (Eds.), Pattern Detection and Discovery. Proceedings, 2002. XII, 227 pages. 2002. (Subseries LNAI).

Vol. 2448: P. Sojka, I. Kopeček, K. Pala (Eds.), Text, Speech and Dialogue. Proceedings, 2002. XII, 481 pages. 2002. (Subseries LNAI).

Vol. 2451: B. Hochet, A.J. Acosta, M.J. Bellido (Eds.), Integrated Circuit Design. Proceedings, 2002. XVI, 496 pages. 2002.

Vol. 2453: A. Hameurlain, R. Cicchetti, R. Traunmüller (Eds.), Database and Expert Systems Applications. Proceedings, 2002. XVIII, 951 pages. 2002.

Vol. 2454: Y. Kambayashi, W. Winiwarter, M. Arikawa (Eds.), Data Warehousing and Knowledge Discovery. Proceedings, 2002. XIII, 339 pages. 2002.

Vol. 2455: K. Bauknecht, A M. Tjoa, G. Quirchmayr (Eds.), E-Commerce and Web Technologies. Proceedings, 2002. XIV, 414 pages. 2002.

Vol. 2456: R. Traunmüller, K. Lenk (Eds.), Electronic Government. Proceedings, 2002. XIII, 486 pages. 2002.

Vol. 2469: W. Damm, E.-R. Olderog (Eds.), Formal Techniques in Real-Time and Fault-Tolerant Systems. Proceedings, 2002. X, 455 pages. 2002.

Vol. 2470: P. Van Hentenryck (Ed.), Principles and Practice of Constraint Programming – CP 2002. Proceedings, 2002. XVI, 794 pages. 2002.

Vol. 2483: J.D.P. Rolim, S. Vadhan (Eds.), Randomization and Approximation Techniques in Computer Science. Proceedings, 2002. VIII, 275 pages. 2002.